Environmental Issues in the Mediterranean

'A timely and highly readable book fusing historic environment change with contemporary pressures affecting the Mediterranean Basin. Read it before you go on field excursions to the region!'
Richard Spalding, *School of Geography and Environmental Management UWE-Bristol.*

Environmental Issues in the Mediterranean reviews both physical and social aspects of this region, in relation to its environment. The Mediterranean has been subject to changing human settlement and land-use patterns for millennia, and has a history of human exploitation in an inherently unstable landscape. It has been at the centre of the development of agricultural technology and major population movements, and is still frequently affected by volcanic eruptions and earthquakes. The region is characterized, and set apart from other areas, by the presence of the Mediterranean Sea, and its current environmental problems can have social and economic impacts for the whole region. The sea affects many of the landscape processes and has long been the principal axis of movement, meaning that social adaptations to the environment have common points throughout the basin.

An introduction to the Mediterranean region, its history, physical and human geography and its environmental problems is given. The book then goes on to examine:

- the dynamic environment – climate variables and fluctuations, vegetation, the hydrological cycle of the basin and its watershed, processes of erosion, fire and of the Mediterranean Sea itself.
- the human impact on the environment – prehistoric and historic land use, traditional agriculture, rural and urban settlement and use of mineral resources.
- the Mediterranean environment under increasing pressure – the present human landscape, changes in agriculture in the twentieth century, the impact of depopulation, pollution, water resources, desertification and potential climatic change.

The authors conclude with a discussion of the region's ongoing environmental issues of water resources, land degradation, agricultural intensification and tourism, and consider how these can be approached using management techniques and national and regional policies.

Environmental Issues in the Mediterranean is ideal for students who are studying a range of environmental issues, but would like to see them linked within one regional context. It also contains a wealth of background information for those going on a field trip to the Mediterranean. It is highly illustrated with line figures and maps, and includes case studies from countries throughout the region. Chapter summaries, suggestions for further reading and topics for discussion are also included.

John Wainwright is Professor of Geography, and **John B. Thornes** is Research Professor in Physical Geography, at King's College London.

Routledge Studies in Physical Geography and Environment

This series provides a platform for books which break new ground in the understanding of the physical environment. Individual titles will focus on developments within the main subdisciplines of physical geography and explore the physical characteristics of regions and countries. Titles will also explore the human/environment interface.

Environmental Issues in the Mediterranean

Processes and perspectives from the past and present

John Wainwright and John B. Thornes

Routledge
Taylor & Francis Group

LONDON AND NEW YORK

First published 2004
by Routledge
2 Park Square, Milton Park, Abingdon, Oxfordshire OX14 4RN

Simultaneously published in the USA and Canada
by Routledge
711 Third Avenue, New York, NY 10017

First issued in paperback 2014

Routledge is an imprint of the Taylor & Francis Group, an informa business

© 2004 John Wainwright and John B. Thornes

Typeset in Galliard
by Wearset Ltd, Boldon, Tyne and Wear

British Library Cataloguing in Publication Data
A catalogue record for this book is available from the British Library

Library of Congress Cataloging in Publication Data
A catalog record for this book has been requested

ISBN 13: 978-1-138-86710-9 (pbk)
ISBN 13: 978-0-415-15686-8 (hbk)

For Katerina and Rosemary

Contents

PART 2

The dynamic environment

57

Figures

Tables

Boxes

Preface

> The 21st century will be one of massive change at local, national and global levels. The key challenge for science and society will be coping with this change in the most effective manner possible ... A greater demand on both the social and natural sciences for information to support and guide policy will emerge ... The challenge of coping with change in the coming century is unprecedented in human history ... The relationship between humans and environment must change – this is now the central challenge for humanity.
>
> (Dovers and Handmer, 1992: 262)

The above statements were made by Australians in 1992 and, between then and the publication of this book, a tidal wave of change has affected the Mediterranean environment. First came the onrush of international concern for sustainability as reflected by the Rio Conference Agreement on Agenda 21 (1992), then the International Convention on Combating Desertification (1996) and the Kyoto Declaration on global warming (1998). Coupled to these has been the progressive improvement of modelling capacity and data availability that seem to show that human-induced greenhouse warming is now a global reality. The European Union has been equally determined in promoting the change from the Common Agricultural Policy (CAP) to the newly proposed, more environmentally friendly Agenda 2000, a post-productionist approach to sustainable agriculture and the extensification of agriculture. The new Framework for Water Directive has highlighted the coming crisis of surface and ground-water quantity and quality that will affect the whole world, and the Mediterranean in particular.

In this morass of change, we set out to try to capture in this book the main challenges and problems facing the environments and peoples of Mediterranean countries and the principles that might be used to address them. As befits our geographical training, we attempt to see the problems from both natural science and social science perspectives. Above all, we approach the issues from a historical point of view, in the firm belief that the past is the key to the present in understanding contemporary environmental issues and their management. What is more, the past usually provides the starting conditions without which it is impossible to attempt such an understanding. The spatial diversity of the Mediterranean land surface and the dynamical complexity of the problems lead us to conclude that the management approaches to the mitigation of these problems will need to be as diverse as the landscape, as varied as the history and as complicated as the cultures. There is no rule-book in this respect and we do not attempt to provide one.

In the long journey from the outline plan to the final delivery of the text to our publishers there have been several encumbrances, not least in sustaining major research projects and publications. For John Thornes there was the obstacle of a major illness and he would like to thank the staff at the Horton Hospital and Rivermead Rehabilitation Centre for their essential care in the early months of this project. Many colleagues have helped us in many ways and we hope that the finished product justifies their efforts, which we deeply appreciate. In particular, Roma Beaumont and Carolyne Megan, from our own department, have improved our rough figure drafts beyond belief and produced the fine illustrations. Francisco Alonso Sarria also provided invaluable help with the figures. Jean Poesen, Juan Puigdéfabregas, Nick Drake and Keith Hoggart all provided references and offprints that greatly helped our search for the broad range of literature required to put the book together. We would like to thank our numerous other colleagues who have provided us with references and offprints that have proved directly or indirectly useful in the preparation of the book. Richard Harrison kindly read through some draft chapters and provided his usual thought-provoking comments.

We would also like to thank our partners – Katerina and Rosemary – for their tireless efforts in all phases of the writing of this book. Without this help, it is unlikely that it would ever have seen the light of day.

John Wainwright and John B. Thornes
October 2001

Acknowledgements

The authors and publishers would like to thank the following for granting permission to reproduce material in this work:

Figure 1.1 and Table 15.2 from J. Margat (1988), reproduced in Grenon and Batisse (1989), *Futures for the Mediterranean Basin: The Blue Plan*. Based on UN Statistics. Bibliothèque de l'Institut de Botanique, for Figure 1.3c. Figures 1.4, 3.8, 14.3. © UNEP-BP/RAC (1989), reprinted from *Futures for the Mediterranean Basin: The Blue Plan* edited by Michel Grenon and Michel Batisse (1989), by permission of Oxford University Press. Figures 2.2 and 2.3 reprinted from D. Ninkovich and J.D. Hays, 'Mediterranean island arcs and origin of high potash volcanics', *Earth and Planetary Science Letters* 16, 331–345, © (1972), with permission of Elsevier Science. Figure 2.4 and Table 2.1 reproduced with the permission of Nelson Thornes Ltd from *Holmes' Principles of Physical Geology*, 4th edition, first published in 1978. Figure 2.10 reprinted from A.E.M. Nairn, W.H. Kanes and P.G. Stehli (eds), 'Post-Miocene depositional patterns and structural displacement in the Mediterranean', in *The Ocean Basins and Their Margins: The Eastern Mediterranean*, Vol. 4A, 77–150 (1978), with permission from Plenum Press. Figure 2.11a and b reprinted from G.B. Vai, 'Palaeozoic strike-slip rift pulses and palaeogeography in the circum-Mediterranean Tethyan realm', *Palaeogeography, Palaeoclimatology, Palaeoecology* 87, 223–252, copyright (1991), with permission of Elsevier Science. Figures 2.11c–e reprinted from L.-E. Ricou, 'Tethys reconstructed: plates, continental fragments and their boundaries since 260 Ma from Central America to South-eastern Asia', *Geodinamica Acta* Vol. 7(4), 169–218, © (1994), with permission from Elsevier Science. Figures 2.11f and 2.11g reprinted from A.H.F. Robertson and M. Grasso (1995) 'Overview of the Late Tertiary – Recent tectonic and palaeo-environmental development of the Mediterranean region', *Terra Nova* 7, 114–127, with permission of Blackwell Science Ltd. Figure 2.14, F.J. Sawkins, *Metal Deposits in Relation to Plate Tectonics* (1990), © Springer-Verlag. Figures 3.1a, 3.2a, 3.3 reprinted from R.G. Barry and R.J. Chorley, *Atmosphere, Weather and Climate* (1992), with permission from Routledge. United Nations Environment Program for Figures 3.1b, 3.1c, 3.2b, 3.2f and 3.2g. European Communities for Figures 3.2c, d, e, h, i, j, k from M. Conte and M. Colacino (1995) 'Climate', in *Desertification in a European Context: Physical and Socio-Economic Aspects*, edited by R. Fantechi, D. Peter, P. Balabanis and J.L. Rubio, pp. 79–109, © European Communities. Figures 15.11 and 18.5 from P.M. Allen, I. Black, M. Lemon, R.A.F. Seaton, C. Blatsou and N. Calamaras (1994) 'Agricultural production and water quality in the Argolid valley, Greece: a policy-relevant study in integrated method', in *Understanding the Natural and Anthropogenic Causes of Soil Degradation and Desertification in the Mediterranean Basin*, Vol. 5: *Agricultural Production and Water Quality in the Argolid, Greece*, edited by S.E. van der Leeuw, pp. 3–166, Final Report on Contract EV5V-CT91-0021, © EU. Figure 6.5 (from J.W.A. Poesen (1995) 'Soil erosion in

Mediterranean environments', in *Desertification in a European Context: Physical and Socio-economic Aspects*, edited by R. Fantechi, D. Peter, P. Balabanis and J.L. Rubio, pp. 123–152, European Commission Report EUR 15415, © EU. Figure 9.4 from P.V. Castro, R.W. Chapman, S. Gili, V. Lull, R. Micó, C. Rihuete, R. Risch and M.E. Sanahuja Yll (eds) (1998) *Aguas Project. Palaeoclimatic Reconstruction and the Dynamics of Human Settlement and Land-Use in the Area of the Middle Aguas (Almería), in the South-East of the Iberian Peninsula.* © European Commission, EUR 18036. Figures 3.6, 20.1, 20.2 and 20.3 from J. P. Palutikof, M. Conte, J. Casimiro Mendes, C.M. Goodess and F. Espirito Santo, 'Climate and climatic change', in *Mediterranean Desertification and Land Use*, edited by C.J. Brandt and J.B. Thornes, pp. 43–86 (1996), © John Wiley & Sons Limited. Figure 3.7 © Bartholomew Ltd (2002) reproduced by kind permission of HarperCollins Publishers www.bartholomewmaps.com. Figure 3.9 from C. Goosens, 'Principal component analysis of Mediterranean rainfall', *Journal of Climatology*, Vol. 5, pp. 379–388 (1985), © John Wiley & Sons Limited. Figure 3.10 reprinted from D. Martyn, *Climates of the World*, copyright (1992), with permission of Elsevier Science. Figure 3.12, M. Puigcerver, S. Alonso, J. Lorente, M.C. Llasat, A. Redaño, A. Burgueño and E. Vilar, 'Preliminary aspects of rainfall rates in the north-east of Spain', *Theoretical and Applied Climatology*, 37, 97–109 (1986), © Springer-Verlag. Figure 3.13, H.J.P.M. Mommersteeg, M.F. Loutre, R. Young, T.A. Wijmstra and H. Hooghiemstra, 'Orbital forced frequencies in the 975,000 year pollen record from Tenaghi Philippon (Greece)', *Climate Dynamics* 11, 4–24 (1995), © Springer-Verlag. Figure 3.14a reprinted from C. Vergnaud Grazzini, J.F. Saliège, M.J. Urrutiaguer and A. Iannace (1990), 'Oxygen and carbon isotope stratigraphy of ODP Hole 653A and Site 654: the Pliocene-Pleistocene glacial history recorded in the Tyrrhenian Basin (West Mediterranean)', in K.A. Kastens, J. Mascle *et al.*, *Proceedings of the Ocean Drilling Program, Sci, Results*, 107: College Station, TX (Ocean Drilling Program), 361–386. Figure 3.14b reprinted from R.C. Thunell and D.F. Williams, 'Paleotemperature and paleosalinity history of the eastern Mediterranean during the late Quaternary', *Palaeogeography, Palaeoclimatology, Palaeoecology* 44, 23–39, © (1983), with permission of Elsevier Science. Figure 3.14c reprinted from R.C. Thunell, 'Climatic evolution of the Mediterranean Sea during the last 5.0 million years', *Sedimentary Geology* 23, 67–79, copyright (1979), with permission of Elsevier Science. Figure 3.16 reprinted from B. Huntley, 'The use of climate response surfaces to reconstruct palaeoclimate from Quaternary pollen and plant macrofossil data', *Philosophical Transactions of the Royal Society B* 341, 215–223 (1993), with permission from The Royal Society. Figure 3.17 reprinted with permission from J. Guiot, A. Pons, J.L. de Beaulieu and M. Reille, 'A 140,000-year continental climatic reconstruction from two European pollen records', *Nature* 338, pp. 309–313, copyright (1989), Macmillan Magazines Limited. Figure 3.18 reprinted from J. Guiot, 'Late Quaternary climate change in France estimated from multivariate pollen time series', *Quaternary Research* 28, 100–118 (1987), with permission from Academic Press. Figure 3.24 reprinted from C. Till and J. Guiot, 'Reconstruction of precipitation in Morocco since 1100 A.D. based on *Cedrus atlantica* tree-ring widths', *Quaternary Research* 33, 337–351 (1990), with permission from Academic Press. Figure 3.25 from F. Serre-Bachet and J. Guiot, 'Summer temperature changes from tree rings in the Mediterranean area during the last 800 years', *Abrupt Climatic Change*, edited by W.H. Berger and L.D. Labeyrie, 1987, pp. 89–97, © Reidel Publishers, with kind permission of Kluwer Academic Publishers. Alysen Huxley for Figures 4.1 & 4.2. CEBAS for Figures 4.3 & 4.6. Figure 4.5 from C.S. Kosmas, N. Moustakas, N.G. Danalatos and N. Yassoglou, 'The Spata field site', in *Mediterranean Desertification and Land Use*, edited by C.J. Brandt and J.B. Thornes, pp. 207–228 (1996), © John Wiley & Sons Limited. Consejeria de Medio Ambiente, Junta de Andalucia for Figure 4.7. Figure 4.8 reprinted with permission from J.-P. Suc, 'Origin and evolution of the Mediterranean vegetation and climate Europe', *Nature*, 307, pp. 429–432. © (1984), Macmillan Magazines Limited. CSIC, Instituto Pirinaíco de Ecologia, Zaragoza, Spain for Figure 5.2. Centro de Estudios Hidrográficos (CEDEX), Madrid for Figure 5.3. F. López Bermúdez for Figure 5.5. Instituto de Geografia, Universidad de Alicante, Spain for Figure 5.6. Figure 5.8 from *Groundwater* by Freeze/Cherry, ©. Reprinted by permission of Pearson Education, Inc., Upper Saddle River, NJ. Figure 5.9 from S.M. Burke, 'Groundwater over-exploitation: a case study in Castilla la Mancha, Spain', in *Atlas of Mediterranean Environments in Europe: The Desertification Context*, edited by P. Mairota, J.B. Thornes and N. Geeson, pp. 82–84 (1998), © John Wiley & Sons Limited. Figure 6.2a reprinted from M.I.L.P. De Lima and J.L.M.P. De Lima, 'Water erosion of soils containing rock fragments', in U. Shamir and C. Jiqqi (eds) *The Hydrological Basis for Water Management (Proceedings of the Beijing Symposium, October 1990)*, 141–147, IAHS Publ No. 197, IAHS Press, Wallingford, UK. Figure 6.2b reprinted from J.W.A. Poesen, 'Mechanisms of overland flow generation and sediment production on loamy and sandy soils with and without rock fragments', in A.J. Parsons and A.D. Abrahams (eds) *Overland Flow Hydraulics and Erosion Mechanics*, 275–305, with permission from UCL Press. Figure 6.3a reprinted from J. Wainwright, 'A comparison of the infiltration, runoff and erosion characteristics of two contrasting "badland" areas in S. France', *Zeitschrift für Geomorphologie Supplementband* 106, 183–198 (1996), with permission from Gebr. Bornträger Verlagsbuchhandlung. Figure 6.3b reprinted from *Catena* 26, J. Wainwright, 'Infiltration, runoff and erosion characteristics of agricultural land in extreme storm events, SE France', 27–47 (1996), with permission of Elsevier Science. Figure 6.4 from C.F. Francis and J.B. Thornes, 'Runoff hydrographs from three Mediterranean vegetation cover types', in *Vegetation and Erosion*, edited by J.B. Thornes, pp. 363–384 (1990), © John Wiley & Sons Limited. Figure 6.6 reprinted from P. Farres, J.W.A. Poesen and S. Wood, 'Soil erosion landscapes', *Geography Review* 6 (1993), 38–41, with permission from Philip Allan Updates. Figure 6.7 from Th.W.J. Van Asch, 'Hazard mapping as a tool for landslide prevention in Mediterranean areas', *Desertification in Europe*, edited by R. Fantechi and N.S. Margaris, 1986, pp. 126–135, © Reidel Publishers, with kind permission of Kluwer Academic Publishers. Figure 6.8 from D. Petley, 'The mechanics and landforms of deep-seated landslides', in *Advances in Hillslope Processes*, Vol. 2, edited by M.G. Anderson and S.M. Brooks, pp. 826–836 (1996), © John Wiley & Sons Limited. Figure 6.11 reprinted from P. Ergenzinger, 'A conceptual geomorphological model for the development of a Mediterranean river basin under neotectonic stress (Buonamico basin, Calabria, Italy)', in D.E. Walling *et al.* (eds), *Erosion, Debris Flows and Environment in Mountain Regions*, 51–60. IAHS Publ. No. 209, IAHS Press, Wallingford, UK. Figure 7.1 in L. Trabaud, 'Postfire plant community dynamics in the Mediterranean Basin', *The Role of Fire in Mediterranean-Type Ecosystems*, edited by J.M. Moreno and W.C. Oechel, pp. 1–15 (1994), © Springer-Verlag. Figures 8.1, 8.2 and 8.3 reprinted from P.E. La Violette (ed.) *Seasonal and Interannual Variability of the Western Mediterranean*, © (1995) by the American Geophysical Union. Figure 9.1 reprinted from G.N. Bailey, G. King and D. Sturdy, 'Active tectonics and land-use strategies: a palaeolithic case study from north-west Greece', *Antiquity* 67, 292–312, by permission of Bailey, King and Sturdy and Antiquity Publications Ltd. Figures 9.3 and 10.7 reprinted from T. Van Andel and C.N. Runnels, *Beyond the Acropolis*, © (1987) by the Board of Trustees of the Leland Stanford Jr. University. Figure 10.1 from O. Bar-Yosef, *Current Anthropology* 27(2), 157–162, © (1986) The University of Chicago Press. Professor James Mellaart for Figures 10.2 & 10.3. Figure 10.6 reprinted from E. Galili, D.J. Stanley, J. Sharvit and M. Weinstein-Evron, 'Evidence for earliest olive-oil production in submerged settlements off the Carmel coast, Israel', *Journal of Archaeological Science* 24, 1141–1150. Figure 10.8 reproduced from *Journal of Field Archaeology* with permission of the Trustees of Boston University. All rights reserved.

Sarah Potter for Figures 10.9 and 10.12 from T.W. Potter (1979) *The Changing Landscape of South Etruria*, London: Paul Elek. Figure 10.10 from J. Wainwright, 'Anthropogenic factors in the degradation of semi-arid regions: a prehistoric case study in southern France', in *Effects of Environmental Change on Drylands*, edited by A.C. Millington and K. Pye, 285–304 (1994), © John Wiley & Sons Limited. Figure 10.11 reprinted from N. Roberts, 'Human-induced landscape change in south and south-west Turkey during the later Holocene', in S. Bottema, G. Entjes-Nieborg and W. van Zeist (eds) *Man's Role in the Shaping of the Eastern Mediterranean Landscape*, pp. 53–67, © Balkema 1990. Figure 10.12 reprinted from A. Gilman and J.B. Thornes, *Land Use and Prehistory in South East Spain*, with permission from George Allen and Unwin. Figure 10.13 *Early Hydraulic Civilization in Egypt: A Study in Cultural Ecology*, from K.W. Butzer, © (1976) The University of Chicago Press. Figure 10.15 reprinted from D.R. Lightfoot, 'Syrian Qanat Romani', *Journal of Arid Environments*, Vol. 33, 321–336, © (1996), with permission from Elsevier Science. Figure 10.17 from D.D. Gilbertson, C.O. Hunt, N.R.J. Fieller and G.W.W. Barker (1994) 'The environmental consequences and context of ancient floodwater farming in the Tripolitanian Desert', in *Environmental Change in Drylands: Biogeographical and Geomorphological Perspectives*, edited by A.C. Millington and K. Pye, pp. 229–252, © John Wiley & Sons Limited. Klett-Perthes for Figure 11.1. Presses Universitaires du Mirail for Figure 11.2 (from J.-P. Métaillie, 'Photographie et histoire du paysage: un exemple dans les Pyrénées luchonnaises', *Revue Géographique des Pyrénées et du Sud-Ouest* 57(2), pp. 179–208 (1986) © Université de Toulouse Le Mirail. Figure 12.1 reprinted from B.J. Kemp, *Ancient Egypt: Anatomy of a Civilization* (1989), with permission from Routledge. Catherine Delano Smith for Figure 12.2, from C. Delano Smith (1979) *Western Mediterranean Europe: A Historical Geography of Italy, Spain and Southern France since the Neolithic*. Academic Press, London. Figure 12.5 reprinted from C. Vita Finzi, *The Mediterranean Valleys: Geological Changes in Historical Times* (1969), © Cambridge University Press. CNRS Editions for Figures 12.6 and 9.2. Chrysalis Books for Figure 13.2. United Nations, Food and Agriculture Organization for Figures 14.1, 14.2, 14.4–14.6, 15.1–15.3, 15.6–15.10. Figure 15.4 reprinted from *Report on Aspects of Land Tenure in Cyprus* (1980), with permission from Ministry of Agriculture and Natural Resources, Land Consolidation Authority, Nicosia. The *New Scientist* for Figure 17.1. Table 2.2 reprinted from F.R. Siegel, *Natural and Anthropogenic Hazards in Development Planning* (1996), with permission from Academic Press.

Every effort has been made to contact copyright holders for their permission to reprint material in this book. The publishers would be grateful to hear from any copyright holder who is not here acknowledged and will undertake to rectify any errors or omissions in future editions of this book.

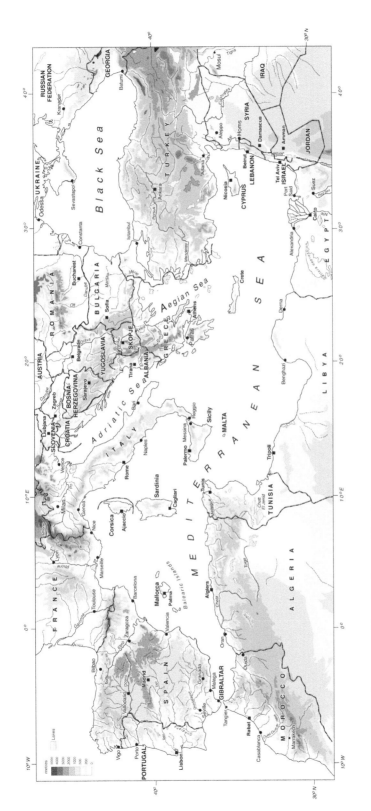

Topography of the Mediterranean with major rivers and locations

Part 1
Background

1 The Mediterranean region in its context

I have loved the Mediterranean with passion, no doubt because I am a northerner like so many others in whose footsteps I have followed.

(Braudel, 1975: 16)

A number of authors have produced geographies of the Mediterranean region (e.g. Walker, 1960; Houston, 1964), historical geographies of the region (e.g. Semple, 1932; Grant, 1969; Delano Smith, 1979) or even what might be described as geographical histories – as in the case of the *magnum opus* of Braudel (1975). The older of these texts fit firmly within the regional geography perspective, in attempting to produce a descriptive explanation of the characteristics and peculiarities of the region. Braudel and Delano Smith, on the other hand, aim to provide explanations of how the landscape has structured human life and history. These explanations are the backdrop of the *longue durée*, to use Braudel's term. Horden and Purcell (2000) have recently tried to expand Braudel's approach over a much longer time span.

However, we should take care not to consider this backdrop as operating in a one-way direction. Equally important is the impact humans have had – and continue to have – on the landscape and other elements of the environment. There are a number of books that deal with this impact, for example, Thirgood's (1981) and Meiggs' (1982) discussions of the link between forests and human activity, and McNeill's (1992) masterly overview of the impacts of settlement on the Mediterranean mountains. In some respects, Hughes (1994) can be thought of as an updating of Semple in that the Classical world is considered in its interaction with the environment. More recently, there have also been two introductory, edited review volumes – King (1997) and Conacher and Sala (1998) – that have dealt with this interaction, but with a focus more on current environments and environmental problems.

It is a central thesis of this book that we are made by the landscapes around us. Our everyday activities are limited by environmental constraints and our attempts to overcome them. Furthermore, modern attempts to live within the environment are controlled by the past development of that environment. This development may be driven by climatic change, internal mechanisms of evolution or the effects of human activity. However, the extent to which any of these are separable on long timescales is indeed debatable. An important consequence of these interlinkages is that understanding where we are today means understanding the ways in which the environment has developed in response to them. To a certain extent, but by no means always, a better understanding developed in this way can help us deal with current and future issues relating to the Mediterranean environment. In this sense, the Mediterranean region is taken as an example of an approach that is generally applicable in any environment. One of the advantages of taking this particular example is that it has had such a long history of environmental interactions, as we will show.

The region is also home to a rather sensitive set of environments, and the timescales over which human and other impacts can affect them vary over a very wide range – from a matter of days to millennia. An appreciation of this sensitivity allows us to take an appropriate viewpoint towards ongoing and future issues of environmental impact. Although human development of the environment will continue to occur – slowly at times, but with periods of very rapid change – we are now more than ever in a position to do so in a responsible way. This is not to say that the past inhabitants had no understanding of how their environment worked – clearly, this is not true – but the level of detail available now, together with an understanding of some of the more important feedback mechanisms and levels of complexity, means that we are much better informed about how environmental systems work. Some of the detail of this understanding in terms of landscape-forming and other environmental processes is therefore one of the key elements of the book. By continually testing our understanding by a process of abduction with past data (Baker, 1998) is one way that we can provide sufficient information from sufficiently extreme conditions to be confident over all appropriate timescales, not just those from our anthropocentric present.

One perspective developing from this view would be to suggest that we should engineer the environment to exploit it in as intensive a way possible while minimizing any 'negative' impacts. Such a perspective would be to miss the point. It must be realized that there are a whole series of issues relating to the quality of life – the aesthetics and other 'intangibles' of our surroundings – that cannot be incorporated into such evaluations. The fact that our actions may affect the environment – and thus our descendants – for millennia to come means that we should proceed with the knowledge of what those actions *could* produce. Thus, the conflicting personal, political, economic and social needs that lead to individuals and societies acting on the environment in a particular way must clearly be balanced in this perspective. One argument that is commonly heard is that we should not worry too much about the environment, as it will probably recover (eventually) from any particular negative human impact. This assertion is almost undeniably true. What the experience of the past shows is that the particular society that caused the problem did not necessarily recover.

It is our aim here, then, to attempt to provide the best explanation possible for understanding the Mediterranean environment as an example of environmental interactions everywhere (some not in as far an advanced state as that in the Mediterranean because of the shorter history of human impacts). This explanation will involve looking at the mechanisms by which the environment operates over timescales ranging from millions of years to the fall of raindrops within individual rain storms. The landscape is a function of this range of slowly operating forces and periodic catastrophic change. We will incorporate human activity and its effects on a range of timescales. The mechanisms and effects of human activity are also spread over a range of rates of operation. The data available necessarily mean that the picture will be more hazy for the earlier periods. Because of the sensitivity of Mediterranean environments, however, they have become an important focus for work by environmental scientists, and thus we have a level of detail on ongoing problems that is almost unparalleled. We can thus aim to let the past help us understand the present, and the present to understand the past.

What and where is the Mediterranean?

As we shall see in Chapter 2, the Mediterranean is a relative newcomer to the Earth's landscape. The closure of the eastern seaway of the Tethys Ocean did not occur until about nine million years ago, bringing about the basic physiography as we see it today. The catchment that drains into the Mediterranean Basin (Figure 1.1) is thus an area that continues to develop and work headwards (see Lewin *et al.*, 1995 for examples of drainage-basin evolution). In one

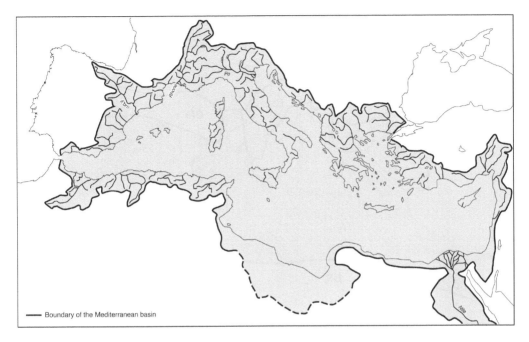

Figure 1.1 The catchment of the Mediterranean Basin

Source: Reproduced in Grenon and Batisse (1989).

sense, this area could be considered as *the* Mediterranean, but the nature of the topography – which is intimately linked with the geological history of the region (see Figure 2.10) – means that such a definition excludes many areas that share important characteristics with the coastal zone. Furthermore, areas of northern Africa, where aeolian activity dominates fluvial, can only be very poorly defined in this way.

It has also been common to define a region of 'Mediterranean climate'. The climate we currently define as Mediterranean is a much more recent phenomenon, as we will discuss in Chapter 3. It has only essentially been in place for the last million years or so, and even then has fluctuated dramatically as a result of the large glacial–interglacial cycles that have affected the globe. The result of these fluctuations has been that present Mediterranean conditions have only been effective for a small proportion of that time. The actual definition of boundaries according to the simple climatic classifications varies quite markedly. Most areas of the Mediterranean fit into the dry or warm, temperate climate zones according to the Köppen scheme (Figure 1.2a). However, it has been demonstrated that the boundaries produced by the Köppen scheme are highly sensitive to the data set used to define it. Emberger proposed simple boundaries to the climate region using an index calculated by dividing the summer rainfall in mm by the average maximum temperature of the warmest month in degrees Celsius. A value of five delimited the Mediterranean climate zone, with seven being the boundary of the sub-Mediterranean zone (Figure 1.2b). Although this scheme works well for the northern part of the basin, it is completely inappropriate for the south, delimiting the limits of the Mediterranean zone in the middle of the Sahara. Daget (1980) defined a compromise based on the ratio of warm-season to cold-season precipitation and the summer concentration of temperatures. The 'Mediterranean Isoclimatic Area' (AIM in French) is thus defined, providing a

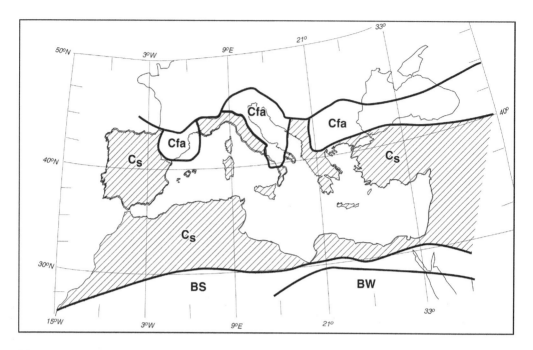

(a)

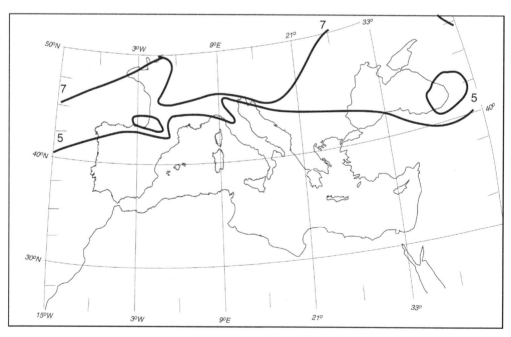

(b)

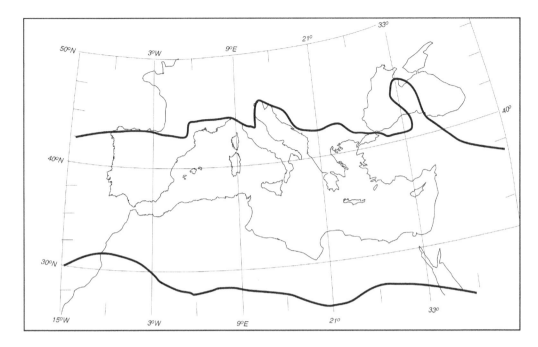

(c)

Figure 1.2 Climatic definitions of the Mediterranean zone: a. Köppen's classification scheme (where B indicates a dry and C a warm temperate climate, f sufficient precipitation in all months, s a dry summer season, S a dry winter steppe, W a dry winter desert, a the mean temperature of the warmest month >22°C, and b the warmest month with a mean of <22°C, but at least four months with means >10°C); b. Emberger's definition of the Mediterranean and sub-Mediterranean zones; and c. Daget's 'Mediterranean Isoclimatic Area'

Note
See text for further details.

better fit to the region as a whole (Figure 1.2c), including the semi-arid zones of North Africa. However, although this is probably the best definition of the Mediterranean climate, there is still a feeling that it is derived empirically, rather than being based on any good *a priori* reasons. Furthermore, as with the other schemes, it extends well into the interior of eastern Asia, and thus incorporates areas that we would tend to exclude from true Mediterranean climates. Overall, then, such simple schemes tend to be unsatisfactory. Perhaps the best definition of the Mediterranean climate is with a set of common factors but with a strong sense of variability over a number of different timescales, and encompassing a range from arid, through semi-arid to sub-humid conditions.

Third, a 'typical' Mediterranean vegetation is often used as a definition of the region, the rationale being that the vegetation should reflect the dominant environmental conditions. If this is so, then such a definition suffers from many of the problems relating to the climate definition. There are also important lags in terms of the response time of vegetation to imposed change, so that employing it as a means of defining climate change is also problematic. As will be discussed in Chapter 4, there are many reasons why the main species around the region are not necessarily distributed at the present time in the same way as they have been during previous interglacials. The tree species that are commonly used, such as holm oak or aleppo pine, tend not to have a

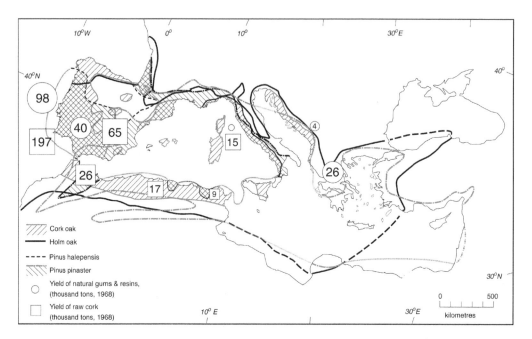

(a)

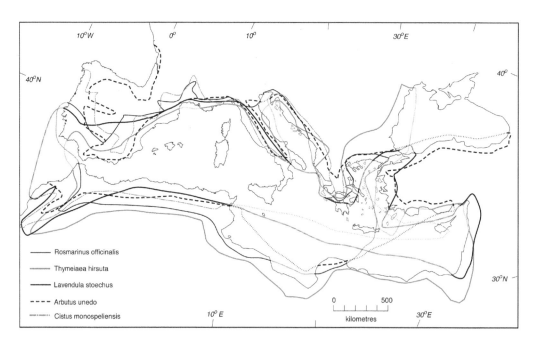

(b)

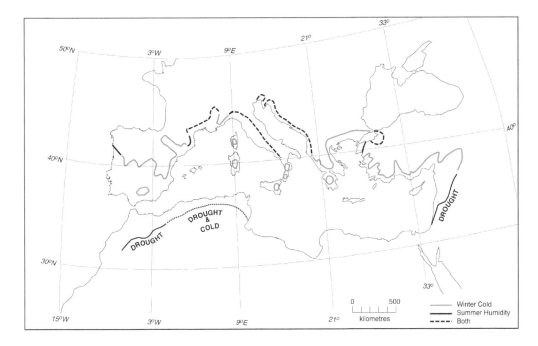

(c)

Figure 1.3 Distributions of 'characteristic' Mediterranean plant species: (a) holm oak and aleppo pine; (b) matorral species; and (c) olive (with indications as to the source of the limits of its environmental range), reproduced by permission of Hodder Arnold

Source: a and b after Beckinsale and Beckinsale (1975); c after Daget (1980).

circum-Mediterranean distribution, and the only species that do are effectively the matorral species, which are generally found as a consequence of disturbance (Figure 1.3). The olive is very commonly used as an indicator species, but, if anything, this use is the most problematic. Its environmental limits of distribution are controlled by a number of different factors, which have different effects in different regions. The distribution of the olive is also very strongly controlled by human activity in the past and present, as we shall see in detail in Chapter 10. An excellent background to Mediterranean vegetation patterns is also given by Allen (2001).

Other authors have chosen to use country boundaries or other political or administrative boundaries (Figure 1.4). Such a definition is also somewhat artificial. If the smallest administrative level is chosen, large areas of the hinterland that are vital to Mediterranean socio-economic systems are cut off. These levels also differ from country to country. If the country level is chosen, then clearly some areas are included that clearly do not share Mediterranean characteristics of any sort – northern France and the south of Libya and Egypt being the clearest examples here. One could also use a *reductio ad absurdum* argument by using past examples – including the parts of India conquered by Alexander the Great, or Britain as a Roman conquest (although the media enjoy the parallels of Britain enjoying a 'Mediterranean' climate under future scenarios of enhanced greenhouse-gas activity!). The administrative units also differ vastly according to the country chosen (reflecting in part the different population densities and their spatial distributions).

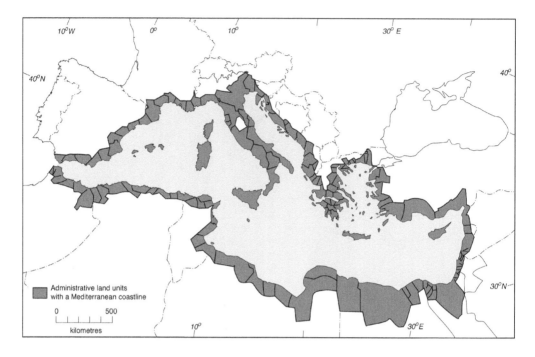

Figure 1.4 Definition of the Mediterranean zone by the Blue Plan according to modern administrative
boundaries with some frontage on the Mediterranean coastline

Source: Grenon and Batisse (1989), reproduced with the permission of Oxford University Press.

In summary, none of these simple definitions of the region are really satisfactory. All include important elements and exclude others. None so far has included Portugal, for example, despite the fact that it shares a number of common traits with the rest of Iberia and the region. The Mediterranean has been a cross-roads, both physically and socially. It has marked a place where plate boundaries have met, air masses mixed, plants combined from different regions and evolved into endemic species, and people have developed new ways of life and spread them to Europe and beyond. Subsequently, the interactions with the wider world have had significant impacts on the peoples and environments of the region. As noted above, it is partly because of this rich and varied heritage that the Mediterranean is an interesting and useful environment to study. In this book, then, we will employ a rather fluid definition of the region, employing climatic, vegetational, environmental and cultural definitions as appropriate. It is by looking at the series of commonalities that arise, and the situations in which they do not work, that we can hope to learn more about the environment and the people that have occupied it.

Current environmental problems in the Mediterranean

As we shall see from Chapter 14 onwards, there are a number of significant problems that currently affect the Mediterranean region. Some of these problems are Basin-wide, while others affect specific regions. The underlying driving mechanisms for these problems are the ever-increasing numbers of people living in the region, and to a certain extent population increases elsewhere in the globe. There is a scarcity of water for drinking, for irrigation and for industry. The source of the scarcity is the underlying nature of the Mediterranean climate, with long periods of hydrologic deficit and high levels of interannual variability in rainfall, but long periods of over-exploitation have exacerbated the issue. Land degradation is occurring because of both intensive use of the land, and its abandonment. The source of these changes are ultimately the long history of settlement and the development of settlement patterns in the region. New problems are also arising, due to the use of new technologies but also the evolution of new political structures, relating to changes on the European and global scales. The attractiveness of the region is also potentially forming part of its downfall. The very large, and continuously increasing, numbers of tourists that visit the region every year have led to a variety of deleterious developments, and increasing pressure on scarce resources such as water, and adding to the pollution of the area. All of these issues are enhanced by the sensitivity of the various parts of the environment, as discussed in Chapters 3 to 8. Given the extent and increasing rate of pressure on these landscapes, it is therefore fundamental that we put our understanding to good effect.

The long history of environmental issues in the Mediterranean region

As suggested in the previous section and the introduction, these issues cannot be divorced from the past. Significant, irreversible changes have been taking place over millennia due to the patterns of human exploitation of an inherently unstable landscape. There has been a continual struggle to grow crops on an often inhospitable soil that is impoverished of water and nutrients. Some of the major technological developments since the Neolithic have been involved in trying to overcome this issue. Once solved, there have often been social revolutions, with an increasing rate of growth, that have often led to further problems with the environment. Thus, there is a continual feedback between growth, pressure and response. Similarly, the control of water has marked a series of fundamental stages in the exploitation of Mediterranean landscapes. Once controlled, there is a tendency to exploit it to its limits, so that the same cycle of growth, pressure and response would have occurred. Not every case in the past provides a success story.

Ancient populations would have had a certain level of understanding of how these systems operated. One only has to look at the often-cited passage from Plato's *Critias*, describing the impacts of deforestation on erosion and the hydrological cycle, to realize this, and other examples can be found in Hughes (1994). Developments such as the plough and terracing can be understood as means of conserving water and soil, as well as being strictly functional in the production process. Thus, the history of Mediterranean land use is also one of discovering problems (often in catastrophic circumstances, as we still find today), and experimenting with means for overcoming them. The benefit may be only short term, though, as both ploughing and terracing can ultimately increase the vulnerability of the landscape, as we shall see in Chapter 11. The more intensive and extensive human modifications of the landscape became, the more sensitive it also became to unintended change.

Factors of change

This idea of sensitivity to change is an important one in the Mediterranean region. The reasons why the region is particularly sensitive will be discussed in detail in Chapters 2 to 8. However, the fundamental factors relate to a number of underlying reasons. First, as noted above, the region is geologically young and tectonically active. Thus, the very materials that make up the land tend to be unconsolidated, weak from tectonic activity, or liable to break up because they evolved in very different conditions deep in the Earth's crust. Because of its youth and the tectonic activity, the landscape is still adjusting to the general geological conditions. The climate is highly variable, and thus can trigger threshold processes very easily. In terms of erosion, the variability of vegetation from year to year that is a response to interannual variability can also trigger significant losses of soil. The general aridity of the climate means that certain areas at least are very sensitive to small changes in climate, with feedbacks again through the vegetation cycle. Fire can also affect this cycle dramatically, and the interannual variability leading to long periods of summer dryness is an important factor in triggering wildfires in the summer. Because of the aridity, flows on the surface tend not to travel very far, and can thus concentrate pollutants. Deposition of sediments can also cause important internal feedbacks, and is one mechanism by which gullying can be triggered in subsequent events. The Mediterranean Sea is also a victim of its own enclosed nature, so that circulation times needed to cycle the entire body of water so that it might become free of pollutants can be on the timescale of a century or more.

Thus far, we have considered sensitivity to change in terms of a series of internal mechanisms. When looking at change, we also need to be aware of externally driven forces. It may be useful to consider these perturbations on three different scales. First, some perturbations act within a strictly local level, for example, deforestation of an area to allow cultivation. The consequence of such activities will generally be important only locally, perhaps providing a 'patchwork' of impacted areas. However, it is also possible that they could have a wider impact. For example, as the patchwork of natural vegetation becomes increasingly sparse, the connectivity between them breaks down, and this can have important effects on factors such as species diversity, and ultimately the health of an ecosystem, as its ability to spread and interact is reduced. Local clearance can also lead to important off-site effects, such as the triggering of gullies over much wider areas, or off-site pollution of water supplies. On a second scale, there are perturbations that have occurred at a pan-Mediterranean scale, but are caused by fluctuations within it. Clear examples of this would be the spread of agriculture, the development of the Roman and Islamic empires, the Crusades and agricultural change following the Renaissance. The impacts of such changes are ultimately more widespread. The change may be diachronous, as in the case of the spread of agriculture, so that there was a relative lack of connectivity in terms of the impact on the landscape, with the connectivity only occurring later and at the local level described above. Alternatively, the change may be relatively synchronous, as with the expansion of the Roman Empire. Third, there are perturbations that are caused by factors outside the Basin, for example, depopulation since the mid-nineteenth century and climate variability or change. It is tempting to think of these changes as always being synchronous, but a detailed observation of their history shows that this is rarely the case. Depopulation has affected different, usually mountainous, areas over various time periods in the past 150 years or so. Climate variability has also tended to affect specific areas rather than being a Basin-wide event, because of the buffering of the Mediterranean climate by the Basin topography, and internal driving of many events. Thus, these sorts of mechanism can have the patchwork pattern of impacts on the landscape. More synchronous effects of climate change do seem to be much more important over longer timescales. Also in this category, developments relating to the expansion of the EU are even more rapid and affect very large areas at a time. We can therefore think of a hierarchy of mechanisms of change, some driven internally, some externally, and operating on a range of spatial and temporal scales.

Thus, we have a complex environment, driven ultimately by the long timescale generative mechanisms of the region. There are a complex series of interacting mechanisms that operate on the environment, whose complexity is enhanced by various interactions and forcing mechanisms, driven by social and other changes. To a certain extent, it could also be argued that social complexity within the region was initially driven by the naturally variable environment in which people found themselves living.

Past and present

Therefore, it is fundamentally important that we look at the relationship between historical and present-day environmental issues. Recent developments can only effectively be understood in the context of changing human-settlement and land-use patterns. For example, modern depopulation of the mountain zones is only really understandable in the context of the historical context of why settlement expanded to occupy these zones, the resulting internal and human-induced impacts on the local environment, and the broader tides of history that ultimately led these marginal areas to become unprofitable or unlivable for many (McNeill, 1992). In turn, those patterns can only be understood by looking at their precursors. In this book, we have chosen the introduction of agriculture as the fundamental starting point of this process for two reasons. First, because its introduction provided a fundamental rupture in the way people lived, and the ways in which they interacted with their environment. Second, its timing is broadly parallel with the post-glacial warming of the early Holocene (whether this is a coincidence or not will be left until Chapter 11), and thus reflects conditions that are comparable with the modern day (Chapter 3). In many ways, the result of the initial spread of agriculture and its current expansion in particular areas (Chapter 15) are the same. However, the context of the causative mechanisms, and thus potential solutions differ. Without the historical and social framework it is impossible to understand these potential differences. It is this framework we hope to provide – or at least an initial version of it – for the Mediterranean region in this book.

Organization of the book

The book is divided into four major parts. Part 1 is the background to the work both in general terms and in long timescale perspective. Part 2 is entitled *The dynamic environment* and presents an overview of the significant processes and factors – climate, vegetation, hydrology, erosion, fire and oceanography – that make the Mediterranean a dynamic and sensitive set of environments. From this base, we move to human action in the landscape, with Part 3 *The human impact on the environment*. The focus is on populations and population movements, agriculture, pastoralism, the impacts of urban environments and the extraction of mineral resources. Finally, we turn to present-day issues and their potential future impacts in Part 4 *The Mediterranean environment under increasing pressure*. We look at the main themes of change, then concentrate on changes in agriculture, depopulation, impacts on water resources, desertification and pollution. These themes are then integrated so that the potential impacts of climate change can be addressed. In the final chapter, we look at the human frameworks within which many of the major problems in the future can be addressed.

At the end of each chapter, we provide a number of key references that develop in more detail the themes covered in an approachable way, together with some topics for discussion. The aim of these topics is to provide stimulus for further investigation, or to act as subjects for seminar discussions. A number of these topics have been used successfully by us in our undergraduate and Master's classes. At appropriate points, we have included information in boxes. The aim of these boxes is twofold: first, to provide supplementary background information at key points for those who may not be familiar in depth with the topic; and, second, to give some more detailed examples or case studies so that the main flow of the text is not interrupted.

2 The geological setting

Introduction

The geological setting of the Mediterranean Basin is significant not only because of the range of active processes which it helps to explain, but also in providing the backdrop for human activities in recent millennia. In further chapters, however, we will discuss how human activities have begun to modify this landscape setting to an increasingly marked level. This chapter will concentrate on the active features of the geological landscape and how they fit into the geological history and plate-tectonic setting. From this, we will progress to the sort of surface and submarine topography that these features have generated, and the types of resources they make available to human populations living within these environments. Because of the extent of the recent geological activity, this chapter will also focus on some of the hazards that these processes cause.

Volcanic eruptions and their effects

On the 24th of August, about one in the afternoon, my mother desired [my uncle] to observe a cloud of very unusual size and appearance.... I cannot give you a more exact description of its form, than by resembling it to that of a pine tree, for it shot up a great height in the form of a trunk, which extended itself at the top into several branches;... it was at one moment white, at another dark and spotted, as if it had carried up earth or cinders....

...[My uncle] ordered large galleys to be launched ... [and] steered his direct course to the point of danger...

Now cinders, which grew thicker and hotter the nearer he approached, fell into the ships, then pumice stones too, with stones blackened, scorched, and cracked by fire, then the sea ebbed suddenly from under them, while the shore was blocked up by landslips from the mountains....

In the meantime Mount Vesuvius was blazing in several places with spreading and towering flames, whose refulgent brightness the darkness of the night set in high relief ... the court which led to his apartment now lay so deep under a mixture of pumice stones and ashes, that if he had continued longer in his bedroom, egress would have been impossible.... [T]he house now tottered under repeated and violent concussions, and seemed to rock to and fro as if torn from its foundations. In the open air, on the other hand, they dreaded the falling pumice stones, light and porous though they were;... They tied pillows on their heads with napkins; and this was their whole defence against the showers that fell around them....

They thought proper to go down to the shore to observe from close at hand if they could possibly put out to sea, but they found the waves still ran high and contrary.... [S]oon after, flames, and a strong smell of sulphur, dispersed the rest of the company in flight.... He raised himself up ... but instantly fell.

(Pliny: Letters VI, xvi: after Melmoth, 1915)

There had been for several days before some shocks of earthquake, which were less worrying for us as they are frequent in Campania; but that night they became so violent that one might think that the world was not being merely shaken, but turned topsy-turvy....

The buildings around us already tottered, and ... there was certain and formidable danger from their collapsing.... Then we beheld the sea sucked back, and as it were repulsed by the convulsive motion of the earth; it is certain at least the shore was considerably enlarged, and now held many sea-animals on the dry sand. On the other side, a black and dreadful cloud bursting out in gusts of igneous serpentine vapour now and again yawned open to reveal long fantastic flames, resembling flashes of lightning but much larger....

Soon afterwards, the cloud I have described began to descend upon the earth, and cover the sea. It had already begirt the hidden Capreae, and blotted from sight the promontory of Misenum.... Ashes now fell upon us, though as yet in no great quantity. I looked behind me; gross darkness pressed upon our rear, and came rolling over the land after us like a torrent ... darkness overspread us, not like that of a moonless or cloudy night, but of a room when it is shut up, and the lamp put out. You could hear the shrieks of women, the crying of children and the shouts of men....

By degrees it got lighter; which we imagined to be rather the warning of approaching fire (as in truth it was) than the return of day: however, the fire stayed at a distance from us: then again came darkness, and a heavy shower of ashes; we were obliged every now and then to rise and shake them off, otherwise we should have been buried and even crushed under the weight....

At last this dreadful darkness was attenuated by degrees to a kind of cloud or smoke, and passed away; presently the real day returned, and even the sun appeared, though lurid as when an eclipse is in progress. Every object that presented itself to our yet affrighted gaze was changed, covered over with a drift of ashes, as with snow.... [We] passed an anxious night ... for the earthquake still continued.

[Pliny: Letters VI, xx: after Melmoth, 1915]

The above passages are eye-witness accounts of the catastrophic eruption of Mount Vesuvius in southern Italy in 79 CE. In the first, Pliny describes the death of his uncle, and in the second the escape of his mother and himself. The eruption led to the burial of the cities of Pompeii and Herculaneum under layers of cinders and ash, preserving them and their inhabitants until excavation by archaeologists in the nineteenth century.

A similar catastrophic eruption took place around 1,700 years earlier on the island of Santorini in the southern Aegean. Again, the city of Akrotiri on the island was buried – up to a depth of seven metres (Doumas, 1983), although the population seems to have had time to evacuate before the destruction of their homes. So powerful was this explosive eruption, that the island was blown away, forming a crater or caldera which now has an area of $84.9\,km^2$ and a depth of about $300\,m$ below sea level, and ejecting an estimated volume of between 13 and $40\,km^3$ of pyroclastics, or volcanic rocks into the atmosphere (Pichler and Friedrich, 1980; Sparks *et al.*, 1983; Pyle, 1990). Ash from the eruption has been found in sediments as far away as the Black Sea (Guichard *et al.*, 1993), eastern Turkey (Sullivan, 1988), and volcanic glasses have been recovered from the Nile delta (Stanley and Sheng, 1986). Historical accounts of the eruption are not available, but interpretations of the distribution of the tephra deposits suggest that the ash clouds were displaced by winds characteristic of the summer (McCoy, 1980).

Perhaps the most significant aspect of the Santorini eruption is the impact it had on the global scale. The eruption plume has been estimated as being about 29 km high (Wilson, 1980), and thus passed into the stratosphere with a potential worldwide distribution, as in the historical eruptions of Krakatoa in 1883, Tambora in 1815, and Pinatubo in the Philippines, which erupted on 14 June 1991, sending a cloud 40 km high. At this level, the dust from the cloud can potentially modify the global climate (Maddox, 1984) by blocking the sun's radiation from the Earth. Tree rings from Ireland show a significant narrowing for the decade following 1628 BCE, demonstrating relatively poor climatic conditions (Baillie and Munro, 1988). In the White Mountains of California, tree rings from the long-lived bristlecone pine show frost damage in the year 1627 BCE (Lamarche and Hirschboeck, 1984; Hughes, 1988). Recent work on tree rings from archaeological sites in Anatolia also show a remarkable growth anomaly which has also been dated to 1628 BCE (Kuniholm *et al.*, 1996). Ice cores from Greenland also show acid deposition correlating to the same event dating to 1645 ± 27 BCE (Hammer *et al.*, 1987). Combined with the less precise radiocarbon dates from the destruction horizons of Thera (Betancourt, 1987; Betancourt and Michael, 1987; Aitkin *et al.*, 1988; Manning, 1988), the suggestion is that Santorini erupted cataclysmically in late 1629 or early 1628 BCE. The consequences were global and regional – possibly leading to the collapse of the Bronze Age Minoan civilization on Crete, possibly by the effects of tsunamis and air percussion waves (Downey and Tarling, 1984) or by the indirect effects of acid leading to crop damage as discussed below. Monaghan *et al.* (1994) demonstrated that a tsunami could have reached Crete from the Thera eruption, and might have been particularly destructive to ships at sea as well as along the coast of northern Crete. The Lisbon earthquake of 1 November 1755 also generated an important tsunami, with sediments from it deposited up to a kilometre inland in the Algarve (Hindson *et al.*, 1996).

Evidence for the regional-scale effects of volcanic activity comes from the historical records of dry fogs in Italy studied by Camuffo and Enzi (1994, 1995). These phenomena were described by chroniclers as

> malodorous fogs which did not wet surfaces, persisting even into the middle of the day, and appearing even in summer, [and] were often accompanied by red dusks, a *weak sun*, solar and lunar halos, which caused damage to the vegetation and brought in their wake epidemics. They were often associated with storms.
>
> (Camuffo and Enzi, 1994: 138)

These dusts are commonly related to non-violent degassing of volcanoes (see below), as well as being produced as the result of gas, dust and aerosols ejected from eruptions. In some cases, the fogs and corresponding acid rains have been related to damage on monuments of the sort produced by industrial acid rains (Camuffo, 1992). The occurrence of rains following the fogs may be because of volcanic particles forming cloud condensation nuclei. Crop damage was noted in a number of occasions, for example, in 936 and 968 CE fogs associated with a south-westerly wind destroyed harvests in northern Italy. This destruction may be due to acidification from aerosols or the formation of calcium fluorosilicate from tephra. Such fog events are also related to illness, particularly relating to respiratory problems, for example, the widespread catarrhal infections in Venice and Padua following the dry fog of 1767. Animals which reportedly died following an occurrence in 1783 may have eaten contaminated pasture. Camoffo and Enzi (1994, 1995) note the occurrence of twenty-eight definite dry fogs in Italy since 1374, with a further twenty events being possible manifestations. This activity is related to increased activity in the Italian volcanoes since 1500, although effects from volcanoes even further afield – notably in Iceland and the Tambora eruption of 1814 – also seem to have been responsible.

Mount Etna on the island of Sicily is one of the most active volcanoes in the world. Rising 3,290 m from the Mediterranean in the Straits of Messina to a summit with four main craters, the mountain has been built up from the first submarine eruptions of subalkaline lavas between 500,000 and 700,000 years ago in a number of phases (Romano, 1982). Recent activity has produced alkaline lavas with intense, effusive activity, although there have been periods of explosive activity about 80,000 and 8,000 to 3,000 years ago. Etna has erupted about sixty times in the twentieth century (Simkin *et al.*, 1981), with a major event in 1991 to 1993 and again in 2001, and provides a continuous stream of gas emissions. It has been estimated that these emissions produce about $26 \pm 6 \, \mathrm{Tg \, a^{-1}}$ of carbon dioxide, approximately half from the summit crater and the rest from invisible, diffuse emissions from the flanks of the volcano (Allard *et al.*, 1991), which are the equivalent of about 21 per cent of the pre-industrial atmospheric concentration of this greenhouse gas, and therefore are probably a major control on the Earth's climate (Caldeira and Rampino, 1992). Etna is also a major source of atmospheric sulphur dioxide, producing an average $0.55 \, \mathrm{Tg \, a^{-1}}$ (Loyé-Pilot *et al.*, 1986) compared to an annual average of $4 \, \mathrm{Tg \, a^{-1}}$ from explosive volcanic events (Bluth *et al.*, 1993; including up to $0.38 \, \mathrm{Tg}$ from the 1977 [Stoiber *et al.*, 1987] and $1.85 \, \mathrm{Tg}$ from the 1991–1993 Etna eruption [Bruno *et al.*, 1994]). These values are significantly smaller than the sudden inputs from the large eruptions such as Tambora in 1835 and Krakatoa in 1883 which are estimated to have produced between up to $175 \, \mathrm{Tg \, SO_2}$ and $15 \, \mathrm{Tg \, SO_2}$, respectively (Stoiber *et al.*, 1987). Again, this production has important climatic impacts, although it has been greatly overtaken in recent centuries by the impacts of industrial pollution (Bluth *et al.*, 1993).

Etna has a relatively large population living nearby, or even on its slopes. The catastrophic effects that overtook the populations of Pompeii and Herculaneum are largely prevented because earthquakes and tremors usually precede the eruption. Before the main 1991 eruption, there was a small eruption in September to October 1989 and the formation of fissures trending along regional fracture patterns in SSE–NNW and NE–SW directions. The fissure leading SSE from the summit filled with magma between June 1990 and July 1991 (Calvari *et al.*, 1994; Rymer *et al.*, 1995). On 14 December 1991 a large number of small tremors were felt, starting at 1.47 in the morning and continuing until 6.30 in the evening. Altogether, 280 tremors with a magnitude of less than 3.3 (see Table 2.1 and below) occurred. The epicentres of these tremors were located along the SSE fissure, and relate to the more rapid movement of magma. Between 8.00 and 10.30 the following evening, another twenty-five events occurred, with a typical mainshock (magnitude 4.5) and aftershock pattern, as faulting occurred on the NE–SW fault under the volcano, to accommodate the passage of the magma (Ferrucci *et al.*, 1993). Over 473 days, $235 \, \mathrm{M \, m^3}$ of lava were erupted, flowing at an average rate of $5.8 \, \mathrm{m^3 \, s^{-1}}$ (Rymer *et al.*, 1995). As the mountain has been unusually quiet since the end of this eruption in March 1993, Rymer *et al.* (1995) suggest that the next major eruption should follow tremors which indicate that magma is moving and opening another dyke.

Attempts were made during this eruption to divert the lava flowing from the summit, which at one point threatened the 7,000 inhabitants of the town of Zafferana (MacKenzie, 1992). This technique was proved during the 1983 eruption of the volcano, when explosives and massive earth ramps were used to contain the lava flows from destroying property. Although this technique is costly, the 1983 operations costing the equivalent of $3M, it is estimated that property between $5M and $25M was ultimately saved. Further experience at containing lava flows using these techniques was gained during the 1991–1993 eruptive phase, successfully

Table 2.1 Comparison of Mercalli earthquake intensity and Richter earthquake magnitude scales, with a description of the principal effects of each intensity

Intensity Mercalli scale	Magnitude Richter scale	Mercalli name	Description of effects	Ground acceleration cm s^{-2}
I	<3.5	instrumental	not felt: only detected by seismographs	<1
II	3.5–3.8	feeble	noticed only by sensitive people	1–2
III	3.8–4.3	slight	like vibrations from passing lorry; felt indoors, especially on upper floors	2–5
IV	4.3–4.5	moderate	like vibrations from passing heavy lorry; felt generally indoors; loose objects, including standing vehicles swing; windows, doors and crockery rattle	5–10
V	4.5–4.9	rather strong	felt generally; sleepers wakened; bells ring; small objects overturned; doors open and close; shutters and pictures set in motion	10–20
VI	4.9–5.5	strong	trees sway; damage from falling loose objects; pictures fall from walls; furniture overturned; cracks in weak plaster and type D buildings*	20–50
VII	5.5–6.2	very strong	general alarm; noticed in moving vehicles; walls crack; plaster falls; serious damage to type D buildings, some cracks in type C buildings; weak chimneys break off at roof level; plaster, loose bricks, stone tiles, shelves collapse	50–100
VIII	6.2–6.5	destructive	steering of cars difficult; very heavy damage to type D and some damage including partial collapse to type C buildings; some damage to type B buildings; stucco breaks away; chimneys, monuments, towers and raised tanks collapse; loose panel walls collapse; branches torn from trees; changes in flow or temperature of springs; changes in water level or wells; cracks on moist ground and steep slopes	100–200
IX	6.5–7.0	ruinous	general panic; complete destruction of type D buildings; serious damage to type B and C buildings, frequently including complete collapse of type C; lifting from foundations or collapse of frame structures; load-bearing members of reinforced concrete structures cracked; underground pipes burst; large cracks in ground (often leading to building collapse); water, sand and mud ejected in alluvial areas	200–500
X	7.0–7.4	disastrous	ground cracks badly; railway lines bent; most masonry and wooden structures destroyed; serious damage to and some destruction of reinforced steel buildings and bridges; severe damage to dams, dikes and weirs; large landslides; water hurled onto banks of rivers, lakes and canals	500–1000 (≈1 g)
XI	7.4–8.1	very disastrous	few buildings remain standing; even large, well-constructed bridges destroyed or severely damaged; underground pipes and cables break apart; great landslides and floods	1000–2000 (1–2 g)
XII	>8.1	catastrophic	total destruction; objects thrown into air; ground rises and falls in waves, leading to large-scale changes in structure of the ground; changes in subterranean and surface streams and rivers; waterfalls created; lakes dammed up or burst their banks	>2000 (>2 g)

*building types from Degg and Doornkamp (1989)

A good construction, mortar and design; reinforced, especially laterally, and bound together using steel, concrete, etc.; designed to resist lateral forces

B good construction and mortar; reinforced but not designed to resist strong lateral forces

C ordinary construction and mortar; no extreme weaknesses such as failing to tie in at corners, but neither reinforced nor designed to resist horizontal forces

D weak material such as adobe; poor mortar; low standards of construction; horizontally weak

Source: After Holmes (1965); Degg and Doornkamp (1989), reproduced with permission of Nelson Thornes Ltd.

containing the lava flow and preventing destruction in the village of Zafferana Etnea (Barberi and Villari, 1994; Di Palma *et al.*, 1994; Vassale, 1994). Earlier attempts to divert lava flows from Etna were unsuccessful, but not simply for technological reasons. In 1669 an eruption took place on the flank of the mountain, forming the secondary peak of Monti Rossi. Part of the village of Nicolosi was destroyed and Catania was threatened with lava flows (Chester *et al.*, 1985).

The hazards at Mount Etna have been summarized by Kieffer and Tanguy (1994), who split them into five groups: cataclysmic events, explosive phenomena, lava flows, secondary effects and seismic activity. In terms of cataclysmic events, Etna no longer produces the explosive eruptions of the type that destroyed Pompeii and Akrotiri. However, it is still capable of producing violent eruptions when groundwater becomes sufficiently incorporated in the magma. Such conditions were thought to have taken place in the eruption of 122 BCE when crops and buildings 25 km away in Catania were destroyed by volcanic 'bombs'. It is thought that this type of eruption is less likely at present because constant activity in the central chimney of the volcano prevents the build-up of the necessary high pressures. Other types of cataclysmic activity could include the collapse of parts of the mountain, or its failure in landslides and debris flows. These might be triggered by the movement of magma, as well as by seismic activity. The second category of explosive phenomena can also be divided into purely magmatic or combined magmatic–groundwater events. The relatively quiet strombolian activity at the summit can be transformed very rapidly into explosive 'fountains of lava' within minutes, as happened in the eruption of 24 September 1986. Similarly, explosive eruptions may take place on the flanks of the mountain, to form secondary eruptive cones, as in the case of the Monti Rossi eruption of 1669 mentioned above. The risks from lava flows themselves on Etna are relatively low, because the rates of magma production are generally low. When lava is produced at rates of several cubic metres per second, the viscous flows of the basalts travel slowly, and they are stopped by surface cooling layers before they can travel more than two to three kilometres. However, if lava production rates are of the order of $100 \, m^3 s^{-1}$, with initial velocities of several metres per second, then eruptions from the summit can reach the tourist facilities built at an altitude of around 2,000 m, as indeed was the case in the 1983 eruption. Eruptions from lower fissures or secondary cones can endanger the populations living much further down the mountain.

The secondary effects of the eruptions are divided by Kieffer and Tanguy (1994) into three groups. First, ash falls may lead to the destruction of crops (being more direct effects of the 'dry fogs' noted above) and be the cause of accidents by making road surfaces slippery. Indeed, on three separate occasions in 1979, 1986 and 2001, Catania airport which is 30 km from the summit has had to be closed because of ash falls. Second, large debris flows have been caused by eruptions in winter and spring leading to very rapid melting of snow and the corresponding release of large volumes of water which entrain sediment and debris on their way down the mountain. Such events were noted in 1755, 1910 and 1985. The third secondary effect is due to phreatic explosions due to lava flows passing over bodies of snow or water, which heat up rapidly, become incorporated within the lava as vapour, eventually building up sufficiently to lead to an explosion, scattering the lava over a local area. Indeed, this type of explosion caused the greatest direct loss of life on Etna, when fifty-nine inhabitants of Bronte were killed while trying to save their crops in 1843. To these secondary effects, we may add the effects of secondary pollution of sulphur dioxide, as noted, for example, on Trajan's column in Rome by Camuffo (1993). Fluoride pollution from the mountain has been measured by Davies and Notcutt (1988), who found concentrations of up to $141 \, \mu g \, F g^{-1}$ on lichens on the mountain, which are comparable to the worst cases of industrial fluoride pollution, and may ultimately have significant effects by concentration in the food chain. It has been noted above that mixing of water and magma can have catastrophic consequences, but there are more indirect consequences in that some elements are concentrated in the groundwater used for drinking supplies for up to 700,000 people, although the main contaminants – magnesium, manganese and iron – appear not to be present in dangerous quantities (Giammanco *et al.*, 1996).

The final category of risk noted by Kieffer and Tanguy (1994) is that due to seismic activity. Large-scale regional earthquake activity will be discussed in the next section. As we have already noted, more local tremors may be caused by the movement of magma, the opening of eruptive fissures, and in certain cases may relate to adjustments of the structure of the mountain to the end of eruptive activity.

The overall effects of volcanic activity can be defined as direct or indirect. The first group includes the fall of tephra and ballistic projectiles, lava, pyroclastic and other flows, and other effects such as phreatic (stream) explosions, the production of gasses and acid rains and the causing of fires (Tilling, 1989). Indirect effects take the form of earthquakes, debris flows, sedimentation problems, atmospheric modifications (Lamb, 1970), tsunami and post-eruption famine and disease (Tilling, 1989). Examples of most of these have been demonstrated for Mediterranean eruptions, and all can be assumed to be important in certain conditions. One of the major problems with the mitigation of these hazards is that their widespread consequences are usually related to short warning periods (Table 2.2). Although prediction of eruptions is improving, these hazards will continue in the volcanically active regions of the Mediterranean.

Earthquakes

> I felt the ground under me begin to dance. I looked up and saw the ceiling start to come away. I made a run for the door while the dust began to cover everything and everyone, and behind me I heard people crying out.
>
> (Hooper, 1997)

The above quotation is from an Italian journalist reporting the effects of the earthquakes that struck Assisi in central Italy on 26 September 1997. Like the description by Pliny cited above of the tremors accompanying the Vesuvius eruption, it gives an indication of the terror that strikes when the Earth starts to move beneath our feet. The Assisi earthquakes caused ten deaths but were more noted for their destruction of historic buildings and priceless thirteenth- to fourteenth-century frescoes. The various consequences of earthquakes are discussed here.

In 1986, a magnitude 5.7 earthquake hit the area of Kalamata in the Peleponnese in southern Greece. Twenty people were killed in the town of Kalamata and several hundred injured, mostly by falling objects. Delibasis *et al.* (1987) suggest many more deaths would have occurred had a large number of the population of the town not been outside attending the opening of a new ferry service in the harbour. Similarly, seven kilometres to the north-west in the village of Elaiochori, there were only two deaths, despite the fact that approximately 30 per cent of the buildings were destroyed, because people were outside attending the dedication of a new icon for the church. The effects on buildings of different types were governed not only by the different soils on which they were built, but also by the construction materials. Only 0.3 per cent of the concrete structures were damaged beyond repair, whereas 18.5 per cent of masonry structures were irreparable (Delibasis *et al.*, 1987).

Earthquakes in the Mediterranean can also have far more catastrophic consequences. For example, the 7.2 magnitude earthquake which hit Messina in Sicily at 4.20 a.m. on 28 December 1908 caused widespread destruction and the deaths of up to 100,000 people (Miyamura, 1988; Kieffer and Tanguy, 1994). The focus of this earthquake was at a depth of 470 km (Console and Favoli, 1988), reflecting the deep structure in this area (see below). Major disas-

Table 2.2 Warning periods, distances reached and frequency and capacity of damage of various volcanic hazards

Hazard	Usual warning period	Distance reached (km)		Frequency of deaths, property damage and other adverse effects with distance (km)					Capacity to cause severe damage	Probability of death or injury
		average	maximum	<10	10–30	30–100	100–500	>500		
lava flows	hours–days	3–4	>100	F	C	VR			extreme	very high
ballistic projectiles	seconds	2	>5	C					extreme	very high
tephra falls	minutes–hours	20–30	>800	VF	F	F	C	R	minimal–moderate	low–moderate
pyroclastic flows and avalanches	seconds	<10	>100	A	F	R	VR		extreme	extreme
lahars	minutes–hours	<10	>300	F	F	R	VR		very high	very high
seismic activity	none	<20	>50	C	C	VR			high	high
ground deformation	hours–weeks	<10	<20	C	C	VR			moderate	very low
tsunami	seconds–hours	<50	>600	A	F	C	R	VR	very high	very high
acid rains and gasses	minutes–hours	<10	>30	F	F	R	R	VR	very low	usually very low

Source: After Siegel (1996).

ters are not uncommon, with the event of 1169 CE causing 15,000 deaths in Catania, across the Messina Straits in eastern Sicily. Between 60,000 and 100,000 people, of whom 18,000 were in Catania alone, lost their lives in the earthquake of 1693 (Kieffer and Tanguy, 1994). The effects of single earthquakes have been felt across half of the entire Mediterranean Basin, as in the events of 20 May 1202 and 26 June 1926 (Ambraseys and Melville, 1988).

On 6 May 1976, at nine o'clock in the evening, a large earthquake hit the area of Friuli, near Udine in north-eastern Italy. The magnitude of this earthquake was 6.5, with an epicentre about five kilometres below the ground surface. Just over four months later, at 9.21 a.m. on 15 September, a second large shock, with almost as great a magnitude (6.3) occurred. By the autumn of the following year, a series of 3,000 aftershocks had struck the region (Wittlinger and Haessler, 1978). The Friuli earthquake is another example where fatalities were relatively limited because it occurred at a time when large numbers of people were outdoors. Nevertheless, 939 people perished and a further 2,400 were injured. Following the first series of aftershocks, the number of homeless rose from 32,000 to 157,000 as weakened buildings successively collapsed or became too dangerous to occupy. Repair of buildings by the second quake in September meant a reduction of the homeless to around 45,000, but this again increased to around 70,000 overnight. This weakened the spirits and resources of the population, and by the end of 1976, despite extensive emergency measures and international aid, 15,000 people were living in camping trailers, 1,000 in tents, and 25,000 in evacuation centres (mainly unoccupied hotels in tourist resorts) on the Adriatic coast. Another 25,000 people were living in prefabs, temporary shelters and railway coaches. In all, damage was estimated at $4.45 million (Geipel, 1982). Again, the region has commonly been affected by strong earthquakes, for example the one that destroyed Gemona in 1511 (Wittlinger and Haessler, 1978).

In the region of Naples, an even larger earthquake – magnitude 6.8 on the Richter scale – took place four years later on 23 November 1980. In this case, aid was also quick in coming but there were problems in administering it due to the flight of people from Naples, and the mistrust of the local population of central government, not least due to the slow reaction (the earthquake took place on a Sunday evening) and poor organization of the authorities. Furthermore, the most intensively affected areas near the epicentre of the earthquake were difficult to access along the narrow mountain roads of the Apennines. In all, there were more fatalities because of the stronger shock but also because of the fact that large numbers were living in poorly constructed urban houses of four or five storeys. More than 3,500 people lost their lives with another 300,000 being made homeless. Damage was estimated at $15.9 billion (Geipel, 1982). Injuries and damage in this event were accentuated by the poor condition of many houses, which was itself a consequence of extensive depopulation of rural areas (Fulton *et al.*, 1987; see Chapter 16). One significant consequence of earthquakes is to trigger mass movements, because the acceleration of the ground from the shock is sufficient to cause instability in the generally steep terrains of the Mediterranean (see Chapter 6). Reconstruction in southern Italy following the 1980 earthquake has been controlled by central laws requiring geomorphologic and seismic mapping to ascertain the potential hazard of building in areas prone to landsliding. Such areas are estimated to cover 26,000 ha in the Basilicata province alone (Fulton *et al.*, 1987).

Given the geological timescale of seismic activity, it is important to observe earthquake activity on as long a period as possible. The range of techniques used to do so is discussed by Vita Finzi (1986). Ambraseys and Finkel (1987) looked at early seismographic and newspaper and other documentary accounts in Turkey and neighbouring regions for the period 1899–1915, and discovered more than 750 new events, increasing the record seven-fold. These sorts of data are invaluable when calculating probabilities of occurrence. Estimates of magnitudes from damage and surface features show that as well as the known areas on the

Anatolian fault zone, other zones have produced large magnitude earthquakes (greater than magnitude seven) over this time period. For the region, an earthquake of such a magnitude can be expected approximately once every five years, on average. Such an average, however, does not imply that there cannot be periods of heightened activity followed by periods of less activity (see Box 5.2 on return periods). Nur and Cline (2000) reviewed the archaeological evidence for the period at the end of the Bronze Age in the Aegean and eastern Mediterranean. They suggest that the destruction of a number of major sites in the period 1225 to 1175 BCE was related to the occurrence of an 'earthquake storm'.

The risk of earthquakes and their consequences depends not only on the frequency, magnitude and location of the activity, but also on a number of other factors (Degg and Doornkamp, 1989). The characteristics of the ground conditions, that is the distribution of rocks, soils, lakes, marshes, reclaimed land and high groundwater tables can affect the passage of shock waves. The surface geology is also important in this respect with a reduction by one intensity class on the Mercalli scale (see Table 2.1) for location on rock compared to firm sediments, but an increase of intensity of one class for loose sediments such as alluvial deposits or sands, or even 1.5 classes for wet sediments and artificial fills. Many of the large Mediterranean cities in sensitive areas are built on recent sediments, if not artificial terrain, and this will accentuate the damage received in these areas. The type of construction will also affect the damage caused significantly with typically 50 per cent of adobe buildings damaged in an intensity VIII event, compared to 40 per cent for unreinforced masonry, 15 per cent for reinforced concrete with seismic designs, 8 per cent for wooden structures and 7 per cent for steel-framed structures. In an intensity X event, the corresponding losses are 100 per cent, 100 per cent, 58 per cent, 23 per cent and 40 per cent, respectively (Sauter and Shah, 1976 cited in Degg and Doornkamp, 1989). It can be noted that traditional building styles in earthquake-prone areas particularly in the eastern Mediterranean have often evolved to include the use of timber frames, to minimize the impact of earthquake damage (see Box 2.1).

Box 2.1 Effects of earthquakes in Israel

An example of the integration of the factors controlling the impacts of earthquakes can be seen in the study of Degg and Doornkamp (1989), who were able to build them into a GIS scheme for predicting intensity of earthquake effects in Israel. Again, long-term historical data were used to build frequencies of occurrence for different areas, and demonstrate that most intense earthquakes occur along the fault zone which follows the Dead-Sea rift–Jordan valley. In many places to the north of the Dead Sea along this valley, events with an intensity of greater than IX can be expected to occur, on average once every fifty years (Figure 2.1). A second zone of high hazard exists along the coastal zone, largely due to the unconsolidated nature of the sediments, where an intensity VIII event can be expected with the same frequency. The associated risk in this area is high because it corresponds with the areas of highest population in the country. The inclusion of secondary information in this approach suggests that earthquake risks are significantly higher than if a simple frequency-based analysis is carried out. For example, Shapira (1983) analysed the relationship between peak ground acceleration and earthquake magnitude for a period of 144 years to 1980 in Israel. The results suggest an intensity VII event could be expected in Tel Aviv and an intensity VIII event in Haifa and Jerusalem, on average, once every hundred years. Intensity IX events in the latter two cities would have a return period of about 1,000 years (see Box 5.2 for a description of return periods).

continued

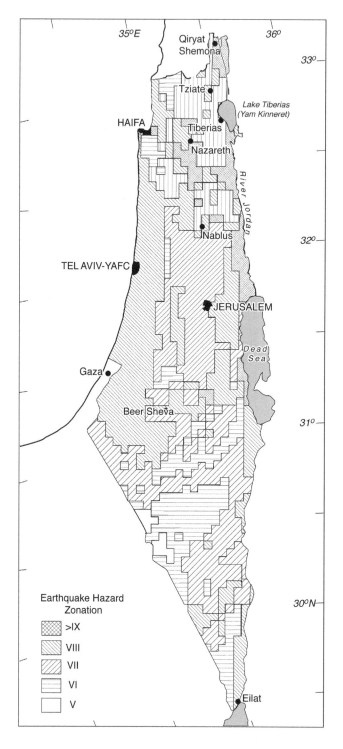

Figure 2.1 Earthquake hazard map of Israel produced by Degg and Doornkamp (1989)

The effects of earthquakes can also be considered as direct – death, injury and damage from the main shock or an aftershock – and indirect – for example, the consequences of landslides and erosion problems, famine and disease caused by infrastructure breakdown. Even with modern technology, the inaccessibility of many remote areas, as demonstrated by the 1980 earthquake in southern Italy, makes relief work difficult. Perhaps the greatest problem is again the unpredictability of events. In some instances, the main quake is preceded by a smaller fore-shock. In these cases, preparations can be made. However, such work is costly to individuals, government and industry, may lead to general panic and often comes to nothing if a major event does not materialize. For example, Thiel (1976) suggests that if three days' warning is given of an earthquake, hazardous areas should be evacuated, and nuclear reactors and petro-leum production pipelines should be closed down. If the warning increases to thirty days, then reservoirs can be drawn down, to prevent the occurrence of potentially catastrophic slope fail-ures and flooding.

The plate-tectonic setting of the Mediterranean

Distribution of volcanoes and earthquakes

Volcanoes

Although volcanic activity has taken place throughout practically all of the Mediterranean through the past 65 million years, presently active volcanoes are limited to two main areas (Figure 2.2a). The first of these occurs to the south of Italy, Sicily and the small islands between. The second is represented by a number of small islands (including Thera) in the southern Aegean, to the north of the island of Crete.

The Calabrian arc of volcanoes (Figure 2.3a) stretches from the island of Pantelleria to the south-west of Sicily and Etna itself to the north through the Lipari (or Aeolian) Islands, to the mainland with Vesuvius, Ischia and the Phlegrean Fields. The Roman god of fire and metal-work, Vulcan, from whose name the word volcano is derived, was believed to live beneath Etna. Many of these volcanoes have received extensive attention, and provide 'type specimens' for different types of eruption. A 'vulcanian' volcano (after the Aeolian island of Vulcano) is one that exhibits moderately explosive activity (Figure 2.4: Holmes, 1965) because of the lava crust that causes gases to build up between eruptions. Another Aeolian island, Stromboli, is used to describe eruptions which are rhythmic or almost continuous. A 'vesuvian' eruption is a more explosive version of the previous types, including the expulsion of magma. Finally, an even more explosive type of eruption, in which clouds of ash and particles are ejected into the upper atmosphere, is termed 'plinian' after Pliny's description of the eruption of Vesuvius, which is given above.

The occurrence of this concentration of volcanoes can be related to the subduction of the oceanic plate beneath the continental plate which carries Italy. Heating of the subducting plate leads to its melting and the formation of magma, which being lighter than the surrounding rock, rises towards the surface. If the magma cools before reaching the surface, it forms a mass of intrusive, igneous rock known as a batholith. If the energy and quantity of the magma are sufficient to allow its rise to continue, it can ultimately erupt in the form of a volcano. The type of magma and therefore the type of eruption are controlled by a number of factors (see, for example, Hatch *et al.*, 1972, for further discussion). Because of the high temperatures involved, these areas can also be important zones of mineralization with considerable economic implications (Sawkins, 1990).

The Hellenic Arc can be explained in a similar way. The actively volcanic islands are found

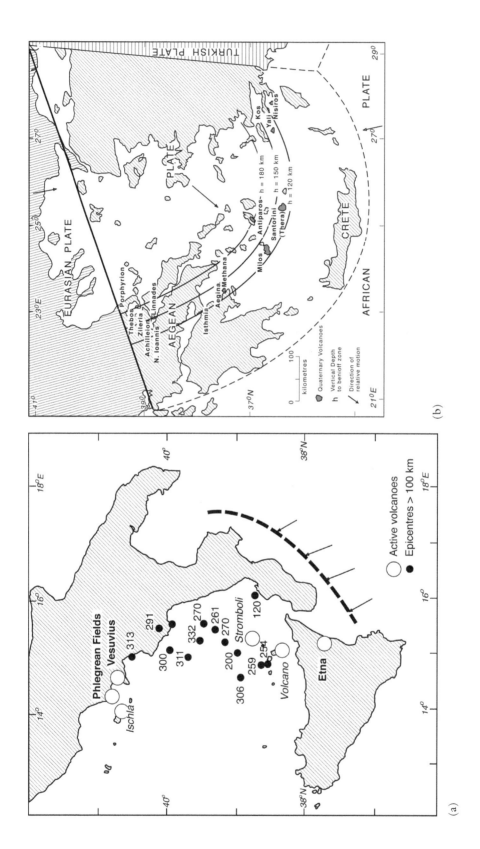

Figure 2.2 (a), (b) Map of locations of active volcanoes

Source: After Simkin *et al.* (1981).

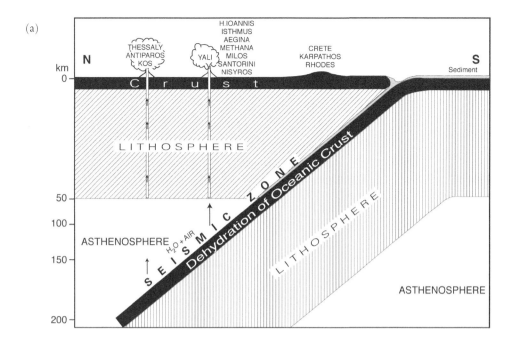

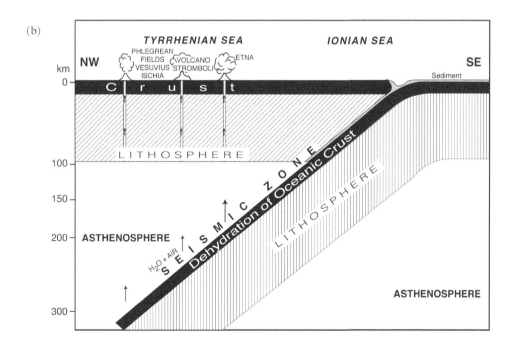

Figure 2.3 Sections through Mediterranean island arcs: (a) Hellenic arc (after Ninkovitch and Hay, 1972; Stanley, 1977; Melentis, 1977); (b) Calabrian arc (after Ninkovitch and Hay, 1972; Stanley, 1977)

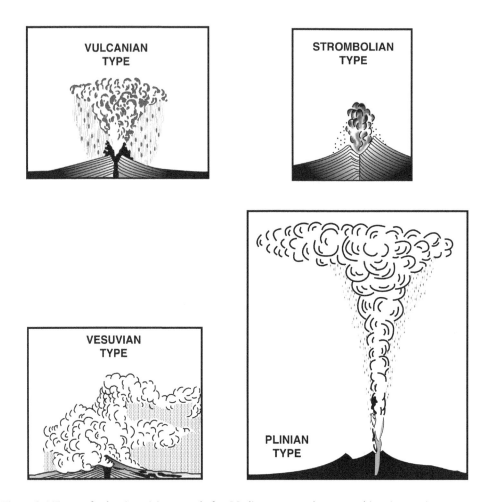

Figure 2.4 Types of volcanic activity named after Mediterranean volcanoes or historic eruptions
Source: After Holmes (1965), reproduced with the permission of Nelson Thornes Ltd.

in two groups. Aghia Theodoro, Aegina, Methana, Melos, Santorini, Nissiros and Ghyali are located about 200 km from the major submarine feature known as the Hellenic Trench (Stanley, 1977: see below for a further description). Antiparos and Kos are located approximately a further 50 km to the north. The Hellenic Trench is thought to be a major subduction zone where the eastern Mediterranean oceanic crust is being consumed beneath the Aegean continental plate. Melentis (1978) divides the Aegean volcanoes into three groups, according to the composition of the magma produced. These groups are the calc-alkaline ('Pacific') group, including the volcanoes of Aegina, Methana, Melos, Santorini, Kalymnos, Limnos, Western Thrace, Halkidiki, North Sporades, Lihades islands and Lokris; the alkaline ('Atlantic') group, made up of Antiparos, Kos, Samos, Mikrothrive and Aghios Evstratios Islands; and the potassic ('Mediterranean') type of Patmos, Nissiros, Ghyali, Samothraki, Mytilini and Imvros (Figure 2.3b). Ninkovich and Hays (1972) relate the different types of magma to the depth to the subducting plate, and consequently the extent to which mantle material is incorporated within.

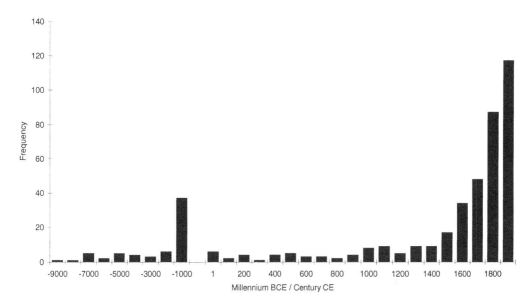

Figure 2.5 Recorded historical eruptions of volcanoes in the Mediterranean
Source: Based on data from Simkin *et al.* (1981).

The history of recent volcanic eruptions in the Mediterranean is relatively well known because of the generally long length of records (Simkin *et al.*, 1981; see also Lamb, 1970). Before the historical period the number drops off rapidly because of the difficulty of producing fine chronologies for small eruptions based on stratigraphic and radiocarbon-dating evidence. The rates of eruption for the Mediterranean as a whole are dominated by the rates for the Italian volcanoes, particularly Etna, which are by far the most active. There seems to have been quite a significant increase in activity in the present millennium, particularly since about 1500, although this may in part be due to the better reporting and distinction of the smaller events. Prior to this, the rates seem relatively constant over the previous two millennia where historical records were noted (Figure 2.5). In the Aegean, where activity in the present arc began about 2.7 million years ago (Horvath and Berckhemer, 1982), there seems to have been a peak of activity in the Santorini group of islands from the 1629/8 BCE eruption until the eruption of Thia in 46 CE. There is then a gap, punctuated only by a subsequent eruption of Thia in 726 CE, until the 1422 eruption of Nissyros. There has then followed renewed activity in the Santorini group since 1570, and three eruptions of Nissyros towards the end of the nineteenth century. Longer-term patterns show evidence for eruptions from a number of other areas. Most notably, the volcanoes of the 'Chaîne des Puys' in the Massif Central of France continued until possibly as late as 80 BCE or even 1050 CE. These volcanoes are now generally considered to be extinct. Volcanic rocks dating from the last 65 million years show activity has been far more extensive, with examples throughout the basin, although with concentrations in the northern basin and north-west Africa (Figure 2.5; Smith and Woodcock, 1982). Given the general relationship of volcanic activity to plate-tectonic activity, this distribution would suggest that extensive and complex plate movements have occurred in the region over this timeframe at least.

Over the longer term, a number of studies have demonstrated a potential feedback mechanism between climate change and volcanic activity in the Mediterranean. Rampino *et al.* (1979) suggested that an apparent correlation of explosive volcanic activity with cold periods over the last 140,000 years might relate to the generation of worldwide stresses concentrated on plate boundaries following the asymmetric mass loading of plates by ice, as in Greenland. A more local mechanism was suggested by Paterne *et al.* (1990), who related an apparent periodicity of 22,000 to 24,000 years in volcanic eruptions in the Mediterranean to glacio-eustatic effects on the plumbing systems of the Campanian volcanoes. McGuire *et al.* (1997) suggested that both rapid sea-level rise and sea-level fall could be mechanisms for increased rates of Mediterranean volcanicity. Sea-level rises of around 100 m result in significant changes of compressive stress within magma chambers at depths of less than 5 km. Unloading due to equivalent falls in sea level reduces radial compressive stresses, favouring the expulsion of magma from island volcanoes (see also Wallman *et al.*, 1988). Secondary effects of sea-level falls could also include slope instability, and structural failures and collapse, again enhancing the likelihood of eruption. In contrast, McGuire *et al.* note that the period of stable, low sea level between 22,000 and 15,000 years ago corresponds to a notably quiet period in the eruptive history of Etna. These results suggest that sea-level changes over the past 2 million years or so (see Chapter 3) may have significantly controlled the timing of Mediterranean volcanism, although Keller *et al.* (1978) have noted that climatic differences may prevent the direct observation of these changes over long time periods.

Earthquakes

Although the spatial occurrence of earthquakes is commonly associated with that of volcanoes, as described above, they are found throughout the Mediterranean (Figure 2.6). However, within this general scatter, there is again a definite pattern. Epicentres pass eastwards into the Mediterranean from the Atlantic. At the Straits of Gibraltar, the trend splits into two, with one group passing through the Baetics in southern Spain and then back into the Mediterranean Sea close to the Balearic Islands. The second group passes through the northernmost parts of Algeria and Tunisia, then continuing across the Mediterranean to Sicily. Here, there is a sharp northward turn, as the concentration of earthquakes passes through the 'toe' of Italy, and up the eastern side of the country. In north-eastern Italy there is a large cluster, passing into a diffuse group in the Alps, but the main group of epicentres swings round, broadly following the Adriatic plate through Croatia, Bosnia and Albania into Greece. The whole of the Aegean is ringed by earthquakes, with one group passing through northern mainland Greece out into the Aegean and continuing through northern Turkey. A second group continues along the Adriatic, passing out into the Mediterranean just to the south of the island of Corfu. This group swings round the island of Crete and continues northwards through western Turkey to rejoin the first. There are more diffuse groups in Cyprus, southern Turkey, the Levant and the Eastern Mediterranean Basin, as well as along the Pyrenees between Spain and France.

When interpreting the significance of earthquakes, it is common to divide them into 'shallow' and 'deep' types, with a threshold between the two usually considered as being about a depth of 70 km. The location of the deeper type of earthquake is much more limited. Apart from a single example in southern Spain, the deep earthquakes are limited to the area to the north of Sicily, with two examples further north in the Adriatic, and in the eastern Mediterranean around the Aegean, Cyprus and southern Turkey. Deep-focus earthquakes are usually interpreted as being related to zones of subduction, where a slab of oceanic crust passes below another plate. The concentrations of deep-focus earthquakes north of Sicily and the Hellenic Trench in the eastern Mediterranean are consistent with the interpretation of the volcanic arcs in these locations. There is probably also some subduction of Mediterranean oceanic crust in

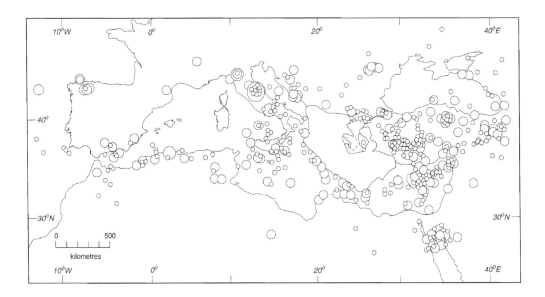

Figure 2.6 Map of locations of present and historic earthquakes

the trench to the south of Cyprus (Robertson and Grasso, 1995). The shallow types are more connected to surface processes. Liritzis and Petropoulos (1992) have suggested that there may be a link between heavy rainfall events and the triggering of large magnitude shallow earthquakes. The mechanism they propose is that the rainfall percolating through cracks which are opening as a result of increasing stress prior to an earthquake can reduce friction and allow the movement of a fault. They found a good correlation between higher than average rainfall and earthquakes in the Athens region.

By analysing the different types of seismic waves as they arrive from an earthquake at different locations, it is possible to define areas of compression and tension in the Earth's crust, and thus the type of movement that has occurred on a fault. From this fault-plane solution, it is possible to determine more details about the structure of the crust, and the type of deformation that is taking place due to plate-tectonic activity. A summary of fault-plane solutions for the Mediterranean shows that the structure is relatively complex. Transform faulting is indicated along the fault passing from the Atlantic into the Mediterranean at or near the Straits of Gibraltar, which is consistent with this feature being an offset from the main spreading ridge in the centre of the Atlantic (Figure 2.7). This pattern of lateral movement is also found along the fault zone in northern Turkey along the Dead Sea rift. Normal faulting, indicating extension of the crust, can be found predominantly in the Aegean (Angelier and Le Pichon, 1978). Thrust or reverse faulting, where compression and shortening of the crust are taking place is common in the Alps and other mountainous areas. Often different types can be found in close proximity to each other, as a result of the complex history and structure of the Mediterranean, as will be seen below.

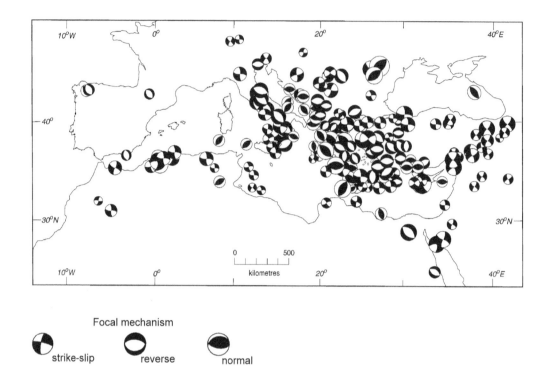

Figure 2.7 Fault-plane solutions of a number of earthquakes in the Mediterranean showing the type of fault movement

Source: Elaborated from McKenzie (1970), Udias (1982) and other sources noted in the text.

Modern geological structure of the Mediterranean

From the evidence discussed above, it is possible to define the modern plate-tectonic structure of the Mediterranean region. Such definitions are based on interpretations of the available data that fall between relatively simple schemes, such as that of McKenzie (1970), and relatively complex ones, for example, that of Dewey *et al.* (1973). Comparing these schemes with those of Biju-Duval *et al.* (1977), Le Pichon (1982), Dercourt *et al.* (1986), Dewey *et al.* (1989) and Robertson and Grasso (1995) shows that there is still major disagreement about the exact number of plates which make up the Mediterranean region as well as the precise locations of their boundaries (Figure 2.8).

The major divisions around the Mediterranean are the Eurasian plate to the north, the African plate to the south and the Arabian plate to the east. Both the Eurasian and African plates have passive (inactive) margins passing out into the Atlantic, where they are separated by a major transform fault. McKenzie (1970) and Dewey *et al.* (1973) suggest this feature passes into a subduction zone which dips under northern Africa, although others suggest the feature is more compatible with compression and thrusting in North Africa and southern Spain (Biju-Duval *et al.*, 1977; Robertson and Grasso, 1995) and combined with more complex transform faulting in the case of Dercourt *et al.* (1986). Stanley (1977) suggests that the Algerian margin is dominated by lateral movement, although profiles through the abrupt boundary indicate the presence of a trench filled with sediments that are being deformed by compression, which may

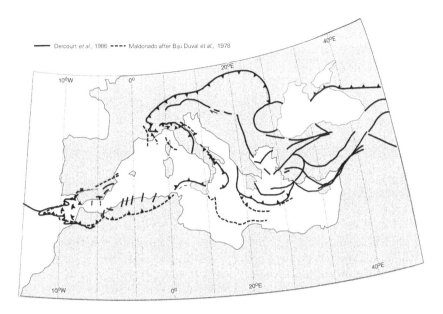

Figure 2.8 Comparisons of different interpretations of the geological structure underlying the Mediterranean region

reflect incipient subduction along this boundary. Continuing east, there is a reasonable agreement that a northward-dipping subduction zone exists (although Dewey *et al.*, 1989, present it as being south-dipping) between Sardinia and Tunisia, passing through the submarine feature known as the Algerian–Tyrrhenian Trough, and curving round south of the Aeolian Islands to pass into Sicily. Continuing through Sicily to the south of Mount Etna, the subduction zone turns in a sharp arc around the submarine Calabrian rise and Taranto Canyon where it continues north as a series of thrust zones along the spine of Italy. The almost-continuous ringing of the Adriatic by thrust zones, together with stratigraphic and palaeomagnetic evidence, suggests that the Adriatic and surrounding areas were formerly part of a separate plate which has subsequently collided with the main Eurasian plate, and the continued thrusting is the representation of the suturing of these two plates.

South of the Adriatic, the subduction zone passes into the Hellenic Trench to the south of Crete and loops up towards Rhodes before swinging back round to the south of Cyprus. Most authors agree on the northward dip of this subduction zone, although Robertson and Grasso (1995) present the Hellenic Trench as a series of offset, southerly-dipping zones opposite the larger northward-dipping zone to the south which is made up of the Mediterranean Ridge and Herodotus Trough. However, this interpretation is more difficult to combine with the location of the Aegean volcanism, and the Benioff zones indicated by earthquake foci (Richter and Strobach, 1978). Biju-Duval *et al.* (1977) interpret this more southerly feature as a gravitational collapse structure. In most interpretations the subduction zone passes into a thrust zone in northern Syria and south-eastern Turkey. Dercourt *et al.* (1986) suggest the boundary passes to the north of Cyprus and into the same zone, although this is incompatible with the evidence of collision of the Eratosthenes sea-mount with a subduction zone south of Cyprus (Robertson *et al.*, 1995). This thrust zone makes up the northern edge of the Arabian Plate, which is bounded to the west by the sinistral transform fault of the Dead Sea Rift. This lateral movement is a consequence of the opening of the Dead Sea, which has taken place over the past 10 million years (Dercourt *et al.*, 1986). The anticlockwise rotation of Arabia also has the effect of squeezing the Anatolian Plate westwards, with the resultant dextral movement along the North Anatolian fault. As this passes into the Aegean, it is translated into an extensional regime, because of the potential of gravitational pull at the subduction zone to the south of the Aegean (Angelier and Le Pichon, 1980).

Background to the Mediterranean Sea

Bathymetry and its relationship to structure

The area of the Mediterranean Basin is about 2.54 million km^2, with an average water depth of 1,500 m and a total length of coastline of *c*.46,000 km, of which *c*.19,000 km are island coastlines (Milliman *et al.*, 1992). The sea contains approximately 3.7 million km^3 of water. The basin can be split into a western and eastern part, divided by the relatively shallow water zone between Tunisia and Sicily (Figure 2.9). The western basin is approximately 0.85 million km^2 in area, and is dominated by the abyssal plains between Sardinia and the Balearics and in the Tyrrhenian, where the depths reach 2,806 m and 3,427 m, respectively. The continental shelves tend to be abrupt in southern Spain, northern Africa, Corsica and eastern Sardinia and the Maritime Alps, where the mountain slopes more or less continue directly into the sea. Larger continental shelves are present from Murcia to the Rhône delta, to the west of Sardinia, north of Tunisia and along most of the Italian coast. In most of these locations the shelves may be due to the thinning of continental crust which took place in the rifting phases when the Balearics, Corsica and Sardinia rotated away from the Eurasian continent, and the Tyrrhenian opened. The abyssal plains are underlain by oceanic-type crust formed during these relatively recent rifting events (see above; Le Pichon, 1982). The western basin forms the only outlet of the present Mediterranean through the Straits of Gibraltar. This narrow zone, 30 km wide and 400 m deep at most forms an important control on the circulation of the Mediterranean and lead to the definition of some of its most significant characteristics, as will be discussed below.

The larger, eastern basin has an area of 1.65 million km^2, and has a much more varied character. There are again two abyssal plains – the Ionian to the east of Sicily and the Herodotus to the north of Libya. While these are deeper than those in the western basin – 4,140 m and 3,219 m, respectively – they are of much smaller extent. Together with a number of submarine rises, valleys and mountains are two roughly parallel features. The Hellenic Trench continues from the submarine Taranto and Otranto valleys south of Italy and the Adriatic and arcs round from the west of the Peleponnese south of Crete, where it splits into three subparallel hollows called the Ptolemy, Pliny and Strabo Trenches, the last of which ends the overall structure in the Rhodes basin to the south-east of the island of the same name. It is in the Hellenic Trench to the west of Crete where the Mediterranean plunges to its deepest point of 4,661 m. As noted above, the features of the trench and the surrounding areas are consistent with its being a subduction zone. In parallel, starting between the Ionian Abyssal Plain and the Peleponnese and arcing round to end between the Rhodes Basin and Cyprus, is the Mediterranean Ridge. This ridge rises from depths of greater than 3,000 m in the Herodotus Trough and Abyssal Plain north of Africa to a high point of 1,298 m below sea level, before descending back into the Hellenic Trench. Initial interpretations of this feature were that it formed a spreading ridge. However, seismic and sea-floor studies have demonstrated that it is a compressional rather than an extensional feature (Hsü, 1978a), and is most probably the result of the compression between the African plate, which is moving northwards, and the extensional Aegean zone, which is moving relatively to the south. In combination with this compression,

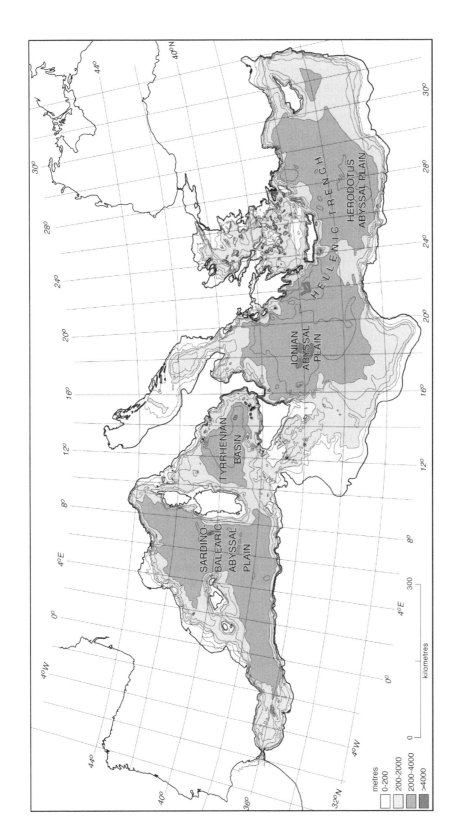

Figure 2.9 Bathymetry of the Mediterranean Sea

the height of the ridge is probably being built up by the accretion of sediments which are scraped off the subducting crust as it descends into the Hellenic Trench (Le Pichon, 1982). Le Pichon believes that the upward motion will continue, and that the Mediterranean Ridge will form a future emergent mountain chain between Crete and North Africa once subduction of oceanic crust is complete within the next 30 million years. The continental shelves in the eastern basin are wide in North Africa, particularly to the east of Tunisia, as well as in the Adriatic, where large areas are less than 100 m deep. Sharp boundaries are common off the Peleponnese, Crete and southern Turkey. In the Aegean, extension of the crust over the past 20 million years or so means that depths are still relatively shallow. The available evidence suggests that the extension is along normal, listric faults rather than by magmatic spreading (Horvath and Berckhemer, 1982). The crust underlying the eastern basin is therefore thinned, continental crust on the peripheries and oceanic in the main basin. The complexities of the basin are probably due to the greater age of the ocean crust, which was largely formed between about 65 and 100 million years ago (Le Pichon, 1982). This greater age means that larger sedimentary covers now mask the oceanic crust. Finally, the ongoing tectonics of compression and subduction structurally modify the crust and its overlying sediments. The eastern basin has two major inflows, from the Black Sea through the Dardanelles and from the River Nile. These inflows are again significant in controlling overall circulation patterns, as will be discussed in a separate chapter.

Messinian salinity crisis

The most marked landscape change in the Mediterranean occurred about 5.75 million years ago. At this time, there seems to have been a marked drop in sea level as there was a complete closure of the seaways between the Atlantic and the Mediterranean, which then existed in the Rif, the Straits of Gibraltar and through the Guadalquivir basin (Weijermars, 1988). Given that the eastern route from the Mediterranean into the remains of the Tethys Ocean (the precursor to the Mediterranean – see below) had already closed about 9 million years ago, this meant that the Mediterranean became a totally isolated basin. Because the temperature remained warm through this period, the net evaporation from the sea continued, and the Mediterranean dried up, leaving behind a series of evaporites. These sediments are made up of chemically precipitated minerals such as halite (rock salt), gypsum and anhydrite. These types of deposit are uniquely found in modern sabkha (coastal flat) environments in arid regions (Hsü, 1983). Given that the major land-based exposures of these deposits are near Messina in Sicily, the period in which this event occurred is known by geologists as the Messinian. During the period up to about 5.32 million years ago, the cycle of closure, desiccation and refilling seems to have occurred a number of times, given the depth of over 1.5 km in places of evaporites (Maldonado, 1983) that are found under the Mediterranean.

A number of points seem to indicate that the Messinian evaporites were deposited in a deep basin which dried up (Cita, 1982). Erosional surfaces created under the sea can be found above and within the evaporites, suggesting that they were subaerially exposed. Deep-sea cores along these surfaces to the west of Sardinia suggest a progression from alluvial to tidal environments. On land, a number of rivers have valley surfaces incised well below present levels and filled with later sediments. Drilling below the Nile during the construction of the Aswan Dam, 1,200 km from the present coastline, showed the presence of a deep canyon 180 m below sea level. The Rhône in France has a palaeovalley beneath it as far upstream as Lyon. This valley is as deep as 500 m below present sea level at the modern coast. Similar features characterize the Var in France and the Ebro in Spain (Clauzon *et al.*, 1996). The presence of these deep incisions, and the deep canyons that can be followed offshore can only be explained by the dramatic change in river base level which resulted from the Mediterranean desiccation. Microfossils in the sediments through the early part of the Messinian show an impoverishment of marine types and the presence of salt-loving types. As the evaporites continue, the faunal remains either die out completely, or are replaced by algal deposits called stromatolites (which are again characteristic of coastal environments: Hsü, 1983) or by salt-loving microfossils, with only occasional brackish water or marine forms interbedded (Cita, 1982). The brackish forms may correspond to the capture of the drainage of inland lakes in the Black Sea area and further east which were the last remnants of the isolated Tethys Ocean in this region. Following the evaporites, planktonic forms characteristic of marine conditions are almost instantaneously found, followed by mollusc and fish. The only types of these which could not have repopulated from the Atlantic are salt-loving forms which probably survived in pockets of saline water within the desiccated basin.

The magnitude of the changes suggested by the desiccation hypothesis for the Messinian salinity crisis led a number of authors to suggest other, less catastrophic models for the deposition of the Mediterranean evaporites (Benson and Rakic-El Bied, 1991; Schmalz, 1991). Selli (1973) suggested that the evaporites could be explained by deposition in shallow, barred basins made up of restricted lagoons. The presence of sabkha deposits that are now in deep basins was explained by subsequent vertical movements of the crust. However, the geological and palaeontological evidence described previously suggests that the Mediterranean was deep before the onset of the crisis. The fact that the evaporites on the seismic profiles from the deep sea follow the modern-day bathymetry of the Mediterranean would also tend to counter the argument that deposition took place in a relatively shallow basin that then foundered (Hsü, 1983; Maldonado, 1983). Furthermore, interbedding of evaporites with deep-water sediments would require unreasonable 'yo-yoing' of the oceanic crust (Schmalz, 1991). A further model suggests that thick evaporites cannot be described with reference to intertidal sabkhas but to other processes which may not be observable, at least easily, in modern environments. The deep-basin model of Schmalz suggests that evaporites may be deposited from brines which concentrate in a barred, deep-water basin such as the Mediterranean became in the Messinian period. If the brine concentrated to an extent that it sank by density difference faster than it could be flushed through the barring sill, then concentration and eventually deposition of evaporites would take place. The possibility of such deposition has been demonstrated in restricted experimental conditions and would continue until circulation or climatic changes altered the sea-water composition. Fluctuations in these conditions could explain the interbedding of evaporites and deep marine sediments. However, this model has difficulty in explaining the observed erosional features and fossil evidence. In all, the deep, desiccated basin model appears the best current explanation for the events of the Messinian salinity crisis.

More recent work by Clauzon *et al.* (1996) has attempted to refine the chronological framework of evaporite formation. They suggest that part of the confusion between the deep- and shallow-basin desiccation models arises from a misinterpretation of the relationship between the evaporites on the continental shelves and the deep-basin evaporites. These have previously been interpreted as synchronous, whereas observation of seismic profiles suggests that the shelf evaporites predate the Messinian erosional surface, whereas the deep-basin evaporites lie above and therefore postdate the erosion. They put forward a two-phase model whereby the shelf evaporites are deposited in an initial period from 5.75 to 5.70 million years ago, as global sea levels dropped by between 10 and 50 m. Although this would have created intertidal conditions on the continental shelves, it would not have been sufficient to isolate the Mediterranean from the Atlantic. During the second phase, from 5.60 to 5.32 million years ago, the deep-basin evaporites were formed as the entire basin dried up. Climatic data suggests there was no marked climatic deterioration during this period which would have caused a sea-level fall, so that the cause of the isolation is more likely to be related to tectonic movements in the Alboran Sea area. There may have been regular overspills from the Atlantic to allow the thick build-up of the evaporites. Evidence suggests that, although the initial drop in sea level may have been glacio-eustatic, following cooling in the southern hemisphere, the removal of large quantities of salt from the oceans of the world may have been responsible for subsequent cooling late in the Messinian period which has been demonstrated from deep-sea cores in the Atlantic and Pacific (Cita, 1982; Weijermars, 1988). The sensitivity of the Atlantic to thermo-haline circulation changes has been demonstrated by the event at the end of the last glacial period, when the injection of large quantities of freshwater into the North Atlantic slowed the Gulf Stream and triggered a cooling phase lasting for up to a thousand years (Broecker and Denton, 1990; Broecker *et al.*, 1990). However, there is little evidence to suggest that this maintained the isolation of the Mediterranean, and the slowing of the relative tectonic-erosion balance at the Straits of Gibraltar seems ultimately to have led to the formation of a gigantic sea-waterfall around 5.32 million years ago. Hsü (1983) estimates that even with an inflow rate of $40,000 \, km^3 \, a^{-1}$ (a hundred times the flow over the Victoria Falls and a thousand times that of Niagara), the Mediterranean would have taken over a hundred years to refill.

Present-day sedimentation

Ongoing sedimentation in the Mediterranean Basin shows a 'remarkable regional variability of depositional thickness and sediment types' (Stanley, 1977). Certain areas, notably off the Rhône, Po and Nile deltas have seen deposition of over a kilometre since the end of the Messinian salinity crisis just over 5 million years ago (Figure 2.10). In these areas, terrigenous materials are deposited by settling in a wedge that thins and contains progressively finer sediments away from the river mouth. These wedges extend for hundreds of kilometres into the deepest parts of the Mediterranean, where the deposition is controlled by variations in the sea-bed due to the salt-tectonics of the underlying Messinian evaporites. In a number of areas in north-western Africa, similar deposits are formed which are dominated by carbonates (Maldonado, 1983). Rates of deposition are summarized by Stanley (1977) who gives values of $22 \, cm \, ka^{-1}$ in the Alboran Sea, $5–9 \, cm \, ka^{-1}$ between the Spanish coast and the Balearics, $15–125 \, cm \, ka^{-1}$ for the Strait of Sicily, $2–4 \, cm \, ka^{-1}$ on the Mediterranean Ridge, $26 \, cm \, ka^{-1}$ in the Hellenic Trench and $>30 \, cm \, ka^{-1}$ in the distal parts of the Nile cone.

Other localized deposition includes organic-rich oozes, sapropel, volcanic ash and aeolian material (Stanley, 1977). Sapropel is a dark ooze, rich in hydrogen sulphide, which tends to form in the deepest parts of the eastern basin where stagnation of the bottom waters can occur. This situation is thought to arise when inflow into the Mediterranean stops briefly, possibly in climatic phases when evaporation decreases relative to precipitation and river inflow (Maldonado and Stanley, 1979), although other explanations have been put forward (see Chapter 8). Because of this large-scale control, the sapropel layers are geographically extensive and can be used in the correlation of sediments from different areas (e.g. Cheddadi *et al.*, 1991). Volcanic ash layers are also useful tools in sediment correlation, as they are again widespread and can be tied to specific sources. Sediments from the Bannock Basin in the central Mediterranean can also be dated by reference to four major Italian eruptions which deposited ash over large areas of the eastern and central basins over the past 155,000 years (Vezzoli, 1991). Similar layers can be traced in sediments on land (e.g. Narcisi, 1996), allowing oceanic and continental long-term sedimentary records to be tied together. This linkage can be important in the study of former climates, as will be seen in the next chapter.

Aeolian deposits are also significant locally, and because of the prevailing atmospheric circulation, tend to lead to a net movement of material from northern Africa into the basin and onto the northern land mass. The role of these deposits is discussed more fully in Chapter 6.

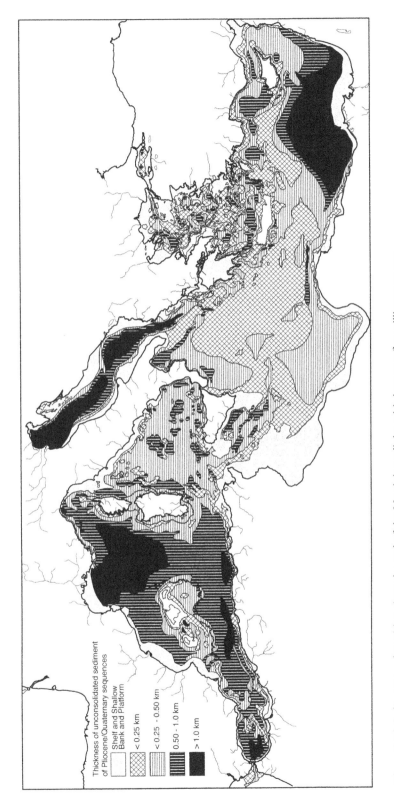

Figure 2.10 Sedimentary deposition since the end of the Messinian salinity crisis just over five million years ago

Source: After Stanley (1977).

Geological history and development – Mediterranean and Tethys

> Like Penelope's canvas, the 'Tethys Sea and Ocean' is a concept incarnated many times.
>
> (Sengör, 1985)

> The Mediterranean, as its impressive size and depth suggest, is no mere trespass of the sea across a continental shelf, as is the North Sea; it occupies a major downfold between two continental masses which had its origin in the Alpine earthstorms of Tertiary times.
>
> (Walker, 1960: 3)

Before the development of technologies for investigating the subsurface and submarine records, the study of Mediterranean evolution was largely equivalent to, and led by, the study of the surrounding mountain chains, in particular the Alps. The Austrian geologist Suess recognized in 1893 on the basis of palaeontological evidence that there must have been a previous seaway running east–west from the Alps to the Himalayas (Jenkyns, 1980). This ocean he named 'Tethys' after the ancient Greek mythological figure who was sister and wife to Okeanos, the god of the sea. The reconstruction of Tethys describes the evolution of the Mediterranean Basin. In recent years, since the publication in 1973 of the classic paper 'Plate tectonics and the evolution of the Alpine system' by Dewey *et al.*, this reconstruction has been carried out within a plate-tectonic framework.

Four main phases can be defined in the geological evolution of the Mediterranean Basin. First, the early history of the Earth's surface follows the development and reorganization of a large number of small plates. Second, by around 505 Ma, three main blocks of crust had

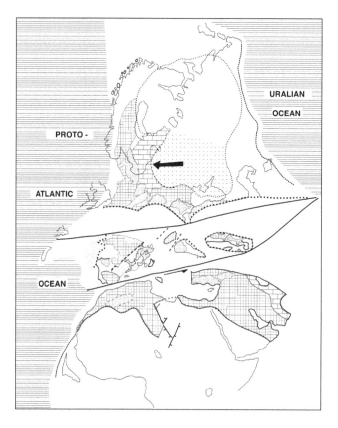

Figure 2.11 Evolution of the Mediterranean Basin: (a) *c.*505 Ma

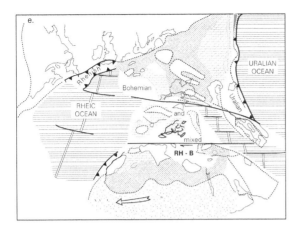

Figure 2.11 (b) *c.*360 Ma

developed in the region (Figure 2.11a). To the north, the main European plate included areas that are now in north-east France, Germany, Switzerland, northern Greece and northern Turkey (Vai, 1991). In the centre lay a block made up of modern Iberia, the rest of France, fragments of Italy and the rest of Greece. To the south, a third block was made up of Africa, Sicily, the rest of Turkey, Cyprus, the Levant and Arabia. Shallow marine conditions were found over large areas of the central and southern blocks. Over a period of about 150 million years, the central and southern areas saw extensive volcanic activity and the rifting apart of many parts of the central block, forming new shallow marine areas. Several phases of rifting and refilling with sediment occurred. With new rifting around 360 Ma moving southern France away from Iberia along the line of the Pyrenees to join the rest of the European plate and an opening up of the crust providing ocean to the south of the present-day Black Sea, the first 'Palaeo-Tethys' Ocean began to form as the first real example of a major, enclosed sea. The 'Palaeo-Tethys' remained as an enclosed sea even after the consolidation of the European, African and American plates into a single plate known as 'Pangaea' (literally meaning the 'whole earth') around 280 Ma.

The third major phase of evolution starts with the rifting of the unstable Pangaea within about 20 million years. This phase led to the evolution of the Tethys Ocean by oceanic rifting along two rift zones – one passing through the area of the Gibraltar Straits to the south of Sardinia, Corsica and the Alps into the remaining Palaeotethys, while the other along northern Africa created the continental blocks that ultimately became the Morocco and Oran Meseta, Sicily and southern Italy and southern Anatolia (Ricou, 1994). Rifting in the central zone of Tethys by around 200 million years ago led to the development of a large number of other crustal blocks – principally the Alboran, Abruzzo-Campanian, Adriatic (Apulian), Karst, Gavrovo, Taurus, Southern Alps, Northern Calcareous Alps, East Central Alps, Moesia and Tisza blocks (Figure 2.11c). The range of marine environments during this period varied from shallow marine carbonates to deep-water oozes, and there was a range of basaltic magma that erupted in association with sea-floor spreading (Dewey *et al.*, 1973; Dercourt *et al.*, 1986; Ricou, 1994). From around this time, the main African block rotated in an anti-clockwise direction in relation to Europe at a rate of up to $1.1 \, \text{cm} \, \text{a}^{-1}$, and this movement accelerated to up to $3.9 \, \text{cm} \, \text{a}^{-1}$ by 130 million years ago (Dercourt *et al.*, 1986; Savostin *et al.*, 1986). Expansion of Tethys by continued sea-floor spreading was accommodated by slower rotation of the European plate. However, during this same period, the actual characteristics of the movement were complex, with more rifting in the northern part of Tethys and a reduction to the south accompanied by the thrusting up of ophiolites (suites of rock indicative of former spreading

Figure 2.11 (c) *c.*200 Ma

centres, with basalt pillow lavas and underlying basic intrusive rocks such as gabbro which are more or less metamorphosed) in the Dinaric and Hellenide Ranges.

Between around 110 and 95 million years ago, the dominant form of movement reversed so that consumption of oceanic crust and collision became the main types of activity (Figure 2.11d). Large subduction zones formed along the eastern part of the southern Eurasian plate, leading to extensive intermediate-type volcanic activity (Dewey *et al.*, 1973; Dercourt *et al.*, 1986). A continental block comprised of land from the Adriatic to the Taurus thus starts to move relatively northwards. At about the same time, lateral faulting along the Pyrenean zone finally brings Iberia to its present-day location in relation to the rest of the European plate (Dercourt *et al.*, 1986; Savostin *et al.*, 1986). A change of direction of the relative movement of the African plate to being dominated by a north-easterly direction by around 65 million years ago meant that the Tethys Ocean began to decrease in width considerably (Figure 2.11e). Thus, the collisional path that led to the Alpine orogeny was set on course – and the ultimate construction of the Mediterranean Basin as we know it today, as indicated by the quotation from Walker given above. By around 50 million years ago, extensive subduction

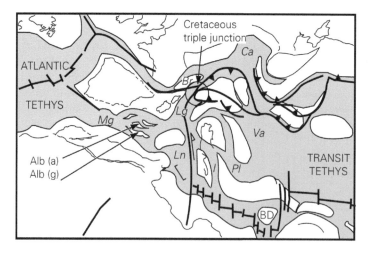

Figure 2.11 (d) *c.*110 Ma

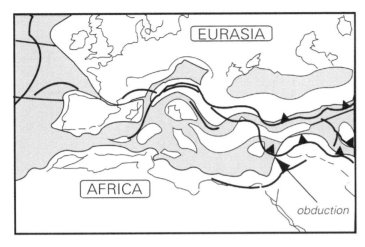

Figure 2.11 (e) *c.*65 Ma

continues in the west of the basin, with the continued development of the Pyrenees, and the onset of deformation and uplift in the Baetics, northern Africa and Pyrenees. Flysch deposition continued around the margins of the Alps, the Hellenides and central Anatolia (Dewey *et al.*, 1973). By around 35 million years ago the present-day land masses were largely in place, although fewer parts were emergent because of the relatively higher sea levels and the early stages of the orogenic process. Iberia became sutured onto the rest of Europe (Figure 2.11f). About this time (Ricou, 1994), or slightly later (Dercourt *et al.*, 1986; Dewey *et al.*, 1973, 1989; Maldonado, 1983), the Corsica–Sardinia block starts to rotate as a result of sea-floor spreading in the present Gulf of Lions. This secondary extension is related to the indentation tectonics of the Adriatic collision with Europe, and is responsible for the rifting of the Rhine–Rhône graben systems. A wide range of sedimentary and igneous rocks were formed at this time through the basin reflecting the great variety of environments. By around 20 million years ago, activity in the Pyrenees ended and there were a relatively complex set of features in

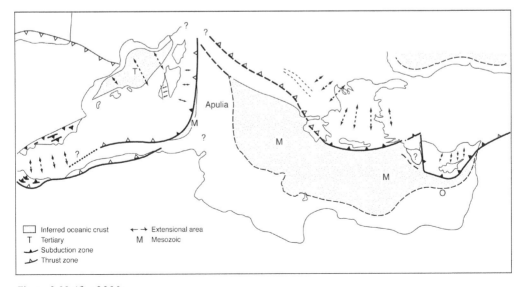

Figure 2.11 (f) *c.*23 Ma

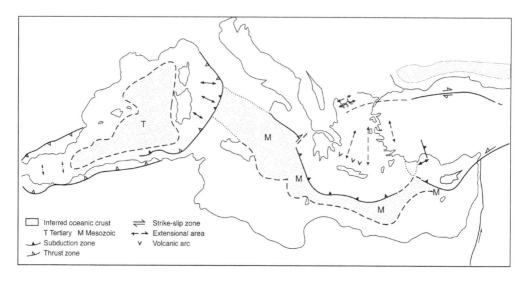

Figure 2.11 Evolution of the Mediterranean Basin: (a) *c*.505 Ma; (b) *c*.360 Ma; (c) *c*.200 Ma; (d) *c*.110 Ma; (e) *c*.65 Ma; (f) *c*.23 Ma; (g) *c*.10 Ma (a. and b. after Vai, 1991; c.–e. after Ricou, 1994; f. and g. after Robertson and Grasso, 1995)

the western basin to permit the oceanic extension of the basin behind Corsica, Sardinia and the Balearics and the rotation of these islands away from the European mainland (Maldonado, 1983; Dercourt *et al.*, 1986; Ricou, 1994; Robertson and Grasso, 1995). Elsewhere, collision continued with the continued uplift of the Alps and other ranges. Around this time, the present extension of the Aegean began due to the presence of the Arabian plate indenting into the Eurasian plate and the development of the subduction zone to the south of the region (Angelier and Le Pichon, 1980; Angelier *et al.*, 1982; Le Pichon, 1982; Robertson and Grasso, 1995). Le Pichon (1982) suggests a later date for the Hellenic subduction, starting around 10 million years ago, with rates of consumption of the oceanic crust of about 4 cm a^{-1}. Depositional environments continue to be increasingly complex. Complex thrusting associated with the collision continues in the Baetics and North Africa to around 10 million years ago. The spreading in the Valencia Trough and Gulf of Lions ceases, leaving Corsica and Sardinia in their present-day locations, but a second spreading centre opened to form the Tyrhennian Sea (Dercourt *et al.*, 1986; Dewey *et al.*, 1989; Ricou, 1994; Robertson and Grasso, 1995). Migration eastwards of the subduction zone led to the collisional association of Sicily, Calabria and the other parts of southern Italy. Thrusting and uplift were extensive from the Apennines to the Alps and Carpathians, and in Cyprus and southern Turkey.

The fourth and final major phase of geological evolution reflects the development of modern conditions. Around 9 million years ago, the link from the Mediterranean in the west to the remnants of Tethys in the east in the Bitlis area of south-eastern Turkey closed (Robertson and Grasso, 1995). This closure was probably related to the initiation of the Dead Sea rift (Ricou, 1994) and the corresponding accelerated northward motion of Arabia at rates of 3 to 3.5 cm a^{-1} (Dercourt *et al.*, 1986; Savostin *et al.*, 1986). From this time, the Mediterranean developed as the enclosed basin we know today. In the past 5 million years, there has been a relative northward movement of Africa at a rate of about 1 cm a^{-1} (Dercourt *et al.*, 1986; Savostin *et al.*, 1986; Dewey *et al.*, 1989). The sutures between the African and Eurasian plates tightened in the west, while the escape tectonics of the Anatolian and Aegean subplates continued in the east (Robertson and Grasso, 1995). The back-arc spreading of the Tyrhennian essentially ceased by about 3 million years ago (Dewey *et al.*, 1989) leaving the ongoing subduction-thrust zone linking northern Africa, Sicily and Italy. The western Mediterranean, underlain by essentially continental crust, is dominated by thrusting from this point, except in the above-mentioned zone between Africa and Sicily. Collision of buoyant fragments of African crust led to the rapid recent uplift of Sicily. Thrusting and uplift also dominate the tectonic activity in the Maghreb and Baetics. Subduction is more dominant in the eastern Mediterranean where the sea is underlain by oceanic or much-thinned North African continental crust (Robertson and Grasso, 1995).

As can be seen from the preceding description, the development of the Tethys and Mediterranean has been a complex process and involves the understanding of the movement, development and modification of a large number of continental fragments. The exact number of these fragments is not known and their origin is often difficult, if not impossible to define with precision (e.g. Ricou, 1994). A much simpler model has been proposed (Hsü, 1978b, 1989; Smith and Woodcock, 1982), which defines the development in terms of only three plates. Between the African and Eurasian plates is a third, made up of the Adriatic, Balkans, Greece and southern Anatolia. Hsü (1989) suggests this third plate rifts from the Eurasian plate around 160 million years ago, during the final stages of the break-up of Pangaea. While this model has much to recommend it in terms of simplicity, it seems that it cannot explain all of the variability of Mediterranean geology, and therefore it has not been adopted here. Among the difficulties encountered with this model are the relative positions of Sicily and the rest of southern Italy; the igneous activity in southern Spain and northern Africa prior to the main collision; the palaeomagnetic evidence amassed by Ricou (1994) and other authors (e.g. Heller *et al.*, 1989); and the stratigraphic data, for example, showing the complexity of the Adriatic area (Ricou, 1994: 188).

Present distributions of structures and rock types

> The five peninsulas of the inland sea are very similar. If one thinks of their relief they are regularly divided between mountains – the largest part – a few plains, occasional hills, and wide plateaux. . . . The Mediterranean is by definition a landlocked sea. . . . It is, above all, a sea ringed by mountains.
>
> (Braudel, 1975: 25)

A look at the topography of the Mediterranean region (Figure 2.12) confirms the virtual omnipresence of mountains around the sea – only the coastal plains of northern Africa, from eastern Tunisia to the Sinai are really free of them. Elsewhere, apart from where large rivers have their deltas, mountains typically plunge straight into the sea, or are separated by only a narrow coastal plain. In a number of locations, high mountain peaks are to be found very close to the coast. For example, Mulhacén in the Sierra Nevada of southern Spain rises to 3,478 m within about 40 km of the coast, the French coastal Alps rise to 3,297 m, Mount Olympus in northern Greece to 2,917 m, the Bey mountains in southern Turkey to 3,086 m and the Moroccan Rif to 2,456 m, all within a similar distance. Mount Etna's 3,323 m are reached directly from the Messina Straits. The Mediterranean topography therefore tends to be very varied and deeply incised, making areas for easy settlement limited, and leading to sharp gradients in climate and vegetation, as will be discussed in the following chapter. The overturning of geological strata due to thrusting and folding in the compression of the Alpine chains means that rocks are unstable and oversteepened, commonly leading to instability.

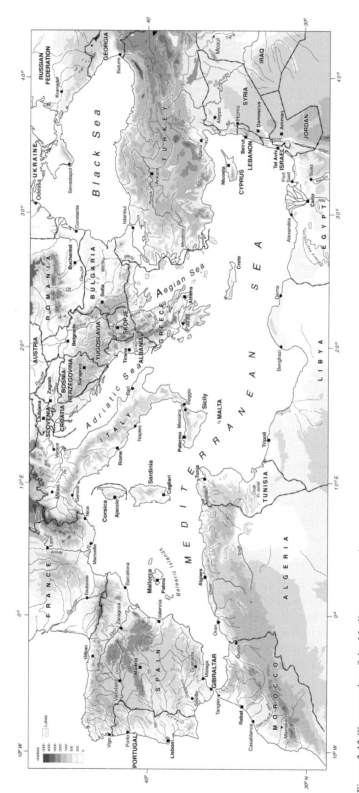

Figure 2.12 Topography of the Mediterranean region

Because of the steepness, which is itself related to the recent uplift of the region, many of the coastal rivers debouch straight into the sea without developing low-energy profiles where they can deposit sediment as coastal plains. This effect is accentuated by the Mediterranean climate, which induces a high-energy 'flashy' flood regime, which again tends to mean that sediment is transported into the sea rather than be deposited beforehand. The very large Mediterranean rivers with better-defined coastal plains and deltas are few in number – the Ebro in Spain, the Rhône in France, the Po in Italy, the Vardar in northern Greece and the Nile in Egypt – are all supplied with water from high mountains which are outside the Mediterranean climatic region, and therefore have more perennial flow regimes (see Chapter 5).

The rock types found around the Mediterranean are also very varied, largely because of the relatively long and complex history of their formation. Particularly within the major mountain belts, the rock type can change from valley to valley quite markedly, or in some cases within the same valley where rocks of significantly different characteristics and ages have been thrust on top of each other. Elsewhere, large normal faults bring rocks of contrasting ages and lithologies next to each other. It is beyond the scope of this book to discuss detailed variations at a finer resolution than regional changes. The major groupings of rock type and their distributions are discussed in the following section.

Resources of the Mediterranean lands

Concern with the resources of the Mediterranean has been important since prehistory, and the relationships between geology and resources have clearly been understood over very long time periods. Indeed, as early as about 1150 BCE, the Egyptians produced maps which related geology to topography (Harrell and Brown, 1992). The 'Turin Papyrus' containing this map includes descriptions of the locations of quarries, in this case 'bekhen' stone, or chloritic sand- and siltstones used for sculptures. The map is believed to have been drawn during an expedition by the Pharaoh Ramesses IV to quarry large quantities of this stone. There are also schematic illustrations of hills which contained gold veins. The descriptions with the map indicate cisterns where water was probably kept for separating this gold by gravity settling.

In the following section, we develop the theme of resource availability in the Mediterranean region. First, the location of mineral resources and soil distributions are discussed on a regional scale, in relation to their main controlling factors. Second, we divide the region into four major geological types – limestone, igneous, metamorphic and marl and unconsolidated sedimentary terrains (Figure 2.13) – and discuss in turn their relief, water and soil resources.

Mineral resources

The Mediterranean region contains a large range of mineral resources which have been exploited economically for millennia. One of the best-known deposits is the Rio Tinto complex in south-west Spain. The 'Red River' is so called for the high iron contents which colour the water. The complex was formed in association with volcanic rocks (Sawkins, 1990) some 350 million years ago. The main part of these deposits are iron sulphides, although copper, lead and zinc are present in appreciable quantities, together with local accumulations of gold and silver (ibid.). The metals were mined from the prehistoric and Roman periods (Domergue and Hérail, 1978: see Chapter 13), although more recent mining has concentrated on extraction of the sulphides to make sulphuric acid (Sawkins, 1990).

Hydrothermal circulation of mineral-rich fluids at oceanic spreading ridges allows the accumulation of minerals in these zones. This process is an important one with regard to the

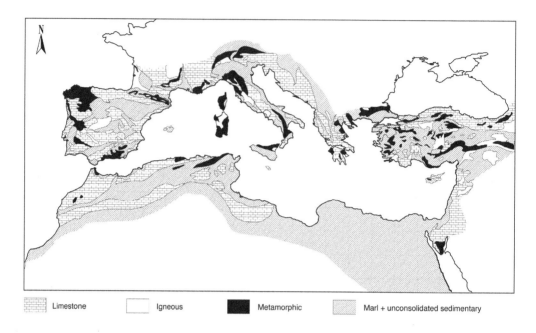

Figure 2.13 Distribution of four major geological types – limestone, igneous, metamorphic and marl and unconsolidated sedimentary terrains

Source: Based on information in Ager (1980), Nairn *et al.* (1978a and b).

Note
See text for discussion.

mineral deposits of the Mediterranean because of the number of spreading ridges which formed during the development of the Tethys and have been subsequently uplifted onto the land. The widespread presence of ophiolite suites as discussed above means that such hydrothermal mineral zones are relatively common. The Tröodos Massif of Cyprus is an important example of this type of deposit – not least because it gives us, via the Roman name for the island, the word copper. Indeed, the Mavrovouni, Limni and Skouriotissa deposits on the island (Figure 2.14) contain some of the world's most extensive copper deposits (Sawkins, 1990). Iron, nickel, cobalt and chrome minerals are also found within the Massif. Chrome is also found in ophiolites from northern Greece, while gold is associated with ophiolites in Italy and Morocco (ibid.).

Minerals associated with collisional tectonic environments are also of obvious importance. In the regions of Portugal, Spain and France which were affected by the Hercynian collisional phase relating to the closure of the precursor to the Atlantic (around 260 to 300 million years ago), there are numerous important deposits of tin and tungsten associated with the emplacement of granites (ibid.). In some parts of the Massif Central of France and west-central Spain, there are also important uranium deposits associated with granites.

Distribution of soils

The general distribution of soils in the Mediterranean region is presented in Figure 2.15. There is a clear distinction between soils in the northern part of the basin, which is dominated by calcic cambisols with localized eutric lithosols and fluvisols (see Table 2.3). Calcic regosols are also important, and cover 30,000 km² in Italy and 20,200 km² in Greece (Louis, 1995). In northern Africa, similar

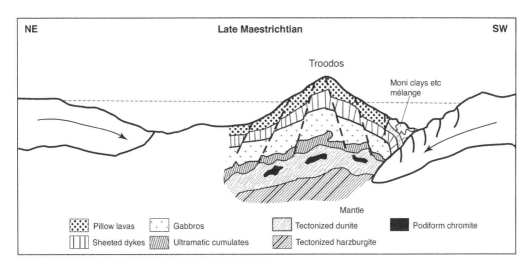

Figure 2.14 Formation of copper deposits in Cyprus

Source: After Sawkins (1990).

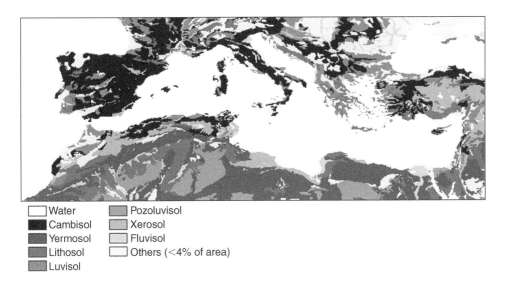

Figure 2.15 The general distribution of soils in the Mediterranean region showing the seven most important soil groups according to the FAO classification (see Table 2.3)

Source: Based on data from the FAO digital soil maps as provided by NOAA (1984).

associations are found in northern Morocco, Algeria and Tunisia. Elsewhere, however, North Africa is dominated by calcic and haplic yermosols, calcic xerosols and orthic solonchaks. The Nile valley area is dominated by calcic fluvisols. The overall distribution is generally controlled, therefore, by local conditions in the northern part of the basin, and by the climate in the southern part of the basin. In general, though, soils tend to be weakly developed and with poor nutrient status. The following sections discuss more specific details of soils on particular terrains.

Table 2.3 Explanation of different soil types used in the text (based on FAO, 1974; and Fanning and Fanning, 1989), together with approximate relative frequency of occurrence in the Mediterranean Basin defined by analysis of digitized soil data (NOAA, 1984) at a resolution of *c.2′*

Soil units	Characteristics	Occurrence %
Cambisol	Soils with at least 25 cm of: (i) moist, dark-coloured surface horizons; or (ii) light-coloured sub-surface horizons altered by pedogenesis	22.8
Yermosol	Similar to a xerosol, but with only a very weakly developed surface horizon	19.6
Lithosol	Shallow, high stone content with a depth to bedrock of less than 10 cm	13.2
Luvisol	Soils with a clay-rich B horizon	9.6
Podzoluvisol	Soils with an irregular, clay-rich B horizon	7
Xerosol	Soils other than those above that have an arid moisture regime and a weakly developed surface horizon	7
Fluvisol	Derived from river sediments with no more than a weakly developed surface horizon	4.2
Rendzina	Having a developed surface horizon overlying a carbonate-rich horizon	2.2
Kastanozem	With a dark-coloured, developed surface horizon overlying calcic- or gypsum-rich layer	2.1
Chernozem	Similar to kastanozems but with a lighter-coloured surface layer	2
Regosol	Dominated by fragments of the parent material with little or no development of horizons	1.9
Vertisol	Little horizon development because of continual shrink-swell behaviour of clays; usually extensively cracked at the surface	1.3
Podzol	Bleached upper horizons with concentrations of salts at lower levels	1.2
Phaeozem	Other soils with a developed surface horizon	1.2
Solonchak	Produced by salinization of the soil profile with no more than a weakly developed surface horizon	1.2
Arenosol	Coarse (sandy) soils, with little definition of soil horizons	0.8
Ranker	Thin, with only a poorly developed surface horizon	0.5
Rock	No soil development – bedrock at surface	0.5
Acrisol	With a clay-rich B horizon but low nutrient content	0.5
Planosol	Soils with poor permeability and hydromorphic characteristics	0.5
Andosol	Relatively poorly developed with a low bulk density	0.3
Gleysol	Soils with simple horizon development and hydromorphic characteristics	0.3
Solonetz	With a sodium-rich B horizon	0.1
Salt	No soil – evaporite at surface	0.1
Histosol	Thick surface horizon of organic material	0.02
Greyzem	Light-coloured surface horizon with light coatings on soil peds	0.01
Nitosol	With a clay-rich B horizon but lacking shrink-swell or iron-rich characteristics	0.01

Sub-units

Calcic	Contains large amount of secondary carbonate by illuviation	
Eutric	Fertile with high content of nutrients	
Ferric	Contains large amount of iron	
Haplic	Simply structured	
Vertic	Has a high clay content and is often cracked	

Limestone terrains

From the discussion of the geological evolution of the Mediterranean above, it can be seen that limestone deposition has been extremely common during the history of the basin, due to its continued location in the subtropics and frequent occurrence of shallow marine environments. Areas such as the *Karst* of the Dinaride range have become type sites for limestone terrains and many of the features are described by terms in Serbo-Croat (as indeed is the term karst, which describes the general landscape dominated by dissolution, with caves, pinnacles, limestone pavements, and so on). Other areas dominated by limestone and karst are to be found in southern France (whence a number of other karst terms are derived), southern Spain, and Greece, among others. Similar features can occur where other soluble rocks outcrop, for example in the gypsum karsts in Almería, Spain.

Relief

Karst landforms are reviewed in great detail in specialized texts (e.g. Sweeting, 1972; Ford and Williams, 1989; Jennings, 1985), and so we will only give a brief overview here. Because of the solubility of the bedrock, these landforms are characterized by the lack of surface water and the presence of hollows or pinnacles relating to solutional processes. Where uplift is important, large, dry plateaux can develop, separated by deep gorges which make movement across the landscape very difficult. The sides of these gorges frequently contain rock shelters or the openings of caves, used for human occupation. The plateaux may contain dolines, dry valleys and limestone pavements. Where the latter have been denuded, very irregular topography results, as seen for example in the areas of 'lapies' in southern France. A variety of swallow holes (ponors) or jamas may connect the surface to great depths. For example, the largest known jama, at Pierre St Martin in the French Pyrenees descends to a depth of 1,971 m from the surface, while another example at Pološka in Croatia descends to 658 m. In areas where rocks have been folded, enclosed basins may be formed because surface water is able to escape to the subsurface. These enclosed basins are termed poljes, and may vary in size from 0.5 km² to 500 km².

Water resources

A full overview of karst hydrology is again beyond the scope of this book (see Bonacci, 1987; Ford and Williams, 1989, for further details). The upper part of the bedrock tends to be very strongly fissured because of weathering processes, allowing the rapid storage of relatively large amounts of rainfall. Below a depth of about 1 m, the bedrock tends to be more massive, and flow tends to concentrate along fissures where the continual flow of water has dissolved the rock along existing planes of weakness. These fissures are the most important pathways in transferring water to the water table.

Karst zones are able to store water rapidly in the subsurface, where it cannot be evaporated to the atmosphere, and subsequently transferred back to the surface via springs. This process is a significant one for water resources in the Mediterranean region, where surface water is otherwise limited. For example, Fabre (1989) notes the usable water supply from karstic sources in Languedoc is of the order of 580 Mm³ a⁻¹. Limestone and dolomite aquifers also supply water to most of Israel (Kronfield *et al.*, 1988). Permanent springs are commonplace in karst zones, providing average flows of $0.1–30 \, \text{m}^3\text{s}^{-1}$ (Bonacci, 1987). Sometimes they may issue from deep vertical shafts, as in the case of the Fontaine de Vaucluse in southern France (after which this type of 'vauclusian' spring is named), where the shaft is of the order of 300 m deep and peak discharges of $150\,\mathrm{m^3\,s^{-1}}$ have been recorded (Fabre, 1989). Other specific types of karst spring include estavelles, which have underwater openings, vruljes (submarine springs found, for example, in the Adriatic), potajnice (a spring with rhythmic flows due to siphoning) and cave springs. An example of the latter, at Dumanli in Turkey has an average flow rate of $25\,\mathrm{m^3\,s^{-1}}$, and a measured peak flow rate of $70\,\mathrm{m^3\,s^{-1}}$ (Bonacci, 1987).

Soils

Because of the widespread nature of the limestones, the soils which tend to form on them are also those which are most commonly associated with the Mediterranean region as a whole. These so-called *terra rossa* can be considered to be a relatively broad group of soils. Yaalon (1997) defines the controls on the terra rossa as being fourfold. First, the Mediterranean climate, in particular the long, dry summer affects the weathering of iron minerals, producing the characteristic colour. Second, the erosional regime only permits the build-up of thin soils given the low amounts of insoluble residue that result from the solution of the limestone. Third, dust input is an important process in building up these soils, as discussed in detail in Chapter 6. Fourth, human activity has been significant in changing soil profiles and hydrologic regime. One specific consequence of this is the very common occurrence of stone fragments at or near the soil surface, which again has significant feedbacks on the soil hydrology and its response to erosion (see Chapter 6).

A *terra rossa* soil is described by Tavernier (1985), developed on a cherty limestone near Rieti in Italy. It is composed of a series of thin horizons, with a 10-cm thick dusky red surface layer, and two horizons continue down to 50 cm and have a reddish brown or dark red colour and incorporating some bedrock fragments (Figure 2.16a). Below this level, the soil is made up of limestone and chert fragments mixed with bright red clays. The texture is dominated by clays and silts. Compared to other Mediterranean soils, the organic carbon and nitrogen content of the surface layer is relatively high, as is the cation-exchange capacity. All of these factors can emphasize soil fertility, and high organic components of the surface layer can help to minimize erosion (see Chapter 6). MacLeod (1980) describes very similar *terra rossa* soils from Epirus in Greece.

Igneous terrains

The igneous landscapes tend to be more restricted due to the nature of the emplacement of the rocks, although very large granite batholiths dominate areas of central and north-western Spain. Other granitic zones can be found in the cores of the collisional mountain chains, such as the Pyrenees or Alps. The granites also dominate certain islands, in particular Corsica and Sardinia. The next most extensive igneous terrain is the ophiolite areas, although these rocks also possess metamorphic characteristics. These areas are made up of basic and ultrabasic igneous rocks which contain lower silica contents compared to the granites, and therefore contain more mafic minerals which tend to be less stable to weathering processes. The ophiolites ring the northern shores of the Mediterranean in relation to the collisional zones described above, with some limited examples in north-west Africa, the Sinai and the Levant (Dewey *et al.*, 1973), although it has been suggested that the Moroccan ophiolites belong to a much older collisional event than the others (Leblanc, 1976). A third group of igneous terrains are the volcanics, which tend to be limited in areal extent because of the nature of their production. Apart from the active volcanoes mentioned above, there are extensive volcanics in east-central Italy, Sardinia, the Massif Central of France, the Balkans, north-east and central Turkey and in the Levant, as well as smaller groupings elsewhere (Smith and Woodcock,

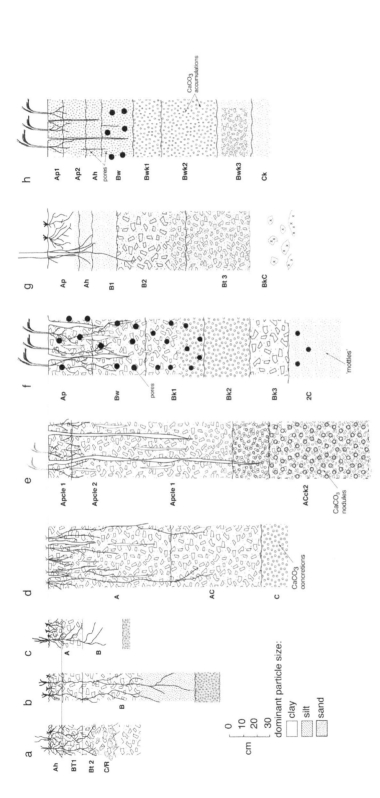

Figure 2.16 Examples of typical soil profiles from the Mediterranean, as described by Tavernier (1995): (a) *terra rossa* on a cherty limestone near Rieti in Italy; (b) a pellic vertisol derived from igneous rocks in southern Portugal; (c) a ferric luvisol derived from schists in Portugal; (d) a 'Black Mediterranean Soil' (pellic vertisol) derived from marls near Florence in Italy; (e) a calcaro-pellic vertisol derived from Tertiary, calcareous lacustrine deposits in Thrace, Greece; (f) a calcaro-vertic cambisol derived from lagoonal clays in south-east Italy; (g) a calci-vertic luvisol derived from Miocene clays near Badajoz, Spain; and (h) a calcic xerosol derived from Quaternary alluvium in Murcia, Spain

1982). Again, the weathering characteristics of the volcanics vary from the other groups, resulting in different landscapes and soil types.

Relief

Because most igneous areas around the Mediterranean are related to recent tectonic activity, they tend to be associated with heavily dissected, mountainous terrain. Areas associated with granites tend to have a more rounded form than the other igneous rocks because of their specific weathering properties. Some of the areas of Italy where recent volcanic rocks are at the surface do not fit into this general pattern, though, and tend to have a more rolling topography. Large areas of south-western Spain and Portugal which are underlain by igneous rocks also tend to have a less marked topography because the rocks in question were formed much earlier, in the Hercynian orogeny, and have thus undergone weathering and erosion processes for much greater lengths of time.

Water resources

In the highest mountains, certain igneous terrains produce perennial or almost-perennial streams, particularly in places where they are overlain by more permeable rocks. Locally, they generate hydrothermal circulation and hot springs, which have been attractions for their curative properties. For example, there are large numbers of hydrothermal springs in the volcanic area surrounding Rome (Bono and Boni, 1996a). Bono and Boni (1996b) noted that some of the largest bottled mineral water productions in Italy derive from aquifers in volcanic rocks, and that there is good evidence that their therapeutic properties were exploited in Roman times. At lower altitudes, unless fed by sufficiently active springs, there tends to be an absence of surface water apart from in flood events.

Soils

Igneous soils also tend to have relatively thin surface horizons, but may be thicker overall compared to the limestone soils discussed above (Figure 2.16b). Because of the nature of the weathering of the bedrock, these soils may contain more sand-sized fragments. Although the cation-exchange capacity tends to be high, the organic carbon and nitrogen contents of the surface layer tend to be low. Yassoglou *et al.* (1997) note that red soils form on basic igneous bedrocks in Greece, although they tend to have a more purple-red hue than other red Mediterranean soils. These soils only have a bright red colour in areas where they are associated with limestones. The authors note that these soils tend to be more resistant to erosion than other red soils.

Metamorphic terrains

Metamorphism, as a consequence of the huge tectonic forces that have developed during the collisional periods of the Mediterranean history, is also widespread in the northern edges of the basin. Deformation and low-level metamorphism are common to the mountain belts from the Rif and Baetics in the west to the Taurus and Pontic ranges in the east. High-pressure metamorphism is more limited, occurring in a small number of zones in the Baetics, Corsica, Apennines, Alps, central Balkans, Hellenides, the Taurus and Pontic zone (Dewey *et al.*, 1973). Marble – the result of metamorphism of limestone – is relatively common and an important resource for ornamental building in northern Italy, Greece, and to a lesser extent in southern France.

Relief and water resources

The general comments with regards to the relief and water resources of the igneous terrains are also largely relevant to the metamorphic terrains. Hydrothermal springs can again be important. For example, at the contact between phyllites and the overlying dolomites in the Alpujarras of southern Spain, there are a number of springs, including some which are used to feed spas. The most important site of springs, at Lanjarón, provides spa water as well as water which is bottled and used for drinking water throughout Spain.

Soils

Metamorphic soils are often relatively thin and dominated by the clays released from the generally unstable mineral assemblages that make up the bedrock. As with other soils with high clay contents, they become very hard when dry. The texture is similar to those found in igneous terrains, but with moderately high cation-exchange capacity, organic carbon and nitrogen contents (Figure 2.16c). Yassoglou *et al.* (1997) note that bright red soils can also form on mica schists, gneiss or marble in Greece. The redness in these soils is due to the formation of haematite in the xeric Mediterranean environments. Indeed, red soils are found elsewhere on marble bedrock, for example, in the marble quarries at Caunes in southern France.

Marl and unconsolidated sedimentary terrains

In this section, two geologically different lithologies have been associated because of the characteristic ways they weather and erode. The marls – essentially muddy limestones of marine or lacustrine origin – tend to be weakly lithified and thus are relatively erodible. Marls were relatively commonly deposited around the edges of the carbonate platforms of the Tethys and Mediterranean, such as in southern France, or in lakes and isolated marine areas such as southern Spain. Recent sediments have similar characteristics because they have undergone relatively weak compaction and diagenesis. They can either be marine sediments which have been uplifted in the recent past – for example, the clays and silts in eastern and central southern Italy – or terrigenous sediments which are the result of uplifted mountains and newly formed basins. Several basins around the Baetics in southern Spain exhibit this pattern, but they can also be found in Greece and Turkey.

Relief

Perhaps the most characteristic feature of this type of terrain is where it has become deeply eroded due to the unconsolidated or weakly lithified nature. These areas typify the 'badlands' – incised gully systems with steep, relatively bare slopes – which are common in southern Spain, eastern and central Italy, Greece and in Morocco (e.g. Scoging, 1982; Alexander, 1982; Imeson *et al.*, 1982). Local examples are to be found in the 'terres noires' or black marls of the French pre-Alps and in Israel (Yair *et al.*, 1982; 1994). In some areas – particularly the more localized example such as the Italian calanchi badlands – the topography can be more rounded (Alexander, 1982).

Water resources

These types of terrain are generally devoid of surface water, except during heavy rainfalls, when flows occur very rapidly. Because of the low infiltration (see Chapter 5), these flows can reach flood proportions and transfer water from the local settings rapidly. However, marls and marine clays will tend to act as aquitards at depth, while the unconsolidated conglomerates will act as aquifers. Indeed, the latter are often important areas of recharge into groundwater because of their linkages with surface watercourses.

Soils

Because of the relatively wide variety of lithologies making up this grouping, there is a wide range of soils that have resulted from it. Soils derived from marls or clay bedrock have very fine particle sizes and thus have a major tendency to crack and bake hard on drying (Figure 2.16d to f). These soils may be moderately deep, tending to grade imperceptibly into the underlying parent material. Cation-exchange capacities tend to be moderate to high compared to other Mediterranean soils, and there are often moderately high organic carbon and nitrogen contents at the surface, if the surface horizon has not been removed by erosion. Some soils derived from underlying clays may actually be relatively sandy due to selective weathering and erosion processes (Tavernier, 1995; Figure 2.16g). In these cases, the soil may have very low organic carbon and nitrogen contents, and much lower cation-exchange capacities than other Mediterranean soils. Soils derived from alluvium on the other hand can be relatively deep, stony and sandy (Figure 2.16h). They may have moderately high organic carbon and nitrogen contents.

Summary

The Mediterranean is still a geologically active region, with constant activity in the form of earthquakes, volcanoes and uplift of the land masses. This activity means that the landscape continues to develop in a highly dynamic way, and this has important consequences on the climate, vegetation, hydrological and erosional characteristics of the region. Such dynamism can create a hazardous environment, both locally from volcanic eruptions or the devastation of large earthquakes, and over larger areas. These larger scale problems can relate to climate modification due to volcanic activity, or tsunamis that follow large eruptions or earthquakes, as well as more pervasive consequences of these events such as landsliding, the destruction of buildings and the diversion or contamination of water courses. As well as hazards, it must be remembered that such geological activity has provided benefits, most notably in the form of mineral resources.

Suggestions for further reading

Geipel (1982) gives an excellent account of the occurrence and aftermath of a large earthquake in the region. The discovery and problems with interpretation of the Messinian salinity crisis are presented in Hsü (1983). Vita Finzi (1986) provides a useful overview of techniques used to reconstruct the history of seismic activity with a number of Mediterranean examples.

Topics for discussion

1 What are the hazards brought by the geological setting of the Mediterranean Basin? Does it offer any advantages?
2 What clues to plate-tectonics are offered by the locations of volcanoes and earthquakes?
3 Explain the locations of the major elements of Mediterranean topography (and bathymetry).
4 What are the main geological controls on available resources in the region?
5 Contrast catastrophic with more slowly acting elements of landscape evolution over the geological time period.

Part 2
The dynamic environment

3   The Mediterranean climate and its evolution

Meteorology

Regional circulation patterns

The general patterns of climate in the Mediterranean are controlled by its location in relation to the general circulation patterns of the Earth's atmosphere. On the largest scale, the climate is controlled by the location of the subtropical high. During the summer months, the Azores anticyclone is typically well established to the west or north-west of the Iberian peninsula, leading to the blocking of frontal systems from the Atlantic. To the south, the low-pressure zone associated with the northward migration of the inter-tropical convergence zone (ITCZ) borders the region from the Sahara to central Asia (Barry and Chorley, 1992; Figure 3.1a). At this time, circulation in the Mediterranean is dominated by two main forms (Dayan and Miller, 1989). In the first (Figure 3.1b), high-pressure zones over central Europe and the west coast of Africa combined with low pressures over northern Europe and northern Africa lead to the development of a high-pressure trough over the Mediterranean, with regional easterly wind patterns dominating. The second type is characterized by a high-pressure cell in the central part of the basin (Figure 3.1c) leading to westerly winds in the north and easterlies in the south of the basin, although northerly winds dominate the Aegean. These patterns can also occur in the winter, when they are responsible for producing foggy conditions (Conte and Colacino, 1995). Depressions that do form in the summer tend to be weak because the regional anticyclonic activity encourages subsidence of air masses. However, weakening of the highs over North Africa can lead to the enhancement of convective activity over the eastern Mediterranean Basin (Kutiel and Kay, 1992). Thermal lows, due to intense daytime heating of land masses can occur in the summer over the Iberian and Anatolian peninsulas, but these produce little rainfall as the relative humidity of the atmosphere is low (Barry and Chorley, 1992; Martyn, 1992).

The winter months are dominated by westerly circulation as the Azores anticyclone is displaced to a more southerly location. The region is typically bounded by high-pressure cells over the Sahara and north-central Europe (Barry and Chorley, 1992; Figure 3.2a). The high to the south is reinforced by the location of the westerly tropospheric jet stream. As the blocking effects of the Azores anticyclone are much less marked in the winter, circulation patterns are much more variable, with up to eleven types of circulation in the winter months (Dayan and Miller, 1989; Conti and Colacino, 1995):

1   The development of a blocking high over the north Atlantic zone leads to the development of low-pressure conditions in a trough along the length of the Mediterranean Basin. Large pressure gradients combined with the strong sea–land thermal gradient lead to

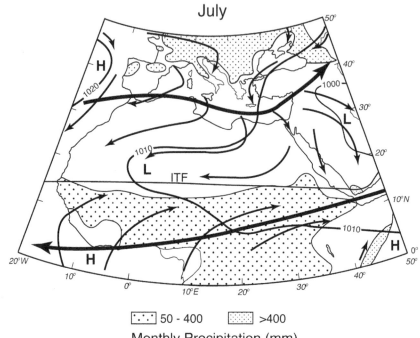

July

50° 40° 1000 30° L 20° 10°N 0° 50°

1020 H

1010

L ITF

1010 H

H

20°W 10° 0° 10°E 20° 30° 40°

[∴∴∴] 50 - 400   [▓▓] >400

Monthly Precipitation (mm)

(a)

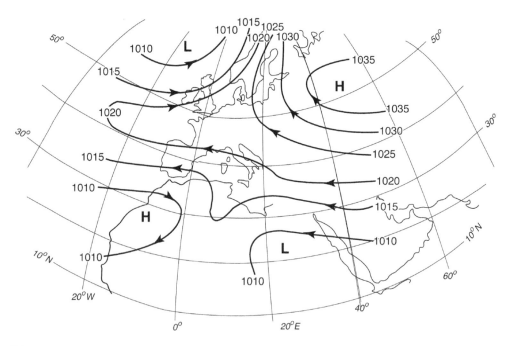

(b)

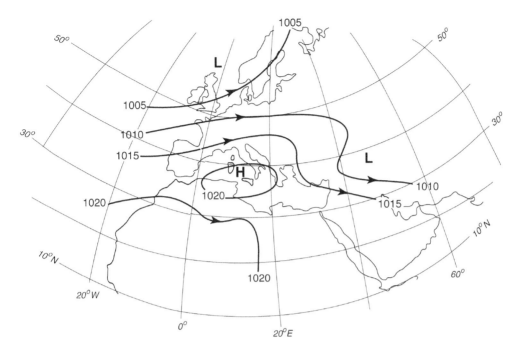

(c)

Figure 3.1 General characteristics of atmospheric circulation in the summer: (a) summer circulation patterns (from Barry and Chorley, 1992); (b) synoptic conditions with a high-pressure trough
over the Mediterranean (after Dayan and Miller, 1989); (c) synoptic conditions with a high-
pressure cell located over the central part of the basin (after Dayan and Miller, 1989)

cyclogenesis in the Gulf of Genoa (Figure 3.2b). The westerly circulation pattern leads to
an eastward movement of these air masses (Dayan and Miller, 1989). Intense cyclogenesis
can also develop in the Gulf of Genoa and surrounding regions when deep lows are
located above the British Isles, and moderate high-pressure conditions occur over Russia.
In these cases, extreme thunderstorms can also be generated at the height of summer,
again as a result of strong thermal gradients (Conti and Colacino, 1995; Figure 3.2c).

2   With the low-pressure area offset between the British Isles and Europe, a north-westerly
circulation pattern can develop. In certain cases, this too can lead to cyclogenesis in the lee
of the Alps in the Gulf of Genoa region (ibid.; Figure 3.2d). This pattern of cyclogenesis is
the most common in the Mediterranean, occurring on average sixty times a year (Barry
and Chorley, 1992; Figure 3.3).

3   A related westerly pattern can also develop in late autumn to early spring when deep lows
over the British Isles and Scandinavia combine with high pressure above North Africa to
channel atmospheric motion along the Mediterranean. In this case, frontal disturbances can
reach along the entire length of the basin (Conti and Colacino, 1995; Figure 3.2e), although
this is relatively rare, occurring on average four times a year (Barry and Chorley, 1992).

4   Another axial circulation pattern is characterized by general low pressure along the
Mediterranean (Figure 3.2f) formed by more southerly incursions of the central European
high-pressure zone (Dayan and Miller, 1989). Deep low-pressure conditions develop over
the western part of the basin, and may lead to strong sirocco winds blowing from North
Africa (Conti and Colacino, 1995).

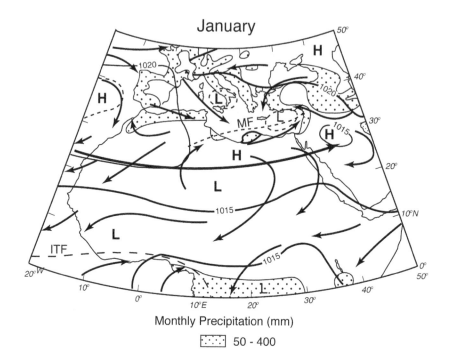

(a)

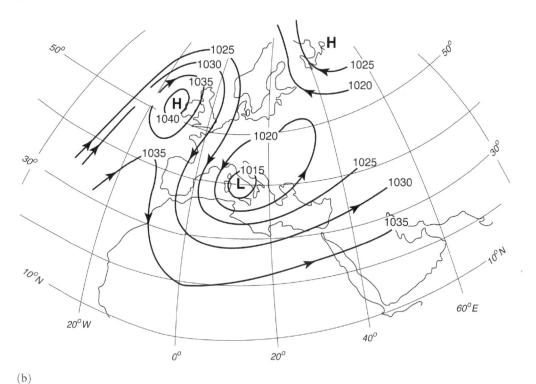

(b)

Figure 3.2 General characteristics of atmospheric circulation in the winter

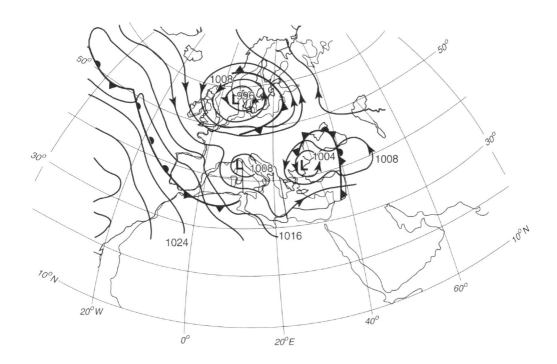

(c)

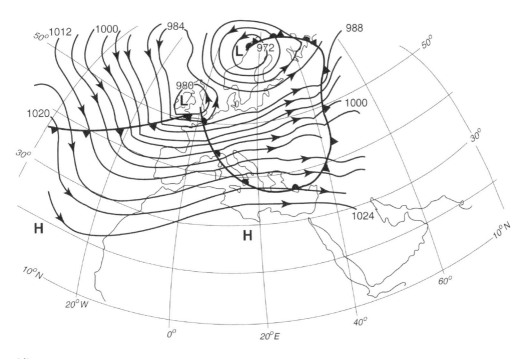

(d)

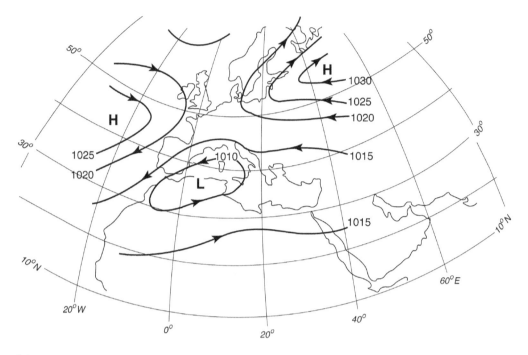

(e)

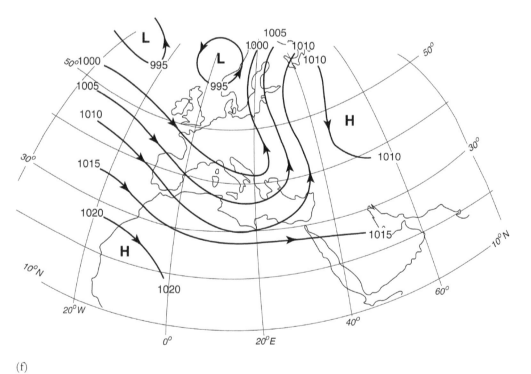

(f)

Figure 3.2 Continued

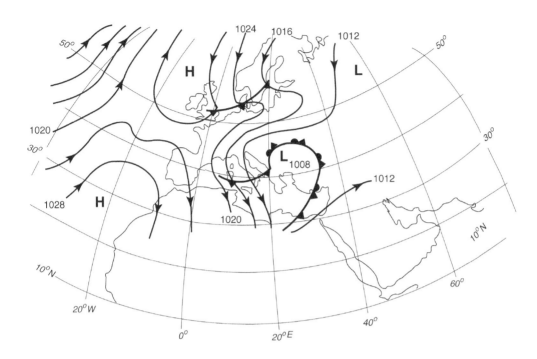

(g)

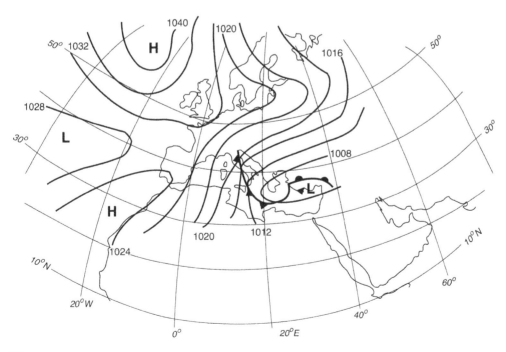

(h)

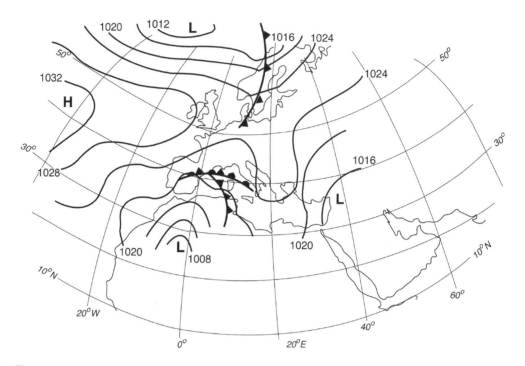

(i)

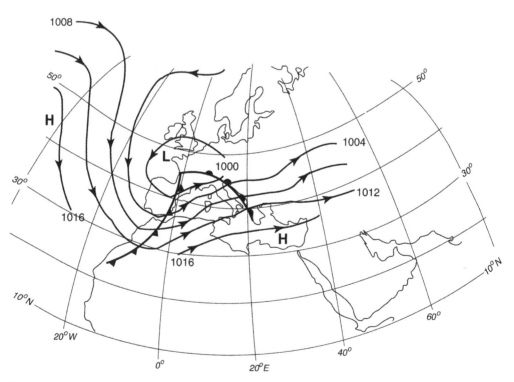

(j)

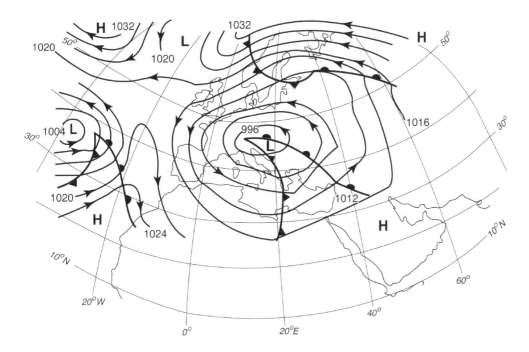

(k)

Figure 3.2 General characteristics of atmospheric circulation in the winter: (a) winter circulation patterns (from Barry and Chorley, 1992); synoptic conditions with: (b) cyclogenesis in the Gulf of Genoa (after Dayan and Miller, 1989); (c) cyclogenesis in the lee of the Alps in the Gulf of Genoa region (after Conti and Colacino, 1995); (d) general low pressure along the Mediterranean (after Conti and Colacino, 1995); (e) strong westerly circulation along the length of the Mediterranean (after Conti and Colacino, 1995); (f) cyclogenesis to the south-west of Greece and in the region surrounding Cyprus (after Dayan and Miller, 1989); (g) Cyprus low depressions in the lee of the Anatolian peninsula (after Dayan and Miller, 1989); (h) depressions forming in the central, western or eastern parts of the basin as a result of northerly airflows originating in northern Europe (after Conti and Colacino, 1995); (i) south-westerly conditions with low pressure above France (after Conti and Colacino, 1995); (j) a large depression expanding from Central Europe in an easterly direction through the basin (after Conti and Colacino, 1995); (k) a strong depression forming in the lee of the Atlas mountains (after Conti and Colacino, 1995)

5   Low-pressure cells in the North Atlantic and north-central Europe lead to the development of strong westerly circulation along the length of the Mediterranean, blocked to the south by the Saharan high (Dayan and Miller, 1989; Figure 3.2g).

6   Throughout the winter, there is a strong thermal gradient between the Saharan high and the eastern basin, with a temperature discontinuity of 12–16°C across the Mediterranean front (Figure 3.2a). This pattern leads to thermal instability and cyclogenesis to the south-west of Greece and in the region surrounding Cyprus, on average fifty-one and twenty-eight times a year respectively (Barry and Chorley, 1992; Figure 3.3). The dominant westerly circulation again means that these storms track eastwards, with landfalls on Greece, Turkey and the Levant.

7   Cyprus low depressions can also form in the lee of the Anatolian peninsula when the circulation is dominated by north-easterly circulation, with incursions of cold air from

north-eastern and central Europe in the winter. In the summer, similar incursions can occur leading to the strong northerly winds known locally as *meltém* (see below; Conti and Colacino, 1995; Figure 3.2h).

8   Depressions may form in the central, western or eastern parts of the basin as a result of northerly airflows originating in northern Europe (ibid.; Figure 3.2i).

9   South-westerly conditions develop over the Mediterranean as a result of high-level low pressure above the British Isles, and ultimately leading to a surface depression above France. This pattern tends to move eastwards and in a number of cases this might develop into the type d circulation pattern (ibid.).

10   A large depression above central Europe can expand to affect many parts of the basin at a time, and generally moves eastwards (ibid.; Figure 3.2j).

11   A strong depression forms in the lee of the Atlas mountains, commonly during the spring. This system tends to move across the Mediterranean in a north-easterly direction, bringing strong sirocco winds to the south of Italy and Greece (ibid.; Figure 3.2k). This type of cyclogenesis is the least common in the Mediterranean, occurring on average fourteen times a year (Barry and Chorley, 1992; Figure 3.3).

In summary, the regional circulation leads to wet, stormy, though relatively mild winters due to the combination of incursions of cold air masses and the relatively high sea-surface temperatures leading to convective instability (Barry and Chorley, 1992). In certain areas, localized

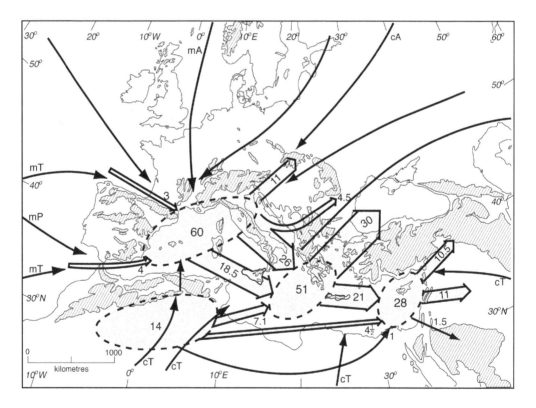

Figure 3.3 Patterns of cyclogenesis in the Mediterranean. Air masses labelled as follows: m = maritime, c = continental, A = Arctic, T = tropical and P = polar. Numbers are average frequencies of depressions

high-pressure conditions develop in response to the position of the major land masses, for example, in Spain (Martyn, 1992). These patterns reinforce the wetter conditions that tend to occur in spring and autumn as the ITCZ position fluctuates in response to the general pattern of circulation. These seasons are typified by the high changeability of the weather – the so-called *tempo matto* (crazy weather) in Italy (Cantù, 1977). By contrast, the summer is typified by long periods of atmospheric stability due to the presence of high-pressure conditions, with only localized thunderstorm activity typically developing later in the season.

Mesoscale circulation features

These general circulation patterns are complicated by two main factors. First, the position of the land masses with relation to the Mediterranean Sea and the dominant air flow. Areas such as Spain and Turkey develop a continental-type climate as moisture-bearing air masses contain gradually less rainfall as they progress eastwards. Second, there is a strong orographic influence due to the extensive mountain ranges surrounding the Mediterranean. The effect of the mountains is to enhance rainfall either by increasing convective activity as air masses are forced upwards as they flow over the mountains, or by increasing cyclonic rainfall by slowing the movement of depressions, or by funnelling air masses and causing convergence (Barry and Chorley, 1992).

The general patterns of circulation mean that the rainfall in the Mediterranean is rarely frontal, and is dominated by mesoscale circulation features. Puigcerver *et al.* (1986) suggest that convective rainfall is the dominant form in Catalonia between May and November. Mesoscale features are described in detail by Atkinson (1981). Convection in the atmosphere leads to the characteristic anvil-shaped cumulonimbus clouds of thunderstorms, and is caused by vertical instability and its release. If the wet bulb temperature near the surface increases significantly above the saturated potential temperature in the middle to upper atmosphere, convection can occur (Carlson and Ludlam, 1968). Atkinson noted that this process is most likely to happen through an increase of the surface wet bulb temperature, which is controlled by the local characteristics of the terrain. In the Mediterranean, this process may be due to the movement of air above the heated land mass, particularly during the summer months, or due to adiabatic heating as air masses moving inland are forced above the coastal mountains. This factor enhances the orographic nature of the rainfall. Following the definitions of Chisholm and Renick (1972), such convective storms can be divided into single-cell, multi-cell and supercell storms. The single-cell storms last for typically less than an hour and tend to be 5 to 10 km in diameter. Multi-cell storms may be 30 to 50 km in diameter and consist of a sequence of cells which form and decay as the storm moves across the landscape. For this reason, such storms can last considerably longer than single-cell storms. Supercell storms are of the same magnitude, but have a single cellular structure, and typically produce more severe weather. They typically occur in regions with strong instability, high subcloud windspeeds, high wind shear within the cloud mass and a large veer of the wind direction with height (Atkinson, 1981). Strong winds and large hailstones are associated with this type of storm (Houze and Hobbs, 1982).

Other significant mesoscale features in the Mediterranean are the winds formed by sea/land breeze circulation and by downslope winds generated by the major mountain chains, most notably the Alps. Neumann (1973) notes the discussion of the causes of sea and land breezes in the classical Greek literature, most notably by Aristotle and Theophrastus. Homer makes several references in the Odyssey to the beneficial effects of using the land breeze to set sail in the evening, because it allows favourable conditions without a large swell. Conversely, Plutarch describes how Themistocles waited for a sea breeze to develop to upset the stability of the boats of the Persian fleet, as a decisive factor in the Greek victory at the Battle of Salamis in 480 BCE. The sea/land breeze circulation is caused due to the differential heating of land and sea masses in a calm, clear atmosphere. During the day, a sea breeze (i.e. from the sea towards the land) develops because the air above the land heats up more rapidly than that above the sea, causing it to rise and be replaced by air over the sea. At night, the land cools more rapidly, leading to the reverse pattern (a land breeze) occurring (Atkinson, 1981). The regular summer sea breezes in Catalonia are known as virazones or garbin, while those on the Atlantic coast of Portugal – the 'Portuguese trade wind' – are common, blowing for twenty-one days in August and fifteen days over the winter. In the Levant, there are regular sea breezes from April to October, and their effects can reach as far inland as Jerusalem or Damascus (Martyn, 1992).

Downslope winds are common particularly in Slovenia and Croatia, where the so-called bora winds are found. Such winds usually start suddenly and are strong and gusty, although there is disagreement as to the exact mechanism of their formation (Atkinson, 1981). They are often defined as bora-type winds if they occur during cold conditions or foehn-type if they occur during warm conditions. The Dinaric boras occur due to the accumulations of continental arctic air in Slovenia and Croatia at the same time as high-pressure conditions develop over eastern or north-eastern Europe, setting up easterly or north-easterly airstreams. The winds cross the Dinaric Alps and descend to the Dalmatian coast, sometimes reaching as far as the Italian coast. Low pressures are found over the Adriatic and Aegean at these times. In the case of the former, the bora is cyclonic and much stronger (50–$60 \, \mathrm{m \, s}^{-1}$) and descends over lower-lying ground towards Gulfs of Trieste, Kvarner, Šibenik and Drin, whereas if the depression is located over Italy, the bora is anticyclonic and the weather dry and cloudless (Martyn, 1992). The mistral of France, the *cierzo* of Spain and the tramontane winds that affect Spain, France and Italy all have downslope-wind characteristics similar to the bora (Barry and Chorley, 1992). Rapid snowmelt can be caused by the Dinaric foehn in Slovenia, Croatia and Bosnia. These winds occur on seventy days a year in Zagreb and 100 days a year in Sarajevo. Foehn-type winds also occur in Languedoc, where they are known locally as the *autan*, and in the Levant, particularly along the Jordan valley (Martyn, 1992).

The direct interaction of the atmosphere and the Earth's surface takes place in the boundary layer which typically extends from one to two kilometres above the surface. Transfer of energy through this zone is dominated by turbulence and convection, derived by the friction of the land mass and differential heating of its surface (Oke, 1987). The energy budget at the surface is an important control on plant productivity, through its control on photosynthesis, and provides an important feedback to the atmosphere in evapotranspiration (Bolle, 1995). These factors are critical in understanding processes of desertification, as will be seen later (Chapter 19). In conditions where the rate of actual evapotranspiration is high compared to its potential rate, stomatal coupling of plants with the atmosphere will be high, leading to high rates of photosynthesis and growth. Conversely, when actual evapotranspiration falls, stomatal de-coupling takes place to control water loss, which limits uptake of CO_2, thus reducing productivity. A resulting loss in leaf mass can lead to less evapotranspiration and the formation of a negative feedback loop which has important consequences for the spatial diversity of Mediterranean ecosystems (see Wainwright *et al.*, 1999a for further details).

Climate

Annual patterns of climatic variables

The average annual rainfall for the basin is shown in Figure 3.4. The rainfall regime is highly variable both throughout the basin and at a regional level. In Spain, the rainfall can vary from 200 mm in the interior of Almeria and Murcia to more than 3,000 mm in Galicia, with the number of days with rain varying from forty-eight in the south-east to 175 in the north

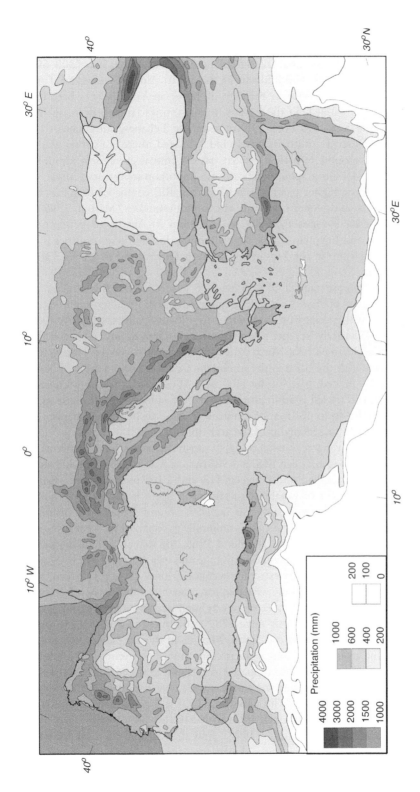

Figure 3.4 The average annual rainfall for the Mediterranean Basin

(Martyn, 1992). In southern France, the gradient is from 511 mm at Le Grau du Roi on the coast on the edge of the Rhône delta to 2,041 mm at Valleraugue at an altitude of 1,567 m in the Cévennes, 80 km distant (data from Météo-France). These gradients clearly illustrate the importance of orographic effects on rainfall. Similar ranges occur within Italy (300–500 mm on the lee slopes of the Apennines to 3,000 mm in the Julian Alps), Greece and Turkey. The maximum in the basin is at Crkvice in the Bay of Kotor in Yugoslavia, where the annual average reaches 5,217 mm (ibid.). In the Levant, the quantities tend to decrease southwards from 600 mm in coastal Syria and even 1,000 mm in the Lebanese mountains to less than 100 mm in the Negev Desert in southern Israel (ibid.). Indeed, there is a strong gradient from the Negev to the area around Jerusalem, which has an annual average of around 550 mm (Striem, 1967; Yair, 1994). Orographic effects are also important in northern Africa, with rainfall up to 1,175 mm in the higher parts of the Moroccan Rif, compared to values of around 400 mm on the northern and western slopes of the same mountains (Martyn, 1992). Moving along the coast, the rainfall varies from up to 1,000 mm in Algeria to between 100 and 200 mm on the Libyan and Egyptian coasts. Rainfall in interiors of these countries is obviously much lower: less than 10 mm in the Libyan and Nubian Deserts and as low as 3 to 5 mm at Aswan in Egypt (ibid.).

The dominance of intense rainfall of a convective nature within this general pattern is demonstrated by the map of days with rain (Figure 3.5). The number again varies widely according to location, with values from forty-eight to 175 in Spain, seventy-five to ninety in France, forty-five to 105 in Italy, seventy to one hundred in Greece, fifty to sixty in the Levant and Israel and as few as ten in parts of North Africa (ibid.). Snow occurs very rarely at sea level, but is important throughout the basin in the mountainous areas, even in areas of North Africa more than 1,000 m asl – indeed, parts of the High and Middle Atlas are snow-covered for six to nine months per year. Dewfall is an important source of moisture, particularly in the more arid regions. For example, up to 150 mm of moisture is obtained by this mechanism in coastal Israel, which occurs on 133 to 255 nights of the year (ibid.).

As well as being spatially variable, rainfall is highly variable between years. Between 1951 and 1960 the annual rainfall varied between 369 and 985 mm at Ajaccio, 388 and 803 mm at Marseille, 275 and 738 mm at Cagliari and between 413 and 1,027 mm at Rome (Estienne and Godard, 1970). The coefficient of variability of rainfall in Spain is commonly around 30 per cent (Mulligan, 1996). In southern France it is between 15 and 49 per cent (median 25 per cent: Wainwright *et al.*, 1999a), and the value commonly exceeds 40 per cent in North Africa (Rumney, 1968). In the period between 1846 and 1965, the rainfall at Jerusalem exhibited a trimodal pattern, with commonly recurring dry, average and wet years and nearly 29 per cent of the years having either extremely dry or extremely wet conditions (Table 3.1; Striem, 1967).

Long sequences of annual rainfall are available for a small number of Mediterranean locations, allowing the study of variability over more extended periods. The monthly precipitation at Rome has been recorded continuously since 1782 (Colacino and Purini, 1986), and seems to show the presence of a long period oscillation. Maxima in this series occurred around 1810 and 1900, with minima around 1830 and at the present time. The twenty-year average precipitation from 1882 to 1901 was 910 mm, compared to 701 mm in the period 1962–1978. These changes seem to have affected the seasonal totals equally. Fourier analysis also demonstrated the presence of significant oscillations with periods of 11.0, 5.3 and 2.6 years. The eleven-year cycle may suggest a link with sunspot activity, which has also been put forward by Thomas (1993), although Mazzarella and Palumbo (1992) show that the effect is weak for Italian data. Camuffo (1984) showed the existence of a similar long-term trend in the rainfall in Padua, while Striem (1967) suggested it is present in the data for Jerusalem, although these do not begin until 1846. Garnier (1974a, b) provides monthly data for a number of locations in

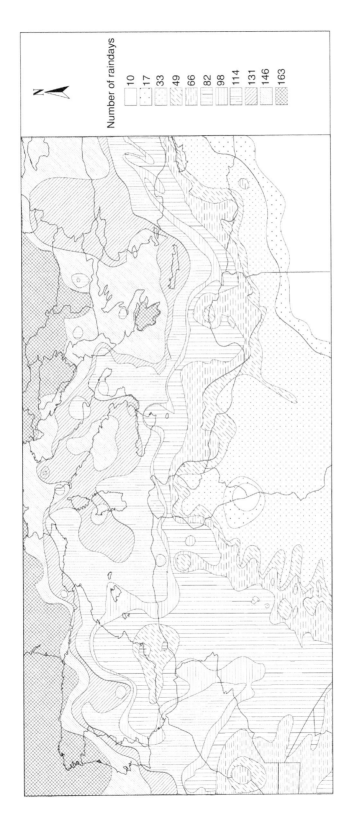

Number of raindays

10
17
33
49
66
82
98
114
131
146
163

Figure 3.5 Map of the average annual number of raindays

Table 3.1 Interannual variability of rainfall at Jerusalem between 1846 and 1965

Years	Range (mm)	Frequency (%)
exceptionally dry	200–380	14.4
dry	380–510	29.7
average	510–600	20.3
wet	600–700	21.2
exceptionally wet	>700 (max.: 1090)	14.4

Source: After Striem (1967).

southern France. Wainwright (1991) used these data to demonstrate that there was no significant difference between eighteenth-century rainfall and that in the late nineteenth and twentieth centuries.

A number of studies of shorter-term variability in rainfall have been carried out. Palutikof *et al.* (1996) looked at standardized anomalies of precipitation for the Mediterranean as a whole since 1960, and specifically for the Alentejo of Portugal since 1940. For the hydrological years 1963/4 to 1967/8 and 1980/1 to 1984/5, the rainfall was consistently below the long-term average in the Mediterranean as a whole. However, the deficits in the 1960s are dominated by the spring deficits, whereas those in the 1980s relate to lower than average rainfall in both winter and summer. In the Alentejo, these authors noted a clear decline in spring rainfall since the 1970s which was not reflected in the annual or other seasonal patterns. Maheras *et al.* (1992) analysed rainfall variability in the central Mediterranean, demonstrating the existence of complex spatial and temporal patterns of variability. Such patterns can occur over relatively short distances, as demonstrated by Gajić-Čapka (1994) for Croatia. Douguedroit (1988) looked at rainfall changes in sixty-two sites in north-western Africa, with the earliest data being from 1922. Her results suggest a complex spatial rainfall pattern, with up to twelve distinct rainfall regimes. Patterns of temporal variability are also complex. In western Morocco, there appears to have been a small peak in rainfall around 1940, and a decline since 1970. The central and northern parts of the studied area show two peaks in the 1950s and late 1960s to early 1970s, with a number of dry years from 1964 to 1967. In contrast, the southern area had dry periods in the 1920s and at the end of the 1950s. Fernández Mills *et al.* (1994) showed that high spatial variability in rainfall regime also occurs in Catalonia, where they defined nine distinct patterns. Dougedroit (1980) has also looked at rainfall records in southern France from the opposite perspective – that of the length of summer drought. Again, there is a certain amount of spatial variability, with four subclasses of behaviour defined. The length of the dry sequence generally increases to reach a peak in July, and follows an exponential distribution.

Palutikof *et al.* (1996) postulated the existence of a Mediterranean oscillation, in which atmospheric pressure distribution fluctuates along the axis of the basin leading to periodic climatic variability. The mechanism proposed is due to the formation of long atmospheric waves over southern Europe due to the presence of the Azores anticyclone, and the thermal discontinuity over the Mediterranean Basin relating to the land–sea interface and orographic effects. The mean annual 500 hPa heights at Algiers and Cairo were found to oscillate inversely, with a dominant cycle of around twenty-two years (Figure 3.6). These authors again tentatively point out the potential link between this variability and that of sunspot activity. The 500 hPa height in the western basin is positively correlated with the existence of more frequent and/or stronger episodes of anticyclonic blocking. These authors subsequently found significant negative correlations between the 500 hPa height and precipitation anomalies and significant positive correlations with temperature anomalies in the western Mediterranean.

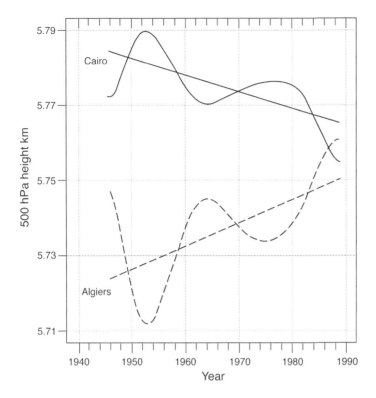

Figure 3.6 Cycles in the mean annual 500 hPa heights at Algiers and Cairo demonstrating the presence of the Mediterranean Oscillation

Source: After Palutikof *et al.* (1996).

The total annual average solar radiation values vary as a function of latitude between 4,190 MJ m^{-2} in the north and 6,699 MJ m^{-2} in the south of the basin (Martyn, 1992). The annual amount of sunshine varies from 1,625 to 1,900 hours in the cloudiest parts of Slovenia and less than 1,750 hours on the coasts of Asturias and the Basque country, to between 2,600 and 2,800 hours on the Côte d'Azur and in Corsica. Elsewhere, there is a minimum of 2,500 hours, reaching up to 2,600 to 3,000 hours on the coastal plains of Morocco and in the High Atlas, 3,100 hours in Lebanon, and 4,000 hours in the southern Libyan Desert (Martyn, 1992). There is a general temperature gradient with annual average temperatures exceeding 18°C in Libya and Egypt and decreasing towards the north and north-west where the annual average is less than 12°C (Palutikof *et al.*, 1996), with even lower values in the mountainous areas (Figure 3.7).

The temperature at Rome has also been measured since 1782, although unfortunately the records for 1793 to 1810 are now missing (Colacino and Rovelli, 1983). Reliable data on minima and maxima are available from 1831. The average annual temperature was relatively stable until 1910 (1782–1794 average 15.4°C, 1891–1910 average 15.3°C), after which there has been an increase to the present average of 16.0°C. The increase has mainly been due to increasing minimum temperatures – indeed, the maximum temperatures declined consistently from the mid-1960s – which Colacino and Rovelli relate to the development of an urban heat-island effect. An increase in the minimum temperature since the 1920s has been demonstrated

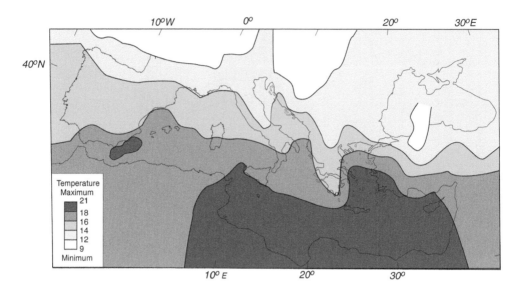

Figure 3.7 Map of average annual temperatures for the Mediterranean

for Athens by Katsoulis (1987), who also demonstrated a significant correlation between the minimum temperature and population of the city. An increase in rainfall in Israel since the 1950s has again been suggested as relating to the development of urban heat islands (Goldreich, 1987; see also Neumann, 1986). In Rome, the period before 1860 showed greater oscillations around the mean compared to subsequent years. A significant oscillation of around eleven years was also found, which Colacino and Rovelli (1983) relate to sunspot activity, and another of about nineteen years, which they believe may relate to the lunar nodal mode. Mazzarello and Palumbo (1994) demonstrated a wider relationship between this cycle and precipitation in the western Mediterranean, which they suggested is due to the correlation between the lunar orbit and atmospheric pressure. Comani (1987) analysed the temperature data from Bologna for the period 1716 to 1774. Until 1741, this sequence oscillates around a mean of 14.2°C. In 1742, the mean temperature was only 11.5°C, after which it seems to have increased in three steps, only reaching the values of the early part of the century from 1765. Comani also presented data from Padua from 1725. This sequence shows a sharp drop in temperature in the year 1740 (but not 1742), and a general decrease in the temperature until 1757. After this time, the temperature remains close to 16°C. Cooler temperatures were noted elsewhere in Europe in 1740 (e.g. Lamb, 1972).

Palutikof *et al.* (1996) also looked at standardized anomalies of temperature. There are a number of positive anomalies in the early 1960s for the Mediterranean as a whole, followed by repeated negative anomalies from 1971 to 1976. These anomalies seem to relate to spring, summer and autumn, rather than winter, variations. The Alentejo series shows a more definite oscillation, with peaks in the late 1940s and late 1980s, and a trough again in the early 1970s. Changes in spring and autumn again bring out the dominant changes. For the period from 1866 to 1985, Maheras (1989) looked at variations in air temperature at a number of stations in the western Mediterranean. Sites in southern France and in Italy showed relatively low temperatures in the period from 1866 to 1910, followed by a pronounced increase and reversion to present levels after the 1940s. Lisbon, Gibraltar, Madrid and Palma demonstrated a different pattern, with an increase from 1866, followed by a decrease from the 1880s to 1930s. There then followed a sharp rise, reaching a peak in the 1960s, which has subsequently been followed by a slight drop. Sahsamanoglou and Makrogiannis (1992) also used data across the basin from 1950 to 1988 to look at temperature patterns. These authors found an average trend of $-0.02°C\,a^{-1}$ in southern Turkey in January and in Turkey, Cyprus and the Levant in July temperatures. This change compares to a general $+0.02°C\,a^{-1}$ trend in July temperatures throughout large areas of the western Mediterranean. These results are consistent with the trends in the Mediterranean oscillation discussed above. Cycles of three, four, five and eight years were found in the central and eastern Mediterranean, compared to significant cycles of about 2.5 years in the west. Katsoulis (1987) found the mean temperature in Athens was higher in the 1920s to 1950s than in the periods before or after. Further north in Greece, a similar cooling trend began earlier – in the 1920s at Larissa and the 1950s at Thessaloniki – with a slight warming at all three locations since the early 1970s (Giles and Flocas, 1984). The overall cooling demonstrated in this study is made up of a complex set of patterns, with spring and winter temperatures making up the decrease in the period to 1944, while summer and autumn decreases are more important subsequently. At Athens, however, there is a continued increase in winter temperatures since 1859, possibly due to the heat-island effect, as noted previously. There is some persistence in these patterns at short lag times, and a complex set of oscillations, with different frequencies at each of the three locations (Flocas and Giles, 1984).

Giles (1984) looked at combined patterns of variability in rainfall and evaporation in southern France since 1931 in order to assess changes in water availability. The period 1971–1980 was relatively wetter in the coastal zones apart from in Rousillon, and relatively drier in the interior in the middle Rhône valley and the southern Alps. These results suggest that there are relatively complex patterns of variability present in space and time in the Mediterranean region.

Seasonal patterns of climatic variables

One of the most noted characteristics of the Mediterranean climate is its seasonal variability. Rainfall tends to occur in the cooler parts of the year – either throughout the winter months, or in two distinct periods in spring and autumn (Figure 3.8). This pattern is intimately linked to the transitional meteorological conditions which arise at these points of the year, as described above. Temperatures tend to be mild in winter and high in summer, peaking in July or August. Rainfall is low to non-existent in summer for the reasons noted above, with either a single peak around December – as in the south and east of the basin – or separate spring and autumn peaks, as occur in Spain, France and Italy. The length of the dry season (i.e. when evapotranspiration exceeds rainfall) can be as short as a month on the Adriatic coast of Italy, but is generally two to four months elsewhere in the north-west of the basin, five to seven months in the north-eastern and eastern Mediterranean and increases from west to east on the southern shores, from five months at Algiers to ten or eleven months in Egypt. Goosens (1985) used principal components and cluster analysis to assess the seasonal patterns of rainfall in the European Mediterranean. Five groups of rainfall were defined with distinct regional patterns, apart from in southern Italy and Sicily, where the characteristics were very variable (Figure 3.9).

Incoming solar radiation in December varies from $168\,MJ\,m^{-2}$ in the northern part of the basin to $419\,MJ\,m^{-2}$ in North Africa, while the corresponding June figures are $796\,MJ\,m^{-2}$ and $921\,MJ\,m^{-2}$, respectively (Martyn, 1992). In winter, the monthly number of sunshine hours can vary from as little as thirty to fifty hours in December in the regions of Ljubljana, Zagreb and Sarajevo because of number of days with fog (between eleven and eighteen) to one hundred hours in southern France and northern Spain, more than 150 hours in the south-west

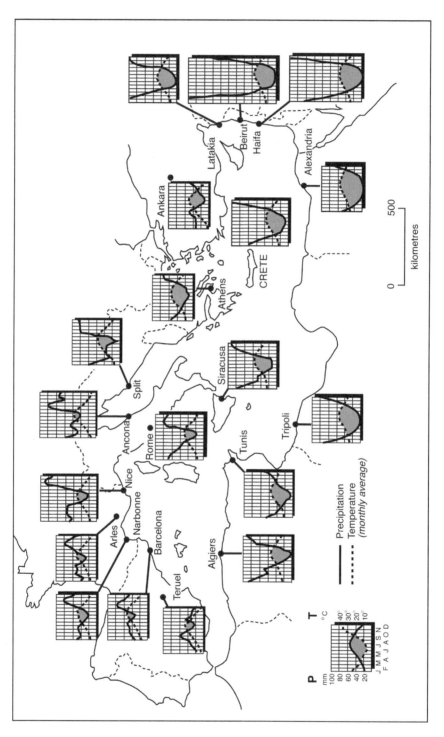

Figure 3.8 Climatograms indicating the length of the dry season at various locations around the Basin (Reproduced with the permission of Oxford University Press)

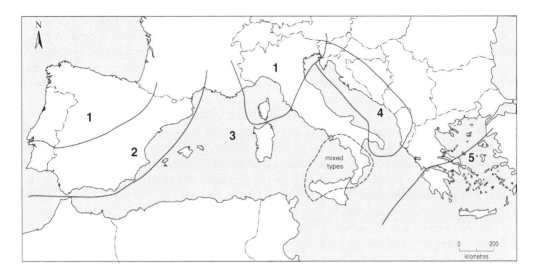

Figure 3.9 Regionalization of rainfall patterns in the European Mediterranean
Source: After Goosens (1985).

of the Iberian Peninsula, one hundred to 120 hours in coastal Dalmatia, Albania and Greece, eighty to one hundred hours in northern and central Turkey, 150 hours in southern Turkey, 150 hours in Lebanon and north-west Syria, and 150 to 200 hours on the North African coast (Martyn, 1992). In summer, the range is from 230 to 370 hours per month in the northern part of the basin, to values in excess of 400 hours per month in the Libyan, Egyptian and Syrian deserts.

The cloudiness in winter varies from 35 to 45 per cent in southern France, 45 to 50 per cent in southern Iberia and 50 to 60 per cent in the Levant, but 70 to 75 per cent over the Lombardy Plain and the northern Adriatic, and 60 to 70 per cent on the Adriatic, Ionian and Aegean seas. In July, the values are less than 10 per cent in southern Greece and Crete, more than 20 per cent elsewhere in Greece and throughout Turkey, between 10 and 20 per cent in central and southern Italy and the Balkan Peninsula, 30 per cent in south-eastern Spain, and 36 to 43 per cent in the Lombardy Plain and Gulf of Genoa (Martyn, 1992).

Corresponding temperatures vary according to similar patterns. The temperatures of the coolest month are between 6°C and 8°C in southern France, between 10°C and 13°C in southern Spain, the Tyrrhenian coasts and the Aegean, between 10°C and 12°C in Syria, reaching up to 17°C in southern Israel and between 20°C and 25°C in North Africa. Exceptions to this general pattern are to be found in the Adriatic coasts where the bora winds depress the temperature to around 6°C to 8°C and in the mountainous areas: temperatures below zero are recorded for the highest parts of the Cantabrian, Castillian and Iberian mountains, the Pyrenees and the Sierra Nevada, average temperatures are around 5°C in the basins of Bosnia, Macedonia and Serbia, but reach a low of −6°C at Cherni Vrukh at 2,883 m in the Balkans, and freezing average temperatures also occur in the High Atlas (Martyn, 1992). Summer temperatures vary from 20°C to 23°C in southern France, between 24°C and 28°C in southern Spain, 28°C to 31°C in Italy, between 28°C and 34°C on the Dalmatian coast, and in Albania and Greece, 27°C on the Black Sea coast and 34°C on the Mediterranean coast of Turkey, 35°C on the Levantine coast and more than 30°C on the coast from Cyrenaica to

Morocco. Temperatures of 40°C or more are recorded in the interior of Turkey, Syria and the desert regions of North Africa. The diurnal temperature range is modified by the effects of the sea, but varies from a range of 9°C in southern France to as much as 24°C in Algeria (Martyn, 1992). There can be a steep gradient from inland to the coast in summer when sea breezes are important: the diurnal range gradient is from 7 to 12°C on the Côte d'Azur and Corsica and in excess of 12–14°C well away from the sea; around 5°C on the Cantabrian coast and more than 14°C in the interior and on the Mediterranean coast of Spain; and between 6 and 8°C in coastal areas, but between 15 and 18°C in central Turkey (Martyn, 1992).

Relative humidities typically vary from a maximum of between 70 and 80 per cent in January in most of the northern basin, although they can exceed 90 per cent in the Lombardy Plains and the area around Milan, but be as low as 60 to 70 per cent on the Adriatic, Ionian and Aegean Seas due to the existence of high temperatures and frequent 'bora winds. On the North African coasts the value falls from 80 to 90 per cent to less than 40 per cent in early morning in January. In the summer, the relative humidities are generally 60 to 70 per cent on the Spanish, French and Italian coasts, but as low as 42 per cent in the interior of Spain and between 45 per cent and 58 per cent in Greece and the Balkans. There are major differences between the coastal and interior areas of Turkey and the Levant, where the values are generally greater than 60 per cent on the coasts and as low as 20 per cent in the interior. North Africa has particularly low values of between 20 and 30 per cent on the coasts and even less further inland (ibid.).

The Mediterranean has a number of characteristic regional winds which are often named (Figure 3.10). The bora winds and their effects have been noted above. Another well-known cold, gusty wind is the mistral which blows along the Rhône valley when central European low pressure systems bring masses of cool, continental air down towards the coast. The mistral is a frequent visitor to southern France, with 103 occurrences per year of wind speeds greater than $11\,\mathrm{m\,s^{-1}}$, and thirty occurrences exceeding $17\,\mathrm{m\,s^{-1}}$ (Barry and Chorley, 1992). In the winter months, similar winds also blow down the Ebro valley, where they are known as *cierzo* (Tout and Kemp, 1985). In northern Greece, mistral-like north-westerly winds – the Vardarac and Struma – blow down Vardar and Struma valleys into the Gulf of Salonika and Strimonic Gulf. During the summer these winds also occur as etesian winds. Etesian winds occur elsewhere in the eastern Mediterranean – in Turkey, where they are known as *meltém*, and lower Egypt – and are northerly winds formed due to the presence of the Azores high ridge over southern Europe and ITCZ over Africa at 18°N (Martyn, 1992). Of the hot winds, the sirocco is the best known. Formed either by the Saharan depressions or those of the western Mediterranean moving along the North African coast, it brings hot, dust-laden air from the Sahara north-wards, particularly in the spring or autumn (Barry and Chorley, 1992). In North Africa, the sirocco is also locally called *chergui*, *sahel*, *chichili*, *chili*, *chihli* and *ghibli*. Its passage over the North African mountains increases the aridity of the air and raises the temperature to 45–50°C. This increase is particularly notable on the coasts of Tunisia and Tripolitania, where sea breezes usually blow during the day (Martyn, 1992). Siroccos reaching the northern shores may still be hot and dry, or may have picked up sufficient moisture on their passage across the sea to cause very humid, warm weather accompanied by fog and rain. In Greece, the lip winds (also known as *livas* or *garbis*) are similar to the sirocco, originating in Libya and sometimes bringing red rain due to the presence of Saharan dust (see Chapter 6). The Levant encounters similar winds – known variously as *samoom*, *simoom* or *chlouk* – originating in the deserts of Syria and Mesopotamia on up to forty-one days a year. Particularly strong instances can occur when these winds develop foehn characteristics when crossing the Lebanon mountains. These can last for a few hours or even a few days when they are known as the 'fire wind', 'breath of death' or *schobe* (Sivall, 1957). The *khamsin* affects Egypt, particularly the Nile Delta and Sinai.

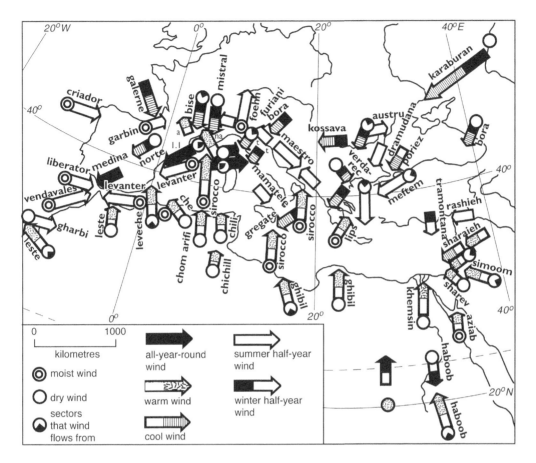

Figure 3.10 Locations of named regional winds

Source: After Martyn (1992).

It is a much more stable but more violent wind, occurring from mid-March and blows for two to three days, four to six times during the fifty days following the vernal equinox, and brings major dust storms (Martyn, 1992). By contrast, cold winds can affect large areas of the north-western Mediterranean. Depressions over the Lombardy Plain, the Gulf of Genoa or the Adriatic lead to cool northern or north-easterly winds call tramontana, which bring cloudless weather to north and western Italy (ibid.). Tramontana winds also blow in the Languedoc, Catalonia and the Balearics (Tout and Kemp, 1985). Other high-velocity winter winds include the *galerna* in northern Galicia and Asturias, and the vendaval and levante winds which affect the Straits of Gibraltar and the south-eastern Spanish coasts, the former blowing from the south-west, and the latter from the east (Tout and Kemp, 1985).

The study by Giles (1984) noted above also looked in detail at seasonal changes in the water balance in southern France in the period from 1971 to 1981. Again, the results demonstrate complex spatial patterns, both in terms of inter-annual variability and in terms of the length and spatial pattern of the dry season in any one year.

Event-based characteristics of climate

Because of the convective nature of much Mediterranean rainfall, there are many events where extremely high rainfall is recorded over short periods of time (Table 3.2). Short-term intensities of more than $720\,mm\,h^{-1}$ have been recorded over periods as long as fifteen minutes in Calabria (Cavazza, 1961), and $110\,mm\,h^{-1}$ maintained for an hour in Catalonia (Llasat and Puigcerver, 1994). The rainfall event of 22 September 1992 in southern France led to up to 300 mm of rain to fall in a period of five hours, with recorded intensities of $190\,mm\,h^{-1}$ over a six-minute duration at surface stations and $370\,mm\,h^{-1}$ over a fifteen-minute duration from radar images (Figure 3.11; Benech *et al.*, 1993; Wainwright, 1996c). Events of 650 mm in eighteen hours have been recorded in southern Spain (Pérez Cueva and Armengot Serrano, 1983), 800 mm in twenty-four hours in 1987 in Valencia (and 200 mm or more over areas of $4,000\,km^2$: Fernández *et al.*, 1995) and events of 909 mm in twenty-four hours have been recorded at Valleraugue in the Cévennes in southern France (Garnier, 1974b), with totals of more than a metre suggested for certain locations (Estienne and Godard, 1970). Valleraugue has a two-year return period, 24-hour rainfall of 151 mm (Wainwright, 1996c). Such short return periods for large rainfalls in many parts of the Mediterranean have important consequences for landscape evolution, soil development and vegetation growth (see below). Llasat and Puigcerver (1994) have noted the meteorological factors which lead to heavy rainfall in Catalonia, pointing out the importance of orography in releasing the convective instability. Different amounts of lifting required lead to different spatial patterns of heavy rainfall, either in the coastal zone, or over the Pyrenees in this case. High 24-hour rainfalls have also been recorded in North Africa, for example, 135 mm at Algiers, 198 mm at Llano Amarillo, 221 mm at Melilla and 154 mm at Susa (Martyn, 1992). The 280 mm that fell in twenty-four hours on Alexandria on 8 December 1888 can be considered quite exceptional when compared to the annual average rainfall of 184 mm over the period 1884 to 1945 (Naguib, 1970).

Puigcerver *et al.* (1986) demonstrated that there exist complex patterns in intensity of rainfall in relation to their duration and the time of the year in which they occur in Catalonia. The highest intensities occur in August and September for all durations. For short durations, the lowest intensities occur in February and May, while at longer durations, the lowest intensities are most commonly recorded in February and March (Figure 3.12). This study also demonstrated that there tends to be an exponential pattern in the distribution of frequencies, and that there are distinct patterns of storms throughout the day. From June to August, the higher intensities are concentrated in the afternoon and evening, while from September to December they can occur throughout the day. This result suggests that local surface heating is dominant in creating convective storms in the summer months, while more general patterns of circulation are important in the autumn and winter. Palumbo (1986) also suggested that there is an important diurnal pattern in Mediterranean rainfall relating to atmospheric fluctuations. Storms in Israel seem to follow a forty- to forty-five-day cycle in the winter months relating to the frequency of generation of Cyprus lows and the time taken for them to move onshore and dissipate (Sharon and Romberg, 1981; Jacobeit, 1981).

Extremes of temperature are equally commonly recorded. Absolute minimum temperatures of $-17°C$ have occurred at Montpellier and Marseille, while temperatures as low as $-30°C$ have been recorded in the interior of Spain at Calamocha (Martyn, 1992). In Italy, the lows vary from $-15°C$ on the Lombardy Plain to $-10.6°C$ in Florence, $-9.6°C$ in Rimini, $-6°C$ in Rome, $-4.4°C$ in Naples, $-2.8°C$ in Bari and $0.4°C$ in Messina, with even lower values in the Alps and Apennines. Lows of $-27°C$ have occurred at Ljubljiana, and $-12°C$ at Thessalonika and in the central Peleponnese. The southern Greek coasts and islands are milder with lows of $-1°C$ to $-4°C$, while $2.2°C$ is the minimum for Crete. Central Anatolia has seen temperatures

Table 3.2 Extreme rainfall events recorded within the Mediterranean for a variety of durations

Location	Rainfall amount (mm)	Rainfall duration (hrs)	Date	Reference
Bejis (Valencia, Spain)	210	1.5	Oct. 1957	Linés Escardó, 1970
	361	24		
Sabadell (Barcelona, Spain)	257	1	25/9/62	Llasat and Puigcerver, 1995
Catalonia (Spain)	866.5	72	Oct. 1940	
	110	1	Sept. 1962	
Giffone (Italy)	160	1	12–13/11/59	Cavazza, S. 1961
	444	6		
	520	24		
Crotone (Italy)	90	1	24–25/11/59	
Pisticci (Italy)	315	24		
Gulf of Policastro (Italy)	>180	0.25	?	
Crkvice (Montenegro)	480	24		Martyn, 1992
Algiers (Algeria)	135	24	Oct.	
Bizerta (Tunisia)	135	24	Oct.	
Llano Amarillo	198	24	Jan.	
Melilla (Spain/N. Africa)	221	24	Jan.	
Susa (Tunisia)	154	24	Sept. + Oct.	
Alexandria (Egypt)	280	24	8/12/1888	Naguib, 1970
Port Said (Egypt)	58	24	24/1/21	
M. Matrouh (Egypt)	99	24	28/12/30	
Cairo (Egypt)	43	24	17/1/51	
Carpentras (France)	19.4	0.1	22/9/92	Benech *et al.*, 1993
	212	24	22/9/92	
Vasion-la-Romaine (France)	14	0.1	22/9/92	
	179	24	22/9/92	
Entrechaux (France)	300	24	22/9/92	
Le Caylar (France)	448	24	21/9/92	
Narbonne (France)	292	24	26/9/92	
nr. Nîmes (France)	420	24	3/10/88	Davy, 1989
Carcagente (Spain)	400	30	20–21/10/1843	Font Tullot, 1988
	154	3	17/5/1850	
	377	24		
	498	24	6–7/12/1853	
Almería (Spain)	158	1.5	14/9/1891	
La Molina (Spanish Pyrenees)	556	48	Nov. 1982	Gallart and Clotet-Perarnau, 1988
Júcar basin (Spain)	650	18	Oct. 1982	Pérez Cueva and Armengot Serrano, 1983
Sorbas (Spain)	200	5	28–29/9/80	Harvey, 1984
Alpujarras (Spain)	250	10	17–18/10/73	Thornes, 1974
Purchena (Spain)	406	36		
Vélez-Rubio (Spain)	150	2		
Vallerauge (France)	520	24	24/02/64	MétéoFrance
	608	24	30/10/63	Estienne and Godard, 1970
	950	24	28–29/9/1900	Garnier, 1974b
Joyeuse (France)	791	22	9/10/1827	Estienne and Godard, 1970
Rousillon (France)	'>1 m'	24	1940	
Barcelona (Spain)	25	0.17	Sept. 1952	Elias Castillo and Ruiz Beltran, 1977
	53	0.5	Sept. 1952	
	143.2	24	Feb. 1944	
Alicante (Spain)	20	0.17	Oct. 1969	
	54.4	1	Oct. 1969	
	125.8	12	Oct. 1962	

(a)

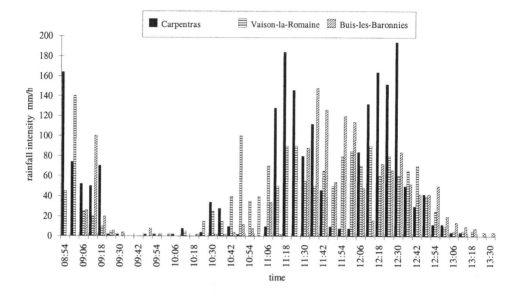

(b)

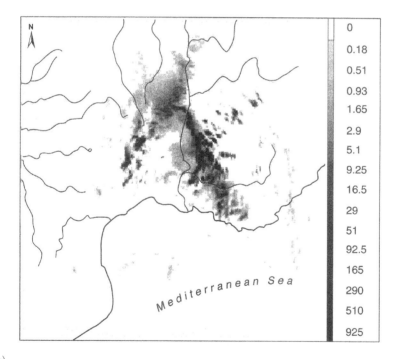

Figure 3.11 Rainfall intensities during the rainfall event of 22 September 1992 in southern France (see Wainwright, 1996; data after Benech *et al.*, 1993): (a) hyetographs of rainfall intensities at Vaison-La-Romaine, Carpentras and Buis-les-Baronnies; (b) radar image of mean rainfall intensity for a 15-minute period from 09:00

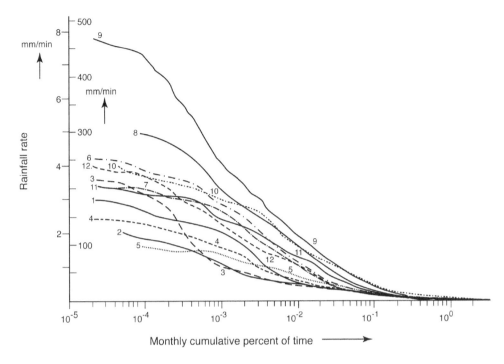

Figure 3.12 Monthly distributions of rainfalls of varying intensities and durations in Catalonia
Source: After Puigcerver *et al.* (1986).

as low as −28°C, with values as low as −10°C in eastern Syria, −1.1°C in Beirut and −3.0°C in Haifa (Martyn, 1992). Absolute maximum temperatures of between 38°C and almost 40°C have been observed in southern France, between 46°C and 49°C in Spain and as high as 50.5°C at Riodades in Portugal. Maxima in mainland Italy range from 38°C in Potenza to 43°C in Turin, while temperatures as high as 43.7°C at Cosenza and 44.4°C at Macomer have occurred on Sardinia and 44°C at Catania in Sicily. In the Balkans and Greece, temperatures of 38 to 40°C are common, with recorded maxima of 45°C at Larissa, and 38 to 41°C on the Aegean islands. Temperatures of more than 40°C recur in central Anatolia and the Mediter-ranean coast of Turkey, as do temperatures of around 45°C on the Levantine coast. Even higher temperatures are of course found in northern Africa, with the highest temperature ever recorded – 57.8°C – at Al Azizia in western Tripolitania on 13 September 1922 (Martyn, 1992). The significance of low temperatures is often most noted in terms of agriculture, particularly in the case of the olive, where crops can be significantly reduced for a period of several years following a severe frost event. High temperatures have a significant impact on human health, with numerous deaths being reported, for example, in Greece in 1996 and 1998. The generally high temperatures are reflected in styles of house construction, which are designed to reduce the amount of sunlight able to reach the interior, and maintain coolness by having thick, insulating walls.

As noted above, high winds do occur on their own in the Mediterranean, associated with bora, mistral or foehn conditions. Speeds of 50–60 m s^{-1} have been noted in the northern Adriatic (Martyn, 1992). High winds are also often associated with severe storms. For example, the 22 September 1992 event in southern France saw windspeeds of 150 to

155 km h^{-1} (Benech *et al.*, 1993). Newspaper reports of the 650 mm storm of November 1963 in the Cévennes recorded velocities as high as 175 km h^{-1} on the peak of Mont Aigoual (1,909 m asl).

Snow is a rare occurrence in the coastal areas, because of the mild winter temperatures, but is frequent in the mountains. The Sierra Nevada – 'snowy mountains' – of southern Spain are only 30 to 40 km from the sea. Heavy snowfalls 'with disastrous consequences' have been noted in Yugoslavia (Martyn, 1992) and even in North Africa a metre of snow persisted for three days at Jebel el Akhdar in February 1949 (Griffiths, 1972).

In summary, although it is common to talk of a single 'Mediterranean climate', this term encapsulates a great amount of variability. This variability is noted on all spatial scales, from the differences between areas separated by ten kilometres or so during a thunderstorm event to those which affect the climate at the basin scale – most notably the north–south and west–east gradients. Temporally, the variability is characterized by significant fluctuations at the storm scale, seasonally and inter-annually. Indeed, perhaps the most unifying characteristic of the Mediterranean climate is its variability.

Reconstruction of climatic evolution

The previous section has provided the context of the present-day climate of the Mediterranean and its inherent variability. In this section, we attempt to define how the Mediterranean climate has evolved to take on its present-day characteristics. This definition effectively takes on three stages. First, the point at which the above characteristics were first developed is assessed, effectively covering the Pliocene and early Pleistocene periods. Second, the variations associated with the glacial–interglacial cycles of the Pleistocene are discussed. Third, the developments associated with the periods of human activity which we will discuss in more detail in subsequent chapters are noted both on the long term, over the last 10,000 years, and over the time period for which historical data are available.

Because no direct evidence obviously exists over such timescales, proxy data for climate change must be used. There are a number of excellent texts discussing the techniques available (e.g. Bradley, 1999; Lowe and Walker, 1997), so we will concentrate in this introduction on the specific problems associated with the use of such techniques in Mediterranean environments.

Oxygen-isotope analysis can give information on the temperature at the time of deposition, albeit modified by the effects of salinity changes as temperatures and circulation patterns change (Vergnaud-Grazzini *et al.*, 1990; Bigg, 1994). The signal from a Mediterranean core will, however, also reflect global changes related to global ice volumes and circulation patterns in the Atlantic, with a variable time lag in the response. A number of studies have demonstrated the potential of oxygen isotopes in marine or terrestrial shells (Letolle *et al.*, 1971; André *et al.*, 1979; Magaritz *et al.*, 1981; Lecolle, 1984), although very few prehistoric samples have been analysed.

Palynology reflects changes in temperature and precipitation less directly through an understanding of the climatic controls on vegetation associations. Despite the generally poor preservation in dry, limestone environments which dominate the Mediterranean, there are an increasing number of studies in a relatively wide range of environments and covering a wide time period of use in climatic reconstruction (see Wainwright, 2003). There is fortunately a move away from trying to interpret climate change in terms of a single 'Mediterranean' species such as holm oak or olive, because of the inherent difficulties involved in this approach, towards an increasing use of vegetation associations and multivariate changes which can more reasonably be used to define changes (see Chapter 1). Many techniques, however, still assume that *stable* vegetation associations are being observed, whereas many changes observed – particularly over the last 14,000 years – are too rapid for such stable associations to develop (see Watts, 1973: 204), and thus interpretations of rapid fluctuations based on pollen data can be questionable. Furthermore, one issue that unfortunately still resurfaces all too often is the use of the Blytt and Sernander zonation to define stages of vegetation development and their timing in the Mediterranean. The Blytt and Sernander scheme was originally developed for Scandinavian peat bogs (Lowe and Walker, 1997) and has no relevance to Mediterranean vegetation sequences as observed from independently dated sequences (see Vernet, 1972). The use of the scheme to define the chronology of Mediterranean vegetation and hence climatic changes is thus based on circular logic. Any study using this scheme without independent dating evidence should therefore be interpreted with extreme caution. This terminology has therefore been avoided in the discussion below.

In limestone environments where pollen data are not well represented, charcoal data from archaeological sites have often been used to provide an alternative means of looking at vegetation change (e.g. Vernet, 1972; Vernet and Thiébault, 1987). While useful, it is important to note the sampling issues raised by this technique. Unless coming from a distinct archaeological unit such as a hearth, the charcoal sampled could have accumulated over a lengthy period of time. The sample numbers tend to be low in comparison to pollen grains because of the state of preservation, and the plants burnt have all been deliberately selected and therefore could be affected by cultural or seasonal biases. For example, Thiébault (1988) demonstrated quite significant differences in the charcoal assemblages at two sites separated by 650 m over long periods of time.

Other techniques used previously include cave, rock shelter and alluvial sedimentology (e.g. Miskovsky, 1974; Vita Finzi, 1969). However, the records produced typically have a low resolution and are difficult to interpret in terms of distinct climatic signals, not least because of anthropic activity in the Holocene (see, for example, Wainwright, 1994; 2000; 2003). Because of the difficulty of disentangling the anthropic from other signals, there is often a great deal of circularity in the climatic interpretations that are derived from such data. Therefore we will consider these data in Chapter 10, rather than use them to discuss climatic signals.

It is important to remember that the Mediterranean climate is dominated by events and great inter-annual variability. These characteristics are often blurred in most palaeoenvironmental records due to factors such as bioturbation, a lack of immediate response of the proxy record to an event, combination of the record for several years, and imprecision in dating. Thus, it is highly unlikely that it will ever be possible to reconstruct these characteristics of the climate. The extent to which general patterns conceal this variability must be borne in mind when interpreting any climate history of the Mediterranean. Specific problems with the use of historical data in climate reconstruction are discussed in a later section.

Patterns in the later Pliocene and early Pleistocene

The origins of the modern-day Mediterranean climate need to be sought in the period since the development of the modern geological configuration and in relation to global patterns of change. Thus, longer term data relating to general changes since the end of the Messinian salinity crisis must be investigated. Oxygen isotope data for the Mediterranean are available at high resolution from 4.7 Ma at the Oceanic Drilling Project (ODP) site 653A (Vergnaud-Grazzini *et al.*, 1990; Thunell *et al.*, 1990), and at lower resolution to *c*.4 Ma at Deep Sea Drilling Project (DSDP) site 132 (Thunell, 1979; Thunell and Williams, 1983), both in the Tyrrhenian Sea (Figure 3.14), and for the period 1.8 to 14.5 Ma for marine sediments now uplifted onto land in southern Sicily (van der Zwaan and Gudjonsson, 1986). These data suggest that climatic conditions would have been up to 3°C warmer than at present in the period until about 3.2 Ma (Thunell, 1979) or 3.1 Ma (Vergnaud-Grazzini *et al.*, 1990). However, it is difficult to attribute changes of oxygen isotopes to changes in temperature because of corresponding fluctuations in salinity (Vergnaud-Grazzini *et al.*, 1990). Moisture regimes, based on vegetation data were either similar to the present, as found in southern Italy and Sicily (Bertoldi *et al.*, 1989) and Israel (Horowitz, 1989), or much wetter, without seasonal drought, as found in southern France and northern Spain (Suc, 1984; for further details, see below). The overall pattern during this time period is one of long-term stability, or even gradual warming (Thunell, 1979; van der Zwaan and Gudjonsson, 1986; Vergnaud-Grazzini *et al.*, 1990), albeit with some oscillations. The first period of cooling occurs around 3.2 Ma, as shown by cores at the DSDP site 132, with a drop of estimated sea-surface winter temperatures by 1.2°C (Keigwin and Thunell, 1979). The timing of this decrease is consistent with the suggested onset of global cooling and sea-level fall from long sequences in the equatorial Pacific (Shackleton and Opdyke, 1977) and the Atlantic (Clemens and Tiedemann, 1997). The downward trend with oscillations around 2.6 Ma and 2.4 Ma, with reversions to warmer conditions, until 2.1 or 2.0 Ma, when there is a general downward trend with onset of large amplitude glacial–interglacial cycles around 900 ka (Vergnaud-Grazzini *et al.*, 1990; Thunell, 1979; Thunell and Williams, 1983). Thunell and Williams (1983) believe the temperature change occurs as three steps, with rapid changes occurring at 2.5 Ma and 900 ka. Comparison of the temperatures from the eastern and western basins suggests that the modern temperature gradient from the west to the east set in about one million years ago. In the period from 2 to 1 Ma, estimates of the summer sea-surface temperatures are 23 to 25°C in the western basin and 22 to 25°C in the eastern. Since 1 Ma, the estimated western basin temperatures have been between 15 and 23°C, while those in the eastern basin have been between 19 and 25.5°C. A similar pattern is observed for the estimated winter temperatures, with values of 13 to 14.5°C for the western basin and 13 to 15°C for the eastern basin in the first period and 7 to 14°C for the west and 10.5 to 15°C for the east since 1 Ma (Thunell, 1979).

Trends in the last million years

The best source of evidence for Mediterranean climate change over the last million years comes from the 196 m cores through the former marshes in the Drama basin in northern Greece at a site known as Tenaghi Philippon (40 m asl). These cores provide a high-resolution record of the pollen deposited in the basin, and have been dated using radiocarbon methods in the upper part and magnetic reversal and oxygen isotope correlation in the lower part (Wijmstra, 1969; Smit and Wijmstra, 1970; Wijmstra and Smit, 1976; van der Wiel and Wijmstra, 1987a, b).

Mommersteeg *et al.* (1995) have studied the detailed changes in the length of the record by grouping the pollen into a number of distinct associations. As well as total tree species and species representing marsh and open water vegetation, and therefore the wetness of the site, the four groups used are of deciduous forest, representing a warm, wet Mediterranean coastal climate; the mixed coniferous–deciduous and mixed evergreen–deciduous forest representative of the modern Mediterranean climate at the site with a dry summer; mixed coniferous–deciduous and steppe vegetation from a montane zone suggesting cooler and relatively wet conditions through the year; and vegetation dominated by steppe species, indicative of cold and dry conditions.

The vegetation, and, by association, climatic record at Tenaghi Philippon show a number of oscillations over the last 975 ka (Figure 3.13). On the most general scale, the forest cover has varied from 0 to 100 per cent, typically in step with changes in global climate reflected by the marine oxygen isotope stages. More specifically, the sub-types have changed dramatically with

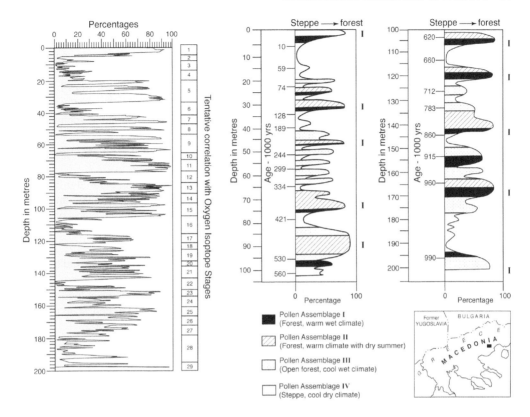

Figure 3.13 Variability in the pollen cores at Tenaghi Philippon

Source: After Mommersteeg *et al.* (1995).

a sequence of oscillations. Before *c.*530 ka there is a tendency for those periods corresponding to the warmer (akin to the modern-day) conditions to be associated with the wet Mediterranean or cooler, wetter montane conditions. After this time, these periods are associated with the modern, Mediterranean type with a dry summer. This shift may be related to the change observed in the Earth's orbital activity some time after 600 ka. Overall, the cold steppe conditions have been dominant over the period of deposition, with 400 out of 975 ka belonging to this type. The cooler, wetter conditions were present in 253 ka, while the two Mediterranean groupings represent a smaller period of time – 200 ka in the case of the dry summer type and 122 ka in the case of the coastal type. Thus, moisture-stressed conditions have been the most important general climatic conditions, with temperatures generally being lower than the present day. Wainwright *et al.* (1999a) noted that the data from the Tenaghi Philippon core suggest that wetter conditions are characteristic of transitional phases between the cold, dry and hot, dry climate types, although the reverse process is not necessarily true.

The study by Mommersteeg *et al.* (1995) also investigated the sequence in terms of the frequencies of the Earth's orbital elements. Although there is a partial problem with this in that the complete sequence is dated by correlation with the marine oxygen isotope sequence, which is itself correlated with the orbital sequence (e.g. Williams *et al.*, 1988), there seems to be an overall match between the orbital elements and the sequence when it is only dated by independent means – radiocarbon in the upper part, and magnetic reversals in the lower part. Variations potentially related to the precession of the equinoxes (Box 3.1), with a frequency of

---

Box 3.1 The orbital variability of the Earth

The orbit of the Earth around the Sun is known to vary periodically in three distinct ways:

1   *Eccentricity* is the extent to which the orbital path of the Earth varies from being almost circular, to becoming increasingly elliptical due to the gravitational effects of the Sun, the Moon and the other planets. The dominant periodicities of the eccentricity are around 100 and 400 ka, but much longer cycles also exist.

2   *Obliquity* is the tilt of the Earth's axis of rotation from the vertical. The angle varies from 22° to 24.5°, with a dominant periodicity of 41 ka.

3   *Precession of the equinoxes* refers to the relative seasonal timing of when the Earth passes closest to, and moves further away from the Sun during its orbit, and is caused by the gravitational pull of the Sun and Moon on the equatorial bulge of the Earth. The precessional cycle varies at periodicities of 19 to 23 ka.

Because the orbital variations affect the quantity and spatial distribution of solar radiation received at the surface, there is thought to be a relationship between them and the quasi-periodic fluctuations observed in the global climate. The so-called 'astronomical theory' or Croll–Milankovitch theory of climate change is discussed in more detail in specialist texts (see Berger, 1978; Bradley, 1999; Lowe and Walker, 1997; Goodess *et al.*, 1992).

---

20 to 24 ka are found in all vegetation-group sequences, but most commonly in that representing a warm, wet Mediterranean coastal climate and in marsh and open water vegetation, and therefore reflect changes in net precipitation. These changes are most probably related to differences in the intensity of the African and Asian monsoonal system, due to relative variations in the position of the ITCZ, which is a dominant control on the weather patterns which reach the Mediterranean. These variations may have led to increased cyclogenesis at the Mediterranean front in the eastern part of the basin, or may have led to more depressions reaching the eastern part of the basin from the west. Changes related to the obliquity of the Earth's orbit, at a frequency of 40 to 44 ka, are clearly represented in the warm, wet Mediterranean coastal climate and dry steppe species associations. Again, these changes may be related to changes in net precipitation. Both of these vegetation groups, as well as the modern-type Mediterranean climate with a dry summer, also vary in association with the eccentricity of the Earth's orbit. These variations have been stronger in the period since 650 ka, and probably reflect global changes relating to glaciation rather than more regional scale variability. In general, conditions seem to have been drier since this time, and this may reflect the longest timescale change in the Earth's orbit, with the 404 ka-eccentricity oscillation. General sequences in the vegetation patterns seem to relate to global changes relating to the Earth's orbit, although due to the complexity of different frequencies of the three orbital parameters, and their interactions (further tonal and harmonic frequencies are found in the vegetation oscillations at 68, 30, 15.5, 13.5, 12.0 and 10.5 ka), the climatic changes in the past reflect the details of these changes.

Temperature fluctuations over long time periods are reflected by the variation of the oxygen isotope composition ($\delta^{18}O$) of foraminifera preserved in deep-sea sediments (Box 3.2). Two cores have been published which reflect the $\delta^{18}O$ changes over the last half million years. TR171-24, located in 2,328 m of water off south-west Crete covers approximately 500 ka, back to Marine Isotope Stage (MIS) 13 (Thunell and Williams, 1983). MD 84641, at a depth of 1,375 m between the Nile Delta and Cyprus, extends back to 461 ka, in MIS 12 (Fontugne

Box 3.2 Marine records of climate change

Due to their steady accumulations over very long time periods, deep-sea sediments provide an invaluable record of long-term climate change. In particular, micro-organisms such as foraminifera which are deposited with the sediment provide information on past temperatures through the analysis of the oxygen-isotope composition of their skeletons. The skeletons are composed of calcium carbonate crystallized from the sea water in which the foraminifera lived. There is a relationship between the temperature of the water at the time of crystallization and the relative proportions of the different stable isotopes of oxygen which make up the carbonate. These proportions are usually recorded as deviations from a modern standard, usually written as $\delta^{18}O$ and measured in parts per mil (‰). In practice, the reconstruction of temperature is more difficult than this would suggest, because values of $\delta^{18}O$ are also affected by global ice-volume changes and sea-water salinity changes (see Bradley, 1999; Lowe and Walker, 1997).

The marine records have been used to define global correlations of climatic variability based on the $\delta^{18}O$ record, which overcomes problems of land-based correlations such as those based on the Alpine glacial or Mediterranean sea-level changes. These correlations are based on marine isotope stages (MIS) which are numbered back from the present. Thus, MIS 1 refers to the present warm stage, MIS 2 the last glacial, and so on, with odd-numbered stages relating to relatively warmer and even-numbered stages to relatively cooler phases.

Table 3.3 Marine isotope chronology and relative stratigraphic names used in northern Europe for the later Pleistocene

Isotope stage	Starting date ka bp	Northern European chronology	Isotope stage	Starting date ka bp	Northern European chronology	Isotope stage	Starting date ka bp
1	12.1	Holocene	16	622	Glacial C	39	1225
2	24.1	Weichselian	17	658	Interglacial III	40	1234
3	59.0		18	695	Glacial B	41	1278
4	73.9		19	729	Interglacial II	42	1295
5	79.3	Eemian	20	743	Helme (Glacial A)	43	1320
5.1	91.0		21	786	Astern	44	1339
					(Interglacial I)	45	1356
5.2	99.4		22	813		46	1370
5.3	110.8		23	831		47	1401
5.4	123.8		24	847		48	1415
5.5	129.8		25	877		49	1430
6	194	Warthe	26	890		50	1445
7	258	Saale	27	917		51	1493
8	279	Drenthe	28	930		52	1535
9	359	Domnitz	29	950		53	1567
		(Holstein)	30	970		54	1582
10	386	Fuhne	31	1007		55	1602
		(Holstein)	32	1043		56	1639
11	430	Holsteinian	33	1081		57	1669
		(Holstein)	34	1106		58	1698
12	486	Elster 1	35	1138		59	1741
13	521	Elster 1/2	36	1156		60	1770
14	544	Elster 1	37	1182		61	1805
15	589	Cromerian IV	38	1201		62	1838

Source: Based on Martinson *et al.* (1987); Williams *et al.* (1988); Lowe and Walker (1997); and Bradley (1999).

and Calvert, 1992). Both sequences show relatively small oscillations of the order of 1.5–2‰ $\delta^{18}O$ between the glacial and interglacial cycles from about 500 to 250 ka (Figure 3.14). After this, the oscillations are much larger, of the order of 4–5‰ $\delta^{18}O$. Both sequences show a peak warmth in the interglacial period about 200 ka (MIS 7), and MD 84641 may also have been warmer in MIS 5 (*c.*120 ka). A peak showing conditions warmer than present in the early part of the present interglacial is found in both these and other sequences in the eastern basin (Thunnell and Williams, 1983; Cheddadi and Rossignol-Strick, 1995), the Tyrrhenian (Paterne *et al.*, 1988) and in the Atlantic off the coast of Portugal (Turon, 1984). Corresponding colder extremes are found in the glacial periods, with minima in MIS 8 (*c.*260 ka) and particularly in the last glacial period (MIS 2) at its maximum extent around 18 ka. The exact temperature fluctuations recorded by these oxygen isotope changes are complicated by changes in salinity of the surface waters, but may reflect conditions of up to 10–12°C cooler than present in the maximum glacial conditions and of 2–4°C warmer in the maximum interglacial conditions (Fontugne and Calvert, 1992). Ariztegui *et al.* (1996) also demonstrated that different methods of estimating sea-surface temperatures produced different results for cores in the central Adriatic. Their results suggest, however, that temperatures may have been up to 8°C colder than present during the last glacial maximum and the height of the Younger Dryas event, and that early Holocene temperatures between 9040–8275 BCE and 7693–7318 BCE were up to 2°C warmer than present (see Box 3.3 for information on dating conven-

---

Box 3.3 Dating conventions

Where dates are given in this book, the following conventions have been used:

> BCE/CE refers to calendar dates before the current era and current era, according to standard practice. If neither are given, then CE is assumed.

Radiocarbon dates have all been calibrated to calendar years (see Gillespie, 1984; Smart and Frances, 1991; Taylor, 1987, for further details on the necessity of calibration), and are quoted as above for direct comparability with historical and instrumental dates. Uncalibrated ages have systematic deviations from real ages, and their use causes confusion at best, and incorrect interpretations of sequences at worst. Calibration has been carried out using the program CALIB 3.0, based on dendrochronological samples for the period to 10,000 bp (before present – by definition, 1950) and interpolated coral dates back to 18,367 bp (Stuiver and Reimer, 1993; Stuiver and Pearson, 1993; Pearson and Stuiver, 1993; Pearson *et al.*, 1993; Linick *et al.*, 1986; Kromer and Becker, 1993; Bard *et al.*, 1993). Dates are presented as a 2σ range (e.g. 24–330 CE), where a full radiocarbon date is reported in the literature, so that the real date has a 95% probability of occurring within the given range. Where an incomplete or interpolated radiocarbon date is given, the calibration is given to the central intercept, and presented as an approximate date (e.g. *c.*1500 BCE). It must be noted that such a date should be interpreted with extreme caution, as the intercept provides a very poor estimate of the actual date of the sample. Dates between 18,367 bp and 45,000 bp are calibrated using the method of Van Andel (1998). At present, there are no techniques for calibrating earlier dates, although in practice this is rarely a problem because of the practical limits of the radiocarbon technique.

Older dates are reported therefore before present, using the terms ka (thousands of years before present) and Ma (millions of years before present).

tions used in the book). However, Bigg (1994; 1995) used ocean general circulation models and flux models to suggest much smaller changes at the last glacial maximum of around 3–4°C colder than present. Cornu *et al.* (1993) have also suggested a temperature of about 3°C warmer than present during the last interglacial period (115–135 ka) based on the $\delta^{18}O$ record from shells in deposits in Tunisia and Mallorca. Sea-level data from Mallorca also suggest warmer conditions during this period (Hillaire-Marcel *et al.*, 1996).

A land-based oxygen-isotope record has recently become available from the Soreq cave site about 60 km from the Mediterranean in Israel (Bar-Matthews *et al.*, 1997). Analysis of spelaeothems from this cave has allowed a reconstruction to be made of the average annual temperature and rainfall at the time of deposition (Figure 3.15). During the last glacial maximum, average temperatures were estimated to be 6–10°C colder than present, with rainfall 50 to 80 per cent of the present average. The reconstruction suggests a general warming trend until around 7,000 years ago, when the present-day range was reached. Estimated rainfall, on the other hand, increases to present-day values between 17 and 15 ka and continues to rise, peaking at 135–190 per cent of the modern average between 10 and 7 ka. From 7 ka, the estimated range is indistinguishable from the modern value.

Other long sequences of pollen records show a general correspondence with the pattern described for Tenaghi Philippon and for the $\delta^{18}O$ sequences. A second long 163 m-deep sequence at Ioannina in north-western Greece extends over the last 423 ka. This sequence oscillates between forest, forest-steppe and desert-steppe vegetation (Tzedakis, 1994). The forested, interglacial periods start with relatively open vegetation, with pine and juniper. Oak woodland then becomes more dominant, although evergreen oak is less common in the earlier part of the sequence, which may correspond to the wetter conditions also recorded at Tenaghi Philippon. The oak occurs in association with ash, pistachio, olive and phillyrea. There is a tendency for the oak to be partly replaced by hornbeam and hop-hornbeam later in the interglacial, before returning to more open conditions with more pine and juniper. The open vegetation relating to colder periods is made up of a number of associations, including open parkland, grassland-steppe and desert steppe dominated by artemesia and/or chenopod pollen (ibid.). A significant feature compared to Tenaghi Philippon is the higher quantities of tree pollen present in the cold periods. Tzedakis (1993) suggests that the existence of these refugial populations may be explained by higher winter precipitation, which may relate to the higher altitude (470 m) and the westerly exposure of the site. The presence of last-glacial forest refugia is a common feature of pollen diagrams in the Balkans (Willis, 1994).

At Valle di Castiglione, near Rome, the last 270 ka are represented by three gaps (Follieri *et al.*, 1988). As with Tenaghi Philippon, vegetation characteristic of dry, steppe conditions is dominant, with 160 out of 237 ka recorded. Submontane and montane groups, indicative of cooler and wetter conditions compared to the present, occur for 52 and 19 ka respectively. The remaining 6 ka are made up of Mediterranean species indicating conditions close to those at present, and occur around 129 ka and in the last thousand years. These data suggest previous interglacials were generally cooler and wetter than the present, in contrast to the Greek sites, which suggest conditions equivalent to the present over a similar time period. Approximately 60 km to the north of Valle di Castiglione is the Lago di Vico sequence, deposited in a volcanic crater lake at an altitude of 507 m asl (Frank, 1969), where a pollen record covering the last 90 ka is preserved (Leroy *et al.*, 1996). The latter part of the last interglacial (MIS 5) is characterized by forest conditions similar to the present day. For part of the early glacial (MIS 4) and the following interstadial (MIS 3), there is a forest vegetation of a much more limited extent, with higher numbers of pine and open species suggesting cooler, wetter conditions. The remaining glacial periods (rest of MIS 4 and MIS 2) are dominated by open steppe species and some pine, together with small numbers of forest species which may represent refugia, as at

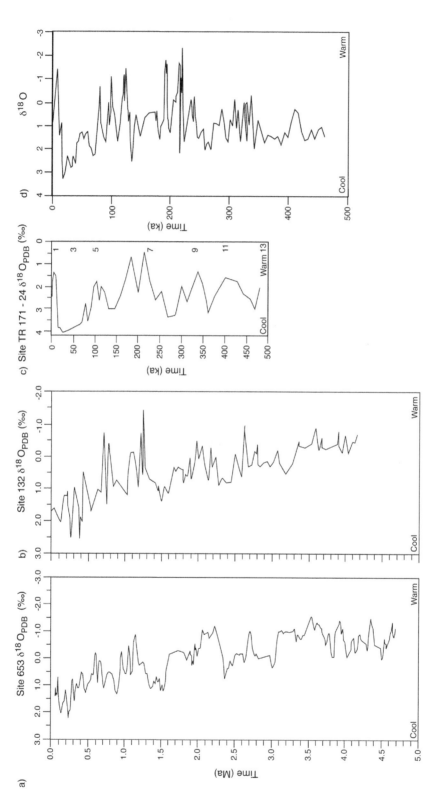

Figure 3.14 Oxygen-isotope fluctuations in four long Mediterranean cores: (a) Oceanic Drilling Project Site 653; and (b) Deep Sea Drilling Project Site 132, both located in the Tyrrhenian Sea; (c) TR171-24, located in 2,328 m of water off south-west Crete; and (d) MD 84641, at a depth of 1,375 m between the Nile Delta and Cyprus

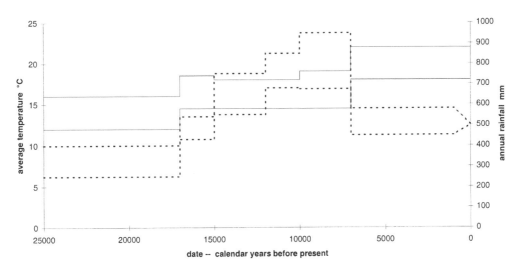

Figure 3.15 Reconstructions of temperature and precipitation in Israel since the last glacial maximum based on spelaeothems from the Soreq cave site

Source: Based on data from Bar-Matthews *et al.* (1997).

Ioannina. Laghi di Monticchio, at 656 m asl, approximately 60 km to the east of Naples, preserves a shorter sequence, beginning around 76.3 ka (Watts *et al.*, 1996a, b; Narcisi, 1996; Zolitscha and Negendank, 1996; Creer and Morris, 1996). There is a higher proportion of tree species throughout most of the record, although even in the last interstadial (MIS 3), the proportion is not as high as present-day conditions. At the last glacial maximum, the vegetation is dominated by grasses, artemisia, chenopodiaceae, pine and juniper, characteristic of dry, steppe vegetation. Response-surface mapping suggests large oscillations in the temperature of the coldest month as represented through the core, with a minimum of 9°C cooler than at present around 15 ka (Figure 3.16; Huntley, 1993). Generally cooler temperatures are represented by the response surface of vegetation growing days (number of days per year above a 5°C threshold), and arid conditions relating to a lack of rainfall between 25 and 18 ka by the response surface of the ratio of actual to potential evapotranspiration (Figure 3.16; Huntley, 1993). Modelling based on mineral magnetic data, however, suggests the period before 35 ka was about 175 mm a^{-1} wetter than at present, and was followed by a drying period where precipitation decreased to only 80 mm a^{-1} wetter than present at 27 ka, before increasing to previous conditions around 20 ka (Creer and Morris, 1996). These changes are earlier than those suggested by the pollen, and conflict with lake-level data elsewhere in Italy (Giraudi, 1989), and with modelling based on pollen and magnetic data in southern France (see below).

In the eastern basin several very long sequences exist along the Dead Sea Rift due to their deposition in this actively subsiding basin (Horowitz, 1979, 1987; 1989; Horowitz and Weinstein-Evron, 1986; Levin and Horowitz, 1987; Weinstein-Evron, 1987). Despite their great depth (1,600 to 2,400 m), the records produced have a lower resolution than the Tenaghi Philippon sequence because sampling intervals had to be much wider because of the origin of the cores from oil and gas drilling. Nevertheless, the sites of Notera 3 in the Hula basin to the north of Israel and Amazyahu 1, to the south of the Dead Sea, both contain records which have been correlated radiometrically and with reference to the marine isotope stages covering approximately the last 3.5 Ma. Horowitz (1979; 1987; 1989) suggests three

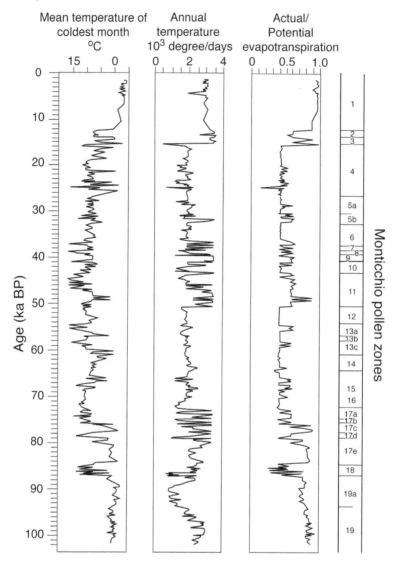

Figure 3.16 Climate reconstructions from Italy based on response-surface mapping of Laghi di Monticchio data

Source: After Huntley (1993).

types of climate occur in Israel: a present-day 'interstadial' type, where climate and vegetation are similar to the present time, a wet 'pluvial' climate where the vegetation is richer in tree species than at the present time, and a very dry 'interpluvial' climate, where the vegetation is dominated by steppe species. The occurrence of these different climate types is controlled by the position of the ITCZ, with a weaker, more southerly position responsible for the pluvial, and a more northerly position leading to the drier interpluvial regime. Data from the Hula basin in northern Israel (Weinstein-Evron, 1987) suggest there is a complex relationship between these phases and the marine oxygen-isotope series. At Ashdod in coastal southern Israel, a discontinuous sequence of 188 m covers possibly the last 850 ka with oscillations between various open steppe communities and those dominated by more wooded communities made up of aleppo pine, together with evergreen oak and pistachio (Rossignol, 1962). Ghab in north-western Syria shows a broad correlation with the Tenaghi Philippon, possibly as far back as *c.*90 ka (Niklewski and van Zeist, 1970). Cheddadi and Rossignol-Strick (1995) presented data from three marine cores off the Nile Delta with both pollen and $\delta^{18}O$ records. There is typically a correlation between warm conditions, as reflected in the $\delta^{18}O$ signal, and the amount of tree pollen, notably oak. In the colder periods, grass, chenopodiaceae and artemisia pollen dominate, reflecting the generally dry conditions. More discontinuous records from the land-based part of the delta are published by Saad and Sami (1967), and record a similar pattern for the late glacial and recent periods.

Extensive work has been carried out on three French sites which lie on the edge of the Mediterranean zone, but provide useful insights into possible changes within it from the latter part of the penultimate glacial period. La Grande Pile (de Beaulieu and Reille, 1992) is located in the Vosges and the Lac du Bouchet (Reille and de Beaulieu, 1990; Williams *et al.*, 1996; Sifeddine *et al.*, 1996) in the Massif Central. More importantly, Les Echets is located in the sub-Mediterranean zone near Lyons (de Beaulieu and Reille, 1984). Guiot *et al.* (1989) use transfer functions based on relationships between modern pollen rain and current climate variables to derive details of possible past climatic conditions. Their results suggest (Figure 3.17) conditions in the last two glacial periods with temperatures 8–12°C cooler than present (present annual average temperature 11°C), and temperatures in the last interglacial between

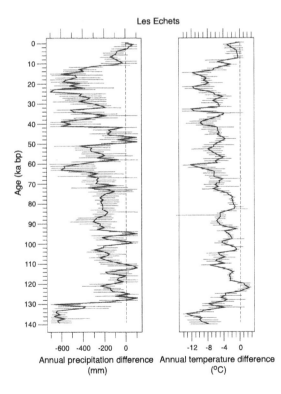

Figure 3.17 Reconstructions of past temperature and precipitation fluctuations in France based on pollen data

Source: Guiot *et al.* (1989; 1992).

0–2°C warmer than present, which are in broad agreement with the ranges suggested for the Mediterranean $\delta^{18}O$ records. Further support comes from the $\delta^{18}O$ signal reported from gastropods in the Pont-de-Mirabeau sequence in Provence, where the general pattern of temperature fluctuation between *c.*30 and 13 ka is in agreement with the Les Echets reconstruction. Precipitation reconstructions are much more variable, but suggest interglacial amounts comparable to the present (800 mm), but glacial periods with between 200 and 700 mm less than the present day, reflecting a very dry landscape. Reconstructions of the seasonal patterns at La Grande Pile suggest that the previous interglacial was broadly similar to the present in this respect also, which has important implications for understanding past landform change in interglacial periods. In the glacial periods, temperature changes are relatively evenly spread through the year, whereas precipitation decreases are concentrated more in the summer, autumn, and, to a lesser extent in the winter months.

The site of Padul in Andalucía is the only published site in southern Spain covering the entire last-glacial cycle (Florschütz *et al.*, 1971; Pons and Reille, 1988). Commencing around 105 ka (MIS 5c), the vegetation is typified by an open mixed-oak forest, reflecting relatively warm, wet conditions. From about 70 ka to 13 ka, the sequence is dominated by a steppe-type vegetation where pine is the only tree pollen found. This sequence represents a phase of cold, dry conditions, interrupted only by a number of short-lived episodes from *c.*55 to 30 ka, where there is a limited re-incursion of oak species. The 30 ka-sequence from Lake Banyoles in northern Spain shows a similar pattern with steppe species dominant until about 14 ka, with some important incursions of pine pollen in the earlier part (Pérez Obiol and Julia, 1994). A similar steppe vegetation is seen in the late-glacial maximum at the nearby site of Pla de Llacs (Pérez Obiol, 1988). Wansard (1996) used the ratio of magnesium to calcium in ostracods from the Lake Banyoles sequence to estimate the summer water temperatures in the period from 24 to 11 ka. The results demonstrated a great amount of variability, from 9–10°C cooler than the present day at 24, 21 and 16–14.5 ka, and peaks reaching modern values (23°C) between 18–17 and 13–11 ka. There are rapid fluctuations, where temperature drops by 8 to 10°C over a century period, showing a great instability in the climate in this period, as recorded in the Greenland ice cores (Johnsen *et al.*, 1992).

In northern Africa, the data which could be used to reconstruct past vegetation and climate change are somewhat rarer, due largely to problems of preservation, but also to the difficulties of carrying out research in these areas. Brun (1989) has compared the data for northern Morocco and Tunisia. Around 30 ka, the vegetation of upland northern Tunisia was dominated by deciduous oak (*Quercus canariensis*) forest, with alder, elm and willow in wetter locations. Oscillations of wetter conditions in the period up to about 28 ka are shown by three major peaks in alder pollen at Dar Fatma 2 (Ben Tiba and Reille, 1982). Similar, relatively humid conditions are shown in the coastal zone of Tunisia, with pine also present. Further inland, steppic vegetation appears to have been more important, although faunal evidence suggests the climate was still significantly wetter than the present day (Brun, 1989). The coastal zone was dominated by artemisia steppe, as shown from cores in the Gulf of Gabès (Brun, 1985). Evidence from alluvial deposits in Morocco, Algeria and Tunisia suggests generally wetter conditions in the period from 40 ka, although the evidence is fragmentary. Variability in these and in palaeosol formation (Rognon, 1987) may also relate to the oscillations which are found in the pollen data. Further evidence for wetter conditions in this period comes from the onset of lacustrine activity at the Chott el Jerid in southern Tunisia dating to around 43 ka, which correlates to a wider-scale wetter period in North Africa (Lézine and Casanova, 1991). White *et al.* (1996) suggest this transition is also related to changes in alluvial-fan activity in the same area. Soil formation in loess sequences in northern Tunisia also suggests much wetter conditions in the period 25 to 28 ka, and generally wetter conditions back to 40 ka (Dearing *et al.*, 1996). Sometime after 28 ka, the sequence at Dar Fatma 2 shows a gradual decline of arboreal pollen, with significant reductions in the presence of alder, and after a short time-lag, increases in pine and evergreen oak, to the detriment of the deciduous oak (Ben Tiba and Reille, 1982). The increasing aridity which this reflects is seen elsewhere in Tunisia, where oak-pine forest dominates from 27 ka until somewhere between 20 and 18 ka, when the steppe vegetation expands relatively rapidly (Brun, 1979; 1989), although the Dar Fatma 2 sequence does show a further humid oscillation around about 20 ka (Ben Tiba and Reille, 1982). Brun (1987) presents data from the Oued el Akarit in central Tunisia which is dominated by desert steppe species which may date to 28 ka, when elsewhere in Tunisia forested conditions were dominant. However, the dating of this site is problematic, and the materials analysed may date to the period after 18 ka, when such an association would be consistent. Dearing *et al.* (1996) suggest the period between 21 and 18 ka is moderately more humid, based on information from soil formation and mineral magnetic data. The increasingly arid conditions from this time led to a break of deposition in the Tunisian pollen sequences until around 12 ka, when Dearing *et al.* (1996) suggest the occurrence of another much more humid period. There is evidence that valley-bottom gullying (arroyo downcutting) begins in a number of places after 20 ka, which, combined with evidence from increased aeolian activity, Rognon (1987) relates to increased occurrence of major storm events in a generally drier environment. This drying is also supported by the smaller number of dates relating to groundwater recharge in the Algerian, Libyan and Egyptian Sahara in the period after 20 ka compared to the period from 50 ka (Sonntag *et al.*, 1980). At Tigalmamine in the Middle Atlas of Morocco, Lamb *et al.* (1989; 1995) have recovered a continuous record starting at *c.*19550 BCE in a 21 m core through lake sediments. At the base of the core, there is a slight decrease in evergreen oak leaving a landscape dominated by steppe species (grasses, chenopods and artemisia), in a similar way to that suggested by the Dar Fatma 2 sequence. Around *c.*14850 BCE, there is an increase in evergreen oak (*Quercus rotundifolia*) together with pine, juniper and ash, suggesting the onset of wetter conditions, probably relating to the onset of high-latitude deglaciation (Lamb *et al.*, 1989) and as seen in the sequences in Spain and elsewhere at a similar time period. After about *c.*11800 BCE, there is a reversion to slightly drier conditions, probably relating to the Younger Dryas event, during which there was a global reversal of the deglaciation, probably relating to circulation patterns in the North Atlantic (Broecker *et al.*, 1989, 1990; Broecker and Denton, 1990; Overpeck *et al.*, 1989; Manabe and Stouffer, 1995). Rognon (1987) suggests there is a similar oscillation in the Maghreb, based on evidence from shells in northern Morocco and charcoal in Algeria. Deposition in Tunisia recommences some time after *c.*12700 BCE with a diverse steppe vegetation, with more grass species in comparison to previously, and with an increase in aquatic species showing generally more humid conditions (Brun, 1989).

The general pattern described by the data above suggests that during periods of glacial activity at high latitudes, the average temperature in the Mediterranean Basin was between 3°C and 12°C cooler than present. These periods generally coincide with drier conditions, with rainfall totals possibly as low as 50 per cent of their present values. Wetter-than-present conditions are often characteristic of transitional periods between cold and warm climate types. In previous interglacial periods, temperatures may have been up to 4°C warmer than present. There is little information on the rainfall regime during these periods, although they may have been relatively similar to that of the present day. It must be remembered that the variability of Mediterranean climates discussed above is not incorporated within these reconstructions, and that there were probably oscillations on a number of timescales through the past.

General circulation model runs since the last glacial maximum at 18 ka can be used to add some detail to this pattern, although it must be remembered that these models only represent the conditions relating to a single glacial maximum and subsequent warming. Kutzbach *et al.* (1993) used the CCMO GCM to retrodict the climatic conditions in January and July at 3 ka

Table 3.4 Average climate reconstructions for GCM model cells representing the Mediterranean region for the period since 18 ka. Data from Kutzbach *et al.* (1993). Dates are given as ka rather than the equivalent calendar years because the model uses solar insolation values for these times as boundary conditions in conjunction with palaeoenvironmental data of the same *uncalibrated* age. There is thus some uncertainty in the absolute timing of these changes (see Box 3.3)

Date ka	Temperature difference °C		Precipitation difference mm day^{-1}		Moisture balance difference mm day^{-1}	
	Jan	Jul	Jan	Jul	Jan	Jul
18	−15.0	−3.9	−0.45	−0.27	−0.07	0.02
15	−11.6	−2.7	−0.21	−0.11	−0.29	0.07
12	−4.4	−1.4	−0.56	−0.13	0.42	0.49
9	−0.24	1.4	−0.08	−0.10	0.32	0.40
6	−1.2	1.1	−0.18	0.02	0.46	0.21
3	−0.2	0.3	−0.24	0.00	0.18	0.07

intervals from the late glacial maximum at 18 ka, to the present day. Boundary conditions for the model were developed using the COHMAP database. These simulations retrodict for the Mediterranean region a mean January temperature difference of 15.0°C colder at 18 ka, increasing dramatically between 15 ka and 12 ka and then remaining fairly stable through the Holocene (Table 3.4). The July temperature differences are less marked according to the GCM runs, with a value only 3.9°C colder at 18 ka, rising to reach a value of 1.4°C warmer than present at 9 ka, after which the values decline again to reach present-day values. Estimated cumulative precipitation shows values generally drier than at present apart from the estimates for July 6 ka and 3 ka. Modelled moisture deficit (precipitation minus potential evaporation) shows a slightly different pattern, with oscillations between wetter and drier in January, and always wetter on average than at present for July. In all cases, there is considerable variability through the region, with both wetter and drier conditions simulated at each point in time for both months. These results should be set in the context of the general reliability of GCM output (see Wigley, 1992, for example), but do at least give some more detail on the seasonality of the changing climates with the transition from late glacial to modern conditions.

The COHMAP project (COHMAP members, 1988; Wright *et al.*, 1993) has suggested how these changes can be explained by general circulation patterns. At the height of the last glacial maximum, there was probably a very strong westerly circulation pattern, with the jet stream diverted northwards away from the Mediterranean Basin. This pattern may have led to the development of less cyclogenic activity in the basin, and hence the drier conditions experienced. The glacial anticyclone over northern Europe and Scandinavia would reinforce the dryness in the summer months by bringing in cooler continental air to the basin. Wetter conditions in the transitional period may reflect the southerly extension of the jet stream in winter, both bringing more frontal rain and increasing the likelihood of cyclogenesis within the basin. In the eastern and southern parts of the basin, the corresponding movements of the ITCZ in relation to jet stream movements have been recognized as being important (Horowitz, 1979; 1987; 1989) in controlling the pluvial regime. When the ITCZ is able to expand northwards, drier conditions are maintained in this area.

The last 18,000 years

As seen above, there was a general warming trend around the Mediterranean starting about *c*.14850 BCE, with a slight reversal between *c*.12050 and *c*.10950 BCE corresponding to the

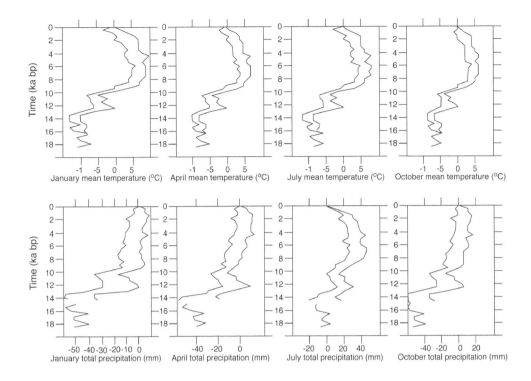

Figure 3.18 Reconstruction of the general climatic patterns in southern France since the last glacial maximum based on pollen data

Source: Guiot (1987).

Younger Dryas event (Pons *et al.*, 1987). The Holocene period comprising the last 10,000 years can essentially be divided into two phases climatically. During the first, there is a transitional pattern which corresponds to the continuing general warming from the last glacial maximum. This phase is followed by a period which is essentially indistinguishable from that of the present day climatically, and commences between *c.*6850 BCE and *c.*3500 BCE, depending on the location (see below, Table 4.1). The oxygen-isotope data from Soreq cave in Israel, discussed above, also suggest conditions similar to the present day for the last 7,000 years.

Guiot (1987) has reconstructed the general climatic patterns in southern France since *c.*19550 BCE using a three-step approach. The method incorporates: (1) multiple regression to derive analogue climates from a number of modern-day pollen sites; (2) canonical correlation and principal components analysis to define 'palaeobioclimate' signals in nine pollen sites; and (3) the application of these analogues to the palaeobioclimate signals using a Kalman filter. This approach permits the definition of past climates within statistical confidence limits. The results (Figure 3.18) are consistent with the last glacial maximum conditions of 7 to 13°C cooler than present, with 20 to 60 mm less rainfall. Warming from *c.*14850 BCE raised the temperatures to 1 to 7°C cooler than present and precipitation to −30 to 30 per cent of present before the reversal to colder, drier conditions in the Younger Dryas. Between *c.*10950 BCE and *c.*10450 BCE this trend is again reversed, with warming leading to conditions between 2 and 5°C warmer than present in the early Holocene, in conjunction with precipitation that is not statistically distinguishable from that of the present day. The exception to this pattern is the

July precipitation which is estimated at between 20 and 40 mm more than at present. This value, together with the temperature values converge on the modern-day values from about 2,000 years ago.

The results of Guiot are consistent with the COHMAP model data (COHMAP members, 1988) discussed above. The reconstructions of Huntley and Prentice (1988) also suggest July temperatures for southern France would have been in the range of 2 and 4°C warmer than present again based on pollen data. However, their results suggest a range of values through-out Mediterranean Europe, with virtually no change over large areas, deviations of more than 2°C cooler than present in southern Italy and southern Greece, and a strong gradient to 4°C warmer than present in northern Greece. By comparison, the range of predicted values by Kutzbach *et al.* (1993) for the same time are from 2.7°C warmer than the present to no dif-ference. Guiot *et al.* (1993) went on to use lake-level data from a number of sites within Europe to refine the picture on a larger scale, suggesting that precipitation was largely unchanged throughout Mediterranean Europe at 6 ka and similarly at 3 ka, apart from in central Italy.

In northern Africa, there is evidence to suggest that the early to mid-Holocene was more humid than at the present time. Fontes *et al.* (1985) (see also Gasse *et al.*, 1987) describe the presence of swamp and lacustrine sediments from the edge of the Great Western Erg in the northern Sahara in Algeria between 9759 and 7507 BCE and 3297 and 1776 BCE. Rognon (1987) presents general evidence for a humid phase from *c*.4850 BCE to *c*.2500 BCE in the Maghreb. Ritchie *et al.* (1985) show the presence of a range of species in lacustrine sediments at Oyo in the Sudan between 7691 and 7312 BCE and 4220 and 3124 BCE, which currently receives less than 5 mm of rain per year on average. Other areas in north-western Sudan and south-western Egypt seem to have had rain-fed lakes which dried up around *c*.4850 BCE. Radiocarbon dates from groundwater in the Sahara in Algeria, Libya and Egypt suggest the existence of relatively humid phases around *c*.6850 BCE and *c*.4850 BCE to *c*.3800 BCE (Sonntag *et al.*, 1980).

Tree-ring data, which can be used to derive climatic signals, have been obtained for a number of sequences around the Mediterranean. However, because of the nature of the tree species growing in the region and the potential for the preservation of fossil wood, most of the sequences are less than 500 years in length and relate to high-altitude conditions (Serre-Bachet, 1991). One interesting exception is the fossil sequence based on about twenty examples of *Taxodion* dating to around 1.3 Ma at Dunnarobba in Italy. This 256-year sequence shows oscillations with frequencies of 85.0, 45.0, 21.3, 6.7, 5.8 and 5.6 years (Rosa Attolini *et al.*, 1988). The sequence of Kuniholm *et al.* (1996) provides a 1,503-year series from 2220 BCE based on wood samples from archaeological sites in Anatolia. The importance of this series in linking to the climatic anomalies following the eruption of Thera around 1628 BCE has already been discussed in Chapter 2. Similar anomalies seem to be present following the Hekla 3 eruption in Iceland in 1159 BCE. The full climatic significance of this sequence has yet to be presented in detail. At present there are few dendrological studies relat-ing to the prehistoric period in the Mediterranean (although see Kuniholm and Striker, 1987). Further studies relating to the historic period are discussed in the following section.

The historical period

The use of early written records can provide useful data regarding past climates, although their interpretation must be made with extreme caution (Le Roy Ladurie, 1983; Flohn, 1985). Such data can be broadly broken down into two categories – continuous and discontinuous. The former are usually based on plant phenological records, recording the date of harvest of a

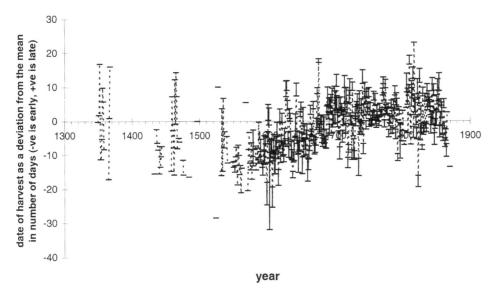

Figure 3.19 Summary of French grape harvest data presented by Le Roy Ladurie (1983; vol 2: 151–169). The graph is a summary of the data for southern French sites only and has been constructed in a different way from the summary graph in Le Roy Ladurie (1983; vol 2: 200). The dates for each location were first transformed by subtracting the long-term average date for the location to remove any spurious trend due to the differential spatial distribution of data through time. These transformed data were then averaged for all locations available in a particular year. The graph also shows deviations from the average in each year as error bars representing one standard error from the mean (note that before the seventeenth century, this calculation is not possible because of the number of records available)

particular crop. In our region, the most useful data come from grape and wheat harvest records (Le Roy Ladurie, 1983). However, these data are not the continuous series they first appear. The date of the grape harvest reflects, at least theoretically, the temperature and hours of sunshine in the growing months, i.e. the summer, whereas the wheat harvest is determined by spring conditions. Therefore these series represent only six months of each year, the missing details needing to be filled in by other means. Furthermore, the date of the grape harvest can also be controlled by economic and social means. For example, from about 1700 to 1840 the harvest in France became progressively later as major landowners applied pressure on tenants to produce grapes with a higher alcohol content to produce a better wine (and also to manufacture *eau de vie* when the wine market became slack). After this period, the increasing independence of small producers caused a relatively rapid return to an oscillation about the original position (Figure 3.19; Le Roy Ladurie, 1983, vol. 2: 200–201). Other data relating to cultivated plants may be difficult to interpret. During the so-called *Little Ice Age*, Le Roy Ladurie notes the extension of the cultivation of the olive, a plant which is particularly sensitive to frost, due to human intervention (1983, vol. 1: 29). Discontinuous data relate to a wide variety of sources such as letters, municipal records, local histories and travellers' memoirs. These data may be highly variable, depending on the interests and perception of the individuals making recordings, and may rely on anecdotal reports or hearsay. The presence of these individuals is also increasingly rare in time and space the further back in time we go, and their presence in particular periods can affect the apparent continuity of a reconstruction (Pavese *et al.*, 1992). Documents may also be incompletely preserved. Where data are missing, it is dangerous to infer relationships between areas, especially in a region with a climate as inherently variable as that of the Mediterranean. For example, Le Roy Ladurie (1983, vol. 1: 80) quotes a peasant of Brie in north-east France:

> The winter [of 1816] was wet. The spring [of 1816] cold and late. The summer wet and late. We only began to cut the wheat on the 20th August, a very late date. Because of continual rain, we were unable to cut any two days running. . . . Frost attacked the budding vines. We harvested the grapes on All Saints' Day. The little wine that was made was undrinkable.

Instrumental rainfall data for this year are available for a number of stations in south and south-west France (Garnier, 1974a, b). Although January rainfall is above average to varying extents everywhere except Montpellier, February and March precipitation is below average almost everywhere – Avignon and Marseille being totally dry – so the winter cannot be said to have been particularly wet. April is exceptionally wet everywhere, but in the south-west especially this appears to represent the *early* advent of the spring rains. Of the summer, July is wetter than average everywhere, but June, August and September have below-average rainfall. If anything, 1816 seems to have been a good year in southern France (Table 3.5).

Data relating to flood occurrence are also frequently used, especially in more remote historic periods (e.g. Lamb, 1977: 144). Interpretation of relatively recent data (Fischer, 1930;

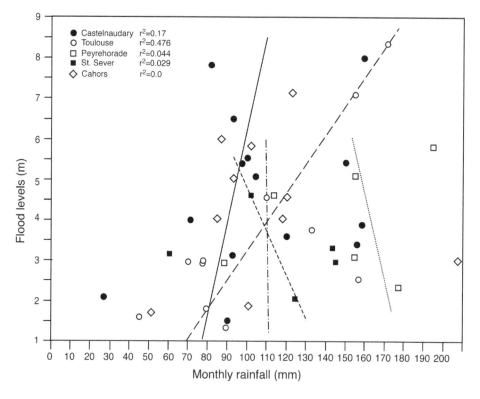

Figure 3.20 Examples of flood events and the corresponding monthly rainfall at local stations in southern France, demonstrating the difficulties of extracting meaningful climatic signals from historical accounts of flood events

Table 3.5 Rainfall records for the year 1816 in southern France, as compared with average values for each station. All values are in millimetres, with figures in bold showing above-average values. *Note*: See text for details

Month	Montpellier		Alès		Joyeuse		Viviers		Avignon		Marseille		St Sever		Toulouse	
	1816	mean	1816	mean	1816	mean	1816	mean	1816	mean	1816	mean	1816	mean	1816	mean
Jan	57	72	**120**	87	**150**	95	**113**	63	**97**	40	**100**	42	**126**	66	**61**	43
Feb	8	46	30	58	62	74	31	48	0	34	0	30	42	75	27	40
Mar	4	69	20	64	19	67	29	54	0	36	0	35	71	51	42	46
Apr	**127**	50	**150**	74	**295**	99	**200**	72	**77**	52	**60**	32	**116**	59	**100**	58
May	79	60	**100**	89	118	139	52	81	**90**	61	**80**	39	36	65	53	82
Jun	27	45	50	50	59	62	55	66	24	27	20	20	16	63	37	62
Jul	**53**	22	**100**	45	**159**	68	**186**	52	**56**	25	**40**	11	**142**	48	**159**	48
Aug	9	29	10	47	23	71	21	61	0	31	0	19	40	44	12	44
Sep	4	76	60	118	77	158	48	117	17	79	15	57	50	59	33	58
Oct	41	110	**150**	126	**211**	198	**131**	124	71	84	70	84	50	71	15	56
Nov	42	102	**100**	93	**202**	162	**115**	114	57	74	50	63	40	78	33	46
Dec	5	89	20	81	12	87	17	66	29	44	30	45	**150**	86	41	45
Year	456	770	910	932	**1387**	1280	**998**	918	518	587	465	477	**879**	765	613	628

Source: Data after Garnier (1974a and b)

Pardé, 1933a and b; 1953) again shows the danger of drawing climatic conclusions. Flood levels are not strictly correlated to the amount of rainfall recorded – none of the relationships tested were significant, even at the 90 per cent level (Figure 3.20) – as the rain may have fallen principally in neighbouring mountainous areas. For example, of the floods of the Tarn at Albi (Monnié, 1931), with a catchment extending into the Massif Central, only that of December 1906 coincides with floods of the Garonne at Toulouse, with a Pyrenean catchment (Pardé, 1953), in the period 1875 to 1930. Although monthly rainfall, where known, is generally above average, respective yearly rainfall can be shown to be below average in 18 to 50 per cent of cases. This example shows the difficulty of using early historical documents on flooding without complementary evidence. Furthermore, runoff production and flooding (see Chapter 5) are also a function of land use, plus continuously occurring extreme rainfall events and numerous other factors, which must be accounted for before drawing conclusions about any climatic change.

Written records

Lamb (1977) believed on the basis of historical records of flooding around the Mediterranean, particularly in Italy, that the region underwent wetter conditions from *c.*500 CE to around 1100, albeit with interruptions from 750 to 850 and in the tenth century. There is also a suggestion that increased cyclonic activity occurred in the northern part of the region from the thirteenth century and particularly in the seventeenth century, with some supporting evidence from lake-level data, although the flood data (ibid.: 144) are not particularly convincing (Figure 3.21). Lamb's study set the stage for a general pattern of historical climate interpretation with conditions deteriorating after the classical period, improving around 1000 CE until the early fourteenth century in the period known as the *Mediaeval Warm Period*, and declining markedly in the period from the fourteenth to eighteenth centuries – the so-called *Little Ice Age* (Grove, 1988; although Le Roy Ladurie [1983] restricts use of the term to the fifteenth to seventeenth centuries). However, a number of more recent studies provide contradictory evidence, or show greater complexities over the Mediterranean which call this general pattern into question.

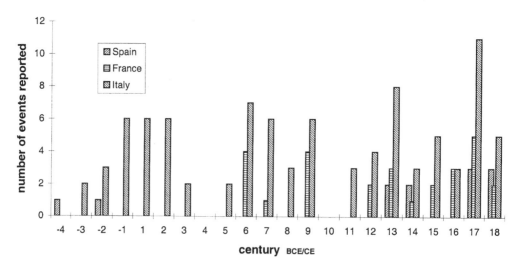

Figure 3.21 Timing of Mediterranean floods according to Lamb (1977)

Note
No account is taken of differential reporting or preservation of records.

Darnajoux (1976) provides information relating to the period up to the end of the tenth century in France. For information relating specifically to southern France, there seems to be a relatively consistent amount of data from the fifth century (Table 3.5). There seems to be little variability across this time period, except perhaps for a slight decrease in the occurrence of severe conditions in the seventh and eighth centuries. Four years between 587 and 593 were also reported as having exceptionally mild autumn or winter weather. The only cold periods noted in Spain by Font Tullot (1988) for this period occurred in 775 and 776, with no abnormally warm conditions being recorded. Drought years were, however, particularly common, especially in the years 680–685, 707–711, 748–754, 846–879 and 976–978, with floods or heavy rain reported only twice (Table 3.6). The Venice lagoon froze in 853, 860 and possibly also in 859 and 864 – these two latter years being the only ones which correspond to reports of colder conditions in southern France – with another severe winter reported in 568 (Camuffo, 1987). In the Bosphorus, extreme cold was recorded in 763, followed by a spring drought in 764 (Telelis and Chrysos, 1992).

The most complete compilation of records for France, Italy and Slovenia for the period from 1020 to 1419 is given by Alexandre (1987), although the sequence is somewhat discontinuous until the thirteenth century. By converting the records into an ordinal scale and calculating moving averages, it is possible to interpret these data in a semi-quantitative way for the period after 1200 (Figure 3.22). These records seem to indicate three phases. The first, starting around 1210, has large oscillations on a twenty-year cycle with hotter temperatures tending to be associated with drier conditions (and vice versa). Around 1255, there is a change to smaller oscillations about a slight downward trend to cooler and drier conditions, with cooler temperatures associated with wetness and higher temperatures with dryness. From 1335, larger oscillations are set in train again, with a reversion to the cold/wet and hot/dry scenario. Data on the French grape harvest are relatively scarce over this period, but suggest values close to the long-term trend in the mid-fourteenth and fifteenth centuries, with a possible occurrence of a cluster of earlier harvests and warmer temperatures in the late 1430s and early 1440s (Le Roy Ladurie, 1983). Spanish data from the tenth to fifteenth centuries (Font Tullot, 1988) suggest much less frequent occurrences of colder conditions, and a number of oscillations between slightly drier and wetter conditions (Figure 3.23). Furthermore, the Spanish sequence tends to be out of phase with the changes suggested by the compound French-Italian-Slovenian sequence, both in terms of temperature and precipitation. In contrast to the often wetter conditions suggested by the Spanish data, the Tiber flood data suggest a lower frequency of flood occurrence until around 1400, followed by a marked increase in the later part of the fifteenth century (Gregori *et al.*, 1988). Records from the Valencian irrigation systems also suggest years with drought were more common in the fourteenth century compared to years with floods (eighteen compared to six: Table 3.5; Glick, 1970). In the fifteenth century, flooding is nearly as common as drought (eighteen drought years compared to fourteen flood years). Complete freezing of the Venice lagoon occurred twice in the twelfth century, once in the thirteenth century and six times in the fifteenth century, with other severe winters noted at relatively constant low rates over these centuries (Camuffo, 1987). In northern Italy as a whole, 'great' winters were noted eight times and severe winters a further nine times in the fifteenth century (Camuffo and Enzi, 1992). The flood-tide data for Venice, which are thought to represent the general persistence of depressions causing sirocco conditions, indicate a period of enhanced occurrence from 1460 to 1500 (Camuffo and Enzi, 1992).

Le Roy Ladurie (1983) notes that for the period from 1491 to 1608, there were eleven very severe winters in southern France with instances of snow, freezing of the Rhône and of olive trees and vines (Table 3.5). Another seventeen severe and cold episodes also occurred over the same period. Summer temperatures, as suggested by the date of the grape harvest, may have

Table 3.6 Summary of historical climate data for the Mediterranean region

Location	Weather type	Years (CE except where noted)	Reference
southern France	severe winters	400, 401, 462, 468, 544(?), 547, 559, 579, 603(?), 604(?), 608, 763, 791, 811(?), 823(?), 842(?), 846, 859, 864, 913(?), 923(?), 939, 964(?), 974(?), 988(?), 992(?), 994	Darnajoux, 1976
southern France	mild autumn/winter weather	587, 588, 589, 593, 993	Le Roy Ladurie, 1983
	very severe winters (snow, freezing of the Rhône and of olive trees and vines)	1494, 1506, 1557, 1565, 1569, 1571, 1573, 1590, 1595, 1603, 1608	
	severe and cold episodes	1491, 1517, 1518, 1524, 1527, 1540, 1543, 1570, 1572, 1581, 1583, 1584, 1587, 1591, 1597, 1598, 1600	
Spain	notable cold conditions	775, 776, 1009, 1010, 1011, 1077, 1110, 1113, 1133, 1161, 1191, 1193, 1201, 1212, 1232, 1233, 1234, 1235, 1258, 1307, 1333, 1334, 1335, 1386, 1419, 1431, 1433, 1434, 1442, 1447, 1458, 1465, 1469, 1476, 1480, 1487, 1503, 1506, 1529, 1531, 1535, 1536, 1539, 1541, 1546, 1548, 1550, 1556, 1559, 1564, 1567, 1570, 1572, 1573, 1574, 1576, 1579, 1582, 1590, 1592, 1593, 1596, 1599, 1600, 1603, 1604, 1607, 1610, 1615, 1616, 1617, 1618, 1619, 1620, 1621, 1622, 1623, 1624, 1626, 1627, 1628, 1640, 1641, 1643, 1644, 1645, 1648, 1654, 1655, 1659, 1663, 1665, 1668, 1678, 1679, 1680, 1682, 1683, 1686, 1693, 1694, 1696, 1697, 1708, 1709, 1716, 1726, 1728, 1738, 1739, 1744, 1750, 1754, 1757, 1763, 1765, 1766, 1774, 1778, 1779, 1782, 1783, 1784, 1786, 1788, 1789, 1797, 1806, 1817, 1819, 1829, 1830, 1838, 1849, 1850, 1862, 1864, 1868, 1890, 1891, 1894	Font Tullot, 1988
	notable warm conditions	1445, 1446, 1449, 1462, 1466, 1518, 1520, 1526, 1527, 1540, 1557, 1565, 1568, 1611, 1612, 1625, 1632, 1633, 1635, 1637, 1646, 1647, 1650, 1651, 1652, 1653, 1656, 1657, 1658, 1718, 1790, 1801, 1802, 1808, 1818, 1839, 1840, 1871, 1878	
	both cold and warm noted	1530, 1543, 1589, 1601, 1605, 1613, 1642, 1787, 1835, 1836, 1837, 1880	
	drought	476 BCE, 427 BCE, 224 BCE–198 BCE, 75 BCE, 680–685, 707–711, 748–754, 846–879, 901, 976–978, 982, 1057, 1058, 1088, 1094, 1213, 1217, 1219, 1255, 1262, 1300, 1304, 1308, 1333, 1346, 1355, 1374, 1376, 1393, 1394, 1401, 1404, 1418, 1421, 1426, 1443, 1455, 1457, 1473, 1484, 1521, 1525, 1539, 1562–1566, 1567, 1569, 1583–1585, 1588, 1598, 1630, 1633, 1640, 1662, 1665, 1675, 1687, 1695, 1699, 1711, 1720–1722, 1724, 1748, 1755, 1757, 1772, 1780, 1781, 1815, 1868, 1869	
	floods and/or heavy rain	500 BCE, 181 BCE, 180 BCE, 49 BCE, 756, 974, 1011, 1084, 1085, 1102, 1138, 1143, 1168, 1172, 1173, 1201, 1203, 1205, 1207, 1211, 1218, 1229, 1236, 1256, 1258, 1264, 1267, 1275, 1283, 1286, 1292, 1297, 1307, 1310, 1320, 1321,	

1328–1330, 1348, 1351, 1353, 1358, 1373, 1379, 1380, 1385, 1388, 1392,
1405–1408, 1412, 1420, 1422, 1424, 1427, 1432, 1434, 1438, 1439, 1441,
1442, 1444, 1446–1449, 1451, 1452, 1458–1460, 1464–1477, 1479–1482,
1485–1488, 1490, 1494, 1496, 1500, 1504, 1505, 1508, 1509, 1512, 1517, 1519,
1523, 1524, 1527, 1529, 1531, 1533, 1535, 1537, 1538, 1541–1544, 1548, 1550,
1572–1582, 1590, 1591, 1594, 1599, 1600, 1602–1604, 1606–1612, 1614, 1619,
1636, 1637, 1642–1646, 1648, 1649, 1651, 1653, 1656–1658, 1660, 1663, 1664,
1667–1674, 1676, 1678, 1679, 1684, 1691, 1693, 1696, 1697, 1700–1702, 1704,
1707–1710, 1713, 1716, 1719, 1726–1729, 1731–1737, 1740, 1742, 1744–1747,
1758–1763, 1766–1770, 1775–1778, 1785–1791, 1793–1795, 1797, 1798,
1800, 1802, 1804–1806, 1808, 1814, 1818, 1819, 1821, 1823, 1825, 1826, 1829,
1830–1834, 1837–1840, 1843–1847, 1850, 1852–1857, 1860–1867, 1871,
1880–1900

| both drought and floods/heavy rain noted | 620, 675, 686, 755, 849, 1302, 1340, 1356, 1402, 1403, 1410, 1414, 1416, 1425, 1450, 1456, 1462, 1472, 1474, 1475, 1489, 1499, 1507, 1511, 1518, 1528, 1540, 1545, 1546, 1551–1561, 1565, 1568, 1570, 1571, 1586, 1587, 1589, 1592, 1593, 1595–1597, 1605, 1615–1618, 1620–1629, 1631, 1632, 1634, 1635, 1639, 1641, 1661, 1666, 1680–1683, 1685, 1686, 1688–1690, 1692, 1694, 1698, 1714, 1718, 1723, 1725, 1738, 1739, 1741, 1749–1754, 1764, 1773, 1774, 1779, 1782–1784, 1792, 1796, 1799, 1801, 1803, 1816, 1817, 1827, 1828, 1841, 1842, 1848, 1849, 1851, 1858, 1870, 1872–1879 | |
| Catalonia | rogations in drought years (number in brackets gives indication of drought severity) | 1500 (20), 1506 (40), 1512 (20), 1520 (20), 1521 (15), 1522 (20), 1526 (80), 1529 (80), 1530 (70), 1532 (20), 1533 (30), 1534 (30), 1535 (20), 1537 (40), 1538 (10), 1539 (15), 1540 (50), 1541 (50), 1542 (20), 1545 (15),1548 (60), 1549 (40), 1550 (70), 1551 (20), 1552 (40), 1556 (15), 1557 (20), 1560 (10), 1561 (100), 1563 (10), 1564 (40), 1565 (60), 1566 (90), 1570 (10), 1571 (50), 1576 (40), 1577 (40), 1578 (50), 1579 (10), 1580 (40), 1583 (120), 1584 (130), 1586 (30), 1588 (20), 1589 (40), 1591 (20), 1594 (30), 1595 (20), 1600 (20), 1602 (60), 1605 (110), 1607 (20), 1608 (20) | Le Roy Ladurie, 1983 |

continued

Table 3.6 continued

Location	Weather type	Years (CE except where noted)	Reference
River Ebro	floods	1421, 1445, 1448, 1517, 1605, 1617, 1775, 1783, 1787, 1826, 1831, 1843, 1845, 1853, 1865, 1866, 1871, 1884	Linés Escardó, 1970
	droughts	1725, 1749, 1751, 1796	
Venice Lagoon	exceptional freezing, including events where ice was thick enough to support people on the lagoon	853, 859(?), 860, 864(?), 1118, 1122, 1234, 1432, 1443, 1475, 1476, 1487, 1491, 1514, 1549, 1595, 1603, 1684(?), 1709, 1716, 1740, 1747, 1755, 1758(?), 1789, 1864, 1929, 1956, 1985	Camuffo, 1987
	partial freezing of the lagoon or internal canals	1561, 1569, 1729(?), 1795, 1814, 1855, 1926	
	severe winters without evidence of freezing	568, 1133, 1216, 1311, 1413, 1419, 1511, 1535, 1596, 1598, 1608, 1731, 1776, 1848, 1849, 1858, 1869, 1879, 1880, 1893, 1901, 1905, 1907, 1940, 1942, 1947, 1954, 1963	
Zamora	rogations in drought years (number in brackets is number of prayers, indicating severity)	1620 (2), 1623 (2), 1628 (2), 1629 (3), 1630 (3), 1631 (3), 1637 (8), 1638 (2), 1639 (3), 1640 (2), 1641 (2), 1643 (5), 1645 (3), 1650 (4), 1662 (4), 1664 (4), 1665 (1), 1666 (4), 1667 (1), 1668 (4), 1671 (1), 1679 (4), 1680 (4), 1683 (2), 1685 (3), 1687 (2), 1689 (1), 1691 (4), 1694 (3), 1698 (4), 1699 (1), 1700 (3), 1703 (5), 1706 (1), 1713 (2), 1715 (3), 1716 (1), 1720 (4), 1722 (4), 1723 (2), 1726 (2), 1727 (1), 1731 (1), 1734 (6), 1737 (1), 1738 (3), 1741 (1), 1742 (4), 1743 (5), 1744 (4), 1748 (1), 1749 (1), 1750 (3), 1751 (2), 1752 (4), 1753 (4), 1754 (4), 1757 (3), 1761 (2), 1764 (4), 1767 (5), 1770 (1), 1772 (1), 1773 (2), 1776 (2), 1779 (4), 1780 (2), 1781 (1), 1783 (1), 1789 (2), 1794 (2), 1797 (3), 1803 (1), 1807 (2), 1808 (2), 1814 (2), 1815 (2), 1817 (1), 1819 (1), 1820 (1), 1824 (1), 1828 (1), 1833 (1)	Alvarez Vazquez, 1986
	rogations in wet years and during floods	1556 (1), 1586 (1), 1597 (1), 1611 (1), 1621 (1), 1626 (1), 1634 (1), 1648 (3), 1654 (1), 1657 (3), 1658 (2), 1677 (2), 1695 (3), 1702 (2), 1707 (1), 1735 (1), 1740 (1), 1766 (1), 1768 (1), 1788 (1), 1799 (1), 1806 (1)	
Valencia	droughts	1313, 1321, 1335, 1341, 1343, 1345, 1351, 1352, 1355, 1358, 1368, 1371, 1372, 1374, 1375, 1376, 1384, 1393, 1400, 1401, 1403, 1404, 1406, 1412, 1413, 1414, 1415, 1420, 1421, 1432, 1435, 1443, 1444, 1445, 1447, 1449	Glick, 1970
	floods	1321, 1328, 1340, 1356, 1358, 1378, 1403, 1406, 1410, 1413, 1415, 1416, 1417, 1427, 1428, 1436, 1453, 1462, 1475, 1487	
northern Italy	'great' winters	1430, 1432, 1443, 1476, 1477, 1483, 1487, 1491, 1511, 1549, 1571, 1603, 1608, 1684, 1709, 1716, 1740, 1755, 1784, 1789, 1795, 1814, 1855, 1864, 1926, 1956, 1985	Camuffo and Enzi, 1992

	Phenomenon	Years	Source
	severe winters	1406, 1408, 1414, 1459, 1462, 1464, 1470, 1493, 1498, 1501, 1504, 1514, 1536, 1561, 1570, 1573, 1595, 1599, 1600, 1602, 1605, 1614, 1665, 1677, 1729, 1747	Gregori *et al.*, 1988
River Tiber	floods	1230, 1277, 1310, 1379, 1415, 1422, 1438, 1467, 1475, 1476, 1480, 1493, 1495, 1498, 1514, 1530, 1557, 1572, 1589, 1598, 1606, 1637, 1647, 1660, 1686, 1695, 1702, 1742, 1750, 1772, 1780, 1805, 1809, 1843, 1846, 1855, 1858, 1870	
River Belbo	floods	1511, 1553, 1646, 1648, 1649, 1651, 1671, 1680, 1684, 1698, 1709, 1742, 1744, 1776, 1792, 1801, 1803, 1839, 1840, 1857, 1859, 1879, 1914, 1926, 1948, 1951, 1957, 1958, 1959, 1960, 1968	
Tanara Valley	drought	1501, 1517, 1562, 1578, 1597, 1611, 1639	
	heavy rain and/or floods	1508, 1510, 1511, 1514, 1515, 1519, 1520, 1521, 1524, 1541, 1545, 1552, 1555, 1556, 1557, 1567, 1568, 1569, 1571, 1579, 1580, 1584, 1593, 1595, 1596, 1598, 1599, 1600, 1601, 1603, 1604, 1605, 1606, 1609, 1612, 1613, 1614, 1616, 1620, 1626, 1627, 1635, 1644, 1646, 1647, 1649, 1653, 1654, 1655, 1657	
	severe winter	1510, 1511, 1515, 1516, 1518, 1523, 1552, 1557, 1564, 1565, 1567, 1568, 1569, 1570, 1571, 1573, 1578, 1579, 1594, 1595, 1599, 1600, 1601, 1605, 1607, 1608, 1610, 1612, 1613, 1614, 1616, 1622, 1635, 1638, 1643, 1650, 1655, 1656	
	unseasonal winter phenomena	1500, 1615	
	mild winter	1593, 1602, 1646, 1647	
	heatwave	1519, 1540, 1555, 1597, 1598, 1603	
Crete	drought	1595 (winter), 1626, 1696	Grove and Grove, 1992
Bosphorus	drought	763 (winter extreme cold), 764 (spring drought)	Telelis and Chrysos, 1992

Note
See text for details of sources.

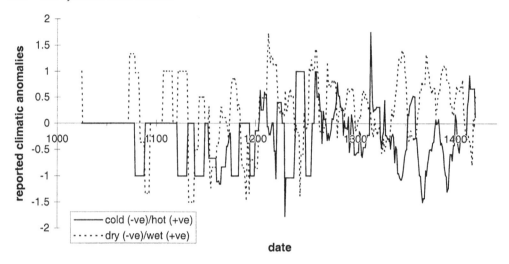

Figure 3.22 Summary of climatic data for southern France, Italy and Slovenia reported by Alexandre (1987) from 1030 CE to 1420 CE. Each record of anomalous conditions in a particular year was converted into an ordinal scale (i.e. cold = −1, warm = +1; dry = −1, wet = +1), summed and the sequence normalized to account for the increase in available records through time. The curves presented are ten-year moving averages

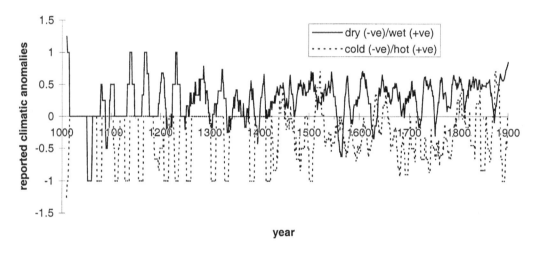

Figure 3.23 Summary of climatic data from Spain (Font Tullot, 1988). The sequence presented here is calculated in the same way as described in Figure 3.22

increased slightly from around 1540 (Figure 3.19; Le Roy Ladurie, 1983). However, in northern Italy, Camuffo and Enzi (1992) only note the occurrence of six 'great' winters and fourteen severe winters over the same time period (Table 3.6). The Venice lagoon froze completely on five occasions and partially another twice. For the Tanara valley in Italy, Gregori *et al.* (1988) note twenty-six occurrences of severe winter conditions, again in the same period. These results indicate the importance of accounting for spatial variability – only the occurrences of 1570, 1571, 1573, 1595, 1600 and 1608 coincide in the French, northern Italian and Tanara valley records, and only 1595 and 1608 are common years if the Venice data are incorporated. None of these years were recorded as cold according to the Spanish data of Font Tullot (1988). These data show a greater incidence in the number of reported floods through the sixteenth century, apart from a period in the 1550s and 1560s, where more mixed or drought conditions are noted. Following a warm peak in the first quarter of the century, there is a slight downward but oscillating trend towards cooler conditions, with a tendency for cold years to occur more frequently with wet years. Again, there is a major contrast with other data sets from Spain, with the number of rogations in drought years noted in Catalonia suggesting relatively frequent and often severe droughts (Table 3.6: Le Roy Ladurie, 1983). Grape-harvest data from Valladolid (Le Roy Ladurie, 1983) also seem to suggest a slight cooling from about 1520 to 1570 followed by the onset of a warming trend. Furthermore, there is only one flood record for the Ebro in the sixteenth compared to the fifteenth century (Linés Escardó, 1970). In Italy, the Tiber flooded six times and the Belbo twice in the sixteenth century, while in the Tanara valley heavy rain or floods were recorded no fewer than twenty-seven times (Gregori *et al.*, 1988). In this location, there were also five recorded droughts, and five heat waves during the century.

In Andalucia, Rodrigo *et al.* (1994) have used historical documents to reconstruct the rainfall patterns for the first half of the seventeenth century. These records mention rainfall-related fluctuations more frequently than temperature fluctuations. Median rainfall deviations are around normal conditions, although Seville shows more instances of wetness than Granada. The patterns of anomalies reported seem to be consistent with actual patterns of rainfall event within the Mediterranean region. Again, the spatial variability of rainfall can be seen, with no overlap between reported floods in the Ebro when compared with either the Seville or Granada records. Indeed, Granada suffered a minor drought in 1604–1605 (Rodrigo *et al.*, 1994). Elsewhere in Spain, Font Tullot (1988) notes an initial colder, wetter phase which represents the last part of the trend starting in the previous century, followed by a more prolonged warmer, slightly drier phase. The number of rogations against drought in Zamora are consistently high from 1620, with much fewer rogations against floods and rain (Table 3.5; Alvarez Vazquez, 1986). The grape harvest in France occurs progressively later through the seventeenth century, but stabilizes by about 1670, whence it oscillates around the long-term average (Figure 3.19; Le Roy Ladurie, 1983). The flood tides at Venice were again enhanced from 1600 to 1620 (Camuffo and Enzi, 1992).

The eighteenth century seems similar to the seventeenth for the most part according to the data of Font Tullot (1988), although the last two decades may have been marked by slightly colder, wetter conditions. Although there is evidence for flooding in the Ebro valley in 1783 and 1787 (Linés Escardó, 1970) and in Zamora in 1788 and 1799 (Alvarez Vazquez, 1986), there is also evidence for drought in both areas during these decades (1796 in the Ebro, and 1781, 1783, 1789, 1794 and 1797 in Zamora). In France, the grape-harvest data for this period are difficult to interpret, as noted above. However, the removal of the trend which is due to the locations for which data are available (see Figure 3.19) suggests oscillation around the long-term average through the century, and the lateness remarked by Le Roy Ladurie as being due to a shift in cultural practices does not appear until the early nineteenth century. The Venice lagoon froze partially in 1795 and totally in 1789. The other five definite complete freezes took place up to 1755 (Camuffo and Enzi, 1992), so there is little evidence to support a later cold phase in Italy. However, there is a final period of enhanced flood tides in Venice between 1740 and 1800 (Camuffo and Enzi, 1992). Other flood data from Italy seem to be evenly spread through the century (Gregori *et al.*, 1988).

During the early nineteenth century, the Spanish data start to show a slight warming, with an approximate thirty-year oscillation between warmer and colder conditions (Figure 3.23; Font Tullot, 1988). There are larger oscillations present in the flood and rainfall data which suggest that the warmer years may also have been wetter. The Ebro flood years are also more frequent than in previous centuries (Linés Escardó, 1970). Notwithstanding, the early part of the century still saw more rogations against drought than rain or floods in Zamora (Alvarez Vazquez, 1986). The French grape-harvest data show the lateness in the early part of the century as noted previously, but return to oscillate around the mean later. There is some suggestion of a slightly later harvest in the 1840s and 1850s (Figure 3.19; Le Roy Ladurie, 1983). Only one instance of a complete freezing of the Venice lagoon is reported, in 1864, with two partial freezes and seven other severe winters, all of which occur from 1848 (Camuffo, 1987). However, only the years 1814, 1855 and 1864 are qualified as 'great' winters in northern Italy as a whole (Camuffo and Enzi, 1992). Flooding on the Tiber seems concentrated in the 1840s and 1850s, but on the Belbo it is remarkably constant through the century (Gregori *et al.*, 1988).

Tree-ring data

Tree-ring data are relatively numerous in the historical period and have a reasonably wide geographical range, but tend to provide sequences of several hundred years at most (Serre-Bachet, 1991). Two floating chronologies exist covering the fourth to first centuries BCE in Israel (Liphschitz, 1986). A larch sequence from 2,600 m asl in the Mercantour Alps in southern France extends back to 933 CE (Serre, 1978), although its direct relevance is limited because of the high altitude of the samples. The larch sequence of Bebber (1990) from the Italian Alps is even longer – beginning in 781 CE – but suffers similar problems. Till and Guiot (1990) have used cedar (*Cedrus atlantica*) rings to identify patterns in annual precipitation since 1100 CE in Morocco. Their results show that large inter-annual variability is a feature of the entire sequence, apart from a period of reduced variability from *c*.1350 to *c*.1540 CE. Despite the large variations, there are very few periods where the reconstructed rainfall is significantly different from the modern-day distribution (Figure 3.24). A marked major departure occurs for two decades either side of 1200 CE, where there is a significant drier period in all the climatic zones of Morocco (Guiot *et al.*, 1982). Samples of *Juniperus phoenica* in the Sinai show a period of favourable growth from 1185 to 1255, probably indicating a period of higher temperatures (Liphschitz, 1986). Years of extreme drought (defined as precipitation of less than two standard deviations less than the mean) occurred on twenty-one occasions, more frequently than extremely wet years, which only occurred six times. There is some clustering of the extremely dry years, with two instances in the late twelfth century, four in the seventeenth century, seven in the eighteenth century, seven in the nineteenth century and one in the early twentieth century. For the extremely wet years, two occur in the third quarter of the thirteenth century and two in the first quarter of the eighteenth century. February, May and June temperatures have also been reconstructed for the Moroccan sequences, but for a more restricted time sequence between 1840 and 1975 (Guiot *et al.*, 1982). There seems to be some correspondence between lower temperatures at Tetouan in the 1840s–1850s and lower rainfall in the same period. However, lower temperatures from the 1910s to 1940s are associated with precipitation close to the long-term mean, demonstrating the existence of complex patterns of climatic variability.

Serre-Bachet *et al.* (1992; also Serre-Bachet and Guiot, 1987) discuss general reconstructions for the period since 1500 using tree-ring series from the Savoy and Maritime Alps, Mont Ventoux, and from Calabria. These series are of pine, larch and fir, and are compared with the cedar series from Morocco. Their reconstructions show a sequence of oscillations around modern mean values on a decadal scale for summer temperatures at Grand St Bernard (since 1585) and Rome (since 1500). The period of the so-called Little Ice Age (*c*.1550 to 1800) shows no difference from the mean values, which is different from reconstructions for more northerly areas, although there is enhanced variability (Figure 3.25). However, this compares

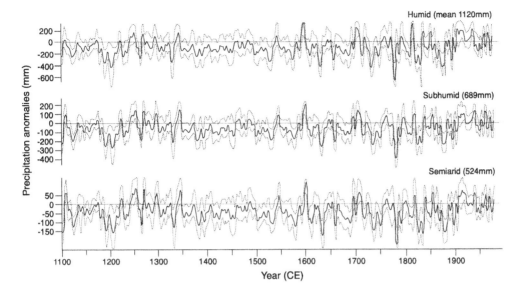

Figure 3.24 Reconstructions of patterns of annual precipitation since 1100 CE based on cedar (*Cedrus atlantica*) tree rings in Morocco

Source: Till and Guiot (1990).

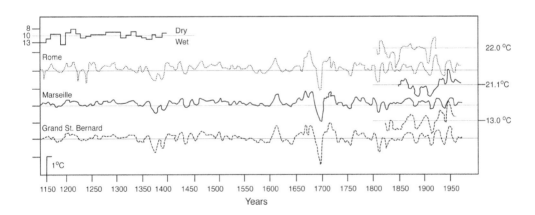

Figure 3.25 General reconstructions of climate for the period since 1500 using tree-ring series from the Savoy and Maritime Alps, Mont Ventoux, and from Calabria

Source: Serre-Bachet *et al.* (1992); Serre-Bachet and Guiot (1987).

with the lower variability of estimated summer temperatures at Marseille during this period (Guiot, 1985). The twenty-year period around 1700 shows a common cold spell with continued departures of −2°C compared with the present values in both locations, although starting slightly earlier in Grand St Bernard. The reconstruction for winter temperatures in north-east Spain, however, suggests that the same period underwent warmer departures of up to 2°C, but with less continuity. Again, the period up to 1815 exhibits greater magnitudes of variability than in the more recent period. The Moroccan data suggest the decade immediately before 1700 was slightly drier than the present mean, while the following decade was wetter (Till and Guiot, 1990). Overall, the tree-ring data suggest that there was no common pattern of change across this period, except for the enhanced variability. Rainfall reconstructions for Marseille and north-east Spain suggest that the period since 1775 has seen oscillations around the present mean, although there is some suggestion of slightly drier conditions at Marseille, starting between 1850 and 1875 (Serre-Bachet *et al.*, 1992). By contrast, a sequence based on bosnian pine (*Pinus leucodermis*) from the southern Apennines in Italy suggests that low growth conditions only occurred for two short intervals – between 1610 and 1620 and between 1685 and 1710 – during this period (Serre-Bachet, 1985). Other Italian sequences show less variability through this period (Biondi, 1992).

The two decades around 1680 are also characterized by low growth in fir trees (*Abies alba* Mill.) on Mont Ventoux in southern France, as is the period from 1706 to 1720 (Serre-Bachet, 1986). Further north in the French Alps, there is some evidence from larch-tree rings to suggest a series of cold periods between 1510 and 1920, corresponding to a 'Little Ice Age', the most important being centred around 1560, 1600, 1700 and from 1770 to 1840 (Tessier, 1986). Larch rings from the central Italian Alps show negative deviations from 1760 to 1767, 1788 to 1840, 1952 to 1966 and 1972 to 1985. Sequences from the eastern Italian Alps show more continuous negative departures from 1760 to 1822, with conditions oscillating around the long-term mean in the later nineteenth century, but a similar pattern in the period from 1952 (Nola, 1994). In Sicily, data from turkey oak (*Quercus cerris* L.) and pubescent oak (*Q. pubescens*) suggest that the middle parts of the eighteenth and nineteenth centuries may have been drier than at present (Martinelli *et al.*, 1994).

Corsican pine (*Pinus nigra* Arnold) tree rings at Cazorla in the Baetic Cordillera have produced a sequence extending back to 1194. Creus Novau and Puigdefábregas (1983) noted two periods of positive growth anomalies in the later sixteenth century and mid-nineteenth century, which they suggested relate to conditions with wet autumn, mild early spring and dry summer weather. Mountain pine (*Pinus uncinata* Ram.) in the Navarro Pyrenees had negative growth anomalies most apparent between 1814–1823, 1894–1903 and 1959–1967, which were interpreted as having warm, dry autumn weather and cool springs (Creus Novau and Puigdefábregas, 1976). Positive anomalies relating to cool, wet autumns and mild springs were greatest between 1729–1738, 1864–1873 and 1924–1933. A further analysis of a longer sequence in the same area was used by Creus Novau (1991–1992) to reconstruct May temperatures. The period from around 1473 to 1817 was characterized by an estimated average temperature about 0.5°C cooler than the long-term mean. Between 1690 and 1725, the estimated average is depressed by 1°C. The same species in north central Spain show positive growth anomalies from 1665 to 1680 and since the mid-1960s, which are interpreted by Génova (1986) to be a function of reduced rainfall values. Negative anomalies characterize the periods 1705–1717, 1745–1760 and 1800–1860, and are interpreted as relating to enhanced rainfall and/or warmer summer and cooler winter temperatures. The period from 1820 to 1850 also has negative anomalies in scots pine (*Pinus sylvestris* L.) growth in Catalonia, with similar climatic implications. In this sequence there is also a consistent, if small negative departure in the 1940s and 1950s, but no trend to suggest current drying (Gutiérrez, 1989). Fernández Cancio *et al.* (1993) used corsican pine in Cuenca to reconstruct precipitation between 1692 and 1988. Apart from an apparent enhanced variability in the first hundred years of this record, the pattern produced is little different from the modern one. Reconstructed rainfall for Navacerrada and Madrid, based on scots and corsican pine sequences, also show little variation from modern patterns (Fernández Cancio *et al.*, 1994). The estimated summer temperature for Navacerrada also showed little difference from modern patterns. There is a notable anomaly in the years between 1770 and 1775, when low temperatures coincide with a peak in rainfall. Reconstructed May temperatures using oak-tree-ring sequences from north-west Spain suggested a higher variability from 1650 to 1870, with some grouping of colder conditions in the 1730s, 1780s and 1815–1825, and of warmer conditions in the 1650s, 1720s, 1760s and 1850s.

Continuous sequences have been recorded for aleppo pine in the Saharan Atlas in Algeria as far back as 1680. Dry episodes are noted for the 1680s, from 1705 to 1720, 1740 to 1780, 1795 to 1810 and 1920 to 1950, whereas wetter conditions were apparent from 1725 to 1735, 1780 to 1790 and 1970 to 1980. There is some variability, with one of the sequences showing wet conditions from 1880 to 1890 and 1910 to 1920 where others are consistently dry (Safar *et al.*, 1992).

In Cyprus, a relationship exists between tree-ring widths in *Pinus brutia* and annual precipitation showing relative dryness in the sixteenth century, followed by great variability superimposed on a continuing dry period in the seventeenth century (Tamari, 1976). From 1830, the Cypriot data show increased moisture. Rainfall has declined on Cyprus since 1910, with droughts in the 1920s to 1930s and in the late 1950s and early 1960s. Analysis of *Pistacia khinjuk* in the Sinai showed periods of wide ring production, reflecting increased moisture, from 1670 to 1712 and 1790 to 1820, and narrow rings, reflecting aridity, from 1715 to 1740 and 1830 to 1860 (Liphschitz, 1986). *Cupressus sempervirens* in the Sinai show wide ring growth from 1808 to 1851 and narrow rings from 1852 to 1917. Cooler conditions are indicated in south Anatolia from narrow rings in *Pinus nigra* between 1720 and 1740 and between 1830 and 1860. Warmer conditions were apparent from 1670 to 1710 and 1800 to 1820 (Liphschitz, 1986).

In the period for which there is a reasonable amount of tree-ring data and those data provide the major source of climatic information, namely the mid-seventeenth to mid-nineteenth centuries, a pattern of great variability emerges. All possible variations of temperature and precipitation are recorded. In the period from 1650 to 1749, the Sinai is reported as having first wet (1670 to 1712) then dry (1715 to 1740) periods, while Algeria first encounters dry (1680s and 1705 to 1720) then wet (1725 to 1735) periods, as does north-central Spain (dry from 1665 to 1680 and wet from 1705 to 1717). A similar pattern of changes in opposite phases between the western and eastern part of the basin emerges in the period from 1750 to 1840. Is it possible that these changes reflect the Mediterranean oscillation as noted from instrumental data in the present century by Palutikof *et al.* (1996: see above)? The compound sequence of Richter and Eckstein (1990) for Teruel in Spain also suggests an approximate twenty-year cyclicity in summer rainfall back to 1350, and there is a similar pattern in the results of Tamari (1976) from Cyprus. A number of sites suggest the presence of cooler conditions at some point during the period 1650 to 1749, but these tend to be located in the Alps, apart from two reconstructions for central Italy. Southern Anatolia has both cool and warm anomalies during the same period, while Spanish reconstructions suggest warmer conditions. In the following century there is some indication of continuing (but less markedly) cool conditions in Italy, while both warmer and cooler anomalies are noted elsewhere. By contrast, the historical records noted above suggest that neither of these two periods were particularly colder than the preceding or following centuries.

Summary of recent climatic fluctuations

The historical variations in climate reflected by the various historical sources, tree-ring data and the instrumental records show that there is no simple pattern throughout the Mediterranean region. The model of climatic decline in the late classical period followed by an 'early Mediaeval warm period', the 'Little Ice Age' and recent warming can no longer be thought of as tenable for the Mediterranean region (see also the discussions by Alexandre, 1987; Camuffo, 1987; Guiot, 1992; Hughes and Diaz, 1994). As with the patterns recorded in the instrumental data, the main characteristics noted are of significant spatial and temporal variability. It seems to be the case that those sites in the high Alps in the area bordering the Mediterranean zone do suggest a cooling which might correspond to the Little Ice Age. However, within the Mediterranean zone proper, there appears to be a complex series of cycles which are not part of this otherwise general cooling trend (see, for example, Grove, 1988). The event-based characteristics of Mediterranean rainfall in general, as well as the high seasonal and interannual variability that characterizes the regional climate, make it difficult to draw firm conclusions from patchy historical records of such events. A number of records have demonstrated a reduction in variability rather than well-defined changes in temperature or precipitation.

Given that it seems to be difficult to pick out sustained periods of climate change in the Mediterranean record, it may be worth concentrating efforts on documenting and understanding shorter-term events and variations. As noted above with the tree-ring data, it may then be possible to relate these to patterns of variability observed in the modern record. We can pick out a number of examples of such events which cover large areas, or periods during which there is great spatial variability. An example of the former might be the period around 1200. The tree-ring data from Morocco suggest a period of dryness from 1190 to 1210, while tree-ring data from the Sinai suggest a longer period of warmer, drier conditions from 1185 to 1255. Dryness is also seen in the historical data from 1184 to 1198 in northern and central Italy (Alexandre, 1987), albeit combined with indications of cooler weather. In contrast, historical data from Spain suggest colder but wetter conditions around 1190 to 1210. A second example of a widespread fluctuation occurred from about 1460 to 1500, when cooler conditions are attested in northern Italy, southern France and Spain from both historical and tree-ring data. The historical data from both Valencia and Venice suggest that the period may have experienced enhanced storminess. The data for the sixteenth and seventeenth centuries clearly show large temporal and spatial variability. For example, in the years 1610 to 1620, tree-ring data from the Italian Apennines and French Alps as well as historical data from Piedmont and Spain suggest cooler conditions, while the southern French grape-harvest data and the Venetian and other north Italian historical data suggest relatively warm or average conditions. Tree-ring reconstructions also suggest a cold phase from 1670 in the French Alps and 1685 or 1690 in Italy to 1710, while historical data from southern France and Spain suggest normal to warmer conditions, and tree rings from southern Anatolia suggest a warm period from 1670 to 1710.

Summary

The Mediterranean experiences a number of recurring weather patterns controlled by its location at the interface between low- and high-latitude weather systems. The interplay of these patterns and the development of weather systems almost entirely contained within the basin leads to the development of a highly variable climate, both in space and in time. Although the simplistic idea of a Mediterranean climate as being one of hot, dry summers and mild, moist winters is in general correct, it ignores a great deal of this variability and therefore care should be taken in employing the term. The dominant characteristic of the Mediterranean climate is variability at all timescales and spatial scales. This ideal also ignores the fact that extreme storms and temperatures, aridity and strong winds also characterize the Mediterranean climate. These features are all highly significant in considering the human settlement of the landscape.

The Mediterranean climate, as we know it today, began to evolve some time around 3.2 million years ago, and continued with a general cooling trend until around 900,000 years ago.

Since this time, there have been a number of relatively large oscillations corresponding to the general climatic cycles seen elsewhere in the world. The relatively colder periods are generally ones with low rainfall, apart from in the far east of the basin, where both cool, dry and cool, wet conditions occurred, in relation to probable fluctuations in the position of the ITCZ. The warmer periods generally experienced conditions similar to those at present, although there is an indication that in the Mediterranean as elsewhere, some of the previous warm periods were several degrees Celsius warmer than present. The transitional warming periods often seem to be accompanied by wetter than present conditions, and there is some evidence to suggest that some, but not all, transitional cooling periods were also wetter.

The present-day climate has been relatively stable for a period of 6,000 to 8,000 years. That is to say that the fluctuations observed at timescales from the storm event to interdecadal variability were also in operation over this time. There is no strong evidence to suggest that there were pan-Mediterranean changes through the middle and late Holocene as proposed by a number of previous authors. The past record, like the present, is one of high spatial and temporal variability. Efforts in future research would therefore be best concentrated in defining the variability at local scales rather than trying to fit the available evidence into increasingly tenuous schemes from outside the region.

Suggestions for further reading

Barry and Chorley (1992) provide the best introduction to meteorology and climate in general, with a specific section on the Mediterranean. Martyn (1992) gives the most thorough review of world climates, while Palutikof *et al.* (1996) present more detail for the Mediterranean region. Bradley (1999) is an invaluable source for understanding methods of palaeoclimate reconstruction. Techniques of historical reconstruction of the climate are dealt with in the classic work of Le Roy Ladurie (1983).

Topics for discussion

1   What makes the Mediterranean climate so unique?
2   Is variability the only real unifying factor in the classification of Mediterranean climates? What causes this variability to occur?
3   Which of the characteristics of the Mediterranean climate would have been most significant in limiting early human populations? How might this situation differ today?
4   Where and when does the Mediterranean climate occur?
5   What are the principal limitations in reconstructing past Mediterranean climates, and how might these limitations be overcome?
6   Can we reconstruct past Mediterranean climates any better from historical data than from other proxy data sources?

4 Vegetation

Introduction

Vegetation is the landscape feature that most characterizes the Mediterranean. It may be described in terms of a strict hierarchical classification of ever-larger spatial units, for example, the divisions developed by the French ecologist Braun-Blanquet (1932: see Box 4.1), in terms of how the plant behaves in response to the controlling factors, or by the functional types, or in terms of its floral content by species, families, etc. Vegetation and flowers are not the same thing. Anyone interested in Mediterranean flowers should refer to a field guide or flora – several English language ones are listed in the further reading at the end of the chapter. Box 4.2 gives the vegetation terms used in this book.

In this chapter we exclude commercial crops, even though they constitute vegetation and much is to be learned from them, for example, the growth of cereals in response to drought (see Chapter 10). Notwithstanding this, the non-cultivated vegetation of the Mediterranean almost invariably has commercial value, for example, the grazing of animals in beech and oak forest. Moreover, the vegetation cover provides valuable protection to the soil and inhibits soil erosion even on quite steep slopes (see Chapter 6). It also provides habitats for the Mediterranean fauna, brushwood for fires, timber for building and chemicals for the health and cosmetics industries.

Mediterranean scale

Controls on plant diversity

Throughout the Mediterranean, there is a high degree of diversity among plant and animal communities due to an 'inextricable web of causes' (di Castri *et al.*, 1981) including the following:

- The Mediterranean is transitional between temperate and tropical regions, mountains and plains, Europe and Asia.
- There has been a complex palaeoclimatic history and the European Mediterranean has acted as a refuge during the periods of advancing glaciation in Northern Europe (see Chapter 3).
- The basin and range topography offers a wide variety of micro-climates.
- The geology (especially lithology) is spatially complex, offering a large and complex set of soils including relict palaeosols.
- There is a correspondingly wide variety of micro-environments
- Plants and animals (including humans) of very different biogeographical and cultural origins have occupied the region from different directions and at different times and there are important residuals from earlier tropical and temperate formations.

Box 4.1 The Braun–Blanquet phytosociology classification

This extract largely follows Tivy (1993). The Braun–Blanquet classification is a complicated system based on the Zurich–Montpellier school of phytosociology, which placed emphasis on the total floristic composition. It takes into consideration the constancy (presence) and fidelity (exclusiveness) of species in the classification of stands in which a species is a measure of the evenness of species distribution. On the basis of 'constancy groupings', species are classified rare, accidental, accessory or constant, as in the list below. The second figure is the percentage of stands in which a species occurs, which is a measure of species distribution.

1	1–20	(rare or accidental)
2	21–40	
3	41–60	(accessory)
4	61–80	
5	81–100	(constant)

Braun–Blanquet's fidelity classes are as follows:

1   Strangers: appearing accidentally.
2   Companions: indifferent species without a pronounced affinity for any community.
3   Preferents: present in several community types, but predominantly in one.
4   Selectives: present particularly in one community type, but occasionally in others.
5   Exclusives: found most exclusively in one community type.

The recognition of plant associations on this basis has been subjected to considerable criticism, because it is theoretically only possible following an intensive investigation of a subjectively defined particular type of vegetation. The system is, nevertheless, widely used by Mediterranean botanists and the *matorral* types are usually classified on this basis.

Box 4.2 Vegetation terms

Sclerophyllous plants have stiff evergreen leaves with thick cuticles, sunken stomata and a dense mesophyll tissue of small thick-walled cells that reduce water loss and maintain leaf shape as the leaves lose water. An example is the oleander (*Nerium oleander*).

   Drought endurers are plants which are specially adapted to reduce water loss during the dry season. Examples are sclerophyllous shrubs (see Box 4.3).

   Drought avoiders include Mediterranean deciduous shrubs, which are adapted to drought conditions by substantially or completely losing their leaves at the beginning of the dry season. An example is *Retama sphaerocarpa*

   Pyrophilous vegetation is fire resistant. An example is *Quercus coccifera*.

In fact, according to di Castri *et al.* (1981), no one single ecosystem can be considered as peculiar to the Mediterranean environment, a gradient of ecosystems being its intrinsic feature. Because the controls of vegetation cover vary at different scales of observation, they are considered at three different scales in southern Spain in the first part of this chapter, following some preliminary generalities. These scales are: the south-facing slopes of the Sierra Nevada, to illustrate the topographic controls; the province of Murcia, to illustrate geological and edaphic controls; and a single valley, the Rambla Honda, to illustrate the hillslope hydrological controls. This approach can be used to understand vegetation patterns throughout the Mediterranean, although specific details of plant types will obviously vary from location to location.

Mediterranean forests

Forests were once widely distributed in the Mediterranean, their type and productivity being largely a function of the temperature and water supply constraints as indicated by Polunin and Huxley (1965) who describe a clear vertical zonation of the main forest types as indicated in Figure 4.1. As with other forest types, there is a great diversity of species, many of which (perhaps as much as 15 per cent) are endemic. We discussed in the previous chapters how the Mediterranean was hotter and more humid than now in the middle Miocene, but with a marked dry period. Woodland at low elevations was of a sub-tropical type. In that period the antecedents of the sclerophyllous plants (adapted to drought), such as oak, occurred in dry rocky areas. From these they spread out in the late Miocene (Messinian) when the Mediterranean was cut off from the Atlantic (see Chapter 2) causing an increase in aridity and potential migration across a shallow Mediterranean Sea and the colonization of the islands. It was also a time of invasion by steppic plants from the Orient. During the Quaternary, alpine and boreal plants were pushed south into the Mediterranean Basin, or died out and became extinct.

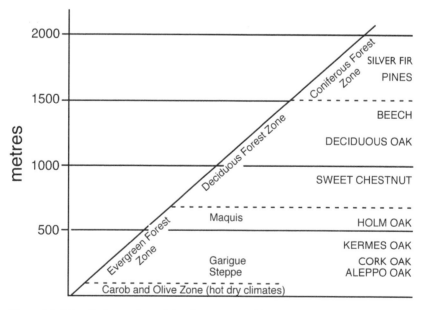

Figure 4.1 Altitudinal zonation of the plant communities in the Mediterranean region
Source: Polunin and Huxley (1965).

Because the snow lines were lower, plant and animal migration across the lowlands was much more feasible.

The destruction of the Mediterranean forests is a result of land redistribution for agriculture (see Chapter 10), but sometimes, as in the Agri valley in southern Italy, the forest was left to occupy the heavier clay lands. In the Agri, conservation practices have produced mature forests dominated by turkey oak and chestnut. Other species are widely distributed, such as maple, hornbeam, cherry and hazel. At higher elevations, the vertical zonation comes into play and, on the slopes of Monte Vulturino, dense beechwoods occur, with an undergrowth of helle-bores, orchids, cyclamen and other perennial flowering plants. The forests are often coppiced, as is the case with chestnut, white hornbeam, maple and oak, on a fourteen-year cycle.

At the scale of the whole Mediterranean, woodland and shrub-winter annual communities (see Box 4.3) account for nearly all the vegetation cover. It is common knowledge that forest once covered the land areas surrounding the Mediterranean Sea and that the retreat of trees is largely attributable to the action of people. Nevertheless, as described in Chapter 2, tectonic activity and hence instability and erosion in the region has led to some dryland habitats being available over much of the last 200 million years (Axelrod, 1993). Even in the wetter periods, when forest communities were dominant, isolated, open, drought-prone habitats existed in which the precursors of the shrub-winter annual communities of the Mediterranean evolved. A dry season was apparently a feature of the evolving Mediterranean climate, with its timing changing from winter to summer during the Pleistocene. Over the last million years, the desta-bilizing effects of the unsettled Pleistocene climate, with extensive erosion and sediment move-ment in the wetter periods, extended the area available for plant colonization. Human activity during the last 10,000 years has further destabilized the landscape (Pons and Quézel, 1985). This activity has resulted in a large increase in the area occupied by the shrub-annual

---

Box 4.3 The composition of the shrub–winter annual community

The shrublands of the circum-Mediterranean region are made up mainly of sclerophyllous shrubs, deciduous shrubs and winter annuals.

Sclerophyllous shrubs, with thick, stiff evergreen leaves are drought-endurers (e.g. *Quercus coccifera*). Their sunken stomata and thick cells ensure that their leaves maintain their shape, even when water is short. Regeneration is from seeds and vegetatively from stools.

Deciduous shrubs in Mediterranean shrubland are drought-avoiders (see Box 4.1). They produce leaves at the start of the wet season (e.g. *Pyrus communis*) that are shed at the onset of the dry season. These 'disposable' leaves have large thin-walled cells with mesophyll tissue consisting of loosely-packed cells separated by air spaces. They are not drought resistant but, in the wet season, they are rapid and efficient at producing carbon through photosynthesis. In some species, the large wet-season leaves are replaced by small drought-resistant ones that persist until the start of the following wet season. The main adaptive trait of the drought deciduous shrubs is the avoidance of the high tempera-tures and drought conditions of the summer by a substantial, if not complete, loss of leaves at the onset of the dry season as revealed by the pattern of litter fall.

Winter annual plants germinate in the autumn and produce mature seed before or at the start of the following dry season, so with their life cycle completed, they avoid drought completely.

(Source: After Clark *et al.*, 1998)

communities, so that they are now the predominant type of vegetation in the Mediterranean Basin.

Mediterranean shrublands

The shrublands are usually separated into *maquis* and *garrigue*. The former is dense vegetation composed of small trees and bushes of medium (2–4 m) height with evergreen leaves, occurring mainly on siliceous substrates, whereas *garrigue* is made up of low, often sparse plant communities on calcareous substrates. The word *maquis* comes from the Corsican dialect word *macchia*, which means mottled. The Spanish word *matorral* fails to distinguish between the two main types except that low, poor or degraded *matorral* is closer to *garrigue* and to the *phrygana* of Greece. The use of *degraded matorral* refers to the concept that the shrublands are derived from the forest by felling, burning and overgrazing, followed by re-colonization by shrubs, as implied by Figure 4.2.

There continues to be a lively debate as to whether the widespread distribution of shrublands is due to highly suitable climatic conditions for shrubs or the history of human intervention. Its ubiquity certainly implies that Mediterranean shrubland is a successful community through time and space. Clark *et al.* (1998) ask: 'Why are these communities so successful in the Mediterranean area, in the face of adverse factors such as fire, heavy grazing, unstable, nutrient-poor soils and drought?' They go on to say that their stability may result from the capacity to keep going through time, by continuing to retain the capacity to regenerate after disturbance (one-dimensional regeneration) or from regeneration by spreading into freshly created spaces resulting from the disturbance, such as plots abandoned after cultivation (two-dimensional regeneration). Two-dimensional regeneration, they claim, is more stable whereas one-dimensional regeneration tends to destabilize the community.

Thornes and Brandt (1993) modelled the effect of long-term variations in rainfall on plant growth and soil erosion, on the assumption that plant growth and erosion compete with each other. More erosion means less soil and so lower water-holding capacity; and more plant cover means less runoff (Francis and Thornes, 1990b). Thornes and Brandt (1993) found that plants died off most rapidly in the shrub communities studied after 'good' growth years in a kind of

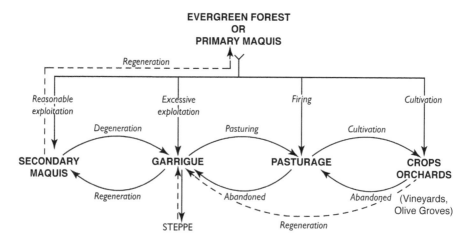

Figure 4.2 Stages of the degeneration and regeneration of plant communities
Source: Polunin and Huxley (1965).

'boom and bust' cycle. Since then a more thorough analysis has revealed the complexity of partial and complete death in shrub communities in the Mediterranean indicating, for example, that it is very species-dependent and soil condition-dependent. Recently it has been shown in the Greek island of Lesvos that there is a minimum soil thickness threshold for the development of a shrub cover.

Controls at different spatial scales

It is impossible to cover all variations in vegetation cover. Rather, we show how different controls operate at different spatial scales using selected examples from southern Spain.

The province of Murcia, Spain

Generally, as we change scale, different controls of the distribution of vegetation appear. However, in Murcia province, Spain (Figure 4.3), which is one of the driest areas of Europe,

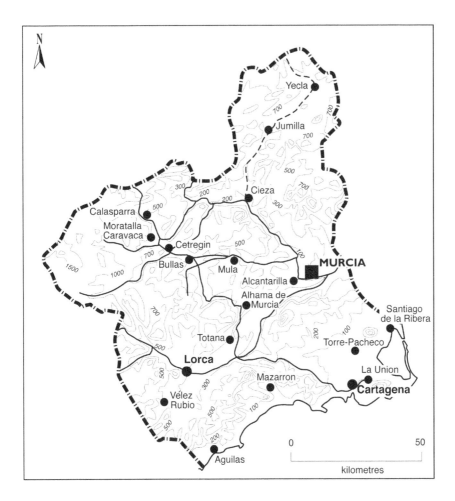

Figure 4.3 Province of Murcia, Spain. Topography and places cited in the text

Source: From Albaladejo Montoro and Diaz Martinez (1983).

the available moisture, topography and soil characteristics continue to dominate. Figure 4.4 shows the monthly rainfall and temperature for four stations. Alhama is on the flanks of the Sierra de Espuna, one of the highest mountains in Murcia. The vegetation and flora of the centre and north of Murcia have been the subject of a detailed study by Esteve Chueca, largely on the basis of the Braun-Blanquet phytosociology classification (see Box 4.1). Not only does the province suffer from intense climatic conditions but, as part of the Baetic Chain, it has sharp relief contrasts, strong variations in aspect and a variety of lithologies, some of which have very poor soils that exacerbate the intermittent severe drought conditions. A long history of human occupation has completely and repeatedly transformed the natural vegetation.

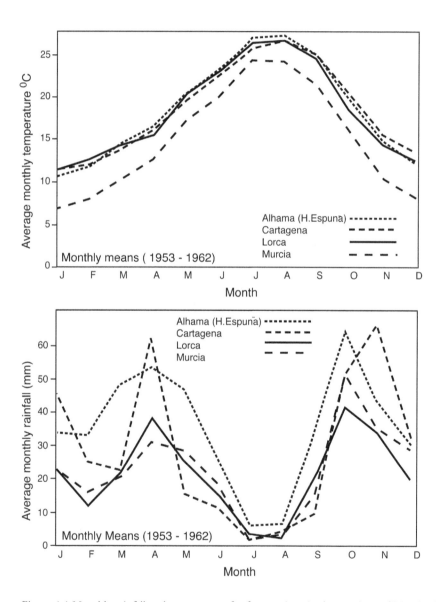

Figure 4.4 Monthly rainfall and temperature for four stations in the province of Murcia, Spain
Source: Esteve Chueca (1972).

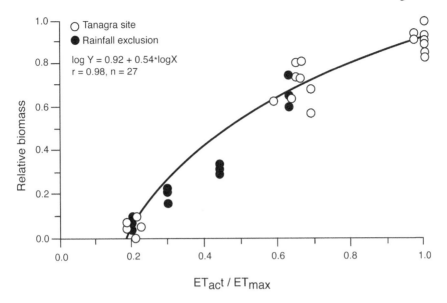

Figure 4.5 Relation between the calculated relative evapotranspiration rate (EA/Etmax) and the relative wheat biomass for different sites and three growing periods

Source: Kosmas *et al.* (1998).

From a management point of view, the species composition of the vegetation may not be the most useful approach. For soil erosion, plant cover (measured as a percentage of plan view obscured by plant material) is important and, in Murcia, this is commonly between zero and 30 per cent. This level of cover is crucial, because a good rule of thumb is that a 30 per cent cover of vegetation reduces soil erosion to 10 per cent of that on bare soil, but this relationship depends on rainfall intensities (see Chapter 6). Another useful indicator of resistance to erosion is the above-ground biomass. Again a useful rule of thumb is that biomass correlates very closely with the ratio of actual to potential evapotranspiration as indicated by Kosmas *et al.* (1998: Figure 4.5). Usefully, the percentage of vegetation cover is also closely related to biomass, so that with more biomass there is less erosion. Since the biomass increases after land has been abandoned from cultivation, the vegetation cover increases and soil erosion decreases after abandonment in the rugged conditions of Murcia.

The regional variability of vegetation in the province of Murcia, indicated in Figure 4.6, reflects essentially the following controls:

- The sharp contrasts in topography leading to strongly juxtaposed basins and mountains. This is seen spectacularly in the vicinity of Murcia city, where the Sierra de Carrascoy, covered with pine and oak, rises steeply above the arid plains about the dry river bed in the great structural graben of the Guadalentín valley.
- The progressive change to less arid conditions from the east to the south-west. Near the coast, at San Javier, the rainfall is about 200 mm per year, but in the high mountains of the south-west it is colder and much wetter. Within 20 km of the coast, the miniature palm (*Chaemerops humilis*) that has a very restricted distribution in south-east Spain and the blackthorn combine to make a very poor, low, degraded *matorral* with cover in the

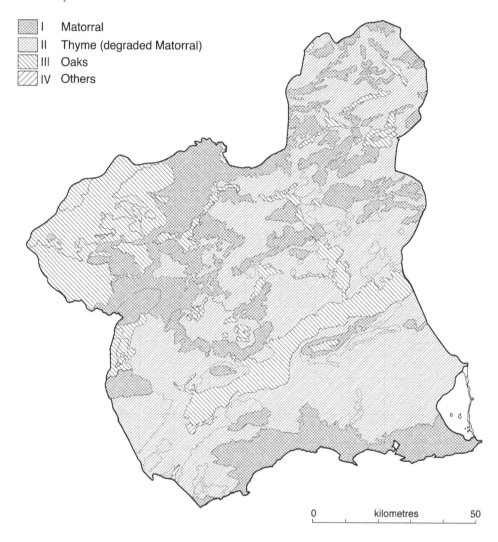

I   Matorral
II Thyme (degraded Matorral)
III Oaks
IV Others

Figure 4.6 Main vegetation types in the province of Murcia, Spain
Source: Redrawn after Albaladejo Montoro and Diaz Martinez (1983).

order of 10 to 30 per cent. Around the farms, the prickly pear cactus (*Opuntia*) – a species introduced from the Americas – is often used as a living fence to keep animals away from buildings. Moving inland to the foothills and the interior basins south of the Sierra de Carrascoy, taller *matorral* with trees begins to appear; this may reach 2–4 m with kermes oak and *Lentiscus* bushes.

Throughout the region, especially in the interior plain lands around Mula, the *matorral* reflects the results of human activity. It includes esparto grass (*Stipa tenacissima*), which once was widely grown as a commercial crop. Where groundwater is within about 10 m of the surface, it is accompanied by the tall willowy broom (*Retama sphaerocarpa*). Phreatophytes, such as the poplar (*Populus*), occur along the alluvial fills forming the floodplain of the main rivers (reflected by the common use of *Alamo* in Spanish place names). Otherwise the woodlands, mainly oak and pine, are confined to the uplands, as in the Sierra de Espuña and the Sierra de Carrascoy, where conditions are cooler.

South flank of the Sierra Nevada, Spain

The Sierra Nevada rises steeply from the valley of the Guadalfeo River in the province of Granada and has a strong distribution of vegetation according to altitude from the valley up to the Pico de Veleta (Figure 4.7). This is a south-facing slope that exhibits all the main Braun-Blanquet divisions (see Box 4.3: Junta de Andalucia, Consejo de Medio Ambiente, 1996). The Vega de Granada floor itself is hot and dry with heavy cultivation. The first stage, up to 600 m, has cultivated olives and *algarrobe* (Verabonia). Also occurring is the small fan palm (*Chaemerops humilis*) that characterizes the dry Mediterranean areas and *pistacia*. A principal bush here is the spiny blackthorn (*Rhamnus oleoides*).

The lower limit of winter snow on this south flank of the Sierra Nevada occurs at about 1,200 m, and from 600 m up to this level the vegetation is characteristic of the Supra-Mediterranean division of Braun-Blanquet (see Box 4.3). Now the bushes give way to oak trees that have been thinned by deforestation for the building of villages. *Quercus ilex* and *Juniperus oxycedrus* are the main tree stands with the local names of *encinar* and *sabinar*. The low-lying oak, *Quercus coccifera* appears with the aromatic herbs rosemary and thyme, with elms and poplars in the rocky gullies. At 1,000 to 1,200 m is a line of villages that have used the upper slopes for grazing and cultivation and substantially changed the vegetation cover. Here the winter cold controls the vegetation and the trees are more northern species such as

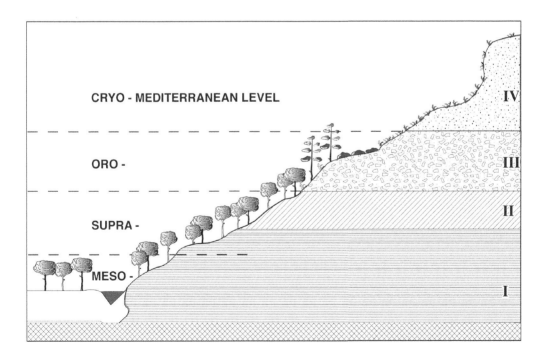

Figure 4.7 Altitudinal distribution of vegetation in the Sierra Nevada from the Vega de Granada to the Pico de Veleta. I *Matorrales* of various kinds; II *Thymus* shrub alliance; III Oaks and poplars; IV Others

Source: From Consejo de Medio Ambiente, Andalucia (1996).

the Pyrenean oak (*Quercus pyrenaica*), the Montpellier maple, beech and chestnut. This suite is characteristic of all the mid-mountain levels right across the Mediterranean. On west-facing slopes the zone is lower, down to about 800 m, where a line of villages is located. Above the villages, conifers give the appearance of temperate mountains and gorse and heather give the environment a familiar feel to North European visitors. The trees gradually die out, to be replaced by alpine plants on the stony, glaciated terrain around the Pico de Veleta.

This sequence is repeated all round the Mediterranean and the human land use is closely adapted to the vertical zonation at the regional and local scale. Aspect and slope play a very fundamental role with small variations leading to strong contrasts in water and radiation availability. These contrasts can be easily observed and predicted.

Local scale – Rambla Honda

At the local scale, the most important controls on the spatial arrangement of plant cover are lithology and soils, microclimate, aspect and topography. The impact of lithology is clearly demonstrated by the study of Puigdefábregas *et al.* (1998) in the Rambla Honda, Almería, Spain.

The Rambla Honda is an ephemeral river that drains a 30.6 km² basin on the Sierra de los Filabres in the eastern part of the Baetic Cordillera in the province of Almería. It is one of several field sites of the European Union MEDALUS (Mediterranean Desertification and Land Use) project (see Box 4.4), where detailed observations have been carried out over ten years into the dynamic behaviour of plant communities and its interaction with soil processes. It is one of the most detailed investigations of its type in the whole of the Mediterranean region and in one of the driest areas. The climate is semi-arid with hot dry summers and rainfall in autumn and spring. The area has dispersed settlements with isolated farms and has been subjected to grazing and harvesting of esparto grass (*Stipa tenacissima*), almonds and figs. There is a strong catena from the top of the western hillslope with eroded slopes and thin stony soils down to the valley floor that is filled with alluvium and has thicker soils. Down this catena are three well-defined plant communities with esparto grass dominant on the drier uppermost slopes, anthyllis (*Anthyllis cytisoides*) a drought-resistant shrub in the mid-slope areas and, in the foot slopes and on the rambla floor, the broom (*Retama sphaerocarpa*). This species is a drought-avoider because it can tap the groundwater at great depths. The three different species show different abilities of water uptake and use. *Retama* has access to water all the year round,

---

Box 4.4 The MEDALUS project

MEDALUS (Mediterranean Desertification and Land Use) was a European Union research programme coordinated by John Thornes, that seeks to investigate the characteristics and causes of desertification in the European Mediterranean, past, present and future. The project is also linked to the International Convention on Combating Desertification (UNEP, 1994) that has a special Annex IV dealing with the Mediterranean countries, including those of North Africa.

The MEDALUS project has established desertification monitoring areas (target areas) in Portugal (Lower Alentejo), Spain (the Guadalentín basin), Italy (Sardinia) and Greece (Lesvos). In these areas, policies are being developed to mitigate the impacts of desertification. The results of the first phase are reported in Brandt and Thornes (1996). See also Chapter 19 in this book.

Stipa is adapted because it loses cell turgor (water pressure) later than either *Anthyllis* or *Retama*, but *Anthyllis* is more susceptible to desiccation, according to the work of Puigde-fábregas *et al.* (1996).

Effects of human activity

Human influence on vegetation can be seen to take three major forms (Wainwright *et al.*, 1999a). First, direct modification of species assemblage involves clearance or deforestation or deliberate planting and introduction. Clearance has been clearly demonstrated in the discussion of vegetation changes noted above, and tends to be a relatively localized phenomenon, at least until about the last 2,000 years. Deliberate introductions obviously include cereals (see Chapter 10) as well as tree crops such as olive, vine, walnut and chestnut. Evidence for the olive and vine can sometimes be confused because of the presence of these species from the early Holocene over large parts of the Mediterranean Basin. Walnut and chestnut seem to be spread deliberately from the Middle East in the last 3,000 years or so, and generally provide a clearer mark of human activity on the landscape.

Second, indirect modifications of species assemblage relate to the development of new habitats due to agro-pastoral land uses. The variety of matorral associations are the clearest indication of this process (see Box 4.3). There is a strong debate about the impact on vegetation of grazing by sheep and goats. It is argued by some that grazing actually encourages growth, thereby increasing productivity of specific plants. Grazing is also spatially concentrated about water holes, as are the beneficial effects of fertilization of the soil. The Mediterranean zones were estimated in 1963 to have 370 million sheep, but the number of sheep in the European Union Mediterranean is now falling quite significantly. It has been estimated that sheep remove in the order of $1500\,\mathrm{kg\,ha^{-1}a^{-1}}$ of dry matter. A recent study (Imeson and Bakker, 1998) found no evidence to support the assertion that soil erosion is correlated with grazing intensity on the island of Lesvos (Greece), and thus there is unlikely to be a feedback to the vegetation type by this mechanism. Given the rates of removal cited above, this is a quite surprising result, and caution is urged in accepting the traditional mantras about Mediterranean erosion in relation to grazing. The reasons for exaggerating overgrazing are clearly summarized by Le Houérou (1981): (i) the information on grazing intensity is in reality very poor; (ii) the vegetation is not protected from livestock in winter, animals remain on the pasture the whole year and pastoralists see no need to establish forage reserves for the unfavourable season; and (iii) the nomadic way of life is culturally intrinsic and cultural beliefs and practices encourage relatively dispersed grazing (see Chapter 11).

Third, there are a number of inadvertent modifications which result from the competitive advantage which is accorded to certain species by environmental modifications. The clearest examples of this are the spread of holm oak, which tolerates the drier conditions at the soil surface resulting in the clearance of other vegetation and is resistant to repeated occurrences of fire (see, for example, Barbero *et al.*, 1992), and of beech in open upland settings (see Reille and de Beaulieu, 1988, for example). The impacts of fire on vegetation and vegetation regrowth are more fully discussed in Chapter 7.

Development of vegetation through time

Pliocene and Early Pleistocene

The climatic and vegetational character of the Mediterranean at the transitions from the Miocene to the Pliocene was of a warmer and wetter nature than at present. The Pliocene background to the lowland Mediterranean vegetation was thus of a tropical nature, with a number of tropical families (Anancardiaceae, Apocynaceae, Burseraceae, Icacinaceae, Malpighi-aceae, Mastixiaceae, Menispermaceae, Nyssaceae, Symplocaceae) and genera (*Calamus, Cin-namomum, Diospyrus, Eurya, Ficus, Lagerstroemia, Mallotus, Musa, Pandanus, Sterculia*) present from fossil remains (Pignatti, 1978). At higher altitudes, there is evidence to suggest that the evergreen subtropical forest gave way to an evergreen temperate forest, dominated by *Ilex aquifolium, Taxus baccata, Daphne laureola, Ruscus* and *Buxus*. Yet higher there would have been an evergreen coniferous forest, with pine, fir and cedar, and at the highest levels, a spiny shrub community dominated by *Astragalus* and *Genista*. These associations can be thought of as having been stable over very long time periods.

The vegetation of the north-western Mediterranean in the Pliocene has been summarized by Suc (1982; 1984; 1989; Suc and Zagwijn, 1983). Before about 3.2 Ma, the vegetation is dominated by forest dominated by *Taxodium*, with *Myrica, Symplocos* and *Nyssa* in the coastal zone, with species requiring less moisture, such as *Engelhardtia, Carya, Rhioptelea, Hamamelis* and *Embolanthera*, in the hinterland (Suc, 1984). These plants seem to have been in place for several million years, as suggested by pollen diagrams relating to the period before the Messinian salinity crisis (Suc and Bessais, 1990) and supporting the suggestion of long-term stability (Pignatti, 1978), although more open, steppic vegetation did characterize the period of the salinity crisis itself (Suc and Bessais, 1990). Suc terms this the P I vegetation zone, which lasted from about 5.4 to 3.2 Ma, with a slightly drier interlude from about 4.7 to 4.2 Ma (P Ib: see Figure 4.8). The end of this zone is marked by a relatively rapid change in the records from the Gulf of Lions, but a more gradual shift off the Catalan coast. Oak and alder pollen becomes much more common, and there are complex vegetation associations with cedar, fir, spruce and beech present. Patterns of complex associations with no modern ana-logues have been demonstrated to be common in periods where climate is rapidly changing (Huntley, 1990). Most importantly, a number of species which tolerated the drier conditions became more important. These species included phillyrea, olive, cistus, holm oak and pista-chio, and are all species which characterize the Mediterranean at the present time. This change probably reflects the greater adaptability of these species to the climatic changes that were ongoing, with the probable first occurrence of a climatic pattern characterized by a strong summer drought, which probably stabilized around 2.8 Ma (Suc, 1984). As a result of these changes, the more tropical elements such as the Taxodiaceae disappear from southern France at this time (Suc and Zagwijn, 1983). In contrast, data from the early Pliocene (4.5–3.2 Ma) in southern Italy suggest that the sub-tropical species are less important, with much higher pro-portions of modern, Mediterranean-type evergreens and pines. Bertoldi *et al.* (1989) suggest from these data that there was a strong climatic gradient across the Mediterranean at this time, from arid conditions in North Africa to semi-arid conditions in southern Italy and finally sub-tropical conditions in southern France and northern Spain. Furthermore, these authors find less seasonal conditions in the globally cooler phase from around 3.2 to 2.4 Ma, in contrast to the findings from southern France. The end of Suc's second phase (P II) occurs around 2.3 Ma, with the increase – again more markedly in the French sequences – of steppe species, including Amaranthaceae-Chenopodiaceae, *Artemisia* and *Ephedra*. Representing a probable drying and cooling, it has been suggested that this phase (P III) is the first of the cycles relat-ing to glacial–interglacial oscillations (Suc and Zagwijn, 1983). Higher resolution data corre-sponding to approximately the same period (2.6–2.2 Ma) at Semaforo in southern Italy suggest a more complex series of oscillations, with up to five cycles of change including deciduous forest, tropical humid forest (which expands to significant levels in the so-called 'glacial' phases), coniferous forest and open vegetation (Combourieu-Nebout, 1993). Data from Capo Rossello B and Gela G1 in southern Sicily are less continuous but would seem to fit into the

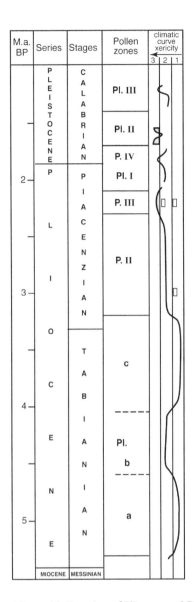

Figure 4.8 Zonation of Pliocene and Pleistocene vegetation types
Source: After Suc (1984).

same general pattern, although deposition in these sections seems only to have occurred in phases where deciduous or modern-Mediterranean-type species were dominant (Bertoldi *et al.*, 1989). Sequences corresponding to the lower complex of the Villafranchian-type area in northern Italy show at least four oscillations of arboreal pollen in the period between 2.9 and 2.2 Ma, with pine and oak dominant together with rare Taxodiaceae and sub-tropical species. Towards the end of the sequence, there is an increase in the non-arboreal pollen (Carraro *et al.*, 1996), which may correlate in part with Suc's sequence.

In Israel, the Pliocene vegetation has been divided into two chronological associations (Levin and Horowitz, 1987; Horowitz, 1989). The first suggests a cool temperate woodland made up of oak and spruce or fir (Horowitz and Horowitz, 1985; Horowitz, 1992). During the second, tree species become scarcer, with less than 10 per cent pollen composed of pine and fir. Oak only makes an appearance at the very end of the phase in a number of locations. Steppe species dominate, and increase in the period from around 3.5 Ma, especially in the south of the country, indicating an environmental gradient approaching that of the present day was beginning to develop at this time (Levin and Horowitz, 1987). In comparison with the western Mediterranean sequences, the Israeli sequences suggest that there are also four cycles of dry to wetter conditions, represented by fluctuations in the amount of arboreal pollen in the period from 2.6 to 2.0 Ma (Horowitz, 1989).

The next vegetation stage in the north-west Mediterranean (P IV–Pl. I) begins at around 2.1 Ma and straddles the boundary of the Pleistocene, continuing until around 1.7 Ma (Suc, 1982; 1984; Suc and Zagwijn, 1983). This phase seems to be the first with a fully Mediterranean-type zonation in southern France, reflected by olive and *Ceratonia* with pistachio, phillyrea and *Myrtus* at the lowest altitudes; phillyrea and holm oak with olive, hop-hornbeam and buckthorn at middle altitudes; deciduous oak and hornbeam with hickory, elm and zelkova at higher levels; and, finally, a mountain forest including cedar, pine, fir, spruce and hemlock. Such a zonation would not be out of place in the Pontic zone at the present time (Suc, 1984). In southern Italy, the site of Camerota shows a clear increase of modern Mediterranean-type species around this time, with a vegetation dominated by *Carya*, together with cedar, pine, hemlock, birch, *Ceratonia*, evergreen and deciduous oak and olive. The sub-tropical Taxodiaceae disappear here during this period (Baggiono *et al.*, 1981), as well as in Sicily (Bertoldi *et al.*, 1989). The Semaforo sequence in Calabria again provides a more detailed record of this phase. Here the Taxodiaceae are still present, but declining, with deciduous woodland species becoming more important. A number of drier sub-phases are represented, where non-arboreal pollen and ultimately artemisia steppe become important. These drier periods have been correlated with cooler conditions reflected by $\delta^{18}O$ values both in Mediterranean and north Atlantic cores (Combourieu-Nebout and Vergnaud Grazzini, 1991). The succeeding phase of Suc (Pl II) again represents a drier period, with a more steppic vegetation composed of pine, Cypressaceae and grass species (Suc, 1978). The length of this phase is uncertain, but may continue until around 1.4 Ma, when there is a further wetter oscillation (Suc, 1984). The upper complex in the Villafranchian-type area would seem to support this, with a semi-open vegetation similar to the present day, with pine and oak as well as grasses and artemisia dating to around 1.4–1.3 Ma (Carrara *et al.*, 1996). Fossil trees from this period show that Taxodiaceae are still present in central Italy (Rosa Attolini *et al.*, 1988). The section from Vrica in Calabria would again suggest the period contains a complex series of wet and dry cycles relating to global changes as reflected by $\delta^{18}O$ records (Combourieu-Nebout and Vergnaud Grazzini, 1991). This sequence would place the wetter phase at around 1.3 Ma, with an extensive dry phase with artemisia steppe around 1.4 Ma. At Saint-Macaire, near Béziers in southern France, the period between about 1.4 and 0.7 Ma is represented by open vegetation, including some steppe species, with little variability through the sequence. A small number of tree species present may represent refugia, which expand twice, near the base and midway through the sequence (Leroy *et al.*, 1996), again perhaps representing more humid phases, possibly around 1.3 and 1.1 Ma. There seems to be little other clear evidence for the northern basin from this time until the start of the Tenaghi Philippon core at around 975 ka, where the basal conditions are indicative of artemisia steppe (van der Wiel and Wijmstra, 1987b).

Later Pleistocene

The general changes at Tenaghi Philippon were described in Chapter 3. The groupings of vegetation types used by Mommersteeg *et al.* (1995) can be traced elsewhere in the Pleistocene vegetation history of the north and west of the basin, with differences due to elevation and local climatic factors. At Ioannina in the Pindus mountains of Greece (alt. 500 m asl), the vegetation record shows nine forested phases between 423 and 74 ka (Tzedakis, 1994). The detail of this record shows a general succession of species during these warmer, forested periods with an early expansion of oak and elm/zelkova followed by hornbeam and hop-hornbeam and finally fir which often occurs with beech. Other modern Mediterranean species such as olive, box and juniper are present in varying, but generally low, quantities through each of these periods. Colder periods are typically represented by three phases: (1) an initial steppe, dominated by grasses, Chenopodiaceae and *Artemisia*, with decreasing trees; (2) a main desert-steppe where *Artemisia* is the dominant species, with grasses less important and a wider variety of shrub species; and (3) a steppe-forest transition where pine and then oak increase relatively rapidly. Similar overall patterns have been seen above at other sites in Italy (Follieri *et al.*, 1988; Watts *et al.*, 1996a, b) and Spain (Pons and Reille, 1988). African data are sparse and relate only to the late glacial period, where desert-steppe conditions seem to have applied (Ben-Tiba and Reille, 1982; Lamb *et al.*, 1989). In contrast, the long Israeli sequences continue, with both forested and steppe conditions characterizing the colder periods and *maquis* dominating the warmer periods (Horowitz, 1987). Similar climatic patterns are confirmed from groundwater, loess and lacustrine data (Issar and Bruins, 1983), as well as from geomorphic, sedimentological, pedological and faunal data (Magaritz and Goodfriend, 1987).

In summary, the Pleistocene vegetation consists of a series of oscillations between forested and open conditions, with a movement from a moist forest to a more evergreen one after about 530 ka. Colder periods tend to be characterized by more arid steppe and desert-steppe conditions. The reverse of this pattern seems to occur in Israel, because of its transitional location, so that colder periods are characterized by forest, due to the incursion of Atlantic frontal systems into the eastern basin when the ITCZ was weaker in northern Africa, or steppe during stronger phases of strong ITCZ activity (Horowitz, 1987).

Holocene

Greece

At Tenaghi Philippon, the Holocene is divided into a number of vegetation phases. The first two represent open oak forest with pistachio, juniper and then pine, in combination with ever-decreasing grass pollen. Between *c.*6850 BCE and *c.*3800 BCE, the dominant vegetation is a holm oak forest, progressively in combination with hazel, ash and hop-hornbeam. After this time, the vegetation becomes much more open (Wijmstra, 1969). These phases represent an initial woodland with warm conditions and relatively humid responding generally to increasing summer drought. In the most recent phase, the oak percentages decrease by 50 per cent, and species characteristic of much more open vegetation structures, such as *Rhus*, pistachio, *Erica arborea* and strawberry tree (Wijmstra, 1969). This change, at about *c.*1850 BCE most likely reflects the impacts of deforestation by humans. At Xinias in Thessaly (Bottema, 1979), deciduous oak woodland starts to increase from around 10865–10424 BCE, although this forest may at least initially have been relatively open. Around *c.*5450 BCE, there is some evidence for agriculture, at the same time as a first, albeit slight increase in evergreen oak, as well as a reduction in deciduous oak at the expense of other trees such as hornbeam, hop-hornbeam, fir and pine. There are another two cycles of deciduous oak expansion and subsequent contraction simultaneously with evergreen oak expansion. The last expansion is again not well dated, but seems to occur after *c.*1250 BCE, and is followed by a later expansion of olive.

At Aghia Galini in Crete, the deciduous oak woodland expands around *c.*8550 BCE (although the exact point is poorly dated). This woodland was relatively open, and there are expansions of grasses both before and after a horizon dated at 7474–7048 BCE (Bottema, 1980). A third open vegetation phase is dominated by Umbelliflorae and Liguliflorae, and ends when the oak forest expands again around *c.*6100 BCE. A final major clearance takes place in the upper part of this core, the exact timing of which is again uncertain, but may lie between *c.*5250 BCE and *c.*3800 BCE. Willis (1994) suggests there is a general pattern through the Balkans of woodland reduction, both in terms of density and diversity, which begins around 4072–3541 BCE at Kopais in Attica (Turner and Greig, 1975) and continues locally throughout the next 1,500 years (to *c.*1850 BCE). This reduction is accompanied by an increase in grass pollen, including cereals, and species indicative of disturbed ground, such as plantain. The expansion of walnut, chestnut, plane, olive and beech trees at this time is interpreted by Willis (1994) as being the result of clearance opening up the landscape and providing niches where these species could be competitive (see also Bottema, 1982, and below). The general onset of more open *maquis* vegetation is clear elsewhere in Greece, for example, at Rezina in the north from *c.*4345 BCE (Willis, 1992b) and at Kleonai in the Peloponnese from at least *c.*3640 BCE (Atherden *et al.*, 1993). Willis (1994) suggests the onset of present-day vegetation occurs through Greece between 4072 and 3541 BCE and *c.*1850 BCE (see Table 4.1).

Turkey

In northern Turkey, Beug (1967a) has studied sediments from two lakes on the edge of the beech and black pine forest zones, showing little vegetation change over the past 7,000 years. There is evidence of human intervention and clearance, if not cultivation, for about the last 2,000 years at Yeniçaga lake (alt. 976 m asl). At Beyşehir Gölü, in south-central Turkey, the lower part of the profile is made up of a cedar forest, which is replaced by pine around 5220–4805 BCE (Bottema and Woldring, 1984). In the levels immediately before one dated to 1618–1437 BCE, the pine is replaced by oak, ash and juniper, and the vegetation becomes markedly more open, with grasses, plantain and Chenopodiaceae. Tree species such as chestnut and walnut are also apparent from this time. In the uppermost part of this sequence, there is a reversion to pine woodland before the onset of a more recent clearance phase. At Karamik Bataklığı, there is also a switch from a cedar to a pine forest around 5573–5287 BCE, although in this sequence the two species subsequently oscillate in their dominance (van Zeist *et al.*, 1975). The clearance phase here is much later, possibly as late as *c.*200 CE. At Pinarbaşi, tree pollen which is dominated by pine but with fluctuating amounts of cedar begins to decline after 6540–4165 BCE (Bottema and Woldring, 1984). Cereals are present during this decline, in quite large numbers after *c.*1900 BCE. From about the same time, there is an expansion of mixed oak forest, although the landscape remains relatively open. Akgöl Adabağ shows a forest expansion after 11196–10576 BCE, dominated first by birch and then by turkey oak. From about 7419–6542 BCE cedar and juniper are also important components of this forest. There is an important clearance phase suggested by a single sample in the later part of the sequence, although a large hiatus makes precise dating of it impossible. At Ova Gölü, the sequence is similarly broken, but shows the same general pattern. Olive shows a large peak, possibly around *c.*2500 BCE, and there is a suggestion of a clearance phase around 397 BCE–72 CE.

The Levant

The Holocene part of the Ghab valley sequence in Syria is not published at a very high resolution. However, oak (*Q. calliprinos*-type) together with some cedar, juniper, pistachio hornbeam, olive and pine increase rapidly from the levels immediately below the one dated at 10110–9059 BCE. Through the Holocene, the vegetation becomes more open, particularly dominated by grass species, albeit with some oscillations (Niklewski and van Zeist, 1970). In the Hula basin of northern Israel, the later glacial period saw a steady increase in oak woodland (*Q. ithaburensis* and *Q. calliprinos*), peaking around 11816–11268 BCE (Baruch and Bottema, 1991; Baruch, 1994). After this, there is a general trend towards a more open, *maquis*-type vegetation, albeit with two slight reversions to more wooded conditions around 10688–9942 BCE and 8824–8029 BCE. At Lake Kinneret, there is a continuous replacement of open oak woodland by *maquis* vegetation from the base of the profile at 5216–2784 BCE (Baruch, 1994). There are two phases of enhanced reduction in woodland, probably representing clearances which occur after 1629–550 BCE and between 759 BCE–242 CE and 778–1286 CE.

Croatia and Slovenia

In Croatia, Beug (1967b) has distinguished four phases in the sequences represented in the lacustrine deposits of the Malo Jezero on the island of Mljet. From the start of deposition (*c.*8030 BCE) until about *c.*6420 BCE, the vegetation is dominated by deciduous oak with some hazel, pine, elm, hornbeam, ash, pistachio and phillyrea. This period is followed by a phase where juniper and phillyrea are most important, the deciduous oaks decline and evergreen oaks start to increase. Beug interprets this as relating to the local onset of modern full Mediterranean conditions. From around *c.*5260 BCE there is a dramatic increase in evergreen oak, with some associated species such as *Erica*. The sequence ends around *c.*10 CE, when pine starts to increase. The latter part of the sequence also shows a marked increase in olive pollen, which probably relates to the expansion of its cultivation. On the nearby mainland, Brande (1973) has sampled six sites along the Neretva river. The dated sequence of Vid I again begins with a forest dominated by deciduous oak together with hazel, beech and some ash. At 6362–5630 BCE, evergreen oaks start to appear, accompanied by an increase in phillyrea. The evergreen oaks increase to the detriment of the phillyrea around *c.*5520 BCE, although the nature of the forest is still mixed. Traces of olive pollen occur throughout this period, and there is a general increase of species indicating open and disturbed conditions. From 764 BCE–370 CE, olive increases, together with other species which are probably being deliberately cultivated, such as vines, walnut and chestnut. Again, there is a general tendency towards more open conditions in the more recent period.

In northern Croatia, Beug (1977) has sampled a number of coastal sites. Here, deciduous oaks, together with some evergreen oak, hazel, elm, hornbeam, ash and box are dominant from the middle Holocene (*c.*5850 BCE to *c.*650 CE). The most recent levels show a rapid increase in *maquis* vegetation, with phillyrea and then juniper dominating, accompanying a general decrease in deciduous oak and elm. At the same time, there is a slight increase in the presence of hornbeam, as well as vines, chestnut, walnut and pistachio. Olives are only sporadically represented in these sites.

In the coastal zone of Slovenia, the earlier Holocene is characterized at Koper by oak–beech forest, which is then overtaken in dominance by hazel with some hornbeam and alder. Between 3760–3102 BCE and 2179–1442 BCE beech becomes dominant again, with two major oak peaks. Following a further oscillation with hazel expanding, oak becomes more important during a phase where the number of herb and shrub species are increasing, and vine, cereals,

Table 4.1 Summary of major vegetation stages derived from pollen and charcoal data. ≫ indicates that this stage is present at the earliest part of the record

Country	Site	Altitude m asl	Woodland development	Evergreen woodland development	Matorral development	Extensive clearance	Reference
Greece	Tenaghi Philippon	40	c.10500 BCE	c.6850 BCE	c.3800 BCE	c.1850 BCE	Wijmstra, 1969
	Kopais	100	c.9050 BCE	≫		4072–3541 BCE	Turner and Greig, 1975
	Litochora					900 BCE	Turner, 1978
	Osmananga					1745–1204 BCE	Turner, 1978
	Terraglu					1858–1323 BCE	Turner, 1978
	Gramousti	285	c.9050 BCE	undiff.		c.2900 BCE	Willis, 1992a
	Rezina	1,800	>9914–8937 BCE	undiff.	c.4345 BCE	c.2800 BCE	Willis, 1992b
	Khimaditis	560	c.9250 BCE			>2845–2331 BCE ?	Bottema, 1982
	Edessa	120	10740–10501 BCE			1680–1423 BCE	Willis, 1994
	Xinias	500	10850–10421 BCE		c.2500 BCE		Bottema, 1979
	Ioannina	500	10362–9151 BCE				Willis, 1994
	Giannitsa	0	≫	3363–3047 BCE		c.3150 BCE	Willis, 1994
	Lerna	0	≫		c.3150 BCE ?		Willis, 1994
	Kleonai	0	≫	3782–3043 BCE ?	c.3640 BCE		Atherden et al., 1993
Greece (Crete)	Aghia Galini	1	≫	>7474–7048 BCE	yes	yes	Bottema, 1980
Croatia	Malo Jezero	0	>c.8030 BCE	c.6420 BCE		? c.10 CE	Beug, 1967b
	Vid 1	0	≫	6362–5630 BCE ↑ c.5520 BCE		764 BCE–370 CE	Brande, 1973
Slovenia	Rovinj	0	≫	mixed oak cont.			Beug, 1977
	Koper	1	≫	mixed oak cont.			Culiberg, 1995
	Ledina	1,100	pre Ho	mixed forest cont.	c.650 CE	>885–1249 CE	Culiberg and Šercelj, 1996
	Ljubljana	300	c.10600 BCE			965–1269 CE	Šercelj, 1966

Country	Site						Reference
Turkey	Yeniçağa	976	≫ c.15400 BCE	undiff.		c.10 CE	Beug, 1967a
	Karamik Bataklığı					c.300 CE	van Zeist et al., 1975
	Beyşehir Gölü		c.13300 BCE			c.1618–1437 BCE	van Zeist et al., 1975; Bottema and Wooldring, 1984
	Akgöl	1,000					Bottema and Wooldring, 1984
	Pinarbaşi	980					Bottema and Wooldring, 1984
	Ova Gölü	15					Bottema and Wooldring, 1984
Syria	Ghab	800	>10110–9059 BCE	>10110–9059 BCE	?	yes but poor resolution	Niklewski and van Zeist, 1970
Israel	Lake Hula		open woodland >19087–17659 BCE	peak 11816–11268 BCE	from early Ho.	yes but poor resolution	Baruch and Bottema, 1991; Horowitz, 1992
	Ashmura					3507–3034 BCE	Horowitz, 1971
	Lake Kinneret					1629–550 BCE	Baruch, 1986
Italy	Cánolo Nuovo	900	≫ 13005–12126 BCE			n/a	Grüger, 1977
	Laghi di Monticchio	656	13249–12306 BCE			2558–2206 BCE	Watts et al., 1996a
	Valle di Castiglione	44	shortly after c.15100 BCE	undiff. >c.4100 BCE ?	>c.4100 BCE ?	1916–1676 BCE	Follieri et al., 1988
	Agoraie	1,328	≫2907–2576 BCE	n/a (mixed oak) 6353–5994 BCE	n/a	182 BCE–73 CE	Cruise, 1990a
	Bargone	831	≫		n/a		Cruise, 1990a
	Prato Spilla	1,550	12843–12209 BCE	never well developed	n/a	2468–2199 BCE	Lowe, 1992
	Lago di Ganna	459	pre Ho			c.10 CE	Schneider and Tobolski, 1983
	Lago Albano		c.15000 BCE	c.8000 BCE ?		yes	Lowe et al., 1996; Oldfield, 1996

continued

Table 4.1 continued

Country	Site	Altitude m asl	Woodland development	Evergreen woodland development	Matorral development	Extensive clearance	Reference
France	Paillon, Nice	100	≫			3986–3547 BCE, ↑ 3029–2485 BCE	Nicol-Pichard, 1982
	Tourves	298	12938–11774 BCE	8011–7422 BCE	6175–5597 BCE		Nicol-Pichard, 1987
	Étang de Berre	1.5		6634–5633 BCE (mixed oak)		758 BCE–390 CE	Triat-Laval, 1982
	Les Frignants	0.7	10562–9089 BCE	4237–3669 BCE	>5571–5222 BCE		Pons et al, 1979
	Augéry	3			4783–4086 BCE		Pons et al, 1979
	Fos-sur-Mer	–5			4455–3548 BCE		Triat, 1975
	Palavas	0		4455–3548 BCE			Aloïsi et al., 1978
	Mauguio	0		>3755–3342 BCE	>3755–3342 BCE	409–164 BCE	Planchais, 1982
	Maguelone	0				>422–786 CE	Planchais and Parra Vergara, 1984
	St Cyprien	0		>7890–7299 BCE			Planchais, 1985
	Canet St Nazaire	–1		>3756–3369 BCE			Planchais, 1985
	Drôme fan				5220–4592 BCE	3016–2496 BCE	Brochier et al., 1991
	Beauchamp Panières	1,200	c.9250 BCE pre Ho	n/a	n/a	n/a	Laval et al., 1991
	Le Bouchet			n/a	n/a	c.1250 BCE	Reille and de Beaulieu, 1988, 1990
	Pyrenees (combined sites)	880–2,080 various	c.10300 BCE	present after c.8300 BCE c.6400 BCE (mixed) ↑ c.5150 BCE to c.3150 BCE	c.5150 BCE to c.3150 BCE	4357–3772 BCE ↑ c.3200 BCE	Reille and Lowe, 1993
	charcoal data (summary)						Vernet and Thiébault, 1987
France (Corsica)	Étang de Sale	1		4929–4043 BCE		c.1300 CE	Reille, 1984

Spain	Lake Banyoles	173	9569–8267 BCE	5482–5148 BCE			Pérez Obiol and Julià, 1994
	Sidera	440	>7241–6610 BCE	?		yes	Pérez Obiol, 1988
	Les Palanques	440	>6365–5989 BCE	c.3550 BCE		c.250 CE	Pérez Obiol, 1988
	Llauset	2,132	10967–9037 BCE	≈4228–3969 BCE		yes	Bergadà et al., 1992
	Quintanar de la Sierra	1,470	10428–9041 BCE	8332–7619 BCE ?		1606–921 BCE	Cristina Peñalba, 1994
	Padul	785	c.12700 BCE	c.12700 BCE	4455–4103 BCE ?	core ends	Pons and Reille, 1988
Portugal	Lagoa Comprida	1,600	≫	n/a		3309–2696 BCE	van den Brink and Janssen, 1985
Morocco	Tigalmamine	1,625	>c.10000 BCE	>c.10000 BCE		33 BCE–608 CE	Lamb et al., 1989, 1995
Algeria	Oumm el-Khaled	500	>4038–3644 BCE	undiff.?	2886–2459 BCE		Ritchie, 1984
Tunisia	Dar Fatma 2	780	present through glacial		4243–2327 BCE	yes	Ben Tiba and Reille, 1982

chestnut, walnut and olive start to increase, with a major peak particularly in olive at 885–1249 CE. From this point, the sequence shows evidence of more open vegetation and continued cultivation (Culiberg, 1995). Culiberg points to the effects of intensive sheep grazing and transhumance in the destruction of vegetation and erosion leaving the karst zone of Slovenia a 'stone desert' by the beginning of the nineteenth century. Further inland at higher altitudes, the sequence is slightly different with clearance occurring somewhat later. At Ledina in the Julian Alps (alt. 1,100 m asl), the vegetation at the Holocene boundary is dominated by pine, spruce and artemisia. This is overtaken by a birch and then a mixed oak–elm–lime forest with hazel becoming increasingly important. By 5972–5620 BCE a beech–spruce–fir forest develops, continuing with oscillating frequencies of the dominant species until the recent period. A major clearance phase peaks around 965–1269 CE with extensive use of trees for charcoal for iron working and for timber (Culiberg and Šercelj, 1996).

Italy

The sequence of Cánolo Nuovo in Calabria (alt. 900 m asl) shows the consistent presence of a deciduous oak forest, with some traces of holm oak from 13005–12126 BCE. The resolution in this sequence is low, but the most recent level shows a slight increase in shrub species, particularly Umbeliferae and artemisia, indicative of clearance (Grüger, 1977). At the Laghi di Monticchio in Basilicata, oak pollen starts to increase at around 13249–12306 BCE, just preceded by birch. There is a slight dip in the oak with birch and artemisia re-establishing themselves during the Younger Dryas event, with the oak becoming dominant by 10610–10177 BCE. Olive pollen is continuously present from this time. By 7831–7500 BCE, the oak forest with elm, birch and hornbeam and some beech is well established, with alder making a first appearance and vine occurring sporadically. From this point, the oak declines until the most recent levels. The alder peaks by 2558–2206 BCE, when chestnut makes a first appearance. At this point, plantain increases, suggesting clearance, as does the continued and increasing presence of grasses to the top of the profile. There is a gradual decline of tree pollen from this point onwards, and holm oak is relatively continuously present. Analysis of diatoms suggests that the present fen at Monticchio originated during the Holocene (Watts *et al.*, 1996b). At the site of Valle di Castiglione 20 km east of Rome, tree pollen starts to increase shortly after *c.*15100 BCE, with deciduous oak increasing more rapidly than evergreen oak, and accompanied with hazel, elm, beech, birch and hornbeam. There is again a short decline in tree species reflecting the Younger Dryas event. In the lower part of the Holocene to *c.*4100 BCE, the vegetation remains relatively open, with high proportions of *Artemisia* and Chenopodiaceae. After 1916–1676 BCE the tree species declines, although holm oak becomes more dominant. This decline is accompanied by a higher presence of holm oak, juniper, olive, chestnut and walnut, all probably reflecting clearance and cultivation. The former is also attested by a more continuous presence of plantain (Follieri *et al.*, 1988).

A number of Holocene sites have been studied in the Ligurian Apennines (Cruise, 1990a, b). At Agoraie, the period from before 2907–2576 BCE to 182 BCE–73 CE is dominated by a fir–beech–oak forest, with small amounts of herb pollen, which is largely made up of *Filipendula* and Cyperaceae. From this time, beech and oak become more dominant, and arboreal pollen percentages drop overall. Walnut and chestnut start to make an appearance, with olive and vine showing a sparse presence further up in the profile. The sequence at Casanova shows an early dominance of pine and fir, with beech and oak developing from 4035–3641 BCE. There are scattered appearances of vine and olive, and walnut and chestnut again appear in an undated upper part of the profile, as non-arboreal pollen – in this case dominated by grasses and Cyperaceae – increases (Cruise, 1990b). At the lower altitude site of Bargone (831 m asl), deciduous and evergreen oaks develop together from after 6353–5994 BCE (Cruise, 1990a). In the same area, the high altitude site of Prato Spilla (1,550 m asl) provides two complete Holocene sequences. The lowest part of the profile shows the final glacial dominance of pine with significant non-arboreal pollen made up of grasses, artemisia and Cyperaceae. By 12843–12209 BCE fir and, to a lesser extent, oak are also present. There is a short phase of oak regression and pine increase centred around 11780–11312 BCE, probably reflecting the effects of the Younger Dryas event. The start of the Holocene is marked by a rapid increase in oak (10643–10252 BCE and 10417–9884 BCE), which remains dominant with fir until 2468–2199 BCE, when there is a rapid decline in fir, and a gradual falling-off of oak to the top of the profile. The vegetation becomes more open from this time, dominated by grasses and Cyperaceae, although beech also increases at this point (Lowe, 1992).

At Lago di Ganna near Varese in the Italian Alps (alt. 459 m asl), a pine-steppe late-glacial vegetation is replaced by an oak–elm–lime woodland, with subsequent increases of ash, holm oak and fir. Beech becomes dominant here from around *c.*3800 BCE, and hornbeam, dock and Cyperaceae develop from a similar time. Holm oak is absent for some of the time between this latter date and around 2,000 years ago, and only makes sporadic appearances in the upper part of the profile. From this time, chestnut and walnut become important, and there is a continual decline in tree pollen (Schneider and Tobolski, 1983). At Lake Albano, near Rome, a deciduous oak woodland expands from about the time of the Younger Dryas event (Lowe *et al.*, 1996). Evergreen oak starts to expand from around *c.*8050 BCE, and there is a slight decline in woodland after 6046–5804 BCE, when wild grass species also become more important, and there is a second decline in woodland after 2393–1975 BCE. The upper part of the profile which probably represents the period of the past 2,000 years suggests the presence of four major clearance episodes. Similar general patterns are seen in the nearby Lake Nemi, although the dating is less secure there (Lowe *et al.*, 1996). In the central Adriatic core PAL94-8, a well-dated sequence shows the expansion of deciduous oak woodland after 12949–12262 BCE (ibid.). Holm oak increases after 7570–7327 BCE, and is accompanied by species such as beech, juniper, hornbeam and fir. There is some decline in tree species after 255–556 CE. Olive, chestnut and walnut increase notably from this date, but are sporadically present from 3493–3035 BCE. At another central Adriatic core, RF93-30, holm oak has started to increase before 4905–4590 BCE, and there are three notable phases in woodland reduction: one after 2587–2284 BCE, one centred on 83–240 CE, and a smaller one after 639–881 CE. Olive is sporadically present from the first clearance, but expands more noticeably from 1520–1264 BCE. At Laggacione di Valentano (355 m asl) in Lazio, the vegetation at the beginning of the Holocene (11047–10588 BCE) is dominated by deciduous oak, including turkey oak, with hazel and elm increasing slightly later (Follieri *et al.*, 1995). Evergreen oaks are present in small quantities. Artemisia and grasses are important initially, but decline rapidly. After 7480–7005 BCE, beech becomes dominant and evergreen oak expands. In the period between 5999–5700 BCE and 2449–1888 BCE, the vegetation was probably dense forest dominated by deciduous oaks, together with hazel, hornbeam, beech and evergreen oak and decreasing quantities of lime and elm. There are sporadic appearances of cereal pollen. After this period, the woodland declines and is replaced by grass, although there a few traces of cultivated plants. The upper part of this profile is, however, poorly defined. The Lago di Vico (500 m asl) is a crater lake, about 60 km north of Rome, and has been studied by a number of authors (Frank, 1969; Watts *et al.*, 1996a). Follieri *et al.* (1995) studied a dated sequence from core V1, in which trees start to increase after 11649–10918 BCE. A more complete forest, dominated by deciduous oak, is present by 10441–9313 BCE, with elm, lime and hazel also present. Beech and evergreen oak take on this secondary role before 5999–5687 BCE. After 976–427 BCE, cultivated species such as chestnut, olive and walnut expand, and there are continuous traces of vine and cereals, although there appears to be no evidence of widespread forest clearance. At the Stracciacappa crater lake (220 m asl, 35 km north of Rome), there is again a rapid increase of tree pollen from 12550–11744 BCE to 9047–8351 BCE, made up of principally deciduous and evergreen oak, together with hazel, elm, beech, hornbeam and alder (ibid.).

France

A number of sites provide vegetation data from southern France, using pollen and the analysis of charcoal from archaeological sites. A relatively short sequence from Nice shows a decrease in arboreal pollen after 3986–3547 BCE, which becomes marked after 3029–2485 BCE. The lower part of the profile is dominated by alder with hazel and deciduous and evergreen oak, the upper part by pine and *Filipendula* (Nicol-Pichard, 1982). To the north of Toulon, the site of Tourves (alt. 298 m asl) provides a more complete sequence from the late glacial period. A pine–artemesia steppe is in place from before 12938–11774 BCE, although there are dating problems in the lower part of the profile. There is a sudden decline in pine before 9927–8540 BCE, when deciduous oak starts to increase, accompanied by juniper. Holm oak shows a greater presence after 8011–7422 BCE. There is a decline in arboreal pollen around 6175–5597 BCE, which remains at a lower level to the top of the profile, with a greater presence of open species, particularly sedges (Nicol-Pichard, 1987). A number of sites close to sea level have been studied in the Étang de Berre. Pine is present at relatively high levels throughout most of the profiles. Deciduous and evergreen oaks start to increase some time after 6634–5633 BCE. Olive shows a significant increase in many of the sequences, and the start of its rapid rise is dated to 758 BCE–390 CE in one of the cores. Walnut appears shortly afterwards. The vegetation appears to have become much more open in the last few hundred years, with the pollen dominated by sedges (Triat-Laval, 1982).

On the Rhône delta, the site of Les Frignants (alt. 0.7 m asl) contains a relatively complete sequence of the first half of the Holocene. The late glacial vegetation is dominated by pine, birch, grasses, sedges and Cichoriaceae. Some time after 10562–9089 BCE, deciduous oak starts to develop, in conjunction with hazel and elm. Evergreen oak is discontinuously present, becoming more constant after 4237–3669 BCE, at the same time as lime. Fir, alder and hazel become more dominant too from this period. The second half of the Holocene is represented nearby at Augery (alt. 3 m asl), starting before 5571–5222 BCE with an assemblage similar to that at Les Frignants. As with a number of other sequences, Augery shows a number of oscillations in the occurrence of holm oak. It becomes relatively more important after the overall reduction of tree pollen just before 4466–3803 BCE, when the landscape and vegetation took on its modern appearance (Pons *et al.*, 1979). At Fos-sur-Mer, a now-submerged sequence shows a similar pattern occurring some time after 4783–4086 BCE (Triat, 1975).

Along the Languedoc coast, the increase of evergreen oak, in this instance accompanied by phillyrea, pine and box occurs just before 4455–3548 BCE at Palavas near Montpellier. Other herbaceous species indicative of a matorral association are also present at this time (Aloïsi *et al.*, 1978). Nearby, a later Holocene sequence is preserved in the Mauguio lagoon. Beech and holm oak dominate the period from 3755–3342 BCE, together with hazel, alder, birch and a number of species characteristic of a matorral association. Walnut, olive and chestnut increase after 409–164 BCE. The vegetation becomes much more open after 645–884 CE, with the presence of a wide variety of shrub and herb species (Planchais, 1982). Olive is dominant by 422–786 CE at Maguelone, with a vegetation otherwise characterized by holm oak, walnut, chestnut and vine (Planchais and Parra Vergara, 1984). The earlier Holocene is briefly represented at the sites of Canet St Nazaire and Saint Cyprien on the Rousillon coast. The pollen at the latter site suggests a relatively open vegetation with matorral elements and a tree pollen dominated by hazel, cork and holm oak from before 7890–7299 BCE until after 5997–5696 BCE. A not dissimilar pattern is present from the start of more continuous deposition at Canet St Nazaire before 3756–3369 BCE. Tree pollen drops in general between 3016–2496 BCE and 1220–1406 CE, when there is a peak in vine and olive, and the start of a rise of chestnut and walnut. However, the resolution in this part of the sequence is low, and the rise of these cultivated species may have occurred somewhat earlier than the latter date (Planchais, 1985).

Away from the coast, there are the longer sequences that have previously been discussed. Les Echets near Lyons contains only a poor resolution Holocene sequence (de Beaulieu and Reille, 1984). The discontinuous sequences through the Drôme fan at its confluence with the Rhône show relatively open vegetation as early as 5220–4592 BCE, with forest species dominated by pubescent oak, together with holm oak, box and juniper and species such as hazel, birch and pine as well as numerous heliophyllous herbaceous species suggesting relatively open conditions (Brochier *et al.*, 1991). The lake at Le Bouchet in the southern Massif Central (alt. 1,200 m asl) shows a rapid initial Holocene increase of hazel and deciduous oak pollen, accompanied with some elm. Lime and ash come in with the decrease of hazel around *c.*4900 BCE. Shortly after the oak peaks around *c.*3800 BCE, beech increases rapidly, together with smaller amounts of alder and fir. The vegetation becomes much more open after *c.*1250 BCE, with species such as dock and plantain indicating relatively disturbed conditions, although clearance may have started as early as *c.*3400 BCE (Reille and de Beaulieu, 1988; 1990).

Numerous sites in the Pyrenees have been studied, with a recent review of sites in the Eastern Pyrenees being published by Reille and Lowe (1993). These authors suggest five phases of Holocene development. An 'Early Holocene Woodland Colonization' took place from *c.*10300 BCE to *c.*8300 BCE with an early, rapid expansion of birch accompanied by taxa such as *Rumex* and *Filipendula*, suggesting tall grass/herb communities, followed by an increase in oak and then hazel. The succeeding 'Mixed Deciduous Woodland' phase lasted from *c.*8300 BCE to *c.*6850 BCE, and was characterized by a maximum in the hazel record, together with elm and oak extension. The presence of holm oak and pistachio suggests relatively warm, dry conditions. The next phase lasts until *c.*3200 BCE and is characterized by forest diversification and the first evidence of human impact on the woodland. Ash, yew, lime, maple, holly and ivy all develop, with fir or oak being the dominant species according to the exposure of particular sites. An initial clearance phase, with reduction of fir, and increase of hazel, Poaceae, heather and *Pteridium* is dated to 4357–3772 BCE. The fourth phase (to *c.*2,000 years ago) is characterized by the spread of beech and continued human disturbance. At Gourg Nègre (alt. 2,080 m asl), the increase of beech at 1944–1120 BCE is accompanied by an increase in holm oak and a more continuous presence of Cyperaceae and olive. The most recent phase consists of a continued reduction in woodland cover and human activity. Walnut expands around *c.*300 BCE, together with olive and followed by chestnut. Clearance seems to peak from the Middle Ages to the nineteenth century, when cereal pollen is highest (Reille and Lowe, 1993). The Holocene sequence from Biscaye (alt. 409 m asl) in the western Pyrenees shows a relatively similar sequence (Mardones and Jalut, 1983).

Other vegetation analyses in the interior depend on the analysis of charcoal from archaeological sites, with associated problems of sampling. However, such studies are useful in filling gaps in regional records where pollen data are poorly preserved or environments unsuited for their preservation. Studies have particularly been carried out in Languedoc (Vernet, 1980) and in the pre-Alps (Thiébault, 1988). Vernet and Thiébault (1987) summarize these data, and suggest that they show four phases of vegetation development. The first continues from the late glacial until around *c.*6400 BCE, and is characterized by juniper, scots pine, almond, some deciduous oaks, rare box and phillyrea. The second phase consists of dominant deciduous oak and the appearance of holm oak, and has a variable final expression, ending as early as *c*.5150 BCE at Dourgne or as late as *c*.3150 BCE at Font Juvénal. The final two phases are characterized by box and holm oak, and suggest the extension of a matorral associated with human activity.

On Corsica, Reille (1984) has studied a number of low-altitude sites on the eastern coastal plain, although there are only two dates available. The earliest phase of the Étang del Sale (before 4929–4043 BCE) consists of pine, alder and deciduous oak forest, with some lime and sporadic peaks of birch. After this date, holm oak expands and the forest species increase in proportion and number. This forest continues until relatively late (Reille suggests to *c*.1300 CE), when there is a large-scale clearance, and the dominance of herb species such as Cyperaceae, Poaceae, Chenopodiaceae, *Cistus*, Asteroideae and Chicioraceae.

Spain

A number of coastal sites have produced charcoal results similar to those in southern France (Vernet and Thiébault, 1987; Guilaine *et al.*, 1982). A number of more continuous pollen sites also exist. In Catalonia, Lake Banyoles (alt. 173 m asl) contains a complete Holocene sequence. Here the late glacial vegetation at 11655–11145 BCE is dominated by pine and Poaceae, with some maple, birch and juniper. Deciduous and evergreen oaks are present in very low quantities. At 9569–8267 BCE, deciduous oak increases dramatically, followed by hazel. Holm oak is present more continuously, but as elsewhere, presents a number of cycles through the Holocene. As the deciduous oak and hazel decline slightly, fir expands, reaching a peak before 5482–5148 BCE. The sequence above this is relatively low resolution, but shows a re-expansion of deciduous and especially evergreen oak (Pérez-Obiol and Julià, 1994). At Olot, near Girona, four sites have been studied which cover various parts of the Holocene. By 7241–6610 BCE at Sidera (alt. 440 m asl) and 6365–5989 BCE at Les Palanques (alt. 440 m asl) a forested vegetation is well developed with pine, deciduous oak and hazel dominant, with smaller quantities of elm.

In southern Spain, the important long sequence at Padul unfortunately does not give a complete record of the Holocene, terminating around 3350–2920 BCE (Pons and Reille, 1988). Mixed oak forest, dominated by holm oak increased rapidly about midway between levels dated at 14256–13250 BCE and 12699–11672 BCE. There are two reversions to more open landscapes, the first representing the Younger Dryas event and the second ending around 7471–7006 BCE. Pistachio is constantly present from 10122–9031 BCE, and cork oak from 7471–7006 BCE. Relatively large quantities of olive pollen are intermittently present from around 7030–6430 BCE.

Roure Nolla *et al.* (1995) identified four pollen phases in Almería. In the first, late glacial conditions with steppe taxa, pine and evergreen oak together with fir and juniper make up the vegetation. The second phase, which is well established at Antas by 8029–7437 BCE, reflects the increase of deciduous and evergreen oak accompanied by other deciduous species together with the development of a regional *maquis* landscape, with pistachio, olive, Ericaceae and holm and scrub oak. After around 5292–5058 BCE at Antas, or 5043–4623 BCE at Roquetas del Mar, a third phase with more sclerophyllous conditions are established with increasing steppe and edaphic species. The fourth phase continues from 3367–2879 BCE at San Rafael or 2553–2145 BCE at Roquetas del Mar. Tree species decline, apart from pine, with the landscape dominated by Asteraceae, Artemisia, Chenopodiaceae and other steppe and edaphic species. A decline in total vegetative cover is also suggested by decreases in the absolute pollen frequencies. Cereal pollen is almost continuously present, with vine and then olive present as cultivated species near the top of the profiles (although vine is absent at San Rafael), which the authors ascribe to Roman occupation, although this is only dated by interpolation. An olive peak may occur prior to *c.*515 BCE at San Rafael, where the main peak of pollen presence is a recent phenomenon, occurring from the seventeenth century. In the eastern Iberian peninsula area, from Alicante to southern Catalonia, the lower parts of the sequences studied by Roure Nolla *et al.* are dominated by pine and some steppe species (e.g. before 7033–6484 BCE at Amposta in Tarragona). This phase is followed by one with mixed deciduous and evergreen oak forest, in which hazel is a significant species at Amposta until after 5566–5079 BCE. Olive appears after 5198–4782 BCE at Torreblanca (Castellón). From this time, evergreen oak dominates with some deciduous oak, and the pine and hazel values decrease. The vegetation becomes more open after 1241–901 BCE at Amposta, 838–539 BCE at Torreblanca, and perhaps as early as 5266–4466 BCE at Almansa (Albacete). Olive starts to increase after 537–766 CE at Amposta, although the authors note its presence in small quantities from 389–8 BCE. There is a greater presence of open vegetation more indicative of *maquis* starting from levels corresponding to the Mediaeval period.

In the Balearics, the end of the last glacial period shows the development of forest with hazel, evergreen oak and birch followed by an increase of deciduous oak and juniper, peaking around 8333–8021 BCE in the marine core of KF14 off Menorca (Roure Nolla *et al.*, 1995). By 6990–6477 BCE at Aljendar (Menorca), the vegetation is dominated by box, juniper and hazel together with low quantities of deciduous and evergreen oaks and a variety of steppe species. A similar association is found at the base of the sequence presented for Alcudia on Mallorca, beginning 6116–5678 BCE. After 5225–4856 BCE at Alcudia, there is a gradual decline in juniper and box together with a corresponding increase of olive and pine. On Menorca, a similar change occurs between 6606–6254 and 5421–5054 BCE at Galdana, where precise dating is difficult because of apparent disturbances in the lower part of the profile, but is delayed until 2875–2465 BCE at Aljendar. Deciduous trees such as deciduous oak, alder, birch and elm tend to decline after 795–404 BCE at Alcudia at the same time as plantain, Poaceae and cereals increase. Similar increases are noted in the Aljendar sequence following the decline of juniper. In the most recent periods, there is some evidence for an expansion and second decline of the mixed oak forest, although this is not well dated. The high-resolution marine records from near Barcelona show complete records from approximately 1050 CE (TG8) and 480 CE (TG9), which can be compared with historical records from the area. There is a general decrease in tree pollen, particularly pine and oak, although with notable oscillations in the evergreen oak results with minima around 1000–1100 CE, 1400 and 1700 CE, which correspond to short-term increases in pine and *maquis* species. The twentieth century shows a general decline in all tree species except pine.

The early Holocene is recorded at Salines in Alicante (Julià *et al.*, 1994), where a pine-steppe community is replaced initially by a juniper–pine steppe, and then more gradually after 10195–9135 BCE by a mixed deciduous and evergreen oak forest, probably of an open character, given the percentages of shrub species. This association is well established by 8013–7623 BCE, when juniper starts to decline finally to a relatively constant but low percentage. The authors suggest that a short-term decline in evergreen oak just after this period corresponds to a short phase of cooler climate (sedimentary data reported below suggest that this is associated with more humid conditions also). At Elx in Murcia, the period before 7850–7495 BCE has a deciduous oak forest, with pine rather than juniper playing a secondary role (Julià *et al.*, 1994). From 7848–7449 BCE, evergreen oak becomes the dominant species in the Salines core, with other species such as olive/phillyrea, *Ephedra*, Cistaceae, pistachio, *Artemisia*, Chenopodiaceae and Poaceae again suggesting a rather open community. Open-ground vegetation such as Chenopodiaceae, *Artemisia*, Poaceae and plantain all increase after 2828–2233 BCE, following a period in which pine was dominant, and there is a slight increase in representation of evergreen oak. This may suggest the development of *maquis* communities in the period before 2201–1830 BCE. There is also an appearance of cereal pollen at this time. Both the Salines and the Elx cores suggest a subsequent oscillation, with evergreen oak again becoming the dominant tree species – dated around 1208–898 BCE at Salines, at the same time as olive/phillyrea and vine pollen start to increase, and cereal pollen is first present. A further increase in pine is marked by a rapid rise in olive/phillyrea pollen, commencing *c.*1200 CE. A similar rise is seen after 1225 CE in the La Cruz core in Cuenca, where there are earlier, smaller peaks *c.*300–400 CE and around *c.*750–800 CE. There are suggestions of similar peaks at approximately the same times in the Salines core. The activity of fires through time is assessed in this work by counting charcoal fragments. These data show relatively high values in the pre-Neolithic levels at Salines, Elx and Ebre, suggesting the ongoing importance of wildfires. In the later periods, there is a good correlation between charcoal content and known historical events in the Ebre core.

Portugal

In Portugal, the site of Lagoa Comprida (1,600 m asl) provides an indication of extensive clearance at relatively high altitudes (van den Brink and Janssen, 1985), where the oak–beech woodland present from the beginning of the sequence was apparently destroyed by fire and replaced by heath around 3309–2696 BCE. At lower altitudes in central Portugal, Diniz (1995) has studied five cores in detail, with more arid, southerly sites unfortunately containing no pollen. The regional vegetation indicated by all five cores is composed of deciduous oak and alder suggesting a pattern similar to the present day for a considerable period – the longest sequence at Junqueira starts at 4930–4577 BCE and continues to at least 24–330 CE. Evergreen oak (*Quercus ilex-coccifera*) is generally present at low frequencies throughout all the profiles. Human influence is suggested as occurring around 24–330 CE at Junqueira, where there is a peak in Poaceae, Asteraceae and Artemisia. At Galeota, it occurs after 1401–1020 BCE, with decreasing pine and alder and relatively high Poaceae, plantain and Asteraceae. Cultivated species occur at various times – olive *c.*3570 BCE at Junqueira, *c.*175 BCE at Galeota, *c.*600 BCE at Pául de Goucha (S13), and from the earliest levels (385–45 BCE) at Pául de Goucha (S20) albeit with breaks; vine from *c.*1830 BCE at Junqueira, *c.*830 BCE at Pául de Goucha (S13), and from the earliest levels (385–45 BCE) at Pául de Goucha (S20), again with breaks; and chestnut between *c.*1825 and *c.*570 BCE at Junqueira, from *c.*175 BCE at Galeota, *c.*1000 BCE at Pául de Goucha (S13), and from *c.*585 CE at Pául de Goucha (S20). The early occurrences at Junqueira may suggest either contamination of the section or major variations of the sedimentation rate.

Morocco, Algeria and Tunisia

In the southern part of the basin, the best representation of vegetation development is again the site of Tigalmamine in Morocco (Lamb *et al.*, 1989; 1995). Some time before *c.*10000 BCE, there is a rapid rise in the tree species represented, dominated by evergreen (*Quercus rotundifolia*) and deciduous (*Q. canariensis*) oaks. Cedar (*Cedrus atlanticus*), which is characteristic of cooler, moister conditions, starts to increase from about *c.*5700 BCE, and overtakes the deciduous oak around *c.*2000 BCE. Grasses are the other dominant species, and Mediterranean species such as pistachio and phillyrea are also present in small quantities. Lamb *et al.* (1995) suggest on stratigraphic grounds that there have been five relatively short-lived periods of aridity in the Holocene period, between before *c.*10530 and *c.*8280 BCE, before *c.*7860 and *c.*5550 BCE, *c.*3060 and 2910 BCE, *c.*1040 and *c.*880 BCE, and between *c.*70 and *c.*290 CE. During the three latter periods, there is a marked decrease in the percentage of deciduous oak, which is less tolerant to dry conditions than the other species.

Similar cycles are difficult to see in the record from Dar Fatma 2 in Tunisia (Ben Tiba and Reille, 1982), which may be a function of the shortened Holocene (starting around *c.*3250 BCE) sequence and the lower resolution. The lower part of the Holocene sequence sees the reduction of the deciduous oak (and alder that dominated in the later Pleistocene), with the increase of cork oak (*Q. suber*) and *Erica arborea*, together with the expansion of grass and shrub species, which become dominant. Elsewhere in Morocco, the Holocene vegetation becomes dominated relatively early by *Ephedra* and *Artemisia* steppe at lower levels, with the appearance of holm oak and aleppo pine, together with olive and pistachio at medium levels (Brun, 1989). The undated Holocene sequence of Daya Tighaslant at 2,197 m asl in the Atlas shows a general progression where pine is overtaken first by deciduous (*Q. canariensis*) and then evergreen oak (*Q. ilex*). More open species such as *Artemisia* and *Ranunculus* become dominant in the most recent part of the sequence (Bernard and Reille, 1987).

At Oumm el-Khaled in eastern Algeria, an oak–pine woodland is dominant at the base of the core (4038–3644 BCE: Ritchie, 1984). From 2886–2459 BCE, there is a continuous decline of tree species, with grasses and shrubs taking over. There is a peak in olive pollen towards the top of the profile, which may represent the extension of Roman cultivation in the region, followed by a further expansion of arid species, with *Artemisia* becoming dominant. This pattern is consistent with the earlier Holocene period of increased humidity, followed by a general drying in North Africa, as discussed above (Chapter 3). Compositae and grasses are dominant in the Capéletti cave in eastern Algeria from levels dated from 5929–4928 BCE to 3613–2459 BCE, with very sparse cedar, pine, holm oak, yew and ash pollen (Beucher, 1979). Similar tree species are recorded in charcoal remains, and this site also produced large numbers of grape pips from a level dated from 4508–3954 BCE to 4468–3828 BCE (Portères, 1979). Brun (1983) shows a generally similar pattern based on data from the Gulf of Gabès in Tunisia, with relatively stable vegetation with some oak and pine, but a dominance and general increase of *Artemisia* steppe.

Summary of major trends since the late glacial maximum

From the detailed regional sequences outlined above, it is possible to elaborate some general trends in terms of the vegetation development. These trends have been outlined in Table 4.1. The late glacial maximum vegetation was dominated by open steppe vegetation in most areas except eastern Turkey, Israel and the lower altitudes of parts of North Africa. In these areas, the vegetation at this time was characterized by open woodland. Elsewhere, deciduous, open woodland starts to appear earliest in Italy, as well as in isolated locations in Turkey, Spain and later in southern France (Figure 4.9a). There is some reversion to more open conditions during the oscillation of the Younger Dryas event. In most places, the onset of woodland development occurs after the end of the Younger Dryas event. There is no clear geographical or altitudinal pattern to this growth of the forest, most likely because the woodland spread occurs from local refugia, although the rate of spread seems to have been much slower in the western part of the basin. The latter pattern would suggest that temperature was a more important control on the dispersal than moisture. The evergreen Mediterranean forest then takes a variable time to develop (Figure 4.9b). In Syria, Israel, North Africa and southern Spain, it is present locally before the start of the Holocene. In the northern and central parts of the basin, the same vegetation takes several millennia to occur, and is often found latest in southern France and parts of Italy, which suggests that migrational factors may be more important than successional ones in the development of this type of vegetation. The migration pattern is often also reversed when compared to the original forest spread, with the initial areas now on the coast (although there are notable exceptions, such as in Catalonia and Languedoc-Rousillon). Once evergreen forest is present, the onset of matorral is then relatively rapid, often occurring within a millennium. From this time on, the history of vegetation is best understood in terms of regional and local human interventions.

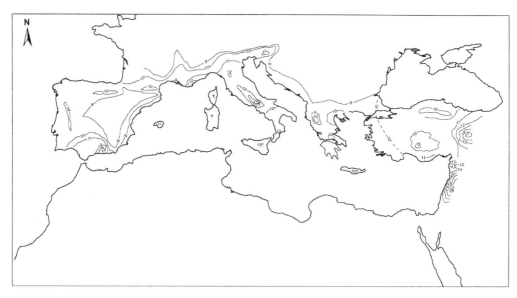

(a)

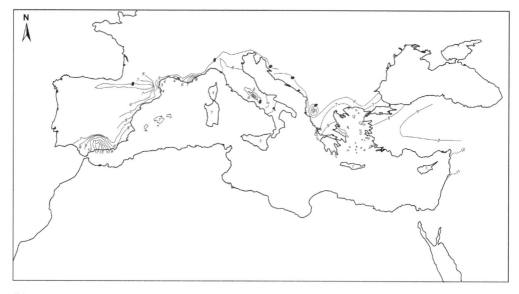

(b)

Figure 4.9 Tentative interpretation of the major changes in late Pleistocene and early Holocene changes in the Mediterranean vegetation: (a) timing of the replacement of steppe vegetation by woodland; and (b) development of evergreen woodland. Note that all dates are in years BCE and that no reconstruction has been attempted for North Africa because of the paucity of data

Timing of human impacts on the regional scale

General records of anthropic modifications to the vegetation in the Mediterranean as a whole have been discussed by Andrieu *et al.* (1995). Wainwright (*in press*) used the data collated by these authors to define the general periodicity of such modifications in the basin as a whole (Figure 4.10). Although deforestation is attested as early as the late seventh millennium BCE, it only begins to show a continuous presence from the middle of the fifth millennium until it

a.

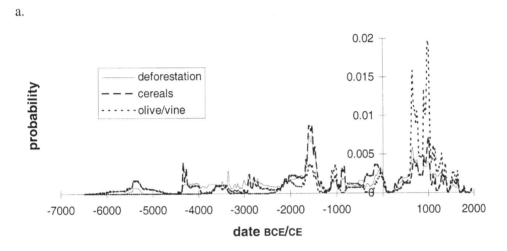

b.

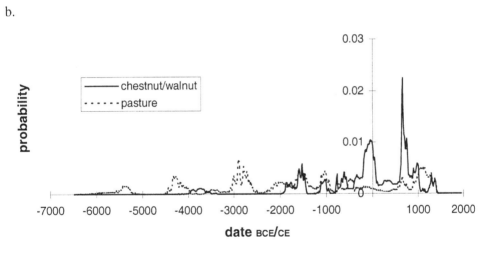

Figure 4.10 Summary of the first occurrences of various forms of anthropic modification of vegetation assemblages within the Mediterranean as a whole. The curves represent cumulated probabilities of calibrated radiocarbon dates for each recorded event, so that for higher probabilities, there is a greater likelihood that the modification was a widespread phenomenon. The reduced number of dates corresponding to the last 1,000 years is related in part to the fact that the changes had already taken place in many of the pollen cores making up the data, and in part to the fact that many pollen sequences are truncated in this period due to changes in land use, etc.: (a) based on indications of deforestation, cereals or olives and vines (see text); and (b) based on indications of chestnut and walnut and pasture

starts to rise at the end of the third millennium, suggesting that it remained a relatively local-ized phenomenon until that time. By contrast, there is a major peak in the middle of the second millennium BCE, suggesting that there was widespread pressure on the landscape during this time. This point will be followed up later (Chapter 10). The second half of the first millennium CE shows a similar peak, coupled with expansion of olive and vine cultivation. The latter is first recorded on a general level from the middle of the third millennium BCE, and follows a second phase of expansion from *c.*1100 BCE. Cereals, on the other hand, follow the same general trend as deforestation, as might be expected. Widespread pasture is first recorded in the pollen record from the mid-sixth millennium, and seems to have undergone phases of expansion at almost 1,000-year intervals from the end of the fifth millennium. As already noted, the expansion of chestnut and/or walnut is a more recent phenomenon, first noted around *c.*1900 BCE. The main phases of expansion are through the first millennium BCE and from *c.*600 CE to 1000 CE. A further example of human impacts on the vegetation in Greece is given in Box 4.5.

Box 4.5 Timing of human impacts: the case of Greece

It is important not to extrapolate widely from general data without evaluating the more local conditions. We therefore present here a brief case study outlining the impact of human populations on vegetation in Greece, based on the work of Bottema (1982). At Edessa in Thessaly, grasses increase dramatically between 4331–3949 BCE and 1968–1682 or 1616–1423 BCE, and there is evidence for cereals, plantain and a number of open-ground species. There is also a large peak in chestnut pollen around the later dates. At Khimaditis, there is an early peak in cereal pollen around 7326–6615 BCE, but indi-cators of open ground and chestnut and vine again increase later still, around 1521–1215 BCE. Further inland, at Kastoria, there are sporadic indicators of open ground from before 2871–2465 BCE, but the dramatic impact on the vegetation does not occur until 825–1035 CE, after which grasses dominate the signal, with important numbers of cereals, plantain and nettles. At Giannitsa, there are probable clearance phases around 6211–5966 BCE and 5661–5283 BCE, as well as more recently, marked by very rapid expansions of grasses and Chenopodiaceae. A similar clearance phase at Khimaditis takes place around 6105–5778 BCE, but the landscape only really becomes open after 2845–2331 BCE. This gradual and localized opening up of the vegetation in Greece is also suggested by Willis (1994: see main text).

Suggestions for further reading

Di Castri and Mooney (1973) and di Castri *et al.* (1981) provide the most useful introduc-tions to Mediterranean ecosystems, encompassing examples from all areas with Mediterranean-type climates. Reviews in Mairota *et al.* (1998) provide more direct material for the Mediterranean region itself. The following are useful flora for Mediterranean species: Blamey and Grey-Wilson (1993); Davies and Gibbons (1993); Huxley and Taylor (1989); Polunin and Huxley (1965); and Polunin and Smythies (1973). Allen (2001) gives a thorough overview of the biogeography of the Mediterranean region.

Topics for discussion

1　Describe how Mediterranean vegetation is adapted to the climatic conditions found there.
2　There are two main reasons why the vegetation we see in the Mediterranean today cannot be thought of as 'natural'. What are they?
3　Outline the main factors that account for the great diversity observed in Mediterranean vegetation.
4　What are the common traits of vegetation evolution in the Mediterranean, and how do they differ spatially and temporally?

5 The hydrological cycle of the Basin and its watershed

Hillslope hydrology in Mediterranean environments

The hydrological cycle of Mediterranean environments is essentially the same as in any environment and the main characteristics have been described by many authors (Chorley, 1968; Eagleson, 1970; Jones, 1997). We concentrate here on those features that are particular to the Mediterranean. These are driven especially by:

- the intensely seasonal characteristics of the rainfall;
- the very strong coupling between climate and vegetation cover and its consequences;
- the massive inter-annual variability.

The hydrological cycle is summarized by the water budget equation (a continuity equation) that is defined as:

input − output = change of storage

This calculation must be for a strict period, if accounting is to be achieved at experimental sites. The water budget is, then, in annual terms:

precipitation − evapotranspiration − deep drainage = change in soil-moisture content

Because all these can be measured in millimetres (i.e. depth per unit area), it is usual to express them all in the same units for hillslope soil work. For larger units, such as drainage basins, they can still be expressed in millimetres, but for very large basins, they are expressed in cubic metres or even cubic kilometres per year. A useful conversion is that $1\,mm$ over $1\,m^2$ is one litre, and in some Mediterranean countries rainfall is expressed in litres (implicitly per square metre).

Precipitation varies enormously in time and space. At the level of a single storm over an area of tens of square kilometres and hours or days in duration, the most important property is rainfall intensity, which is the total amount falling in a unit of time (usually $mm\,h^{-1}$). If the rainfall intensity is greater than can be infiltrated by the soil, overland flow occurs, at a rate that is proportional to the intensity once a threshold of soil water storage has been passed. In extreme cases, floods and intensive erosion will result. There is no extensive compilation of rainfall intensities for Mediterranean environments, but individual countries have their own compilations. Thus, for example, given the very wide variety of Mediterranean environments within Spain, the ICONA publication *Precipitaciones maximas en España* (1979) gives clear information on the maximum rainfall for different time periods and over different time periods for the

Table 5.1 Precipitation (mm) of storms of different durations occurring with different return periods for Almería city

Return period	Storm duration					
	10 min	*30 min*	*60 min*	*2 hr*	*6 hr*	*24 hr*
2 year	7.1	13.4	18.1	21.6	27.5	36.6
5 year	10.5	22.2	30.4	37.6	43.5	54.3
10 year	12.8	28.0	38.5	48.1	43.5	66.2
20 year	15.0	23.9	46.4	58.3	64.6	77.6
50 year	17.9	48.8	56.5	71.5	77.9	92.3
100 year	20.2	46.2	64.0	81.0	87.6	103.1

Source: ICONA (1979).

whole country. Of course, for shorter durations and longer return periods, the intensity is greater. Table 5.1 gives an example for Almería, for different durations and for different return periods.

Thus, for example, $64 \, mm \, h^{-1}$ can be expected on average once in a hundred years, and $63 \, mm \, h^{-1}$ for a ten-minute period once in five years. The figures also illustrate that artificial rainfall simulators, typically producing 48 to $120 \, mm \, h^{-1}$, often have intensities that are far too high for these kinds of environment and thus may exaggerate the actual magnitude of rainfall intensity effects on processes that depend on it (such as runoff and erosion) if interpolated to lower rates. The ICONA manual also gives the formulae for calculating the maximum amount of rain for any duration (up to two hours) and any return period (up to five years). The spatial distribution of rainfall intensity for the Iberian Peninsula clearly indicates that there is a different intensity of rainfall regime in the Levant and the south than in the interior and the north. One of the most intense rainfalls recorded contributed to the Biescas disaster discussed below. According to García-Ruiz *et al.* (1995), the rainfall in the Barranco de Betes, the catchment area in the Pyrenees contributing to the flood, estimated on the basis of peak runoff, was in excess of $500 \, mm \, h^{-1}$. Shaw (1983: 233) gives a relationship for the world maximum rainfall intensities as follows:

$$R = 425 \, D_h^{0.47}$$

where R is the precipitation in mm and D_h the duration in hours of the rainfall. Thus the world maximum in fifteen, twenty and thirty minutes are $221.5 \, mm$ ($886 \, mm \, h^{-1}$), $253.6 \, mm$ ($761 \, mm \, h^{-1}$) and $306.8 \, mm$ ($613.7 \, mm \, h^{-1}$) respectively, very close to those estimated for the Aras sub-basin of the Barranco de Betes.

Rainfall amounts are also strongly influenced by relief. Lautensach (1971) calculated the gradient of rainfall in mm per 100 m of altitude for various massifs in the Iberian Peninsula. For Sierra de Espuna in Murcia, he obtained 36 to 48 mm per 100 m in the period 1906–1925, at Alhama and Totana respectively for the annual rainfall totals.

A second major component affecting the hydrological balance is soil texture. Sandy soils have high infiltration rates, so a great deal of water can get into the soil, but high hydraulic conductivities so the water drains easily from the base of the profile and can be extracted easily by evaporation to the atmosphere. Fine soils (silts and clays) have the opposite properties, so have better water retention. Table 5.2 gives examples of infiltration rates for different soil types in southern Spain. According to studies in Israel: 'The breakpoint between the advantage of

Table 5.2 Infiltration rates as measured on different lithologies using cylinder infiltrometry at Soria and Ugijar in Spain

Lithology	Saturated hydraulic conductivity $mm\ h^{-1}$
decalcified marl	2.7–21.2
wealden sandstone	9.4–113.2
marl	33.0–162.0
limestone	69.6–166.6
sandy silt	157.2
sandy marl	170.3
gravels	190.1
sand	253.2
limestone breccia	337.7
sands and gravels	401.8–650.6

Source: Data from Scoging and Thornes (1979).

coarse-textured soils over fine-textured soils versus its inherent disadvantage lies somewhere between a mean of 200 and 500 mm of precipitation' (Noy-Meir, 1973). Where there is a lot of surface rock, the runoff is greatest. As a consequence, quite small amounts of soil covering slopes can have an important effect on overland flow in Mediterranean dry environments, even where rainfall amounts and intensities are constant. Even small amounts of loess soils, deposited on slopes, can change runoff amounts quite significantly without the need to argue for climatic change in dryland areas. Thornes (1990) has detected and modelled similar controls on the spatial distribution of vegetation types on Mediterranean hillslopes.

Catchment hydrology in Mediterranean environments

Why flash floods occur

Floods occur when the rainfall intensity produces so much water that the capacity of the landscape to get rid of it is exceeded. In the Mediterranean, flash floods are not uncommon because rainfalls are intense and cover small areas. There are two main causes for this. One is the closeness of mountains to the coastline. If warm humid air is forced over the land and then has to rise steeply and quickly, violent instability sets up in the atmosphere, leading to torrential downpours, often on to steep land. On higher ground, short, very intensive storms (one hour) occur more often than longer, less intense storms. Storms at the coast usually penetrate a long way inland. Even though the French Mediterranean coast extends only 400 to 500 km, its climatic influences penetrate far north into the interior. A typical meteorological situation in southern France consists of a classical Atlantic depression coming in over Western Europe, but subject to some blocking by high pressure over Central Europe. Then the flow over the warm Mediterranean Sea results in a strong moisture-loading of the lower layers and sometimes a secondary cyclogenesis over Spain and the western basin. This situation produces very active convective rain systems, both within the warm sector and in the vicinity of the almost stationary undulating cold front (Obled, 1988).

Another cause is the very low vegetation cover. Since the proportion of rainfall lost as runoff rises significantly as the percentage of vegetation cover falls below about 30 per cent, bare soils produce a large runoff coefficient. Moreover, soils in the Mediterranean often have low infiltration capacities (see Box 5.1) because the soils dry out in the summer, following the winter

Box 5.1 How runoff is generated

We can imagine the soil as represented by a tank with a perforated lid and an outlet at the bottom (Figure 5.1). The storage capacity (C) of the tank depends on its depth and porosity. Deep soils with a high porosity have good storage capacity. Rain falling on the perforated lid passes through the surface as infiltration at a rate in mm h^{-1} and moves down through the unsaturated soil to the saturated zone. If the rainfall intensity (mm h^{-1}) is higher than the infiltration rate, then the rainfall runs off. This process is called Hortonian overland flow (HOF). After time, the spare capacity in the tank (C − S) reduces as more infiltration occurs until (C − S) = 0. Spare capacity approaches zero and this slows down the entrance of water into the soil. When all the storage is filled, the arrival of more water causes saturated overland flow (SOF). The relative role of these two mechanisms varies downslope because throughflow, if it occurs at all in Mediterranean soils, brings the lower parts of the slope to saturation earlier. This is called the partial contributing-area concept.

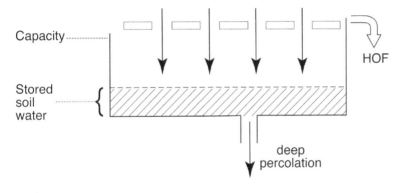

Figure 5.1 Tank model of soil hydrology

moisture. Fine surface 'seals' develop, perhaps only 1–2 mm thick, and these increase the runoff rates. Sometimes evaporation leads to chemical as well as mechanical seals (often of calcium carbonate) that again reduce the infiltration rates very dramatically. Mediterranean soils with high erosion rates also often have stones exposed at the surface. In this position, the stony soils create higher runoff rates and lower infiltration rates, whereas soils with a large proportion of buried stones may have higher infiltration rates (Poesen and Lavee, 1994).

Some examples of flash floods

In the historical records there are some remarkably intense rainfalls in Mediterranean areas. In the 18–19 October storm of 1973, that traversed from Malaga to Benidorm in south-east Spain, daily totals of 250 mm were commonly recorded; and in the flooding of the Biescas campsite in the Spanish Pyrenees on 7 August 1996, 250 mm fell in the Barranco de Arías in the valley of the River Gallego (Figure 5.2). The basin is 18.8 km², with the highest elevation at 2,189 m and its lowest point at 940 m, and mainly lying between 1,200 and 1,600 m. It is largely covered with meadows and pine forests. Storms developed in the Ebro depression to the south of the Pyrenees and when the humid hot air from the Mediterranean arrived at the

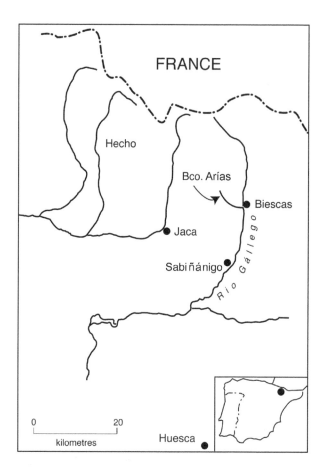

Figure 5.2 Location of the Barranco de Arías
Source: García-Ruiz *et al.* (1996).

mountain front, 160 mm fell in the village of Biescas. This rain fell on soil already saturated because the previous six months had been wetter than usual (García-Ruiz *et al.*, 1995). In one of the sub-basins, the discharge estimates indicated an intensity of 515 mm h^{-1} and the very highest locally estimated value was 997 mm h^{-1}, which is a record intensity for this location in the period since 1940. A series of check dams was destroyed by the flood, which carried down sediments from behind the check dams on to the alluvial fan where eighty-seven people were killed in fifteen to twenty minutes. Between 122,000 and 136,000 tonnes of sediment were released from the check dams, filling up the main channel across the alluvial fan. The specific flow was estimated at 27.7 m^3 s^{-1} km^{-2}.

Pérez Cueva and Calvo (1984) estimated a flood of between 11 and 38 m^3 s^{-1} km^{-2} in the Valencian storm of 1982 and García-Ruiz *et al.* (1988) a massive 58 m^3 s^{-1} km^{-2} in a small catchment, the Najerilla, in the Iberian System in north-central Spain. Such intense runoffs cause immense damage, through the debris transported, and the debris flows triggered (super-concentrated flows or sometimes mudflows). Modern flood protection, ironically, often leads to greater loss of lives, because the population at risk feels safer, and this was certainly a factor at Biescas, when the 144-year flood occurred. Of course, the use of return periods is very misleading because they are extremely difficult to estimate, especially in semi-arid mountain environments and also because they do not indicate that it will be 140 years before a similar flood magnitude occurs.

Flash floods and land-use changes

Because historical data are so inadequate, and because there is a tendency for insurance claims to rise steeply over time, it is not possible to know if flood damages are increasing through time. However, a knowledge of hydrology suggests that flooding is intimately related to land use and that the progressive history of the devastation of plant cover in the Mediterranean is likely to enhance the flood risk. These factors are a major reason for concern about desertification (see Chapter 19). Moreover, thicker soils can hold more water if they are porous and as the soil is eroded, this storage is reduced to produce a downward spiral of more runoff, more erosion and more flooding (see Chapter 6).

The removal of vegetation by fires has a similar effect (less interception, less soil storage) and this has been experimentally documented after major fires (see Chapter 7). In the first few years runoff is usually observably higher because the soils are made hydrophobic (water-resisting) by fire and because there is less vegetation to store the rainfall.

Flash flood predictions

A major problem in these environments is being able to predict the occurrence of storms and the intensities associated with them (see Box 5.2). Storms covering perhaps $100 \, km^2$ may have within them very intense cells, so that even if a storm track of an existing or historical storm is known, it is practically impossible to say what rainfall they will produce, or have produced, over what time. The French Ministry of Agriculture attempted to investigate this problem, based on south-east France, where they recorded the sixteen-year maximum daily rainfall per year in 300 measuring stations. This study showed that, typically, a ten-year storm of 1-hour duration was fairly uniform (autocorrelated) over about 50 km, and the 24-hour rainfall could be regionalized over 80 km; but for a more intense 100-year storm, the corresponding figures are about 40 km and 60 km (Slimani and Lebel, 1986). The experiment also supported the observation that, in mountain areas, extreme and often convective rainfalls, over short time periods, often occur preferentially in the lower parts of relief; and large amounts over long time periods tend to occur higher in the mountains.

The spread of radar technology in Mediterranean countries is providing fresh insights into variations in intensities within storms. Figure 5.4 shows the intensities for the 1982 storm in the provinces of Murcia and Alicante. Nevertheless, until radar and ground calibration are combined, the errors in radar forecasting of rainfall are still too large to permit 'real-time' estimation of the rain falling on a given catchment, so that flood warning has still to rely on traditional methods. Flood hazard is, alas, likely to continue to be a major hazard in Mediterranean mountain environments.

Regimes of Mediterranean rivers

The regimes (long-term flow patterns) of Mediterranean rivers reflect:

- high inter-annual variability of rainfall;
- strong seasonality of rainfall within the year;
- rare extreme events as outlined above.

Box 5.2 Return periods

The concept of return period is very important. The essential idea is that floods of a given size occur an average distance apart in time. This is not the same as saying that they do occur this fixed time interval apart, though. Bigger floods are less frequent, smaller floods more frequent. Size is usually measured by total rainfall, rainfall intensity, or flow peak volume. A plot of size against return period for the Sierra Nevada in Spain (after Cirugeda, 1973) is shown in Figure 5.3. From the graph one can read the return period for an event of a given magnitude or, conversely, the magnitude of an event of a given return period. In designing flood-prevention works, engineers have to decide what event they are working to. Often this is the one hundred year flood, i.e. the flood peak flow that occurs *on average* once every one hundred years. The quality of these estimates improves as the history of flows increases.

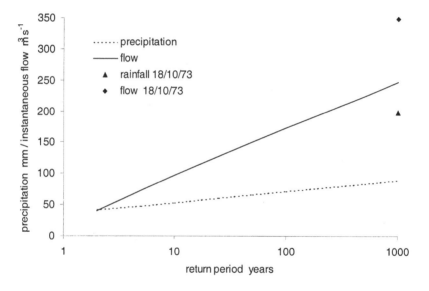

Figure 5.3 Graph of size against return period for the Guadalfeo River, Sierra Nevada, Spain
Source: Cirugueda (1973).

For example the Río Segura in Murcia Spain has been studied in detail by López Bermúdez (1976). At Murcia city, the river has a catchment area of 6,960 km². The monthly flows of the Guadalentín, the main tributary (2,780 km² at Totana) are shown in Figure 5.5 and range from a mean monthly flow of $0.4 \, m^3 s^{-1}$ in July to $2.58 \, m^3 s^{-1}$ in January. The average annual precipitation in the basin is 271.6 mm. The maximum recorded flow is $547.1 \, m^3 s^{-1}$ in the flood of October 1948, the minimum $0 \, m^3 s^{-1}$ in August 1945. The rivers in this region are broad, braided, gravel-bedded streams known by the name *rambla*. In smaller catchments, flow is rare and even intermittent and these are called ephemeral channels. Again, even these smaller channels can be subject to very great variations in flow. Because they are an important source of irrigation and building material (gravel and sand) for roads and concrete structures, they have been studied quite intensively (Thornes, 1977, 1993a, b). At tributary junctions, there are big

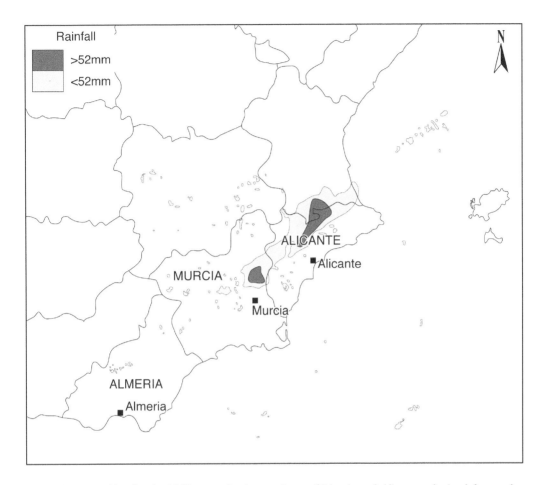

Figure 5.4 Intensities for the 1982 storm in the provinces of Murcia and Alicante, obtained from radar imagery

alluvial fans, where material brought in by tributaries piles up in the main channel, only to be moved by later exceptional storms in a downstream direction. It is these large moving volumes of bedload that cause most damage to life and property in large flood events.

Mediterranean rivers that rise in high mountains, such as the Apennines and the Pyrenees, still show the summer drought, but often show a late spring peak, such as the Ripoll, a tributary of the Rio Besos that drains the Catalonian Pyrenees. As shown in Figure 5.6, there is a high flow in November, reflecting the autumn rains and a peak in April and May, reflecting the snow melting under the onset of spring rains and warmer temperatures in the mountains.

By concentrating on floods and the damage they cause, we have emphasized the negative aspects of Mediterranean rivers, but they are above all critically important for their water supply. Often dry river beds hide the fact that water is flowing beneath them, partially from groundwater that drains to the lowest point, partly from direct infiltration into the bed by flows across the surface. Human activity tries to maximize this source by encouraging flow directly into the river bed, diverting it from hillsides and terraces. These ancient technologies are discussed in Chapters 10 and 18. The Arabs built brick-lined tunnels beneath the river

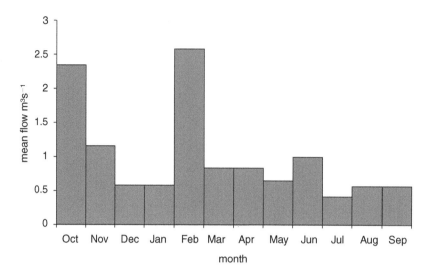

Figure 5.5 Mean monthly flows of the Guadalentín River at Totana, Murcia, Spain

Source: López Bermúdez (1976), reproduced by permission of Hodder Arnold.

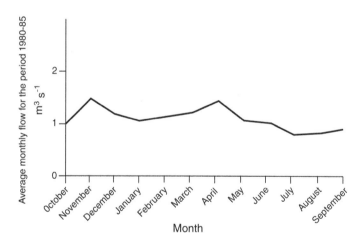

Figure 5.6 High flow in the Ripoll, a tributary of the Rio Besos (Pyrenees) in November

beds, so that they could catch water during heavy storm periods for later use. Such a system of underground tunnels is found in the lower Río Guadalfeo in the province of Granada where, even today, 1,000 years after its construction and despite the enormous floods that have occurred from time to time, it provides water to the irrigation system of the *Vega* of Motril. At the present time, hardly any Mediterranean rivers have 'natural' flows because nearly all are used for irrigation, urban water supply or waste disposal (see also Beaumont, 1993; and Chapter 17 on pollution)

Channel- versus catchment-flood prevention

In Italy, a country that has experienced a number of severe flood disasters, research is alive and well, but legislative and regulatory activities and the efficiency of the responsible state agencies are rather poor (Rossi and Siccardi, 1989). This problem arises partly because the responsibility has been devolved from central state organizations to local regional governments. The difficulties are compounded by the Mediterranean character of the environment – high intensities with up to 30 to 40 per cent of the annual rainfall occurring in a single day. Mean response times of the hydrographic networks are short because of the mountainous terrain. High sediment production from erosion results in channel sedimentation and avulsion (rapid changing of the course of the river).

Italy has a long history of flood protection and soil conservation that dates back to the Roman period. Like other Mediterranean countries, flood control has mainly been carried out by structural passive measures aimed at improving the hydraulic capacity of major rivers, usually by walls or similar concrete or masonry structures. Along the Po River between 1861 and 1876, the levees gave way on 214 occasions; since the beginning of the twentieth century, such collapses have been reduced to single figures. The Po floods of 1951 highlighted the inadequacy of traditional structural defence measures and shifted the emphasis to catchment measures including soil conservation. Just two years afterwards, Luna Leopold addressed the American Society of Civil Engineers on the need for catchment rather than channel measures and pointed out that any system of embankments cannot represent a formal and safe solution to the problem of protection, because reinforcements and raising of levees caused disasters to be less frequent but more disastrous. A new cycle of flood disaster control measures followed the Florence flooding of 1966, the emphasis this time being on flood warning measures (e.g. telemetry: Rossi and Sicardi, 1989).

Throughout the Mediterranean there is now much more attention to risk, hazard-zone management and formal flood risk assessment. The intensive use of floodplains for irrigation agriculture and urban development, especially in coastal tourist areas, coupled with the high risks from debris slides and debris flows triggered by intense storms and tectonic instability, ensure that flood/hillslope hydrology occupies a pivotal role in the Mediterranean environments. These factors help to explain the key role of foresters in Mediterranean environmental problems, where afforestation is often regarded as a palliative to all environmental protection and conservation problems.

Wetlands

Although not generally associated with the Mediterranean, wetlands do form where flat, poorly-drained land collects enough water for the surface to be submerged or saturated most or all of the time. The wetlands are characterized as much by their vegetation as by the continuity of water. Their great importance is as areas where migrating birds can feed and replenish their survival prospects on the long journey between Africa and Europe. They are protected by international conventions. There are two main types of wetland: inland depressions, such as the lakes in the Ebro valley and the Tablas de Damiel in the headwaters of the River Guadiana on the meseta of Castilla la Mancha; and coastal wetlands, such as the coastal lagoon, Lac de Bizerte in Tunisia and the Coto Doñana wetlands in the Guadalquivir delta.

The Tablas de Damiel (Sánchez Soler and Fernandez del Rincon, 1991) are formed at the junction of the seasonally flowing Ciguela River and the 'permanent' freshwater Guadiana River. Because of groundwater withdrawals, the Guadiana is rarely flowing and has led to serious changes in the water balance of the Tablas de Damiel, such that they are in danger of being lost completely. They are protected because they are a good example of the interaction between wetlands and aquifers and because they are representative of an ecosystem that once covered a much larger area of Castilla la Mancha. The vegetation essentially comprises aquatic species adapted to the saline conditions (halophytes) that form islands separated by creeks, whose genesis is related to the seasonal rise and fall of the water levels. They are under threat because of the strong economic base of irrigated agriculture in Castilla la Mancha.

For centuries people lived in harmony with the wetlands, using the resources sustainably. However, in the 1970s, canals and drainage were introduced and the courses of the Rivers Ciguela and Guadiana were re-aligned, with the objective of increasing cultivation of the neighbouring flatlands by improving irrigation. The river channels were also deepened and the supply of water to the upper reaches was re-directed by means of the inter-basin Tajo-Segura canal that takes water to south-east Spain, to provide water for the citrus orchards of Murcia and Almería. In a study by the Geological Survey of the Spanish government, it was estimated that the recharge of the aquifer should be $335\,\mathrm{Hm^3a^{-1}}$. The annual extraction in 1987 was estimated at $520\,\mathrm{Hm^3}$. This annual extraction of a volume of water, far greater than the recharge, has de-watered the aquifer by an estimated $2{,}000\text{--}2{,}500\,\mathrm{Hm^3}$. In 1974, an area of 30,000 ha was irrigated from subsurface aquifers. By 1987 this figure had increased to an estimated 130,000 ha. For a fuller discussion of controls on the recharge of this aquifer, see below.

In 1984, the Ministry of Public Works developed a plan for the hydrological regeneration of the National Park, involving:

- in the short term, construction of wells which could be used at critical times to augment flow into the Tablas de Damiel;
- in the medium term, a review of the abstraction of flow to the Tajo-Segura aqueduct;
- in the long term, construction of a dam in the River Bullaque, a tributary of the River Guadiana upstream of the Park, that will take any excess water from the basin back into the Park.

But the main problem will be to restore the groundwater inflow to the lakes, by limiting over-exploitation of the groundwater. In 1987, a measure was introduced by the Guadiana water authority to limit over-exploitation of the main aquifer that is causing depletion (Aquifer 23).

The Tablas de Damiel are not only a resource for Spain, but for the whole of Europe, because of their position on the major bird migration routes and the great areas of drylands that surround them. Unless the urgent actions in train are implemented vigorously and accepted seriously, this important wetland could be lost for ever. Even now, spontaneous combustion of organic matter through drying out is further reducing the area of halophytic vegetation quite dramatically.

The Lake Garaet El Ichkeul is one of the most studied wetlands in the Mediterranean and one of the most regulated (Hollis *et al.*, 1992; Figure 5.7). The Ichkeul lake drains into Lake Bizerte to which it is connected by Wadi Benhassim. It has experienced important human modifications over the centuries and is characterized by strong coastal interactions. It is now threatened by climate change that will probably increase salinity and reduce inflow indirectly through the greater expected demand for irrigation water. Hollis *et al.* (1992) estimate that the average storage of irrigation water will fall by 26 per cent and that the wetlands will be nearly empty for up to 19 per cent of the time. The combined effect of new dam schemes and the rise of temperature will effectively turn the Ichkeul National Park Reserve into a saline *sebkha* and it is likely to lose all its food plants for over-wintering and breeding waterfowl. Agriculture is likely to change to even more intensive irrigation and foundation problems

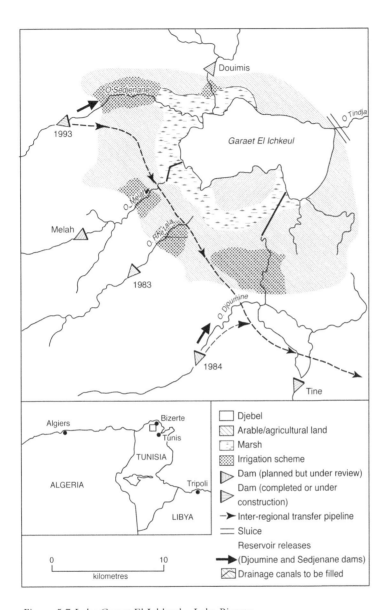

Figure 5.7 Lake Garaet El Ichkeul – Lake Bizerte
Source: Hollis *et al.* (1992), reproduced by permission of Hodder Arnold.

could occur in roads and buildings from saline groundwater. Clearly the management of this wetland is fraught with great difficulties now and in the twenty-first century.

Hollis *et al.* (1992) built a large computer-simulation model of the controls on the lake level and lake salinity on a daily basis using daily weather data to produce daily hydrological data under the various management scenarios. They show that rising temperature will increase the actual evapotranspiration and thereby reduce surface runoff. The second year of modelling suggests a reduction of water surplus (runoff and groundwater recharge) from 56.4 mm to

10 mm, a fall of 82.4 per cent. This is equivalent to only 9.3 per cent of the rainfall appearing as runoff.

Groundwater

Groundwater is the water that is stored in rocks below the surface, some of it dating back to prehistoric times. The groundwater store is recharged by rainwater and by water infiltrating from river beds.

Near the coast, deep gravel-filled canyons or wide coastal plains get water from the rivers flowing across the surface and from infiltration by sea water. The fresh water is controlled by surface runoff. When fresh water is pumped out for irrigation, a pressure gradient is set up, because sea water is denser than fresh water. The sea water invades the aquifer in a process known as sea-water intrusion. Changes of the fresh water infiltration process (for example, by damming surface rivers) affect the balance and can actually increase intrusion. Once it has occurred, pumping for irrigation pulls up brackish water. This is the case in southern Spain, where progressive reservoir construction for flood protection and for irrigation has caused less river flow. Salt-water invasion (with brackish wells) has become more common, as in the Andarax in the province of Almería (see Box 5.3). Very large floods can bring the water level nearly to the surface, making wells easier and cheaper to dig and more productive for irrigation.

In large interior basins, aquifers often comprise a mosaic of permeable and porous rocks, sometimes at great depths. The great interior meseta of Castilla la Mancha, Spain, is underlain by aquifers that are used to keep agriculture irrigated in times of drought (Figure 5.9). In more restricted basins, such as the great rift zone that is occupied by the Guadalentín and the Segura, in Murcia, the aquifers are at even greater depth under the progressive build-up of sediments in the floor of the rift over geological eras. The rocks forming the aquifer may outcrop many miles from the points where water is extracted. Burke (1998) investigated the controls on aquifer recharge in a large area of Castilla la Mancha. She found, using computer simulation, that rainfall was the main control of recharge, but also that the rock type at the surface and soil characteristics, such as depth and permeability were important. An important result is that, under the kind of irrigation practised, the farmers often over-irrigate, i.e. they put on more water than the plants need or can use. Of course the water is recycled. In fact, the main changes in aquifer budget over the past forty years have been caused by increase and construction of the areas under irrigation, after rainfall changes have been taken into account (Burke, 1998).

In small islands, all irrigation and about one-third of all potable water in the public supply comes from groundwater, so the recharge of aquifers is crucial and is of concern in light of expected future climate changes. In Malta, Gozo and Comino, there are both perched and mean sea level aquifers (Figure 5.10a, b). The perched aquifers are found in the higher parts of the islands, but private extraction by farmers has almost completely exhausted them.

Rivers and channels

Because of the seasonal and sporadic nature of rainfall in Mediterranean environments (except for mountainous regions), rivers are characterized by having no flow for long periods of the year and/or flow occurs as flash floods. As a consequence, Mediterranean rivers are often ephemeral, i.e. there is no flow for months or sometimes even for years. The large rivers that are responsible for the volume and pollution in the Mediterranean Sea are *allogenic* (the flow comes from outside the Mediterranean region in the climatic sense). These rivers, such as the Rhône, Nile and Po are usually perennial.

Box 5.3 Saltwater intrusion

The groundwater is fresh at the landward side and salty at the seaward side. The freshwater/seawater interface is quite abrupt and is usually close to the shore. Its position depends on the rate at which the freshwater can drain towards the sea. The seaward flow of freshwater prevents the landward flow of saltwater and there is an equilibrium position, which has slight variations due to tides and season. If water is pumped out of the freshwater side, the interface shifts and the result is saline water encroachment into the aquifer. This can reverse the flow, bringing salty water up into the wells, as shown in Figure 5.8. Alterations to the freshwater regime, through dam construction and land-use or climate change (e.g. more or less moisture) may change the position of the interface.

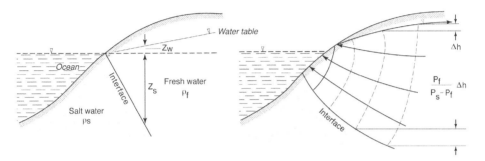

Figure 5.8 Marine intrusion shown schematically
Source: After Freeze and Cherry (1979).

A good example of saltwater intrusion has been studied on the alluvial plain of Capoterra in southern Sardinia, west of Cagliari (Collin *et al.*, 1998; Barrocu *et al.*, 1994). The coastal aquifer is jeopardized by saltwater intrusion in the area exploited by the ever-increasing demands from agriculture, industry and domestic use. Groundwater is drawn from the aquifer through some 300 wells, scattered over an area of $60\,km^2$. The aquifer is in multi-layered recent alluvial deposits in the upper part of the Capoterra plain and the confirmed aquifer is at a depth of 40–50 m in the older alluvial deposits. Continuous monitoring by Barrocu *et al.* has indicated a rise in salinity particularly in the central part of the plain near the lagoon, probably reflecting leaching from the lagoonal deposits and spray transported from the salt extraction pans by the wind.

There is a large literature on ephemeral channels and their behaviour (Thornes, 1977; Butcher and Thornes, 1979; Thornes, 1993a and b). The main features are:

- The channel bed is usually determined by the last flow in the channel, which might have been days, months or even centuries ago. In contrast, perennial flows continually adjust their bed characteristics to the flow passing through.
- Water in the flow infiltrates into the bed, so that flow may decrease downstream, through transmission losses into the sub-alluvial aquifer.
- Coarse sediments in large volumes may occur on the stream bed. There is always something to be eroded in these channels, although the exact amount may depend on the time passed since the last extreme flood event.

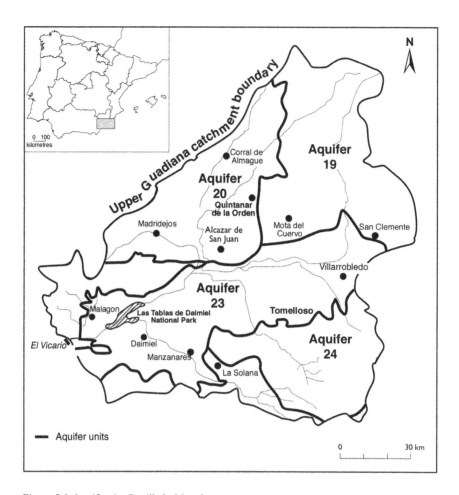

Figure 5.9 Aquifers in Castilla la Mancha
Source: Burke (1998).

- Tributary flows may not coincide with flows in the main channel, especially if the storms are highly localized over a tributary catchment.
- The bed may be characterized by material moving as waves of sediment rather than as individual particles, because alluvial fans are formed at channel junctions. These are called 'sediment slugs'. The significance of the large sediment yields comes mainly from the need to construct water storage dams that may be quickly filled with sediment and so lose their efficiency. Small dams (called check dams) are specifically constructed to limit sediment movement to reservoirs or large dams, or from damaging fields or infrastructure (see Figure 5.11).
- Sometimes flows only pass part way through the channel as a result of transmission losses, so that sediments may be 'dumped' by the flow in unexpected places. As a result, the patterns of sediment behaviour normally found in channels tend not to occur in ephemeral channels.

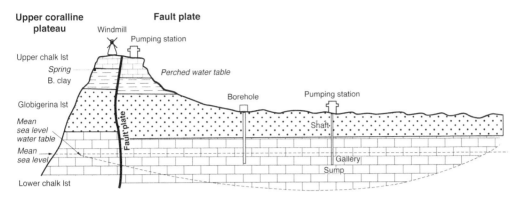

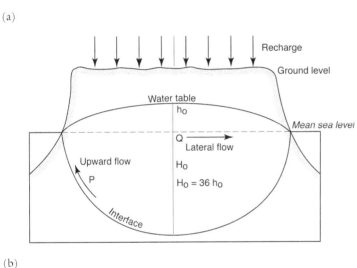

(b)

Figure 5.10 (a) Perched and (b) mean sea level aquifers in Malta

Source: After Attard *et al.* (1996).

Another feature of ephemeral channels is the ubiquitous occurrence of vegetation that can use the water in the sub-alluvial aquifer for its growth. Such plants are called *phreatophytes*, and examples are oleander and *retama* (Spanish broom). They can also survive buried by sand and gravel and re-sprout to give a more permanent look to ephemeral channels than might otherwise be expected. Even these hardy shrubs find it difficult to survive flash floods and often they can be used to identify and date extreme events, producing earlier flood flows.

Suggestions for further reading

The papers in the Perugian Symposium, edited by Siccardi and Bras (1989) are all in English. An account of the impact of the September 1973 flood in the Alpujarras, southern Spain is found in Thornes (1974; 1976). Wainwright (1996c) gives an account of the September 1992 event in southern France in the context of other extreme rainfall events.

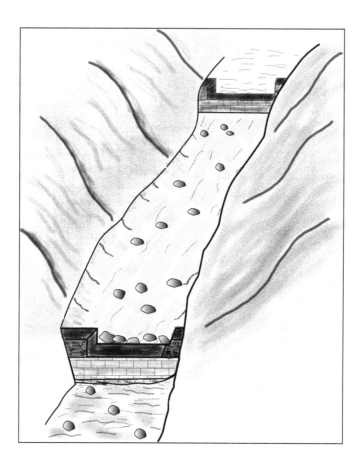

Figure 5.11 Check dams constructed to limit sediment movement to reservoirs or large dams

Topics for discussion

1   Discuss the factors that lead to recurrent significant flood disasters in Mediterranean environments.
2   Compare and contrast channel and catchment actions as methods for flood risk remediation and explain their relative advantages and disadvantages.

6   Erosion processes

Introduction

The Mediterranean landscape as we see it today is denominated by features that are related to erosion in its various forms. In the majority of areas, water erosion and mass movements are the dominant processes because of the high relief which recent tectonics have imparted to most of the region. Elsewhere, the erosion from uplands may be redeposited forming distinctive coastal plains – this process will be discussed in more detail in subsequent chapters. Finally, large areas to the south of the basin and in some coastal zones are dominated by aeolian erosion and deposition. Thus, the interplay of climate, lithology and weathering, and vegetation defines the location and extent of the different types of erosion in the Mediterranean Basin. These controls are discussed in this chapter.

Controls on erosion by water

Erosion can be considered as the balance between erosivity – the forces causing movement – and erodibility which is the resistance of the material to being moved. In the water-erosion process, the erosivity is derived from the energy of falling raindrops and any subsequent overland flows. The erodibility of the soil surface is related to the physical and chemical characteristics of the soil. In both cases, the presence of vegetation can be very important in modifying the response of a particular surface to erosion.

Erosivity

The energy possessed by a raindrop impacting the surface is a function of its mass and its velocity. The mass is related to the diameter of the raindrop, while the velocity is related to both the drop diameter and the height from which it falls. The diameter of raindrops in any particular storm varies with the rainfall intensity. For example, Feingold and Levin (1986) showed that for a series of storms on the Mediterranean coast of Israel the modal diameter is around 0.4 mm in a low-intensity storm ($5.8\,\mathrm{mm\,h^{-1}}$), increasing to 1.2 mm in a higher intensity event of $39\,\mathrm{mm\,h^{-1}}$. The maximum recorded drops in these events are 3 mm and 4.6 mm, respectively. They also demonstrated that the size distributions vary dramatically through time in an event. This variation is also significant spatially due to the often localized nature of rainfall events in the region (e.g. for Spain: Alonso Sarria and López-Bermúdez, 1994). The fall velocities for drops of different sizes are presented as a function of the fall height by Epema and Riezebos (1983). Terminal velocities – i.e. the maximum reached during free fall through the atmosphere from a sufficient height – range from $6.68\,\mathrm{m\,s^{-1}}$ for a 2 mm diameter drop to $9.24\,\mathrm{m\,s^{-1}}$ for a 6 mm drop. For drop diameters of less than 3 mm, the terminal velocities have effectively been reached over a fall height of 13 m, whereas larger drops require increasingly larger heights because of the increased frictional resistance to their fall. However, 90 per cent of the terminal velocity is reached between fall heights of 4.5 to 5.5 m, and 95 per cent between 7.0 to 8.5 m.

Given information on the drop-size distribution during an event and the terminal velocity, it is theoretically possible to estimate the rainfall kinetic energy directly (e.g. Brandt, 1990). In most cases, this is not practical because measurements of drop-size distribution are either costly or time-consuming, and thus not commonly available. A number of studies have related the rainfall kinetic energy to the rainfall intensity during an event (Zanchi and Torri, 1980; Renard *et al.*, 1991; Sempere Torres *et al.*, 1992; and Coutinho and Tomás, 1994: Figure 6.1). The results of these show that there is generally an asymptotic increase to a peak energy per millimetre of rainfall, and that the Mediterranean measurements are all higher than the US average. The latter is probably related to the frequent occurrence of convective events and is one of the reasons that soil-erosion potential can be high in the region.

The effect of vegetation is generally to reduce rainfall erosivity by the reduction of the fall height and thus the velocity of the drop on impact. As noted in Chapter 4, much of the Mediterranean vegetation is low and thus intercepted rainfall will only drip from relatively short heights. Even where tree species are present, they are generally less than 5 m tall, but most matorral areas will be dominated by plants less than one or two metres in height, with a consequent reduction in the achievable velocity. However, the process is complicated by the fact that intercepted water may concentrate on a plant and can thus drip as much larger drops than those in the original rainfall. Brandt (1989) has demonstrated that for tall canopies of greater than 10 m in height, this process can lead to an increase in rainfall energy beneath vegetation and thus much lower erosion rates in these areas must be attributed to other factors (see below). Although this could also occur under Mediterranean vegetation, it is less likely,

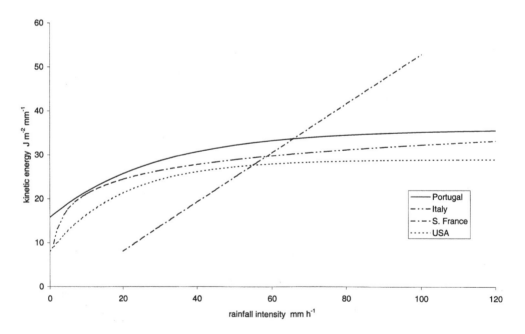

Figure 6.1 Relationship between rainfall intensity and rainfall kinetic energy measured in a number of Mediterranean locations as compared to average calculations for the USA

Source: Elaborated from Poesen and Hooke (1997), based on data from Zanchi and Torri (1980), Renard *et al.* (1992), Sempere Torres *et al.* (1992) and Coutino and Tomás (1995).

given the canopy height (see Parsons *et al.*, 1992; Wainwright *et al.*, 1995, 1999b and 2000, for examples from shrub canopies in the American Southwest). Protection of the surface in this way may be relatively unimportant because of the sparse vegetation *cover*, particularly in the drier regions, or in areas where the vegetation has been cleared or disturbed.

The erosivity of runoff can be considered as a function of its velocity and the slope of the surface. The velocity is often highly variable across a slope because of the irregularities of the surface due to the presence of stones, irregularly distributed vegetation, animal activity and, not least, previous erosion of the surface. The irregularities often act to channel flow between them, leading to faster, more erosive flows which can concentrate the pattern of erosion at various locations across the surface. This process will be considered further below. The velocity is related to the depth of flow and thus generally the overall volume of flow produced, which has been considered above in Chapter 5. Given that many Mediterranean slopes are steep because of the active uplift discussed in Chapter 2, flow erosivity will again tend to be high. Again, we see that erosivities will be generally high, and thus the soil-erosion potential will also be high.

Erodibility

At a first approximation, the erodibility of a soil can be related to the parent material from which it is derived. The weathering characteristics of the parent material (see Chapter 2) affect the particle size of the soil as well as its chemical composition. For example, marls will commonly weather into silt and clay-sized grains ($<63\,\mu$m), whereas granite typically weathers into sand-sized grains ($63\,\mu$m–2 mm). Limestones and dolomites, on the other hand, contain numerous stone fragments due to the break-up of the rock along bedding and fracture surfaces.

However, there is not a clear relationship between the particle size and the erodibility of the soil. Clays may have a low erodibility because of cohesion relating to the structure of the minerals making up the clay (Poesen, 1992). It is also possible that the soil particles are arranged in clods or aggregates, which are stable when wet, and thus more resistant (Poesen, 1995), although after periods of extended drying cracking of the surface may increase the erodibility of these aggregates. However, certain clays will disperse on contact with water, reducing their cohesion and the stability of aggregates. This process appears to be significant in generating highly erodible surfaces in the Vallcebre badlands in the Spanish Pyrenees (Solé *et al.*, 1992). Ternan *et al.* (1994) found low aggregate stability for soils developed from Tertiary sediments in central Spain, whereas soils formed from limestone bedrock were much more stable. These authors also found great variability between different land-use modifications. Imeson *et al.* (1982) also demonstrated the importance of this parameter in the development of pipe erosion (see below) in certain badlands in Morocco. Sandy soils are typically highly erodible because they lack cohesion and exert insufficient resistance from their own mass. These factors are exacerbated by the low organic matter content of most Mediterranean soils (Chapter 2) which means there is little aggregation of soil particles (Poesen, 1995). Stony soils show a decreasing erodibility due to the mass of the stones relative to the erosivity of most overland flows. Poesen (1992) demonstrated that there was an exponential decline of erodibility with increasing surface cover of stones, which is also consistent with the findings of de Lima and de Lima (1990) for soils in the Alentejo (Figure 6.2). The importance of such soils in the Mediterranean is due to their spatial extent, with over 60 per cent of Spanish, Italian and Portuguese soils and 82.4 per cent of Greek soils characterized as being stony (Poesen and Bunte, 1996; the lower proportion – 36 per cent – in France is probably due to the inclusion of non-Mediterranean areas). Through time, continuous erosion is likely to lead to the formation of stony soils by the selective removal of finer particles. The presence of a gravel lag on the surface

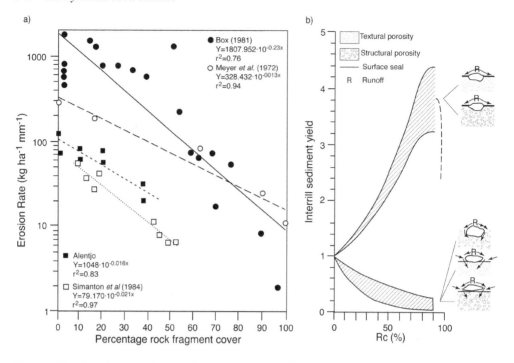

Figure 6.2 Erosion rate at a given surface stone cover relative to the erosion rate for a surface with no
stones

Source: Based on de Lima and de Lima (1990) and Poesen (1992).

may ultimately act as a negative feedback to reduce overall erosion rates, both due to the
increasing grain size and the protection of the finer particles by stones sitting on the surface
(Poesen and Bunte, 1996). However, this effect may be counteracted by the concentration of
flow between the surface stones (ibid.) leading to increasing flow erosivity as described above
(see also Bunte and Poesen, 1993).

A further complicating factor is the surface soil-moisture content. Govers *et al.* (1990)
demonstrated experimentally that as soil moisture increased from 8 to 21 per cent, the rill
erosion decreased by more than an order of magnitude, leading Poesen (1995) to suggest that
Mediterranean soils would be most erodible in autumn when the first large storms encountered dry soils. However, this effect may vary between different erosion types, as soil cohesion
is also reduced with increasing soil moisture, which would increase the erodibility with respect
to splash (Truman and Bradford, 1990) and possibly also in the case of rill and gully erosion
with side walls becoming more likely to collapse.

Vegetation can reduce the erodibility of the soil both directly and indirectly. Perhaps the
most important effect is the presence of a root system which adds strength to the surface. The
significance of this mechanism is more pronounced under vegetation such as grasses where
there are large numbers of lateral roots near the surface (Dissmeyer and Foster, 1981; Poesen,
1995). Indirectly, the vegetation cover tends to produce higher surface stability by the presence of leaf-litter layers and increased organic matter contents, although again these factors
may be relatively unimportant given the low rates of litter and organic matter production in
many Mediterranean ecosystems (see Chapter 4).

Slope-erosion processes

Water erosion on slopes is due to three distinct processes. Before runoff is generated, erosion occurs by splash due to the action of raindrops only. After runoff generation, the irregular and unconcentrated flow on hillslopes can transport sediments. Such a mechanism is termed inter-rill erosion. As flow accumulates, there is a tendency for it to concentrate in more hydraulically efficient areas. These are termed rills or gullies, depending on their size. A fourth erosion process is common in certain areas under the surface, leading to the formation of soil pipes.

Splash

Splash is ultimately controlled by the raindrop energy impacting the surface. When the drop encounters the surface, it dislodges or detaches sediment particles. These particles may remain on the surface, or be splashed upwards in rebounding water droplets. Soil can thus be moved in any direction from the point of impact, although there is usually a net downslope transport (e.g. Savat, 1981; but see Torri and Poesen, 1992). The distance of transport is usually less than a metre (Savat and Poesen, 1981). Splash most commonly transports material of fine sand size (0.125 mm: Savat and Poesen, 1981), although particles as large as *c.*12 mm have been

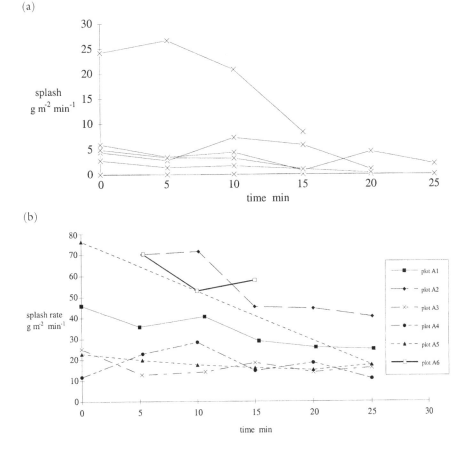

Figure 6.3 Splash erosion rates measured by Wainwright (1996a, b): (a) on a badland surface; and (b) in the intercrop part of a vineyard

observed moving (Kotarba, 1980). Coarser particles are also usually transported for much shorter distances (Wainwright, 1992; Wainwright *et al.*, 1995). Once flow begins, splash can continue, but at an exponentially declining rate as the water depth builds up because the rain-drops are increasingly less likely to impact the surface (Torri *et al.*, 1981).

Splash rates have been measured in southern France by Wainwright (1996a, b) on two dif-ferent types of surface under extreme rainfall intensities of around $100\,\mathrm{mm\,h^{-1}}$ (Figure 6.3). On agricultural land under vine cultivation with soils derived from Quaternary sediments, splash rates varied between 0.4 and $76.3\,\mathrm{g\,m^{-2}\,min^{-1}}$, with higher values on unvegetated (median $25\,\mathrm{g\,m^{-2}\,min^{-1}}$) compared to vegetated ($5\,\mathrm{g\,m^{-2}\,min^{-1}}$) interrow surfaces. Rates on bare marl surfaces in badlands were lower, varying between 0.4 and $26.7\,\mathrm{g\,m^{-2}\,min^{-1}}$, reflect-ing the higher cohesion of the clay compared to the coarser agricultural soils. Van Asch (1980) found similar results from rainfall-simulation experiments in Calabria, with the highest rates found on freshly ploughed agricultural surfaces and the lowest on those with a dense herb and shrub cover.

Interrill erosion

Unconcentrated overland flow generally has insufficient energy to detach soil from the surface (Abrahams *et al.*, 1991) and thus tends only to transport sediment that has already been detached by the action of raindrop impacts. For example, in the experiments described above, the proportion of interrill erosion compared to splash had a median value of only 11.6 per cent for the agricultural soils and 67.8 per cent for the badland marls (Wainwright, 1996a, b). In general, the transport capacity of these flows can be thought of as a function of the amount of flow present. There is again a tendency to transport soils of finer sand size preferentially, although the upper size limit is higher than for splash, being between 5 and 10 cm, depending on the particle density (Abrahams *et al.*, 1991; Wainwright, 1992; Parsons *et al.*, 1996). Transport distances are also typically further than for splash, although Parsons *et al.* (1993) found that they rarely exceeded 3 m in an individual storm event.

Francis and Thornes (1990b) carried out rainfall simulation experiments in Murcia, Spain, to compare the runoff and sediment produced under different vegetation covers for different intensities of rainfall. They found that for rainfall intensities of around $100\,\mathrm{mm\,h^{-1}}$, mean erosion rates were $2.24\,\mathrm{g\,m^{-2}\,min^{-1}}$ on sparsely vegetated matorral slopes compared to $0.32\,\mathrm{g\,m^{-2}\,min^{-1}}$ on denser matorral slopes and $0.04\,\mathrm{g\,m^{-2}\,min^{-1}}$ in aleppo-pine forest. They found a strong exponential decrease in the erosion rate with increasing vegetation cover (Figure 6.4). This exponential decrease has been found in a number of studies (e.g. Elwell and Stocking, 1976), although there is some indication that the spatial pattern of the canopy cover is also important (see Wainwright *et al.*, 1999a, for a fuller discussion). A particular example of this is the formation of patchy vegetation types in more arid areas. Sanchez and Puigdefábregas (1994) discuss how patches of *Stipa tenacissima* grass in southern Spain affect the spatial pattern of erosion, particularly in the deposition of sediment on the upslope and erosion on the downslope side of a clump. This irregular pattern leads to surface instability to which the plant adapts by growing outwards.

The impacts of stones on interrill erosion rates have been outlined above. However, many Mediterranean soils – particularly those developed on marls or clays – also develop surface crusts which can have important impacts on the erosion process. This crusting is often related to the high clay content, although Poesen (1992) has illustrated the importance of the sand fraction in the development of the crust. Where the process is by clay shrinkage on drying, the resulting surface with desiccation cracks can provide an important hydrological control or erosion, allowing water to travel to depths along cracks. If there is sufficient moisture during a

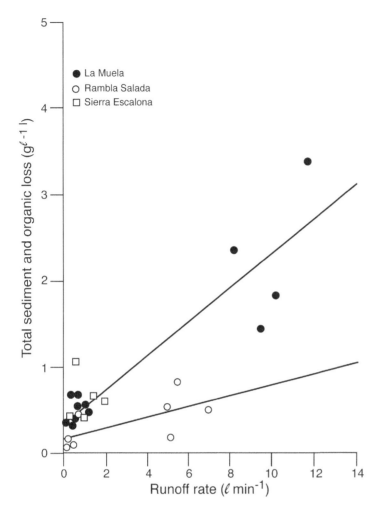

Figure 6.4 Erosion rates as a function of different vegetation covers based on rainfall-simulation experiments of Francis and Thornes (1990b) in Murcia, south-east Spain

storm to allow the cracks to close, then runoff can increase dramatically due to the relative impermeability of the clays (see e.g. Hodges and Bryan, 1982; Howard, 1994).

Rill and gully erosion

The onset of flow concentration and thus rill initiation is difficult to define because of the continuity between deeper threads of interrill flow and the first rills that incise into the surface (Parsons and Wainwright, *in prep.*; see also Parsons *et al.*, 1996). For this reason the definition of a specific threshold can be difficult. Several authors have defined an empirical relationship for the onset of rilling as a basis of the shear strength of the soil and the shear velocity of the flow. However, whereas Rauws and Govers (1988) define a single threshold, Torri *et al.* (1987) suggest a much more diffuse zone of initiation (Figure 6.5) which is consistent with the idea of continuity. The key distinction is that the rill flow possesses sufficient energy to

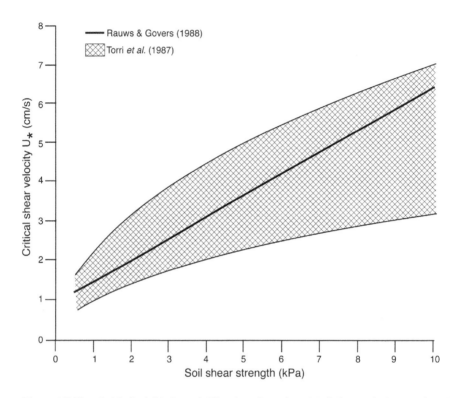

Figure 6.5 Thresholds for initiation of rilling based on the critical shear velocity as a function of the
soil shear strength

Source: After Poesen (1995).

detach sediment and can thus incise a channel bed. Consequently, larger sediment sizes can be
transported for greater distances than in interrill flows, and erosion rates are correspondingly
higher for the same slopes.

The distinction between rills and gullies is also partly subjective. Common definitions
include a cross-sectional area greater than a square foot (900 cm² : Poesen, 1995) or larger than
that which can be removed by ploughing, although the latter is hardly a useful definition for
non-agricultural slopes. Confusion also exists because of the terminology used, in that not all
gullies are formed by continuity of incision due to overland flow, as others can be formed by
the collapse of channel banks or incision into valley floors (Figure 6.6). The latter are commonly termed *arroyos*, after the Spanish term. The processes leading to gully formation are
thus complex and often poorly understood (for fuller reviews, see Bocco, 1991; Howard,
1994; Campbell, 1997).

The significance of gully erosion can be demonstrated by recent studies by Poesen *et al.*
(1998) in southern Portugal and Spain (Table 6.1). Gully erosion made up 80 per cent of the
total erosion in south-east Portugal and 83 per cent in south-east Spain. However, Cosandey
et al. (1986) demonstrated that change in gullies in the Cevennes was episodic, with alternat-
ing phases of erosion and deposition. Deposition was the dominant factor in this system,
particularly in areas protected from grazing.

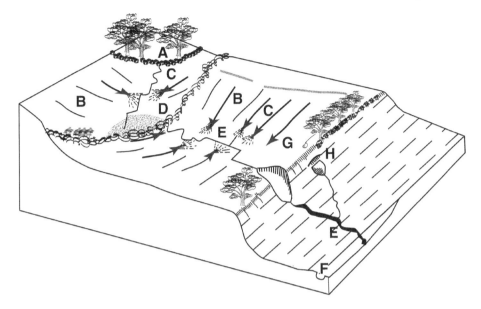

Figure 6.6 Different types of erosion within Mediterranean landscapes, demonstrating the various locations where gullies may form. A: stable and terraced wooded landscape; B. interrill erosion; C. rill erosion with localized depositional splays; D. deposition at field boundary; E. gully; F. valley-bottom gully (*arroyo*); G. soil pipe; and H. sapping/incipient headcutting in field or channel boundary

Source: After Farres *et al.*, 1993; Schumm *et al.* (1983).

Table 6.1 Estimated sediment production by gully, interrill and rill erosion in south-east Portugal and south-east Spain

Location/method	Period of observation years	Ephemeral gully t km^{-2} a^{-1}	Interrill and rill t km^{-2} a^{-1}	% gully
SE Portugal aerial photo runoff plots	3	142.9–688.3	26.0–168.8	
mean	20	415.6	103.9	80
SE Spain field mapping	10	1259.7	259.7	83

Source: Data from Poesen *et al.* (1988).

Piping

Erosion can occur under the surface by both solution and mechanical transport, forming hollows or pipes. Harvey (1982) defines a number of ways in which pipes formed in south-east Spain. Pipes here are both shallow – several centimetres below the surface and relatively small relating to a disaggregated layer below a surface crust (see also Hodges and Bryan, 1982) – and deep, reaching depths of several metres. The latter form either as the result of the removal of weaker subsurface material, or due to solution as a result of different layers of porosity, cementation or solubility (for example, if a calcrete layer is reached below the surface), or as a result of the presence of deep tension cracks which preferentially transport water and thus become enlarged. These pipes can become enlarged, reaching diameters of over a metre, and may collapse, often providing a focus for gully development. Haigh (1990) has also noted the importance of pipes below valley-floor gullies in that their collapse can lead to channel entrenchment. This mechanism has also been noted in wadi-floor gullies in Morocco (Imeson *et al.*, 1982) and Spain (Martin Penela, 1994).

Rates of water erosion

Erosion rates measured using standard erosion plots on uncultivated and abandoned surfaces show great variability and range from approaching zero to $28,000 \, t \, km^{-2} a^{-1}$ (Table 6.2). The latter case is for a clay badland in Basilicata, southern Italy (Rendell, 1986). Most rates measured are below around $500 \, t \, km^{-2} a^{-1}$, although values of over $1,000 \, t \, km^{-2} a^{-1}$ are not uncommon. Rates for cultivated slopes tend to be higher, but reach neither the minimum nor the maximum for uncultivated slopes (although Tropeano, 1984, measured rates as high as $7,018 \, t \, km^{-2} a^{-1}$ on vineyards in northern Italy), because the vegetation cover is always poorer in the less erosive regimes and the more erosive slopes would be too unstable for cultivation. However, the rate of erosion can be seen to be a function of the type of crop and cultivation system. Kosmas *et al.* (1997) compiled the results of erosion rates on 175 standard plots through the region and found the highest rates on average under vines, most likely because of the large bare areas between the crop rows, and the lowest under olives, where most cultivation is undertaken on terraced slopes (Table 6.3). Issues of higher erosion rates on agricultural land will be followed up later (Chapters 10 and 15).

Measurements of erosion rates can also be made using erosion pins or repeat microtopographic surveys to estimate the amount of ground lost to erosion. By measuring or assuming the soil bulk density, these rates can be converted to sediment losses (Table 6.4). Examples of these measurements show that they tend to produce higher estimates (Table 6.5), with equivalent losses as high as $50,000 \, t \, km^{-2} a^{-1}$, although such high values may only be for very limited areas. There is also a tendency to use this technique on highly erodible surfaces, where measurements will be easy to make.

Mass movements

Mass movements occur when the gravitational force on a rock or soil mass is great enough to overcome the resistance of the rock or soil. The major controls on landsliding in a Mediterranean context have been discussed by van Asch (1980; 1986: Figure 6.7; see also Almeida-Teixera *et al.*, 1991). The resistance of a rock or soil is related to the cohesion of the material and the friction between the separate grains making it up. Cohesion is a function of the mineralogy and texture of the rock and soil and may be reduced by high pore-water pressures as the material becomes saturated. Frictional resistance is related to particle size, shape and sorting and can also be reduced under wet conditions when the water acts as a lubricant. Steeper slope angles reduce the effectiveness of frictional resistance.

Van Asch (1986) suggests that a number of geological factors are significant in making landsliding an important process within the Mediterranean landscape. First, clays are relatively common, particularly in Tertiary and Quaternary deposits (see Chapter 2). Second, many of these more recent deposits are weakly consolidated. Third, these deposits tend to be heavily fractured due to tectonic activity (see Mariolakos, 1991, and Sorriso Valvo, 1991, for a fuller discussion). Fourth, many of the older rocks in the region are of a plastic nature. This characteristic

Table 6.2 Erosion rates measured on standard erosion plots throughout the Mediterranean Basin

Location	Cover	Year	Comments	Rock fragment cover (%)	Mean annual soil loss $t\ km^{-2}\ a^{-1}$	Reference
agricultural plots						
Alentejo, Portugal	fallow/wheat rotation	1961–1991		30	55–196	Roxo, 1993
Alentejo, Portugal	wheat	1989–1990	extremely wet	30	933–1,015	Roxo, 1993
Douro, Portugal	vineyards	1979–1988		50	39	Figueiredo and Ferreira, 1993
Douro, Portugal	vineyards	1989–1992	extremely wet	50	110–280	Figueiredo and Ferreira, 1993
El Ardal, Spain	barley–wheat	1989–1992		?	33–99	Lopéz Bermúdez, 1993
Masquefa, Spain	olives	1983–1985		?	81–2,400	Marqués, 1991; Marqués and Roca, 1987
Var, France	vineyards		chemical weed control	10–20	275–5,385	Viguier, 1993
Var, France	vineyards		conventional tillage	10–20	210–1,500	Viguier, 1993
Aix-en-Provence, France	ploughed	1965–1969		0	2,220	Clauzon and Vaudour, 1971
Potenza, Italy	horsebean	1984–1987	34.4% clay	0	199–620	Postiglione *et al.*, 1990
Potenza, Italy	durum wheat	1984–1987	34.4% clay	0	133–200	Postiglione *et al.*, 1990
Tuscany, Italy	wheat				260–1,980	Chisci *et al.*, 1981
Albugnano, Italy	vineyards	1981	RC 3.4%		≥7,018	Tropeano, 1983
Mongardino, Italy	vineyards	1981	RC 2.9%		3,255	Tropeano, 1983
Santa Victoria d'Alba, Italy	vineyards	1981	RC 0.04%		32	Tropeano, 1983
Santa Victoria d'Alba, Italy	vineyards	1981	RC 0.1%		78	Tropeano, 1983
Santa Victoria d'Alba, Italy	vineyards	1981	RC 0.1%		273	Tropeano, 1983
Petralona, Greece	wheat	1991–1992		>40	20–100	Diamantopoulos, 1993
Spata, Greece	vineyards			3–44	38–253	Kosmas, 1993
abandoned and 'natural' plots						
Alentejo, Portugal	abandoned agricultural land/natural vegetation	1988–1991		30	18	Roxo, 1993
Alentejo, Portugal	abandoned agricultural land/natural vegetation	1989–1990	extremely wet	30	28	Roxo, 1993
Alentejo, Portugal	cistus	1988–1991		30	81	Roxo, 1993
Alentejo, Portugal	cistus	1989–1990	extremely wet	30	217	Roxo, 1993
El Ardal, Spain	natural matorral	1989–1992		?	32	Lopéz Bermúdez, 1993
El Ardal, Spain	fallow land with rock fragments	1989–1992		?	1	Lopéz Bermúdez, 1993

continued

Table 6.2 continued

Location	Cover	Year	Comments	Rock fragment cover (%)	Mean annual soil loss t km^{-2} a^{-1}	Reference
Tabernas, Spain	natural vegetation, retama	1991–1992		14–77	6	Puigdefábregas, 1993
Tabernas, Spain	natural vegetation, anthyllis	1991–1992		14–77	22	Puigdefábregas, 1993
Tabernas, Spain	natural vegetation, stipa	1991–1992		14–77	34	Puigdefábregas, 1993
Tabernas, Spain	abandoned agricultural land	1991–1992		14–77	16–40	Puigdefábregas, 1993
Sierra Valencia, Spain	matorral	1988–1989		?	9–21	Rubio et al., 1990
Sierra Valencia, Spain	bare	1988–1989		?	5–16	Rubio et al., 1990
Murcia, Spain	matorral	1985–1992	0% OM	>20%	39	Albaladejo et al., 1991
Murcia, Spain	matorral	1985–1992	2% OM from added refuse	>20%	0.6	Albaladejo et al., 1991
Aisa, Central Pyrenees, Spain	abandoned, dense shrubs	1990–1991		'high'	51	García-Ruiz et al., 1991
Aisa, Central Pyrenees, Spain	abandoned – grass and open shrub canopy	1990–1991		'high'	663	García-Ruiz et al., 1991
Aisa, Central Pyrenees, Spain	abandoned – low cover	1990–1991		'high'	1,400	García-Ruiz et al., 1991
Aisa, Central Pyrenees, Spain	abandoned, stone pavement	1990–1991		'high'	3,800	García-Ruiz et al., 1991
Aisa, Central Pyrenees, Spain	abandoned, bare	1990–1991		'high'	4,600	García-Ruiz et al., 1991
Santa Fe, NE Spain	beech woodland	1982–1985		?	152	Sala and Calvo, 1990
La Castanya, NE Spain	evergreen oak forest	1982–1985		?	203	Sala and Calvo, 1990
Aix-en-Provence, France	matorral	1965–1969			0.5	Clauzon and Vaudour, 1971
Santa Lucia, Italy	natural vegetation	1992–1993		?	34	Aru and Baroccu, 1993
Santa Lucia, Italy	burnt matorral	1992–1993		?	31	Aru and Baroccu, 1993
Santa Lucia, Italy	eucalyptus plantation	1992–1993		?	66	Aru and Baroccu, 1993
Tuscany, Italy	bare		Pliocene clay		1,347	Pannicucci, 1972
Basilicata, Italy	bare		Plio-Pleistocene clay		up to 28,000	Rendell, 1982
Petralona, Greece	maquis	1991–1992		>40	25–218	Diamantopoulos, 1993
Spata, Greece	olive grove with extensive grass cover			3–44	0	Kosmas, 1993
Nahal Yael, Israel	debris slope	1970–1971		high	3–18	Yair and Klein, 1973

Source: After Poesen and Hooke (1997), with additional sources as indicated, reproduced by permission of Hodder Arnold.

Table 6.3 Average sediment losses from standard erosion plots in the Mediterranean

Land use	Number of samples	Average sediment loss $t\ km^{-2}\ a^{-1}$	Standard deviation $t\ km^{-2}\ a^{-1}$
wheat	65	17.6	26.1
vines	9	142.8	157.5
eucalyptus	12	23.8	34.2
shrubland	95	6.7	5.6
olives	3	0.8	3.3

Source: Kosmas *et al.* (1997).

Table 6.4 Conversion chart giving equivalent soil-erosion rates measured in different units assuming a dry bulk density of $1,300\,kg\,m^{-3}$

Units	Rate
$kg\ m^{-2}\ a^{-1}$	1
$kg\ ha^{-1}\ a^{-1}$	10,000
$t\ ha^{-1}\ a^{-1}$	10
$t\ km^{-2}\ a^{-1}$ or $g\ m^{-2}\ a^{-1}$	1000
$m^3\ ha^{-1}\ a^{-1}$	7.7
$m\ a^{-1}$ ground loss	0.00077
$mm\ a^{-1}$ ground loss	0.77

Source: After Poesen (1995).

Table 6.5 Erosion rates based on measurements of ground lowering

Location	Parent material	Date	Measured ground lowering mm	Equivalent soil loss $t\ km^{-2}\ a^{-1}$	Reference
El Barranco,	Ho sediments	1987–1990	2–6	2,600–7,800	Benito *et al.*, 1991
NE Spain	Ho sediments	1987–1990	3–5	3,900–6,500	Benito *et al.*, 1991
La Charca,	Tt clay	1987–1990	5–15	6,500–19,500	Benito *et al.*, 1991
NE Spain	Tt clay	1987–1990	4–12	5,200–15,600	Benito *et al.*, 1991
Propiac, France	marl	1993–1996	2–14	2,600–18,200	Wainwright, 1996b
St Genis, France	marl		10–20	13,000–26,000	Bufalo *et al.*, 1989
Vallcebre, Spain	mudrock		9	11,200	Solé *et al.*, 1992
Ebro basin, Spain	shales and Ho sediments		5–9	6,500–11,200	Benito *et al.*, 1992
Ugijar, Spain	sands, marls and conglomerates		6–27	7,800–35,100	Scoging, 1982
Basilicata, Italy	marine clays:				
	calanchi		5.3–13.6	6,900–17,700	Alexander, 1982
	biancane		22.8–39.7	29,600–51,600	
Zin, Israel	loess	(over 75 ka)	0.75	970	Yair *et al.*, 1982

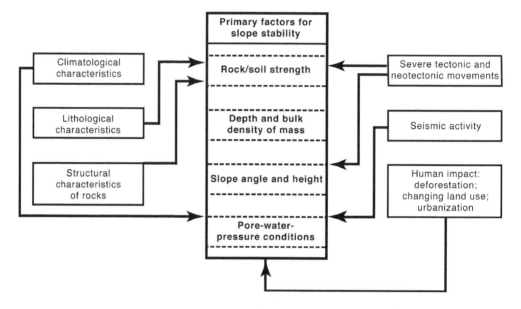

Figure 6.7 Factors causing slope instability in the Mediterranean
Source: After van Asch (1986).

is accentuated in some areas, for example, in the Pyrenees, where overfolding has placed the older rocks above marls, upon which they can slide. Fifth, in the higher mountains, relict periglacial and glacial material is present in thick deposits above impermeable surfaces (see also Wainwright, 1996c). Sorriso Valvo (1991) adds that important weaknesses can also form along karstic or pseudo-karstic features, which are again widespread in the region.

Climatic conditions control landsliding principally by affecting the pore-water pressure. The rainfall regime is thus moderated by the characteristics of the land surface which affect infiltration rates, as discussed above (Chapter 5). Although high-intensity, short-duration rainfall events can cause mass movements, they tend to be of a shallow nature because the infiltrated water cannot travel to depth during the time of the event (see Wainwright, 1996c). Larger landslide events tend to be related to more prolonged periods of lower intensity rainfall, where saturation can develop more readily (Capecchi and Focardi, 1988; Ergenzinger, 1992; Gallart and Clotet-Perarnau, 1988; Sorriso Valvo, 1991; Wainwright, 1996c). In the high mountains, particularly the Alps, high rainfall combined with snowmelt often leads to significant landslides in spring (Tricart, 1974; van Asch, 1986).

Seismic activity is also significant in accentuating the gravitational force, and is thus a significant part of the landslide process in many parts of the Mediterranean, although particularly around the Aegean (see Chapter 2). The higher magnitude earthquakes can be responsible for significant slope failures (van Asch, 1986). Mariolakos (1991) describes the effect of the earthquake on 13 September 1986 near Kalamata in southern Greece (magnitude 6.2, followed by a 5.6 magnitude aftershock two days later), where a large number of rockfalls occurred at or near reactivated fault lines. In the mountain village of Elaeochori, these rockfalls destroyed all but two of the houses. Sorriso Valvo (1991) also points out the importance of slower movements of more brittle rocks along joints and other fractures. These slower rock flows can often develop into large-scale rock avalanches if the continued movement leads to progressive weakening of the rock material. Tectonic activity is also important indirectly, as continued recent uplift has led to river incision and significant oversteepening of slopes (van Asch, 1986). The fact that tectonic activity has produced many deposits which now dip at high angles also increases the likelihood of slope failure as the plane of weakness can be orientated to counteract frictional resistance.

As a final consideration, van Asch (1986) notes the importance of vegetation in improving the stability of slopes. Vegetation increases slope stability by the effects of roots adding strength to the soil mass and by interception and evaporation of rainfall, which will tend to lead to lower pore-water pressures. However, he notes that the weight of trees can increase the stress on the slope, so that a grass cover is often used in remedial situations. Land use and land-use change can thus be thought of as important human controls on slope stability. In Calabria, landslides have been noted far more commonly on areas with agriculture or sparse vegetation than on areas with a closed vegetation cover (van Asch, 1980). On the other hand, protection by vegetation is not likely to be important for deep-seated failures because roots only penetrate to a relatively shallow depth.

Mass movements can occur in a variety of forms, from shallow, translational slides which affect the uppermost tens of centimetres of the soil, through rotational slides and slumps to rock topples, large-scale rock avalanches and slides, as well as mud and debris flows. The formation of these different types is covered in detail elsewhere (Carson and Kirkby, 1972; Crozier, 1986; van Asch, 1980), so only a small number of examples will be dealt with here.

The landslide at Vaiont in northern Italy is one of the world's best known, not least because of the 2,500 lives that were lost following the failure on the night of 9 October 1963 (Figure 6.8). The slide followed the construction of a dam on the Piave river (Petley, 1996). After an initial movement in November 1960, involving 700,000 m³ of slope material, the filling of the dam was carried out slowly and carefully, in the hope that by allowing the slopes to creep as a response to increasing pore-water pressures, they would stabilize. Despite this, rates of slope movement rose rapidly in late 1962 and again in August–September 1963. By this time, cumulative movement had been up to 3.4 m. Although water levels behind the dam were lowered as

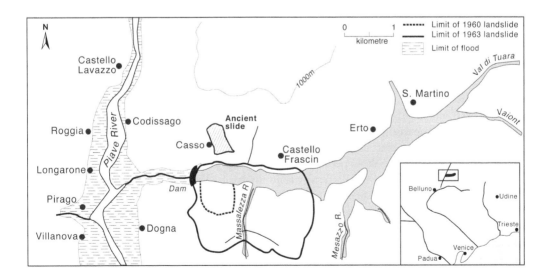

Figure 6.8 The Vaiont landslide of 9 October 1963

Source: After Petley (1996).

a consequence, this was not enough to prevent the sudden failure of $250\,M\,m^3$ of sediment travelling at velocities of up to $30\,m\,s^{-1}$. This sediment displaced around $30\,M\,m^3$ of water over the dam, causing extensive downstream flooding and leading to the fatalities.

A further consequence of landslides are the mudflows which can develop from the saturated, mobilized material. Again, these can have disastrous consequences, as was seen in southern Italy on 6 May 1998. The villages of Sarno and Quindici, between Naples and Salerno, were engulfed by mudflows leaving thirty-seven dead and over 2,000 homeless (Hanley, 1998; Hooper, 1998). Large mudflows are not uncommon in Italy, another significant example being that of Tessina in the pre-Alps, which is almost 2 km long, 400 m wide, extends from an altitude of 1,150 m to 675 m, and has been episodically active since 1960 (dall'Olio *et al.*, 1988).

Large landslides in bedrock are often features which follow deeply entrenched river valleys. Along the Guadalfeo valley in the Alpujarras of southern Spain, these are particularly pronounced. Thornes and Alcántara-Ayala (1998) used field and modelling studies to investigate the controls of these landslides. Their results suggest that interannual rainfall variability is insufficient to cause failure by the modification of the water table. They point to the probable causes as being the long-term evolution of the slope morphology (e.g. increased weathering) or the effect of extreme events causing channel flooding and slope undercutting, as, for example, in the very large flood event of 1973 (see Chapter 5).

Slope failure is also a significant process in relation to other erosion processes. Collison (1996) demonstrated the effect of tension cracks above gully heads in marls in southern Spain in promoting preferential flow. In turn, this increased pore-water pressures generating failure, and subsequent gully extension. Indeed, he suggests that almost all of the observed 70 m extension over a ten-year period could be due to headwall collapse. Van Asch (1980) also notes the importance of rotational and shallow debris slides in the formation and extension of gullies in badland systems in southern Italy.

Catchment-scale erosion rates

Erosion at the catchment scale incorporates the balance of all erosion and deposition due to fluvial transport, and may thus include material mobilized by mass movements as well as by surface and subsurface erosion, and sediments reworked by fluvial activity. Inbar (1992) collated data for erosion rates in Israel and Spain and demonstrated that, with some exceptions, they tend to follow the pattern originally proposed for the USA by Langbein and Schumm (1958). The plot studies of Kosmas *et al.* (1997) also demonstrate a similar relationship for plots under matorral, with a peak erosion rate at 280 to 300 mm (Figure 6.9). However, the maximum for the plot experiments was $21.5\,t\,km^{-2}\,a^{-1}$ at 282 mm of rainfall compared to between 310 and $840\,t\,km^{-2}\,a^{-1}$ in Israel at 300 mm precipitation (Figure 6.10, Table 6.6; Inbar, 1992). Comparison of a wider sample than that of Inbar (1992) also demonstrates a greater variability than that suggested by the Langbein and Schumm (1958) curve (Figure 6.10; Table 6.6). This result is perhaps not surprising given the inherent variability in the hydrology, geology and land-use patterns within the basin. This comparison also raises important questions of scale of measurement, as the catchment scale measurements are clearly incorporating more processes than the plot studies, and imply longer-term imbalances between slope and fluvial erosion. Parsons *et al.* (*in press*; Wainwright *et al.*, 2001) outline the major problems of making comparisons of this type. In terms of overall sensitivity, however, major land-use changes seem to be more important at changing the average erosion rate than moderate shifts in the mean rainfall. It may be more important in the Mediterranean context, though, to consider the impacts of extreme climatic events at controlling erosion rates rather than average conditions (Wainwright, 1996c).

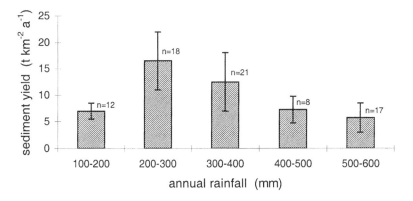

Figure 6.9 Erosion rates as a function of annual rainfall for four sites with matorral cover
Source: After Kosmas *et al.* (1997).

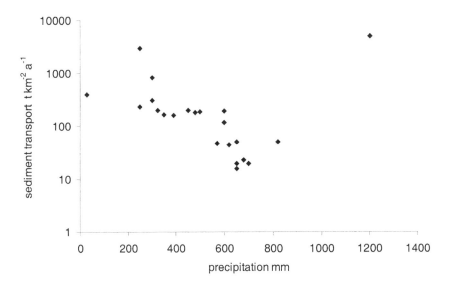

Figure 6.10 Measurements of catchment erosion rates in the Mediterranean
Note
Sources as in Table 6.7.

Ergenzinger (1992) demonstrated the interactions of various processes for the Buonamico Basin in Calabria, Italy. In this study, the most frequent slope events produce relatively small amounts of sediment by surface erosion and shallow landsliding, although there are also a number of significant large landslide events that modify the landscape dramatically (Figure 6.11). Most river activity moves more moderate amounts of sediment, implying that it is limited by supply from the hillslopes except when major landslides occur. The relative rates suggest the landscape takes around twenty-five years to re-equilibrate after the occurrence of the latter. Michaelides and Wainwright (2002) consider the extent to which interactions between slope and channel processes are significant in modifying the hydrology and sediment transport in Mediterranean catchments at a variety of spatial and temporal scales.

Table 6.6 Erosion rates based on sediment-transport measurements at catchment scale

Catchment	Area km^2	Annual rainfall mm	Sediment transport $t\,km^{-2}\,a^{-1}$	Reference
Segura, Spain		250	230	López Bermúdez, 1979
Segura, Spain		250	3,000	Romero Diaz *et al.*, 1988
Tordera, Spain		650	20	Sala, 1982
Bradano, Italy			1,159	Rendell, 1986
Sinni, Italy			2,458	Rendell, 1986
Crati, Italy			1,003	Rendell, 1986
Tiber, Italy			377	Rendell, 1986
Arno, Italy			250	Rendell, 1986
Yael, Israel		30	390	Schick, 1977
Hillazon, Israel	158	680	23	Inbar, 1992
Netofa, Israel	121	600	190	Inbar, 1992
Qishon, Israel	470	480	180	Inbar, 1992
Qishon, Israel	224	650	50	Inbar, 1992
Snunit, Israel	65	620	45	Inbar, 1992
Alexander, Israel	544	650	16	Inbar, 1992
Ayalon, Israel	160	600	117	Inbar, 1992
Eqron, Israel	62	500	185	Inbar, 1992
Soreq, Israel	80	570	47	Inbar, 1992
Pelugot, Israel	200	450	200	Inbar, 1992
Shiqma, Israel	746	390	160	Inbar, 1992
Adorayim, Israel	86	350	165	Inbar, 1992
Lahav, Israel	16	300	840	Inbar, 1992
Shoval, Israel	15	325	200	Inbar, 1992
Gerar, Israel	54	300	310	Inbar, 1992
Jordan, Israel	1,492	820	50	Inbar, 1992
Meshushim, Israel	160	700	20	Inbar, 1992
Morocco		1,200	5,000	Heusch and Millies-Lacroix, 1971

The significance of erosion rates can be defined in absolute terms. For example the LUCDEME (Lucha contra la Desertificación en el Mediterráneo: Pérez Soba and Barrientos, 1986) project in Spain uses a six-fold scheme as outlined in Table 6.7. Rates greater than $1,000\,t\,km^{-2}\,a^{-1}$ are considered unacceptable (ICONA, 1982). Such an area is estimated to cover 76 per cent of Spain (Pérez Soba and Barrientos, 1986): the implications of this estimate will be discussed further below (Chapter 19). However, such an absolute definition can be misleading because replacement of the soil varies significantly according to the underlying

Table 6.7 ICONA classification of soil loss rates in the LUCDEME anti-desertification project

Erosion rate $t\,km^{-2}\,a^{-1}$	Class
<1,000	negligible
1,000–2,500	low
2,500–5,000	moderate
5,000–10,000	marked
10,000–20,000	high
>20,000	very high

Source: After Pérez Soba and Barrientos (1986).

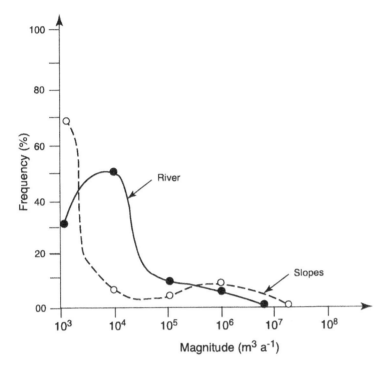

Figure 6.11 Magnitude and frequency of fluvial and slope erosion events in the Buonamico Basin, Calabria, Italy

Source: Ergenzinger (1992).

lithology (see Chapter 2). In this case, it is common to use a soil-loss tolerance or 'T-factor' (Poesen, 1995) to define the acceptability of erosion rates in a given location, although it may be difficult to measure weathering rates accurately. The T-factor may be partly misleading because the soil lost at the surface will be generally richer in nutrients than the soil replaced by weathering. In this case, an eroding soil may become progressively less able to support vegetation and thus more susceptible to erosion, ultimately reaching a catastrophic positive feedback (Thornes, 1985; 1990). The loss of nutrients with eroded soils has been demonstrated as significant by Andreu *et al.* (1994).

The measurements of erosion, if converted to ground loss, suggest the potential dynamism of the Mediterranean landscape under present climatic conditions. For example, a rate of $500\,\mathrm{t\,km^{-2}\,a^{-1}}$, if maintained for a century, would lead to a ground loss of 38.5 mm. Although this scenario might be unlikely because of the feedbacks noted above, not least because many Mediterranean soils have become thinned and enriched in stones, it does give an indication of how rapidly surfaces may evolve. Such rates should not be extrapolated over very long periods, however, because of changes in the controlling factors of erosion (e.g. Yair *et al.*, 1982; Yair, 1994), and because of limitations in the ways that such rates are calculated (Parsons *et al.*, *in press*; Wainwright *et al.*, 2001; see also Trimble and Crosson, 2000).

Controls on erosion by wind

As with other forms of erosion, we can consider aeolian erosion as the balance between erosivity – in this case, that of the wind – versus the erodibility of a surface. Patterns of wind in the

Mediterranean region have already been discussed in Chapter 3. Of the wind types generating aeolian transport noted by Pye and Tsaor (1990), thermally induced circulation can be important in the coastal regions, while the subtropical, stable anticyclonic systems are significant in the southern basin. The more localized mechanisms of thunderstorm squalls and dust devils can also occur in the region as a whole. The coastal regions have equivalent wind energies to other coastal zones in mid-latitude parts of the world (Eldridge, 1980, cited in Pye and Tsaor, 1990). A further important type of wind can be the descending orographic winds discussed in Chapter 3, particularly where they occur on the rain-shadow (typically eastern) side of mountains where lack of moisture may lead to the presence of more erodible soils. During hot foehn winds, this effect may be accentuated. The presence of aeolian landforms such as dunes is therefore a function of the strength and directional variability of the winds (Fryberger and Dean, 1979), as well as other factors controlling erodibility, which are discussed below. It must be remembered that even in the Sahara and Libyan Deserts, only 28 per cent and 22 per cent respectively of landforms are made up of dune fields, compared to 43 per cent and 39 per cent of mountains (Thomas, 1997).

The erodibility of sediments with respect to wind transport is discussed in detail elsewhere (e.g. Cooke *et al.*, 1993; Lancaster and Nickling, 1994; Lancaster, 1995; Pye and Tsaor, 1990; Thomas, 1997), so it will only be summarized here. Since the early experiments of Bagnold, inspired by his travels through the Libyan Desert, it has been recognized that grain sizes of 0.07 to 0.1 mm are most easily entrained by wind. The finest particles (<0.07 mm) may be transported in suspension over relatively long distances as dust. The significance of this process is discussed in more detail below. Particles coarser than around 0.5 mm tend only to be pushed along the surface in a creep motion. Between these two limits, motion is by saltation.

The basic erodibility due to the particle size may then be modified according to a number of criteria. These include the vegetation cover, surface slope, surface moisture and salt content, organic matter content, and the presence of surface crusts. Vegetation cover has a number of complex effects depending on the plant height, flexibility, density and presence of surface litter (Cooke *et al.*, 1993). Wasson and Naninga (1986) demonstrated that for shrub-cover densities of less than 30 per cent, sand-transport rates increased dramatically, particularly at higher wind speeds. The relationship with slope is more sensitive if the wind is blowing in the upslope direction, as it must counter an increasing proportion of the gravitational force with increasing slope angles, whereas there is a rapid decrease of energy required for winds blowing downslope. A number of studies have demonstrated that velocities required to entrain particles increase exponentially for surface moisture contents of 2 to 10 per cent, becoming stable beyond this point. A similar exponential relationship has been found for increasing salt contents. In both cases, this is due to the increased forces bonding particles at the surface together. Organic matter content at the surface can bring stability in the same way as with respect to water erosion, by helping the aggregation of the finer particles. Surface crusts may reduce wind erosion by binding the surface, as in the case of algal or bacterial mats. In extreme cases, the selective removal of finer sediments by wind or water erosion can result in an armour layer or 'desert pavement' which minimizes further movement.

The dominant areas for aeolian activity in the Mediterranean are the coastal zones and inland dunes of the Sahara, Libyan, Sinai and Negev Deserts. However, there may be locally important wind erosion elsewhere in the basin. For example, Quirantes *et al.* (1991) estimate wind-erosion rates of 429 t km^{-2}a^{-1} in the Rio Gualchos catchment of southern Spain compared to water-erosion rates of 390 to 1,560 t km^{-2}a^{-1}. Similar wind-erosion rates in two other catchments made up 26 per cent and 19 per cent of the total annual erosion.

The transport of dust over long distances across the Mediterranean has long been recognized as an important process. Correggiari *et al.* (1989) compared dusts sampled from the atmosphere at a number of locations in the central Mediterranean, with reference to the wind direction at the time of sampling. They found average dust concentrations of $56.4\,\mu g\,m^{-3}$ for S–SE winds, compared to $7.1\,\mu g\,m^{-3}$ for NW winds, $3.8\,\mu g\,m^{-3}$ from NE winds and $22.3\,\mu g\,m^{-3}$ from mixed sources. Not surprisingly, this also leads to a geographical gradient of concentrations, with an average of $56\,\mu g\,m^{-3}$ in samples from the south of their study area, compared with only $5\,\mu g\,m^{-3}$ in the north. This study and those of Tomadin *et al.* (1984, 1990) demonstrate that the dusts from different sources have distinct geochemical signatures which allow the possibility of detecting different source areas and thus deposition mechanisms. The results of these studies demonstrate the importance of deposition from atmospheric dust both of sediments (Tomadin *et al.* (1990)) and, locally, of trace metals (Correggiari *et al.*, 1989).

Pye (1992) measured dust-deposition rates on the island of Crete. The number of days with dust deposition varied from twenty-one at Sitia in the east of the island, to forty-one at Iraklion and fifty-nine at Souda in the west. Mean annual rates of dust accretion varied from 10 to $100\,t\,km^{-2}\,a^{-1}$, although rates as high as $1.42\,t\,km^{-2}\,a^{-1}$ were recorded over a two-month period between February and April 1989. Similar rates of deposition have been measured elsewhere in the eastern Mediterranean (Kubilay *et al.*, 1997). The process is important both spatially and temporally – Bücher (1989) notes 417 dust-fallout events in southern France of a sufficient size to be recorded in historical documents between 1500 and 1983, with a marked lower frequency of records between 1640 and 1800. Yaalon and Ganor (1973) recognized the importance of dust in the formation of Mediterranean soils, distinguishing between three different types. First, soils with small amounts of aeolian material, which has acted as a modifying agent. This category includes the widespread *terra rossa* soils (see Chapter 2), which a number of authors have suggested are reddened by the minerals incorporated within the dust (e.g. Jackson *et al.*, 1982, although this is disputed by some authors, for example, Boero and Schwertmann, 1989; Moresi and Mongelli, 1988). MacLeod (1980) suggested that terra rossa soils in Epirus must have had significant aeolian deposition to produce observed thicknesses in the time available. Pye (1992) suggested that the lower rate of deposition in Crete $(10\,t\,km^{-2}\,a^{-1})$ would lead to the accumulation of 0.77 m of dust in 100,000 years. Second, Yaalon and Ganor define soils with a significant amount of aeolian input, which considerably affects the processes of pedogenesis. In this category, they include coastal plain soils which are often interbedded with large amounts of sand from coastal dunes, as well as soils with well-defined calcrete horizons. In the latter case, calcareous dust allows the formations of such horizons in locations where the underlying bedrock is not itself calcareous, and has been shown to be important in areas of Israel and North Africa. Third, they define soils with a surficial cover of aeolian material. Continued and contemporaneous deposition and pedogenesis leads to the intercalation of loess with palaeosol horizons. Again, these are important in areas of Israel, as well as Tunisia and Morocco (Thomas, 1997).

Issar *et al.* (1989) note that although loess deposits are common in the Negev, modern dust accumulation rates range from 26 to $108\,t\,km^{-2}\,a^{-1}$ – in other words, very similar to those recorded by Pye in Crete – which are again well below the threshold deposition rate of $325\,t\,km^{-2}\,a^{-1}$ that would be required for loess to be actively forming (Pye, 1992). This fact is confirmed by radiocarbon dates of *c.*13500 BCE, *c.*24950 BCE and *c.*37650 BCE on carbonate horizons in the upper three of seven palaeosols contained within the loess. These authors link the formation of the Israeli loesses to northward fluctuations in the past location of the ITCZ (see Chapter 2). In many parts of the Negev, the loess has been covered by encroaching dune sands during the Holocene (Issar *et al.*, 1989; Yair, 1994). Analysis of heavy minerals suggests that these sands relate to the extension of the Sinai Desert rather than from the coastal dune areas. In the latter case, the sands are dominantly made up of sand from the Nile brought to Israel by longshore drift (Issar *et al.*, 1989).

The study above demonstrates that different mechanisms of erosion may have been important at different times in the past. This situation may particularly have been the case during the cold, dry stages which dominated large parts of the Pleistocene climate of the Basin. A further example is given by Ambert (1974), who discusses the importance of aeolian hollows in forming small lakes in southern France in the late Pleistocene, which have subsequently provided useful palaeoecological data.

Suggestions for further reading

Morgan's book (1995) remains the best introduction to soil-erosion process from an applied perspective, while Selby (1993) provides a thorough overview on slope processes in general. The review by Poesen and Hooke (1997) on erosion rates in the region is a very useful overview.

Topics for discussion

1   What are the critical controls on erosion rates in the Mediterranean?
2   Under what conditions will erosion proceed slowly and what thresholds need to be crossed to cause catastrophic erosion?
3   How do human impacts affect the erosivity and erodibility of soils?

7   Fire

Introduction

Often seen simply as a destructive force, the role of fire in the Mediterranean landscape has been reappraised in recent years. The destruction is seen in monetary terms with respect to the prevention of loss to land, property and valuable forests, the costs of fire fighting, and of rehabilitation of the landscape after a fire has taken place (Le Houérou, 1987). However, work on the ecological impacts of fire since that of Naveh (1975) has demonstrated the stability and diversity of Mediterranean ecosystems. This chapter focuses on the controls on forest fires followed by their impacts on vegetation, and the movement of water and sediments through the landscape. Issues relating to more recent impacts are dealt with in a later chapter.

Controls on forest fires

Naveh (1975) pointed out the long history of the use of fire in the later Palaeolithic of Israel and Greece up to 80 ka, although there is evidence for the organized use of fire as early as 380 ka at Terra Amata in southern France (de Lumley and Boone, 1976). However, it is extremely difficult to demonstrate a clear link between the knowledge of fire and its use in the landscape. At this point in history, it was far more likely to be generated by lightning strikes – indeed, up to 4 per cent of wildfires at the present time are still generated by this mechanism (Naveh, 1975; Vafeidis, 2001) – and locally by volcanic activity (Sauer, 1961, cited in Naveh, 1975; Trabaud, 1994). The generation and maintenance of wildfires in dense, dry woody vegetation led Naveh to propose the existence of Mediterranean 'fire bioclimates' – namely those with 'long, hot and dry summer seasons with maximum average daily temperatures around 30°C, average relative humidities of 50 to 60 per cent and frequent heat waves of "sharav" or "sirocco" in the beginning and end of the dry season, when temperatures rise above 40°C and relative humidities drop below 30 per cent'. In this way, the frequency of occurrence of fires (or indeed the likelihood that fires will spread once started) can be seen to relate to both local climatic conditions – dominantly the length of time with insignificant rainfall – and more regional controls on hot, dry winds and humidity levels in general. Microclimatic effects may also be important in controlling humidity levels, for example, beneath a forest canopy.

However, the climatic control is not the only one that is significant. The structure and composition of the canopy, understorey, litter layer and soil may determine the nature of a fire and the mechanism by which it spreads. Common types of fire include canopy fires, ground fires and soil fires. Of course in any actual fire, more than one of these forms may occur at different points in space or time. The mechanism of spread is significant in that it can control which elements of the vegetation are most affected by a fire (Díaz-Fierros et al., 1994). Certain plant species, such as conifers and some shrubs, contain inflammable oils which make them more susceptible to burning (Clark, 1996). Margaris *et al.* (1996) note the importance of biomass accumulation in the ecosystem in controlling the intensities of fires. Very high temperature fires as a result of large biomass accumulations can slow the post-fire recovery of the ecosystem by destroying seed banks. Thus, in regions where the climate increases the susceptibility towards burning, frequent, small fires may be beneficial to the vegetation community as a whole. This has led some authors to suggest that Mediterranean ecosystems are 'perturbation-dependent' systems (e.g. Naveh, 1994). Of course, in most cases at the present time, human activity is the dominant cause of this perturbation (Métaillie, 1981; Le Houérou, 1987; Amouric, 1992).

Impacts on vegetation patterns

Trabaud (1994) notes that fire leads to rejuvenation of certain stands of vegetation, and to the development of mosaics of plant communities. Thus, the action of fire promotes diversity in space, time and species composition of Mediterranean vegetation. Numerous authors have noted the selection of fire-tolerance in numerous Mediterranean species as a result of repeated exposure to the action of fire. Naveh (1975) suggests that this selection occurs by one of two mechanisms. First, a positive feedback mechanism favours those plants which increase their physiological activity after the occurrence of fire. Such activities include resprouting from roots, suckers or stumps (as with *Quercus ilex* or *Q. coccifera*), fire-stimulated seed germination (as with *Pinus halepensis*) and post-fire flowering and seed germination. Certain matorral shrubs have evolved more than one method – for example, *Thymus*, *Rosmarinus* and *Cistus* can regenerate either by resprouting or by fire-stimulated seed germination (Naveh, 1975). Le Houérou (1973) calls the species which use this mechanism 'active pyrophytes'. Second, a negative feedback mechanism favours plants which avoid fires. This mechanism includes those plants which have a direct tolerance, for example, by having extremely thick bark (*Q. suber* or *Q. ithaburensis*) or resistant seeds or basal organs (for example, certain types of grass: Naveh, 1975). These species are termed 'passive pyrophytes' by Le Houérou. It has been suggested that within Mediterranean ecosystems, repeated burning may favour evergreen rather than deciduous oak species (Naveh, 1975; Barbero *et al.*, 1990), although Métaillie (1981) notes that *Q. pubescens* can behave as a passive phyrophyte where its bark is sufficiently thick, and can resprout from the stump as with the evergreen oaks.

Numerous studies have been carried out on the post-fire regeneration of vegetation. Trabaud (1994) gives a summary of recent findings. He suggests that there is a commonly recurring pattern. Immediately following the fire, few species remain, but between around one to three years, there is a sudden increase of species represented (Figure 7.1). These species often include rapidly growing annuals that are not typically found in the ecosystem, their growth being encouraged by the opening of the vegetation cover, destruction of litter, and the post-fire enrichment of soil nutrients. After this period, many of these species disappear as the original vegetation becomes dominant. As this occurs, the structure of the vegetation becomes more complex, with a greater number of layers present. This feature can clearly be seen in the study of Mansanet Terol (1987) in Alicante, Spain, where the heights of regrown plants after a period of twelve years ranged from 65 cm to 155 cm (Figure 7.2). In terms of biomass present, there is also a rapid growth in the first years after a fire, followed by a general slowing of the rate of growth, typically following a logistic growth pattern. For example, Specht (1969) found that biomass on a burned garrigue in southern France was $374 \, \text{g m}^{-2}$ after one year, and $874 \, \text{g m}^{-2}$ after three years. In *Sarcopoterium spinosum* phrygana in Greece, Papanastasis (1977) found biomasses of $120 \, \text{g m}^{-2}$ in the first year, $160 \, \text{g m}^{-2}$ in the second and $280 \, \text{g m}^{-2}$ in the third, while Arianoutsou-Faraggitaki (1984) found rapid increases in such an ecosystem,

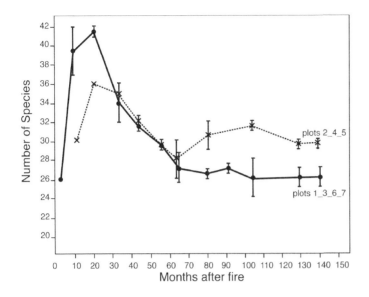

Figure 7.1 Changes in the richness of species present in a dense *Quercus ilex* coppice after fire
Source: After Trabaud (1994).

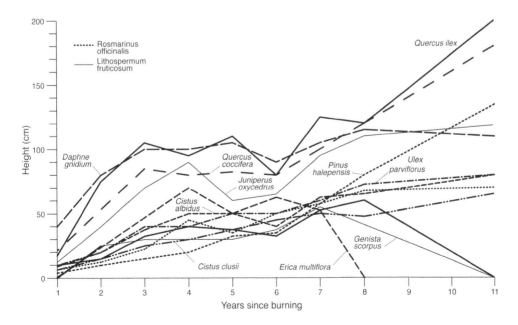

Figure 7.2 Evolution of plant height through time following forest fires in the region of Alicante, Spain
Source: Mansanet Terol (1987).

with $500\,g\,m^{-2}$ after five years. As a summary, these results confirm Naveh's (1975: 206) conclusion that: 'Contrary to the present view of fire as simply destructive,... fire ... [has] favoured genetical as well as ecological diversity.'

Effects on soil properties

The impacts of burning on Mediterranean soils have been intensively studied by Giovannini and his colleagues. Their initial studies involved deliberate burning of a 100-m² plot under matorral in Sardinia and measuring a variety of soil properties over a three-year period (Giovannini and Lucchesi, 1983; Giovannini *et al.*, 1987). They found that both organic matter and aggregate stability decreased markedly in the week following the fire in the surface horizon of both soil profiles measured (Figure 7.3). After three years, the organic matter had recovered to its pre-fire level, whereas the aggregate stability recovered after two years. Some of the lost organic matter was translocated into the B horizon where it remained after three years. A similar increase in aggregate stability in the lower horizons was interpreted as being due to the movement of organometallic cementing agents. This movement meant that the surface horizon of the soils lost its slightly hydrophobic character in the period immediately following the fire, and this property was transferred to the horizon about 10 cm below the surface. Few changes were noted beyond about 20 cm below the surface, although the surface temperature during the fire was probably between 650 and 850°C.

Because different fires can result in a wide range of surface temperatures, according to the rate of spread and the amount of available biomass as fuel, Giovannini *et al.* (1988; 1990) carried out a series of experiments subjecting soils to different temperatures. Both the silty clay and the sandy loam soil behaved similarly, with slight decreases in organic matter content,

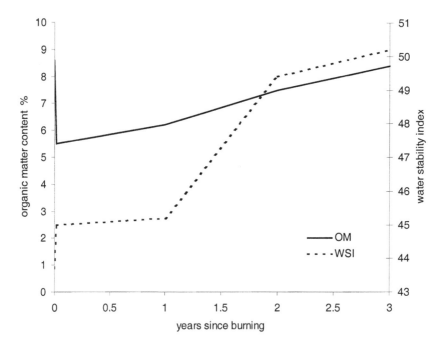

Figure 7.3 Evolution of soil organic matter and aggregate stability in a burnt soil

Source: After Giovannini *et al.* (1987).

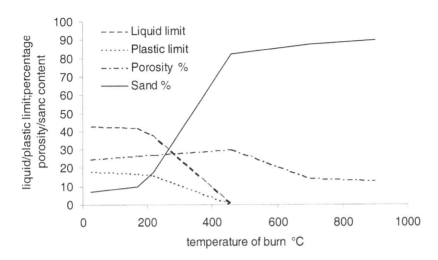

Figure 7.4 Relationship between fire temperature and various soil properties according to the experimental results of Giovannini *et al.* (1988)

liquid limit and plastic limit at temperatures up to 170°C (Figure 7.4). By 220°C, there was a greater decrease in these properties, and by 460°C there was very little organic matter remaining, and the soils had entirely lost these plastic and elastic properties. Porosity increased up to a temperature of 460°C in the silty clay, and then decreased to about half its initial value, whereas the sandy loam constantly decreased, albeit with a large jump between 220°C and 460°C. The aggregate stability increased constantly in both cases, in parallel with the increasingly sandy texture. This change is suggested as being due to the fusion of clay-sized particles into sand-sized particles. These results demonstrate that the response of soils to fires can be complex in terms of the characteristics which affect soil hydrology and erodibility. Nevertheless, Giovannini *et al.* (1988) define two important thresholds: 460°C, beyond which all organic matter is combusted, and 700°C, when the structure of clay minerals in the soils changes due to the loss of hydroxide cations from the clay lattices. The observations of Imeson *et al.* (1992) and Lavee *et al.* (1995) suggest that changes over this range of temperatures can occur at very small spatial scales in forest fires, thus leading to a great heterogeneity in surface response to fires.

Experiments by Giovannini *et al.* (1990) showed a range of responses in terms of soil nutrients according to temperature. Although total nitrogen decreased with increasing temperature, quite dramatically so between 220°C and 460°C, NH_4-N increased up to 220°C before decreasing again (Figure 7.5). Organic phosphorous decreased constantly, while inorganic P increased to reach a constant value above 460°C, while plant-available P increased to 460°C, before decreasing again. Potassium, on the other hand, increased to 700°C, before decreasing rapidly. Calcium, magnesium and sodium showed a variety of complex responses. The pH increased to become highly alkaline at temperatures of 700°C and greater, while the cation-exchange capacity showed a constant decrease. They used the burnt soils to grow wheat in controlled conditions. Biomass (shoot and root), plant height and root extension were highest in those soils heated to 220°C and 460°C, and lowest in those soils heated to 700°C or more (Figure 7.6). Giovannini *et al.* related these results to the increase of NH_4-N at 220°C and of available P at 460°C. At higher temperatures, the plants suffered disease and nutritional disorders because of the highly alkaline conditions and the toxic levels of calcium and potassium ions.

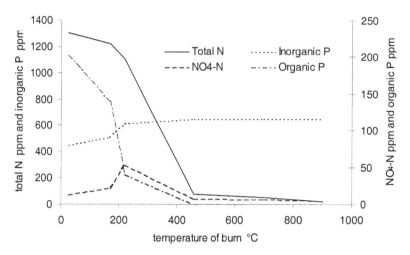

Figure 7.5 Relationship between fire temperature and soil nutrients according to the experimental results of Giovannini *et al.* (1990)

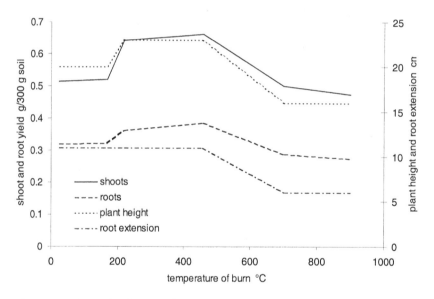

Figure 7.6 Relationship between fire temperature and subsequent growth of wheat according to the experimental results of Giovannini *et al.* (1990)

Note
The results refer to the same soil as shown in Figure 7.5.

Effects on hydrology and erosion

A large number of studies have been carried out on the effects of fire on water and sediment yield in the Mediterranean. Table 7.1 outlines the results from monitoring experimental plots. Most of these results demonstrate an increase of at least an order of magnitude in both runoff and sediment yield (e.g. Vega and Díaz-Fierros, 1987). The only exception to this pattern is the study of Kutiel and Inbar (1993), which was carried out in an aleppo and brutia pine plantation in Israel. The fire which occurred was relatively light, lasted for a short duration, and only affected the understorey without affecting the canopy. This result is consistent with those of Giovannini *et al.* (1988), which suggested that there could be a small increase in the infiltration rate following a low-temperature fire. The study of Kutiel and Inbar also noted a marked increase in available nutrients in the surface soil during the first four months following the fire. Sanchez *et al.* (1994) also found a significant increase in organic matter and nutrients immediately after a fire on matorral plots in Alicante. By six months after the fire, these values were beginning to decrease again. The study by Soler *et al.* (1994) in holm oak forest in Catalonia also showed an increase in solutes in the runoff coming from a burnt plot compared to both unburnt and cleared plots (see also Soler and Sala, 1992), as did that of Díaz-Fierros *et al.* (1990) on a *Ulex europaea* matorral. The results of Giovannini *et al.* (1990) discussed above suggest the role of fire in producing nutrients is by recycling material held within the vegetation biomass, rather than by releasing material incorporated within the soils. These results also suggest that these nutrients could be lost in runoff and erosion events, making the year immediately after the fire a very sensitive one within Mediterranean ecosystems. Field observations also suggest that this immediate post-fire period is one where aeolian activity can be important in transporting the nutrients away from sites in the form of ash.

These results have also been supported by a series of experimental studies. Imeson *et al.* (1992) carried out 150 rainfall-simulation experiments in Mediterranean oak forests in Catalonia. Their results showed that the lowest final infiltration rates occurred on the burnt plots, with an average value of $17.1\,\mathrm{mm\,h^{-1}}$, compared to $36.7\,\mathrm{mm\,h^{-1}}$ on forest plots and $22\,\mathrm{mm\,h^{-1}}$ on cultivated plots in the same area. They also noted that hydrophobic materials were present in both burnt and unburnt sites, mainly composed of fresh or decaying leaves. The significant difference leading to the difference in infiltration rates seemed to be the exposure of the hydrophobic layer at the surface due to the removal of the litter layer in the fire, which also lessened the water-holding capacity of the surface. However, these authors also point to the importance of macropores which can locally bypass this hydrophobic layer. Kutiel *et al.* (1995) showed a continued decrease in final infiltration rate from $29.3\,\mathrm{mm\,h^{-1}}$ immediately after a fire on a *Pinus halepensis–Quercus calliprinos* forest, to $28.8\,\mathrm{mm\,h^{-1}}$ after two weeks and $23.5\,\mathrm{mm\,h^{-1}}$ after a year. These values compare to a rate of $31.2\,\mathrm{mm\,h^{-1}}$ for unburnt plots. There was a corresponding but more marked increase in erosion rates – the unburnt site averaged $13.5\,\mathrm{g\,m^{-2}}$ in the simulated event, compared to $32\,\mathrm{g\,m^{-2}}$ immediately following the fire, $40.8\,\mathrm{g\,m^{-2}}$ two weeks later, and $39.3\,\mathrm{g\,m^{-2}}$ on the burnt sites. Although there was an increase in erosion in this study, it was not as marked as with other studies. This can probably again be explained with reference to the experimental studies of Giovannini *et al.* (1988), because the fire temperatures were only 200 to 400°C in this case. In a further paper, Lavee *et al.* (1995) also point to the effect of variability caused by heterogeneity within the fire. The development of rough, unburnt patches means that there are areas where little runoff is generated close to areas where large amounts of runoff are generated. In some cases, the runoff generated from the former may be reabsorbed in the former.

Summary

The range of studies discussed suggest the importance of fire in Mediterranean ecosystems. In terms of vegetation, it has been suggested that fire adds to genetic diversity and thus stability of ecosystems, and has played a significant role in the development of the species groupings that we observe in the present landscape. This effect is due to the development of various fire-tolerance strategies, which has been accentuated in the more recent past by the deliberate use of fire by humans to modify the landscape.

Table 7.1 Monitored rates of runoff and erosion before and after fires in a range of locations in the Mediterranean

Location	Land use	Annual rainfall mm	Annual runoff mm	Runoff coeff. (%)	Erosion rate $t\,km^{-2}\,a^{-1}$	Solute load $t\,km^{-2}\,a^{-1}$	Organic matter $t\,km^{-2}\,a^{-1}$	Measurement period	Reference
Barrosa, Portugal	mature pine forest	1,156.8 1,231.8	1.6 –	0.1 –	0.7 3.3			Oct 1989–Sept 1990 Mar 1989–Aug 1991	Terry, 1994 Shakesby et al., 1994
Barrosa, Portugal	pine seedling regeneration after 1986 fire	1,172.6 1,240.0	42.8 –	3.7 –	4.9 6.5			Oct 1989–Sept 1990 Mar 1989–Aug 1991	Terry, 1994 Shakesby et al., 1994
Barrosa, Portugal	rip-ploughed and planted with eucalyptus 1–2 years previously	1,172.6 1,240.0	195.4 –	16.7 –	449.1 358.5			Oct 1989–Sept 1990 Mar 1989–Aug 1991	Terry, 1994 Shakesby et al., 1994
Lousa, Portugal	burnt and cleared pine	1,156.8 1,231.8	125.3 –	9.2 –	118.3 110.3			Oct 1989–Sept 1990 Mar 1989–Aug 1991	Terry, 1994 Shakesby et al., 1994
Serra de Cima, Portugal	pine seedling regen. after 1986 fire with natural *Eucalyptus* encroachment	1,172.6 1,240.0	47.0 –	4.0 –	12.0 10.1			Oct 1989–Sept 1990 Mar 1989–Aug 1991	Terry, 1994 Shakesby et al., 1994
Serra de Cima, Portugal	*Eucalyptus* stump regrowth	1,172.6 1,240.0	136.8 –	11.7 –	1.5 2.1			Oct 1989–Sept 1990 Mar 1989–Aug 1991	Terry, 1994 Shakesby et al., 1994
Serra de Cima, Portugal	rip-ploughed and planted with *Eucalyptus* <1yr previously	1,172.6 1,240.0	316.9 –	27.0 –	4,146.7 2,100.9			Oct 1989–Sept 1990 Mar 1989–Aug 1991	Terry, 1994 Shakesby et al., 1994

Location	Description							Period	Reference
Pousadas, Portugal	*Eucalyptus* plantation since 1980	1,065.2	18.4	1.7	3.0			Oct 1989–Sept 1990	Terry, 1994
		1,231.8	–	–	7.4			Mar 1989–Aug 1991	Shakesby *et al.*, 1994
Prades, Spain	burnt oak woodland	441	1.3	–	28.3	0.2	6.5	Oct 1988–Mar 1991	Soler *et al.*, 1994
Prades, Spain	felled oak woodland	441	0.2	–	3.4	0.02	1.1	Oct 1988–Mar 1991	Soler *et al.*, 1994
Prades, Spain	oak woodland (control)	441	0.1	–	1.5	0.01	0.2	Oct 1988–Mar 1991	Soler *et al.*, 1994
Galicia, Spain	*Ulex europaea* shrubland – unburnt	1,400			56.7			Sept 1988 (four months only)	Díaz Fierros *et al.*, 1990
	– burnt				81.6				
Galicia, Spain	means of 29 plots including pine and eucalyptus forests, heather and grass	1,400			1,450			2 months	Díaz Fierros *et al.*, 1987
					192.7			6 months	
					101.0			12 months	
					184.0			>12 months	
Yoqneam, Israel	mixed aleppo and brutia pine forest – unburnt	331*	0.3		0.1			Oct 1988–	Kutiel and Inbar, 1993
	– burnt	331*	0.2		0.04			*(163 days only)	

Moderate fires can have beneficial effects, in some cases by reducing the erodibility of soils and by increasing the porosity and infiltration rates. The nutrients released during moderate-to high-temperature fires can also increase the fertility of soils, again promoting plant regrowth. Higher-temperature fires can have significant effects in accelerating runoff, soil erosion and nutrient removal. These effects are particularly important in the period of up to about three years following the fire, when the soils have not had a chance to recover, and the protective cover of vegetation has yet to grow back. Ballais *et al.* (1992) demonstrated this point clearly with the example of the erosion which followed the Montagne Sainte-Victoire fire of August 1989. Rates of deposition increased rapidly in the year following the fire by a factor of about 650, but decreased back to the pre-fire conditions in the following year due to the regrowth of grasses and *Quercus coccifera*. However, in September 1991, there was a high-intensity storm which caused a second phase of rapid erosion from the massif. The stability of Mediterranean ecosystems with respect to frequent, low-intensity fires should not therefore be underestimated. Correspondingly, great damage can be caused by infrequent, high-intensity fires.

Suggestions for further reading

The best general introductions to the effects of fire in the region are by Naveh (1975), Le Houérou (1987) and Trabaud (1994). The work of Giovannini has been fundamental in understanding the impacts of different temperatures.

Topics for discussion

1   Define the negative and positive impacts of fire in Mediterranean ecosystems.
2   How different might Mediterranean vegetation patterns look if they were not affected by intermittent burning?
3   What is the length of time over which fire affects surface characteristics? How might this vary and what are the principal controls of the variation?

8   The Mediterranean Sea

Main issues

Because of the ubiquitous, continuous and high rates of growth of tourism, petroleum transport and advances in agricultural technology, there are genuine fears that the Mediterranean Sea, as an inland sea, cannot escape pollution on a large scale. In fact, as the evidence shows, these fears are probably wildly exaggerated, in part due to the significant improvements in prevention over the past twenty-five years (see Chapter 17). Nevertheless, the precautionary principle prevails: action taken now may offset severe and irreversible problems later. A series of protocols over the years have worked to prevent the balance from tipping to disaster. The 1953 London Convention sought to prevent pollution of sea water by oil spills. In 1973 this was extended to forbid the discharge of hydrocarbons 50 miles from the shore. In 1976 the Barcelona Convention produced three legal protocols:

- to prevent pollution by immersion operations of ships and aircraft;
- to encourage co-operation between Mediterranean countries (north and south, east and west) in the event of a critical situation (e.g. a large oil spill);
- to protect the Mediterranean against land-based pollution.

So strong are these intentions that Albaiges *et al.* (1984) described the Mediterranean as 'one of the world's most protected seas' (in writing at least). There are, nevertheless, some serious 'hotspots', which are covered in detail in Chapter 17.

The real reasons for understanding the behaviour of the Mediterranean Sea are its local and global significance. For centuries it has been the main axis of commercial and cultural exchange for the people on its shores. The civilizations that evolved on its shores have shaped the history of humankind.

Today, the heat balance and changing water salinity in the Mediterranean might even affect the flow patterns and thus warmth of the North Atlantic and hence ultimately the risk of triggering climate change (Johnson, 1997). But the sea is also the source of the weather that affects and afflicts the surrounding land masses at a scale below that of the general circulation. The autumn depressions in the Gulf of Genoa are essentially located where the polar jet stream, guided by the Rhône depression, is forced east towards northern Italy and interacts with the sub-tropical jet to generate strong vorticity and cyclonicity. The real damage created by these cyclones occurs when they move on to the Mediterranean coast of Spain and are forced upwards by the Baetic Mountains. As the global circulation changes (reflected by changes in the Atlantic atmospheric pressures, the Atlantic oscillations), the incidence of severe storms in autumnal Spain is also expected to change. It is also thought that the prevailing conditions in the Mediterranean Sea affect the timing and intensity of the Asian Monsoon. As the climate changes caused by global warming come into play over the next sixty years, because of the changes of the heat energy budgets, the salinity and circulation patterns of the Mediterranean Sea are also expected to change.

All oceans are complicated and complex. They are complicated in that their behaviour involves many interacting components: the bathymetry, the atmosphere, the variations in inputs and outputs. They are complex in that small changes can produce very substantial responses through the actions of positive feedbacks. They are complex also in that the changes are very difficult to predict numerically and the outcomes are highly uncertain. To this complexity are added the complications that arise from the huge size of the phenomena and the inputs from the Atlantic, from the rivers draining into the sea and their transported loads and not least from the precipitation and evaporation that determine the budgets of water mass, salinity and energy that drive the circulation at many different scales. Finally, we should add to these the complications of the nutrient budgets and the food chains that start with planktonic life and end up with the tuna, anchovies and bream that find their way to the markets of northern Europe.

The knowledge of this complicated and complex ocean has been greatly enhanced in the last half century by a series of major research projects supported by national and international funding. The Physical Oceanography of the Eastern Mediterranean Project (POEM: Rother *et al.*, 1996) and the Western Mediterranean Circulation Experiment Consortium Project have vastly improved the detailed knowledge of both the horizontal and vertical variations and the effects of the strong seasonal variability.

The morphology

Comprising $2.54 \, M \, km^2$, the sea is divided into a series of basins, the main being the western and eastern basins separated by the Italian peninsula and the broad, shallow sill of the Straits of Sicily. Both main basins have a number of smaller, regional basins, each with its own topography and entrances and exits. These include the Adriatic, Aegean, Levantine and Ionian Seas in the east Mediterranean and the Balearic Basin, Alboran, Ligurian Sea and Tyrrhenian Seas in the western Mediterranean (Figure 8.1). This morphology emerged from the complex interactions of the European, African and Middle East plates formed around the Tethys Ocean after the break-up of Pangaea (Chapter 2).

Effects of circulation on temperature and salinity

In general, because the sea lies between 30–40°N, there is a mean westerly circulation and local variations in weather types. The variability is less in summer, when land and sea breezes predominate. The hot dry summers lead to strong open water evaporation, estimated to be over $1 \, m \, a^{-1}$ in the western Mediterranean Basin. Bethoux (1980) estimated the flows of water into and out of the Mediterranean at Gibraltar. The input from the Atlantic was estimated to be $53,000 \, km^3 \, a^{-1}$. The Atlantic water is fresher (not so saline) and lighter, so it flows through the Straits on top of the denser out-flowing Mediterranean water. The outflow was estimated at $50,500 \, km^3 \, a^{-1}$, the difference being the water lost to evaporation. Evaporation is affected by wind as well as sunshine, so the high rates continue throughout the winter months.

Because of the strong 'twin tub' configuration and variations in temperature, densities and salinities of the water, the waters in the Mediterranean are divided into several water-mass types (analogous to the air masses used by climatologists). The incoming Atlantic water is called Modified Atlantic Water (MAW). This is a layer of relatively low density, between 100 and 200 m thick that can be traced eastwards right across the Sicilian sill to the Levant coast-

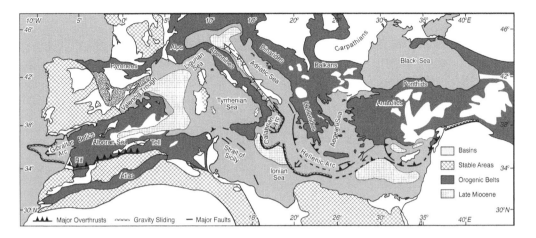

Figure 8.1 Morphology of the Mediterranean Basin

line. As the Atlantic water moves eastwards, it is heated and mixes with the saltier underlying Mediterranean water and the fresh waters coming from Europe and Africa.

During winter, mainly in the eastern Mediterranean Basin, there occurs an intermediate layer of relatively high temperature and salinity, called the Levantine Intermediate Water (LIW). It has the highest salinities in the Mediterranean Sea (between 38.8 and 39.1 ppt). In both basins, below the other layers is the Mediterranean Deep Water (DW) as shown in Figure 8.2(a) and (b). In the west Mediterranean Basin, this has lower temperature and salinity. In the west, the deep water (DW) is generally moving westwards, but in the east it moves from the Sicilian sill towards the Levantine coasts at depths exceeding 1,500 m.

The flow to and from the Atlantic is controlled mainly by the differences in water conditions on either side of the Gibraltar sill. The Straits are about 60 km long, 12 km wide at Punto Cires and approximately 300 m deep at the minimum sill. It is one of the most measured straits in the world, but the flow is complex and difficult to measure, except in special marine projects. As a consequence, a long time series of flows is not available.

The flow of water at known temperature can also be used to check the heat budget of the whole Mediterranean Sea. In a study of the heat flux through the Straits of Gibraltar, a net gain of heat is indicated, mainly due to temporal variations in temperature and velocity, because heat coming from the bed is negligible compared with other fluxes (MacDonald *et al.*, 1995). The largest contribution to this gain is probably generated in the upper 100 m of the water column since both velocity and temperature increase towards the water surface.

According to Manzella (1995), salinity, current and depth of upper water stream show strongly seasonal variations in the Straits of Sicily. The Straits are wider and shallower than at Gibraltar, about 200 m deep between Sicily and Tunisia, except for two narrow channels about 400 m deep between Sicily and Malta and Malta and Tunisia. The Mediterranean Atlantic Water is coming from the west and the more saline Levantine Intermediate Water flows westward underneath, in the relatively deep channel between Sicily and Libya. The west to east flux is 0.65–0.85 Sv (Sverdrup units, named after the Swedish oceanographer: $1\,Sv = 1 \times 10^{6}\,m^{3}\,s^{-1}$); from east to west is 0.6–0.8 Sv. This is not the true balance it appears to be, for there are strong variations in the salinity of the Atlantic Water, whose salinity increases as it moves through the Straits. The water in the eastern basin is generally more saline than in the

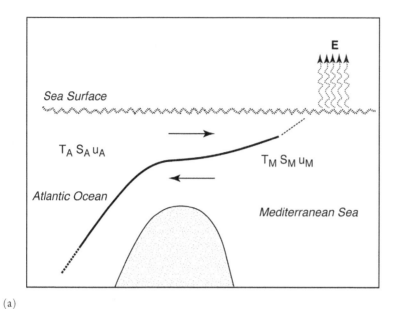

(a)

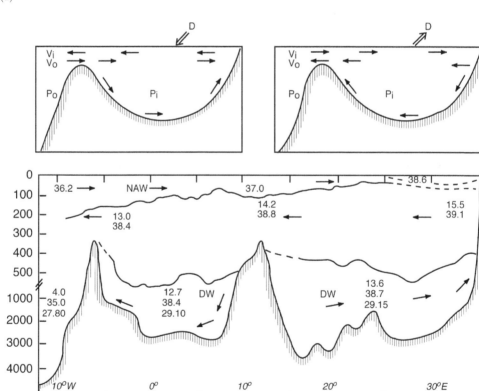

(b)

Figure 8.2 (a) Vertical structures of water types passing through the Straits of Gibraltar. (b) Cross-section of the Mediterranean showing the vertical structure parallel to the main flows in the West–East direction

west and the Levantine Intermediate Water flowing west carries more salt into the western Mediterranean.

Bethoux and Gentili (1994) also report that, over the past thirty years, temperature and salinity appear to have increased in the deep water in the western basin, suggesting that perhaps recent climatic changes have been responsible for changes in the heat and water budget through the surface. In particular it is hypothesized that this might be evidence of a decrease in rainfall. This issue is a hotly debated, but evidence from rainfall gauges also seems to support this hypothesis. The discovery, in the eastern basin, of deep organic muds named sapropels are called upon to link deep water formation with global climatic change and tropical climate oscillations (Cita *et al.*, 1991) and this links with the modern knowledge of the MEDALUS Project (Box 4.4) that the oscillations in rainfall in the Mediterranean land regions are coupled to the oscillations of the pressure conditions in the North Atlantic (see Chapter 3). These relationships are quite significant because they show that Mediterranean oscillations of climate can be coupled to global climate change. The sapropel deposits also show that there must have been interruptions in the deep water conditions in the Adriatic and Cretan Seas over the past 5,000 years.

The organic mud deposits record eleven separate changes in biological production and/or circulation over the last *c.*400 ka. These fluctuations occurred during glacial as well as temperate climates. They would require either a four- or fivefold increase in the present rate of plankton production or a reversal of flows over the Gibraltar sill, both considered by Bethoux and Gentili as 'inconceivable'. Recent explanations have concentrated on density differences between the inflowing Atlantic waters and the deep waters leaving the Mediterranean over the Gibraltar Sill, the onset of sapropels relating to less dense inflowing waters, and thus coupling with the global circulation system. Other explanations have focused on high flows in the Nile basin affecting density gradients and circulation patterns in the eastern part of the basin.

Attempts have been made to consider the impact of glacial maxima on the entire Mediterranean, using models. The most probable scenario is that solar radiation was reduced by 1 per cent, water temperatures by about 5 per cent and that humidity and cloudiness increased. Water loss and evaporation would have been dramatically reduced. Precipitation and runoff were almost certainly lower than that coming from fresh water today. Consequently, in a glacial Mediterranean, as well as the present one, evaporation losses linked to the heat budget remain much greater than fresh water gain. With a water deficit of about $1\,m\,a^{-1}$, the basin water is concentrating. Added to this, in glacial periods, the lower sea level changed the geometry of the Straits and the Atlantic salinity was greater. The sapropel deposits correlate well with the African summer monsoons, suggesting the Nile was a key component in their production.

Bethoux and Gentili also reported the budget for the western basin. Inputs and outputs are not balanced. The input is mainly rainfall and the inflow at Gibraltar and Sicily. The losses are the outflows at these two places and evaporation. Because the budget is negative (about $-1.15\,Sv$ in 1960–1961 and about $-0.8\,Sv$ in 1985–1986) it implies that the western basin is becoming more saline and that there is a high transfer of water and energy to the atmosphere. There is an indication of a warming trend in the period 1967–1987 of about $0.0027°C\,a^{-1}$. Another study examined data from 1909 to 1990 and found clear evidence of increases in temperature, salinity and density. These increases have been linked to the eastern basin's fresh water inputs, resulting mainly through damming of major in-flowing rivers. The closing of the Aswan High Dam in 1964 led to an almost total suppression of Nile discharges (Agnew and Anderson, 1992). Without the Aswan Dam, there would be 2 to 4 cm more fresh water over the eastern basin. There have also been changes of discharges into (and therefore out of) the Black Sea resulting from human activities. The average yearly discharge of the Nile from 1912 to 1942 was about $6.2 \times 10^{10}\,m^3$, but since 1969 it has only been about $0.4 \times 10^{10}\,m^3$. Nevertheless, calculations show that the Nile effect cannot explain the salinity changes in the western basin. The draining of rivers in the Black Sea watershed from 1947 to 1985 diverted $4.5 \times 10^{10}\,m^3\,a^{-1}$ from the eastern basin and this would have doubled the effect of the Nile on salinity.

Local circulation

We have so far discussed the effects of the topography and the prevailing westerly air currents on the conditions of temperature and salinity. However, just as the atmosphere can be generalized in air mass terms, hiding the details of local storms, so the broad classification of the Mediterranean can mask the locally strong circulation in the sea. The density variations and barometric gradients force the development of local currents that can be damped by the vertical structure. Because of the complexity of these motions, they are summarized by averages and studies by computer-simulation models that attempt to solve the partial differential equations of water movement subject to the complex bathymetry.

As the North Atlantic Water comes in through the Straits of Gibraltar, it is exposed to a large anticyclonic gyre occupying the Alboran Sea. The flow is strongest on the Spanish side, moving at almost $1\,m\,s^{-1}$ eastwards (Figure 8.3). The centre part is deeper (at approximately 5,220 m) and perimeter parts are shallower. This configuration leads to some up-welling on the Spanish coast. Near the Moroccan coast, at Centra, the gyre is strongly curved. Hopkins (1999) estimated that 65 per cent of the water coming from the Atlantic is involved in the Alboran gyre. Water escaping the gyre moves off towards the north-east, passing on the main-

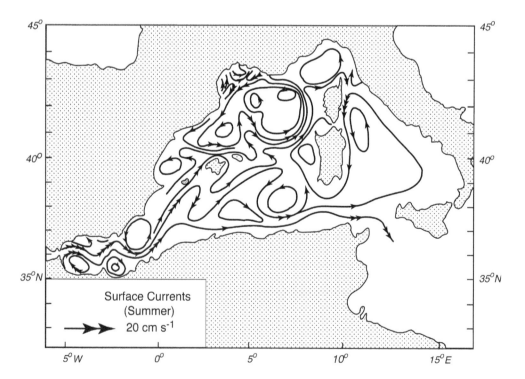

Figure 8.3 Map of the Alboran gyre

land side of the Balearic Islands, or moves along the North African coast. The remaining water moves north-west of Sardinia to join the complex circulation in the Gulf of Lions. This circulation is essential in maintaining the Mediterranean Sea conditions in steady state.

The living Mediterranean

When the Mediterranean is viewed from space, satellite imagery can easily detect plankton blooms on the sea surface (López García, 1991). Plankton live in the photic zone down to about 100 m deep in the summer. Although the surface of the Mediterranean receives about 1.5 m kcal m^2 a^{-1}, only about 44 per cent of this is used in primary production in the western basin (Margalef, 1984): 'The water is full of light, but the organisms can hardly use it.' Phytoplankton are basically eaten by zooplankton and so starts the food chain. There are few pelagic (surface) fish and those that exist are rather small, such as sardines and anchovies. By contrast, estuaries, lagoons and brackish and hypersaline environments are more biochemically active than the open sea. These are places where nutrients from the land come to the sea in fresh river water. There are few up-wellings that bring nutrients up from deep water. In bed sediments, dissolved organic matter is about ten times that of the overlying water. Chemical reactions in the bed sediments lead to a progressive loss of carbon that converts the sediments into the major reservoir of organic carbon in the sea.

According to McPherson and Sarda (1984), the Mediterranean Sea as a whole, and the western basin in particular, has a relatively low productivity rate for fish, compared, say, with the Atlantic. On the one hand, this paucity partly reflects the failure of the fishing community to take the industry as a serious economic enterprise: the most useful indicator is catch per unit of effort. On the other hand, such paucity has been recorded over long periods. As noted by the tenth-century Arab geographer Al-Muqaddasi:

> When God created the Mediterranean he addressed it, saying, 'I have created thee and shall send thee my servants. When those will ask for some favour of me, they will say "Glory to God!" and "God is Holy!" and "God is Great!" and "There is no God but God!" How will thou then treat these?' 'Well, Lord,' replied the Mediterranean – 'I shall drown them.' 'Away with thee – I curse thee – I shall impoverish thy appearance and render thee less fishy!'
>
> (Cited in Horden and Purcell, 2000: 7)

In pelagic fisheries, the most important fish are sardines and anchovies, the latter replacing pilchards in importance in south-east Spain and Morocco. Other pelagic fish include horse mackerel and mackerel. Hake, too, is important and it exhibits a typical inverted U-shape pattern of catches through time: as the fishing effort increases, the catch increases steeply to peak at a maximum where it levels off; then, with a further increase in effort (more boats, more hours, more technology), the catch can fall quite sharply and dramatically, indicating overfishing. According to McPherson and Sarda (1984: 300), 'overall landings show very little variation despite the constant increase in fishing power'.

Where the continental shelf is wider, as in the Gulf of Lions and off south-east Spain, red mullet and young hake are the most important pelagic fish. Overall, pelagic fisheries are more important than demersal fish. Sardines and anchovies account for two-thirds of the pelagic catch, mainly from inshore areas, such as Catalonia and Castellon. The main spawning area for sardines is off the Ebro delta. They are usually found near the coast in the winter spawning season.

Demersals are bottom-living fish, of which hake is the most important, occupying depths from 40 to 400 m. Hake yield is low compared with the North Atlantic. On the other hand, blue whiting is the species with the highest biomass in the Mediterranean Sea and certainly gives the highest yield, but it is less popular and so commands a lower commercial value. Lobsters occur at depths of 300 to 400 m and the Norwegian lobster is the most abundant and highly priced and has been over-fished. In the very deepest water, the economically important harvest is from shrimp. Shell fish are sensitive to pollution, as reported in Chapter 17.

There has to be a constant trade-off in fisheries between optimizing yield for effort on the one hand and ensuring sustainability on the other. It is widely accepted that, because of the low productivity of the waters of the Mediterranean Sea and of the western basin in particular, there is a dynamic equilibrium among the components of the ecosystem. These two factors, taken together, imply that, for a sustainable future, the target of any attempt to regulate exploitation means that maintaining the equilibrium is of utmost importance. There already exists the legal framework for this and the instruments to implement it in EU directives and in changes to fishing practice and fishing fleets. What is needed is what O'Riordan and Voisey (1998) call the transition to sustainability, that is, a commitment in spirit, ethos and stewardship in the form of policy integration, indicators and targets and local initiative. The position with fisheries is similar to that in agricultural pollution (see Chapter 21). They are both environmental issues, but not yet environmental crises and the precautionary principle needs to prevail. Fishing is economically irrelevant compared to tourism, but it remains critically important to the individuals and families in the many small villages and towns on the Mediterranean coastline. The traditional fisheries of the Mediterranean Sea are fast disappearing as technologies change and competing demands for employment create new opportunities.

Suggestions for further reading

Margalef (1984) is a very readable text in English by a distinguished ecologist and is part of the Key Environments Series. Arenson (1990) is subtitled 'The Mediterranean Maritime Civilization' and is a broad level coverage of parts that other books do not reach. It contains some fascinating well-illustrated topics, such as the fish described in wall paintings at Pompeii. It first appeared as a Channel Four television programme. It deals not only with the sea, but also history and culture of the region.

Topics for discussion

1   With the Mediterranean Sea as an example, discuss the main political instruments for the protection of oceanic environments.
2   Criticize the analogy that describes the Mediterranean Sea as an enormous bathtub.
3   Outline the major impacts of the Mediterranean Sea on the adjacent land masses.

Part 3

The human impact on the environment

9 Background to prehistoric and historic land use

As a poet of the human consciousness I suppose I am bound to see landscape as a field dominated by the human wish – tortured into farms and hamlets, ploughed into cities. A landscape scribbled with the signatures of men and epochs. Now, however, I am beginning to believe that the wish is inherited from the site; that man depends for the furniture of the will upon his location in place, tenant of fruitful acres or a perverted wood. It is not the impact of his freewill upon nature which I see (as I thought) but the irresistible growth, through him, of nature's own blind unspecified doctrines of variation and torment.

(Lawrence Durrell, *Justine*, 1969: 95)

Introduction

In this section, we look at the human impacts on the Mediterranean landscape from the earliest human settlements to the early modern period. The more recent developments of the modern period will be dealt with separately because the impacts of industrialization and mechanization have caused a range of different processes to be important within the landscape, as well as accentuating a number of the issues to be discussed here. Although rural landscapes and their problems are the dominant focus of this early period, problems relating to urban settlements also become significant in the last few millennia. As a background to these important issues, the present chapter will discuss the general population changes in the Basin. Such changes give an impression of the periods in which greater or lesser human pressure was exerted on the landscape, although for a number of reasons. However, we will see that in certain instances such a correlation is too simplistic, and that problems can arise from both too dense and too sparse human populations within the landscape.

The time-span of the periods covered in Chapters 10 to 14 is a large one, covering possibly as long a time as a million years. A length of time as great as this obviously leads to the use of a wide range of techniques to develop the chronology of the patterns and changes observed within the landscape. In the earliest phases, this chronology depends on geological and radiometric dating techniques (see Box 9.1). Archaeological data become more important in the medium term, particularly when coupled with radiocarbon dating to provide absolute chronologies (see Box 9.1). These chronologies are often classified according to technological or material culture groupings. The latter have a wide variety of, usually regional or local, names which are often used for correlation where radiometric data are not available. We have tried to keep the use of these names to an absolute minimum, but those which are used are explained in Box 9.1 (refer back to Box 3.3 for dating conventions). Historical data become important in the last two to three millennia, depending on the location, but archaeological and other environmental data are still useful for this period because of the selective nature of the historical record. Some aspects of this issue have already been addressed in Chapter 3. It must be

Box 9.1 Chronologies and archaeological periods

Archaeological periods are usually defined in terms of assemblages of artefacts (stone tools, pottery, metal artefacts, etc.) and are therefore usually *diachronic* in their spatial distribution. Commonly used periods and synonyms that we will encounter are given in the table below, along with approximate dates in different parts of the Mediterranean Basin. We will talk about refinements and deviations where appropriate in the text.

Table 9.1 Summary of archaeological periods and their relative timing in different parts of the Mediterranean Basin

Archaeological period	Meaning	Approximate range of dates in Eastern Mediterranean	Approximate range of dates in Western Mediterranean
Epipalaeolithic *or* Mesolithic	'Middle Stone Age'	⩾10000–between 9000 and 7000 BCE	9540–7000 BCE
Neolithic	'New Stone Age'	between 9000 and 7000–3500 BCE	7000–2200 BCE
Chalcolithic *or* Æneolithic	'Copper Age'	3500–3000 BCE	2500–2000 BCE
Bronze Age		3000–900 BCE	2000–600 BCE
Iron Age		900–600 BCE	600 BCE–start of Roman influence

It is also common to find archaeological periods subdivided, usually into 'Early', 'Middle' and 'Late' phases, and even further divided into A, B, etc. The exact usage of these will depend on local sequences. These phases may themselves be subdivided, if enough information is available, although this stage of subdivision is a common source of argument amongst archaeologists.

remembered that both archaeological and historical data are incomplete records of the human impact on the landscape, and in many cases, indirect evidence is used.

Settlement and land use in the Palaeolithic and Mesolithic

The first human settlement of the Mediterranean Basin

Evidence for the first human settlement of the Mediterranean Basin is by its very nature sparse. This sparseness leads to a number of problems, not least in pinpointing the initial time and location of settlement. Therefore, the data and interpretations based upon them tend to be controversial and may change rapidly as new discoveries, analysis and interpretations are made.

In Israel, the site of 'Ubeidiya contains the earliest known hominid remains in the Mediterranean Basin. Fossil remains belonging to *Homo erectus* are represented (Wymer, 1982; Bar-Yosef, 1987), which are associated with mammal bones and stone tools which suggest a date as early as 1.4 Ma (Tchernov, 1987). 'Ubeidiya has been interpreted as evidence for one of the first migrations of hominid populations out of Africa.

Recent evidence for very early hominids from Spain is reviewed by Dennell and Roebroeks (1996). The site of Atapuerca near Burgos in northern Spain contains fossil hominid remains and artefacts which were originally thought to be around 500,000 years old, but have since been redated as being possibly more than 800,000 years old (see also Carbonell *et al.*, 1995). At Orce in Andalucia, there are a number of remains which have been variously interpreted as being between 800,000 and 1.8 million years old, although the younger date seems much more likely. A number of bones have been suggested as being of hominids, but the evidence for this is quite tenuous. Isernia La Pineta, south-east of Rome, is more securely dated at around 730,000 years ago (Coltorti *et al.*, 1982). The evidence for a colonization of the northern Mediterranean Basin via the Gibraltar Straits, while feasible, is, however, ruled out by Dennell and Roebroeks (1996) because of the paucity of evidence before 800,000 years ago in north-west Africa. These authors prefer a hypothesis based on a reflux of *Homo erectus* populations from Asia, although the evidence of early sites in eastern Europe, Greece or Turkey which would support this is also lacking at present.

At Petralona in Greece, a skull which is dated to between 240,000 and 160,000 years has been noted as having affinities with both *Homo erectus* and *Homo sapiens* (Wymer, 1982), although the fact that there are earlier fossils of *Homo sapiens* in northern Europe implies that there may be earlier fossils of this species still to be found in Mediterranean Europe. The *Homo sapiens* skull fragment from Tautavel in southern France has been dated to a similar time period (de Lumley, 1976). Remains of neanderthals (*H. sapiens neanderthalensis*) are widely reported for the period after around 130,000 years ago, from Israel, Morocco, Spain, France, Italy and Greece (Wymer, 1982).

Modern human populations

An anatomically modern human (*H. sapiens sapiens*) has been found recently in the Nile valley in Egypt, dating to around 55,000 years ago (Vermeersch *et al.*, 1998). This important discovery lends support to the theory that modern humans migrated from Africa to replace the existing neanderthal populations. By about 35,000 years ago, modern populations are common throughout the Mediterranean and Europe, as exemplified by the burials at Cro Magnon in the Dordogne region of southern France, where the type fossils for the early modern populations were discovered. (It is not our intention here to enter into the 'neanderthal debate' – for further details, see Shackley 1980; Stringer and Gamble, 1993; Trinkaus, 1989.)

Burials of modern humans are again known from a wide range of locations, and often associated with grave goods and ornaments (Gamble, 1986). There is an evolution of the stone-tool technology, so that tools based on blades become more common, and art objects, including human and animal figurines are widespread (Bordes, 1984; Wymer, 1982; Champion *et al.*, 1984; Gamble, 1986). Cave paintings like those found in the Dordogne are also found in caves on the Mediterranean coast of France, such as the Grotte Cosquer near Marseille (Clottes *et al.*, 1992). This site is of interest as its entrance is now 37 m below the present sea level, clearly illustrating the effects of the lower glacial sea levels (charcoal from the cave has been dated to *c.*21640 BCE). Animals represented include bison, ibex, deer, chamois, horse, great auk and seals, and pollens from the cave are dominated by the steppe species of the Late Glacial Maximum landscape (see Chapter 4).

Developed hunter–gatherer communities at the Pleistocene-Holocene boundary

The Levant and Cyprus

An increasing number of studies of the subsistence patterns of the hunter–gatherer communities of this period demonstrate that the economies of these groups were becoming increasingly sophisticated. There was a move away from a reliance on more opportunistic food supplies towards attempts to control the food supply. In the Levant, there are a number of important changes which foreshadow the development of agriculture (see Chapter 10). By around *c.*21400 BCE, the first sites belonging to the Kebaran culture appear (Byrd, 1994). The Kebaran sites are concentrated along the Jordan rift and often show patterns suggestive of seasonal migration, with winter camps in the lowlands, and summer camps in the mountains to the north (Bar-Yosef, 1987). Sites such as 'Ein Gev I have evidence of the hunting of gazelle, fallow deer, wild goat, and some roe deer, wild boar, aurochs, equids and various birds, as well as mortars and pestles suggesting that plant remains were being processed for food. A burial in a shallow pit dwelling at this site is dated to 17547–15812 BCE. A wide range of animal species seem to have been hunted, although some specialization is apparent. The fauna at Wadi Madamagh is dominated by 82 per cent goat and that at Nahel Oren by 74 per cent gazelle bones (Mellaart, 1975). Legge pointed out that the age structure of the gazelles at Nahel Oren was similar to that found in domesticated species (Noy *et al.*, 1973), although domestication itself has been rejected by subsequent authors because the behavioural characteristics of gazelle make it unsuitable. The site of Ohallo II on the shores of Galilee has evidence for the exploitation of fish as a food source around *c.*20220 BCE, and the excavation also revealed the presence of a number of plant species such as barley and other cereals, wild grape and olive pips (see Chapter 11 for further details) and almonds (Kaufman, 1992). Emmer wheat and barley first appear in the Kebaran at Nahel Oren (Noy *et al.*, 1973). In the succeeding Geometric Kebaran period, starting around 15699–15137 BCE (Byrd, 1994), the subsistence base appears similar, although the distribution of sites suggests an extension of social interactions, including the exchange of Mediterranean shells over long distances – some being found on sites in the Sinai (Kaufman, 1992). Kaufman suggests that the social interactions implied by these exchanges allowed the development of social agglomeration and ultimately of sedentary lifestyles. These developments should also be seen in their environmental context. An increasing aridification of the climate (see Chapter 3) led to the development of a more open landscape, which would have necessitated the development of different subsistence strategies (Bar-Yosef, 1987). These changes may also have led to the selective replacement of perennial by annual grass species, which were more readily useable as a food source, and ultimately adaptable for domestication (Henry, 1991). There is some evidence from the coastal plain of Israel that more intense occupation in the Kebaran period led to a period of enhanced erosion, which often led to a slight shift in site location by the later Geometric Kebaran period (Schuldenrein, 1986). Similar patterns are also possibly seen in the Negev (Goldberg, 1986).

Sites of the Natufian period (14042–12691 BCE to 10088–9040 BCE: Byrd, 1994) are much larger than those seen before. Ain Mallaha covers an area of at least 2,000 m², and includes rounded houses built on terraces (Bar-Yosef, 1987). There is an expansion of the territory covered by the Natufian, with large sites in or on the edge of the recently developed Mediterranean vegetation zone, and much smaller base camps in the interior shrub or desert-steppe zones (Bar-Yosef, 1987), a development which occurred in the later part of the Kebaran (Kaufman, 1992). Furthermore, there is evidence to suggest year-round occupation of the larger sites, which are generally located near to perennial water sources (Henry, 1991). Seasonal movements are still suggested though by locations of site territories (Vita Finzi and Higgs, 1970; see also Vita Finzi, 1978). Human teeth show evidence of the consumption of stone-ground carbohydrates, and Sr/Ca ratios of bones imply that there were high levels of plant foods consumed (ibid.). More recently excavated sites have direct evidence of plant remains – grasses, légumes and lillies at Wadi Hammeh 27; barley, almonds and possibly peas at Hayonim cave; and over 150 different species at Tell Abu Hureyra in Syria (Kaufman, 1992) – to complement the indirect evidence from plant-processing tools such as mortars, pestles and flint blades with a silica gloss which are usually indicative of their use in hafted tools as sickles

(Bar-Yosef, 1987). Means of storing plant foods also first appear in small quantities in the Natufian (Edwards, 1989a). A further important development that has been suggested is that of specialized hunting of gazelle, as many Natufian sites are often dominated by the bones of this species, and Beidha has 76 per cent of its fauna made up of goat bones (Mellaart, 1975). Hunting strategies also seem to have been more specialized, with evidence from Hayonim cave and Ain Mallaha that male gazelle were being selectively culled, leading to increased sexual dimorphism in the species (Cope, 1991; Tchernov, 1991). Campana and Crabtree (1990) propose that some form of communal hunting took place at Salibiya, possibly using nets or fire, to trap whole herds of gazelle, as the age profiles of bone assemblages are close to those of modern herds. They go on to suggest that the coordination and leadership that are required for communal hunting may have been a precursor to the mobilization of the labour forces required to carry out agriculture. However, this model has been challenged on ethnographic grounds (Edwards, 1989a). It is clear, though, that social interactions in the Natufian are greatly enhanced compared to the previous periods, with greater quantities of shells from the Mediterranean being exchanged, and even obsidian from central Anatolia is found in the later part of the period in Levantine sites (Bar-Yosef, 1987).

The Natufian has examples of stressed populations, for example the dental hypoplasia seen on increasing numbers of skeletons through time and the possible evidence for female infanticide (Henry, 1991). Population stress may also be implied by the territorial expansion at this time (see Cohen, 1977, for a fuller discussion). It has been suggested that the use of a wider subsistence base (the so-called 'Broad Spectrum Revolution') was significant in the development of the domestication of plants and animals (Flannery, 1969), although this has been challenged for the Levant by Edwards (1989b) who points out that there has been a very broad subsistence base since at least the Middle Palaeolithic. The use of a broad spectrum of resources does, however, suggest the buffering of subsistence stress on populations. A number of authors (e.g. Henry, 1989) also suggest that acorns may have been used as a food supply in the developing Mediterranean oak forests, although direct evidence for this is sparse (probably in part due to excavation techniques – remains have been discovered in the Kebaran site of Wadi Hammeh 26: Kaufman, 1992). Ethnographic parallels suggest that acorns tend to be used in periods of resource stress, although McCorriston (1994) points out this does cause problems with some of the Natufian population estimates (see below; balynophagy is discussed further in Chapter 11). The Natufian seems to represent a period of expanding population which is exploiting a wide range of techniques to sustain itself, and developing a greater understanding of the interactions between plants, animals and the landscape. As we will see in the next chapter, this understanding ultimately led to the development of agriculture in the region.

At Aetokremnos on the southernmost tip of Cyprus, there is evidence as early as 12343–11187 BCE for human colonization of one of the larger Mediterranean islands (Simmons, 1991), and thus for sea-faring, as the deep waters surrounding the island would still have been present even at the lowest sea levels of the LGM. Other evidence for sea travel at about the same time includes the discovery of obsidian from the island of Melos at the cave of Franchthi in the southern Argolid. Van Andel and Runnels (1987) suggest that the location of the obsidian on Melos means that it is unlikely that it was discovered by accident, and may relate to quite extensive exploration of the Aegean islands by peoples fishing from the mainland. The Aetokremnos finds are all the more important because of the association of the human settlement with large numbers of bones of now-extinct species – mainly pygmy hippopotamus, but also some pygmy elephant – which were previously thought to have died out before the first colonization of the island (Simmons, 1991). Although there is no direct evidence for the extinction as being due to 'overkill', there is a suggestion that the wide range of bones represents a mass kill, possibly by driving the animals over the edge of the cliffs, which reach 70 m in places. However, the 'naïve' character of the pygmy species, which had hitherto not had competition from predators on the island, does not necessarily imply any great strategy on the part of the hunters (e.g. Simmons, 1991).

Greece and Italy

In northern Greece, Upper Palaeolithic settlement patterns have been studied in detail by Bailey *et al.* (1993). Compared to the Levant, it must be remembered that the European Mediterranean at the Late Glacial Maximum was a much more desolate place, vegetated by desert steppe (see Chapter 4), and with relatively sparse animal as well as plant resources. This study develops on the earlier research of Higgs, who believed that Upper Palaeolithic sites such as Asprochaliko and Kastritsa were the winter and summer transhumance camps on a seasonal migration route, following herds of red deer (transhumance is discussed in greater detail in Chapter 11). The analysis of a number of sites (Figure 9.1) showed that although the coastal zone was occupied for over 100,000 years, the mountains of the interior were only occupied after the Late Glacial Maximum, possibly as a result of competition from other carnivorous species in the lowlands for increasingly sparse resources. Thus the severity of the environment may have led to the development of a subsistence pattern which necessitated the exploitation of a range of resources distributed over the landscape, in which seasonal mobility was required to exploit the resources to the full. Bailey *et al.* suggest the mountainous topography was an important component of this system, forming steep limestone ridges to constrain movement to a set of predefined routes. Combined with a series of disturbed flysch soils in the uplands which would have been unattractive to the browsing animals, and the blocking of the higher mountain areas by glacial moraines, it becomes clear that the Upper Palaeolithic sites of the Epirus are located preferentially to exploit a range of seasonally migrating species – ibex, chamois, wild boar and horses, as well as red deer. Analysis of the local site locations suggests that they were often placed to be away from main migration routes, but in good positions to control the movements of animals into enclosed topographic areas where they might more easily have been hunted.

Seasonal exploitation of herds of red deer has also been proposed for the Upper Palaeolithic of Lazio in central Italy (Barker, 1981). As with the Epirus, there is evidence for an expansion of occupation into the uplands in the Upper Palaeolithic, with specialized hunting of red deer and steppe horse being carried out, although a wide range of other species were also hunted. There is again evidence for the concentration of sites at gorges, allowing the control of the movements of animal herds. The age structures of faunal remains show specialized predation of animals killed either in their first year, or at the time of their optimal body weights, after four to eight years. There also appears to have been some specialization of stone tools between lowland and upland sites, possibly reflecting the different activities that occurred on them. Similar site patterns also seem to be found in Campania and Toscana, and although the evidence is less definite, Barker (1995) has also suggested a similar pattern occurring in the Biferno valley of east-central Italy. Movement between the coastal zone and the interior, with a broad usage of resources including hunting of large ungulates and other animals, the use of fish and shellfish, and plant remains such as acorns, wild olives and strawberries combined with grinding stones, also seems to have been a characteristic of the period after the LGM in Portugal (Ferreira Bicho, 1993).

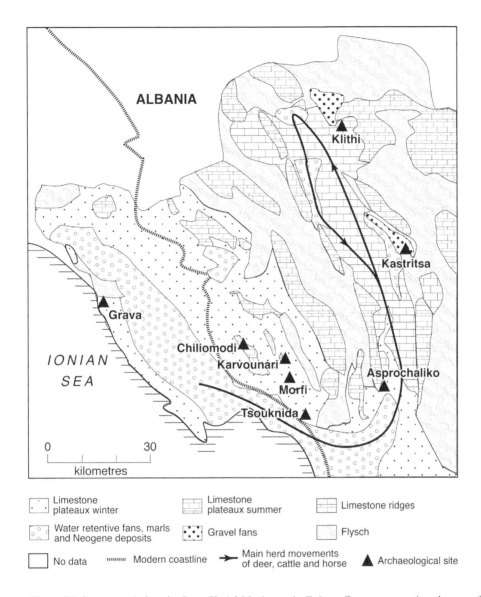

Figure 9.1 Sites occupied at the Late Glacial Maximum in Epirus, Greece, suggesting the use of seasonal migration patterns to follow herds of animals

Source: After Bailey *et al.* (1993).

Spain, France and North Africa

In the Valencia region of Spain, the sites of Parpalló and Les Mallaetes seem to have been occupied almost synchronously from around *c.*23350 BCE to *c.*12050 BCE, together with a number of more sporadic occupations at other nearby cave sites, and the site of Volcán, 21 km away near the present coast (Bailey and Davidson, 1983). Evidence from the site territories of Parpalló and Les Mallaetes, which are located at similar altitudes on the same massif, suggest that although a two-hour radius from the sites gives essentially overlapping territories, a half-hour radius would allow the exploitation of mutually exclusive zones. There is some evidence from faunal remains (Davidson, 1983) that Les Mallaetes was a summer camp, and Parpalló a winter camp, which would also accord for their locations. Les Mallaetes is north facing, making it unsuitable for occupation during the Late Glacial winters, with their temperatures of 8–9°C cooler than at present, and would have been ideally located for the hunting of red deer and ibex in the summer, while Parpalló could have been used to control access of the red deer as they moved from winter to summer pastures. A number of other nearby sites could have been used for the blocking and observation of red deer herds in a controlled hunting strategy (Bailey and Davidson, 1983). Parpalló may have been a winter focus for a number of groups that were more widely scattered during the rest of the year, its special status possibly being indicated by the fact that 6,000 engraved or painted plaques with animal and human figurines were retrieved from the site (Davidson, 1991). Unfortunately, the faunal remains from these sites precluded detailed analysis, although Davidson (1983) suggested that there is an increased specialization in red deer through time at Parpalló, Les Mallaetes and Volcán, a pattern which does not seem to be explicable by environmental change. Long-distance connections between the sites are also attested by the presence of marine shells in the inland sites, and possibly also by the import of flint from other areas, as there appear to have been no local sources.

In northern Spain, relatively long-distance connections of over 20 km are again attested by the presence of marine shells in upland sites. At Castillo in Calabria, there are also Mediterranean shells in the later Upper Palaeolithic levels, suggesting contacts were established over distances of about 500 km (Bailey and Davidson, 1983). Sites are found in two locations in Cantabria and the Asturias: in clusters towards the rear of the coastal plains, where they could control access to the valleys, and in the uplands close to steep terrain. The latter groups of site tend to be dominated by faunal remains of ibex, although red deer were again clearly hunted in both groups of site. At La Riera in Cantabria, there seems to have been an evolution in hunting strategies through time (Clark and Strauss, 1983). Horse, aurochs and bison were originally preferred animals, followed by a period of ibex specialization, and finally a concentration on red deer. After 19490–17464 BCE, hunting of the red deer tends to be concentrated in the summer and autumn months, and from 10868–10290 BCE there is a marked increase in juveniles, suggesting further specialization. Ibex also seem to have been killed shortly after birth. The bones of red deer also show similar age structures to those described for central Italy, which Clark and Strauss suggest may indicate the use of drives to trap large numbers of animals at the same time. They also suggest that the highly fractured nature of the bones from the site may indicate the maximum exploitation of all the resources from the carcasses (or a high degree of trampling in the cave!), which often seem to have been transported whole from the kill site to the cave.

On the edge of the Montagne Noire in southern France, the site of Abeuradour shows an interesting evolution in the subsistence base from the latest Palaeolithic to the Mesolithic (Vaquer and Barbaza, 1987). Even the earliest levels, suggested as being contemporaneous with the Younger Dryas or slightly earlier, contain evidence for carbonized grains. The immediately postglacial deposits contain a wide range of remains of animals – chamois, wild boar, deer, aurochs, rabbit, hare, birds and salmon – albeit in small quantities. More significantly, large numbers of carbonized plant remains have been found, including hazelnuts, cherries, grapes, lentils and other pulses. Later levels (including one dated to 8002–7540 BCE) are almost totally made up of grains and flint microliths. The latter are common throughout the sequence at the site, and may represent the remains of hafted tools used to collect plant remains. Some of the later levels also include stones apparently used for grinding plant food-

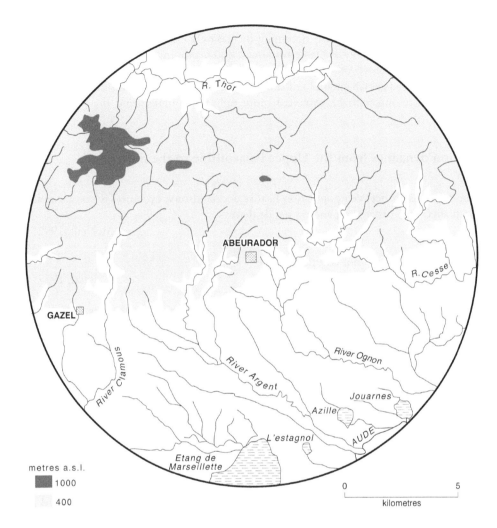

Figure 9.2 Site-catchment analysis of the Mesolithic site of Abeuradour in southern France, showing the wide range of territories exploited

Source: After Vaquer and Barbaza (1987).

stuffs. The variety of remains suggests that a large territory was being exploited from the site (Figure 9.2), although it is not known whether this was on a seasonal or a permanent basis.

Capsian sites of the Maghreb seem to have focused on the extensive use of land snails to complement other hunted resources. Sporadic or rotating use of a number of midden sites over an extended period may have resulted in the conservation of resources (Lubell *et al.*, 1976), although it is impossible to tell whether this was deliberate or as the result of local resource exploitation.

While the direct impact on the landscape of Upper Palaeolithic and Mesolithic hunter–gatherers was almost certainly relatively minor, the examples described above illustrate two significant developments. First, there was an increasing awareness of the wide range of resources available within the landscape, both plant and animal. This awareness included an understanding of how these resources might be exploited and their distributions modified, which were important precursors to the development of agro-pastoral systems. Marine resources (and transport) were also becoming increasingly important. Second, there seems to have been a use of seasonally available resources, and a movement through the landscape to follow resource availability. Although this was by no means a transhumant system as developed in many places later (see Chapter 11), it does represent an understanding that mobility allows relatively scarce resources to be exploited more fully while minimizing the pressure on the resource.

Population dynamics from the Upper Palaeolithic to the nineteenth century

The relationship between population history and landscape evolution is not a simple one. Each affects the other in an endless pas de deux.

(McNeill, 1992: 147)

Palaeo-demography is not an exact science.

(Reed, 1977: 553)

It is not our intention in this section to give a summary of population changes with the aim of demonstrating a link between population change and environmental change. As McNeill notes, such an aim would be particularly fruitless and contradictory. Nor do we aim to generate population estimates for the entire Mediterranean world. Such a task would involve an enormous amount of effort for questionable results. The errors involved with such an aim would again make the results rather meaningless (as noted by Reed) and may lead others to draw unwarranted conclusions about local cases based on local evidence. Cohen (1977) notes the difficulties with making population estimates based on archaeological evidence. Typically, evidence is collected based on the number of houses or sites, or the amount of food resources under production or consumed. This evidence is then converted to absolute or relative population changes assuming complete (or consistently incomplete) recovery of the evidence, that there is a link between the measured evidence and the population numbers it would support, and that all the sites which are assumed to be contemporaneous were in fact occupied at the same time (rather than seasonally or as a result of longer-term movements). Unfortunately, these assumptions are rarely ever achievable in a given location. Cohen also notes that there can be a significant over-estimation of the population immediately after the adoption of agriculture because the permanent sites tend to be much better preserved than mobile camps. Conclusions drawn from this that agriculture was the result of population pressure (or led to greater population pressure) can therefore be over-stated. Thus, the figures presented below for prehistoric population changes should be taken as being illustrative of particular ranges rather than as being absolute values. Historical data are on the whole more representative, although they can still contain significant sources of bias, depending on the source of the information. The data may be infrequent, unreliable, based on censuses for taxation (causing some of the population to become 'missing'), be rough estimates, use changing administrative boundaries, or only include information on numbers of males or of households. Finding a reliable, consistent multiplier from the household to the population total may be difficult, even in the same region (see McNeill, 1992, for further discussion of estimating the historical populations of a number of Mediterranean mountain zones).

Thus, our aim here is to demonstrate a number of general trends in population change by means of giving the background to some of the issues to be discussed in the following chapters. By focusing on specific case studies at the regional and local level, we can also demonstrate the importance of variability compared to the general trends observed.

The Levant and Egypt

Perrot (1962, cited in Cohen, 1977) estimated that there was a population build-up during the Upper Palaeolithic in Palestine, based on the number of sites discovered in the coastal zone. A similar conclusion was reached by Hours (1982) for the period from *c.*21330 BCE in the northern Levant. Reed (1977) suggested that the population of Palestine (*c.*27,000 km^2) may have been around 10,000 at *c.*10968 BCE, based on estimates of the density of Natufian sites and their probable population (see also Bar-Yosef, 1991). Assuming a population increase of 0.1 per cent per year based on figures from the literature, the estimates become 56,800 at *c.*9230 BCE (at the very end of the Natufian) and 188,500 at *c.*8030 BCE (Reed's actual figures were 10,000 at 11000 bp, 27,000 at 10000 bp and 74,000 at 9000 bp, the differences being caused by the variability in the radiocarbon timescale). McCorriston (1994) suggests that the figure of 10,000 would actually be too *low* to have caused the population pressure implied by stress-related features seen in the Natufian (see above), based on ethnographic parallels with pre-Hispanic Californian populations. Hours' (1982) data show a rapid increase in the number of sites after the development of agriculture in the northern Levant, albeit with a reversal between around *c.*8800 to *c.*7575 BCE.

As discussed in Chapter 10, late Palaeolithic populations in Egypt were very low, but increased rapidly in the Neolithic. Butzer (1976) estimated the population of Egypt to have been 866,000 in the predynastic period, 1,614,000 during the Old Kingdom and 1,966,000 in the Middle Kingdom. During the period of the New Kingdom (*c.*1570–1069 BCE), Egypt may have had a population of around three million, with the populations of major cities such as Memphis or Tell el-Amarna reaching 50,000 and 30,000, respectively (Brewer and Teeter, 1999).

Greece

Van Andel and Runnels (1987) attempted to reconstruct the population of the southern Argolid – an area of about 2,000 km^2 – based on extensive field surveys, covering all periods (Figure 9.3). The population of the Upper Palaeolithic was probably small, between around twenty-five to fifty people, and there was no significant change until the Neolithic (from the early sixth millennium BCE), when the population may have been between one hundred and 450 people. Peaks were reached in the early and late Bronze Ages, with population estimates of 1,200 in both these periods, separated by a period of much lower population. There is very little evidence for occupation in the so-called 'Dark Age' between the end of the thirteenth century and the tenth century BCE. From the seventh century BCE, though, there is a rapid increase to about 6,800 to 7,300 in the Classical period, followed by a subsequent decline and dark age, increasing steadily from the thirteenth century CE, reaching 1,500 to 2,000 during the early eighteenth century, 7,000 just after independence in 1832, 10,000 in 1851 and 12,000 at the start of the twentieth century (compared to the present population of 12,000 following a number of oscillations).

In comparison with the Levant, we can see that the population density was much lower in the Argolid in the Upper Palaeolithic and Neolithic periods, and that the initial increase in population was not as significant following the adoption of agriculture: at least for perhaps three thousand years. This is not to say that the same pattern was found all over Greece. In Messinia in the south-west of the Peleponese, intensive surveys have been carried out over an

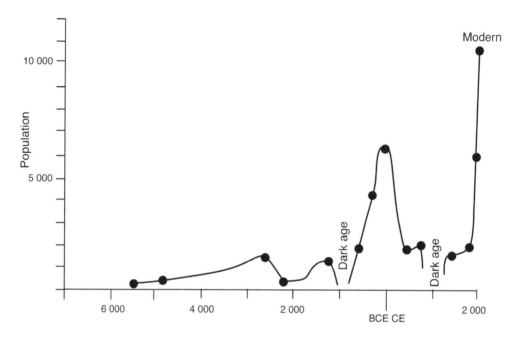

Figure 9.3 Population reconstruction based on archaeological data for the Argolid, Greece
Source: After van Andel and Runnels (1987).

area of about 3,800 km² (McDonald and Rapp, 1972). Palaeolithic inhabitation of the area is uncertain, but is probably of the same order of magnitude as in the southern Argolid, and there are only five Neolithic find spots (implying a population of 500–2,000 using the same figures for extrapolation). There is a rapid increase in the Bronze Age, notably due to the development of regional centres such as Pylos: the Early Bronze Age population is estimated at 3,500, the Middle Bronze Age at 10,000, and the Late Bronze Age at 50,000 (McDonald and Simpson, 1972: the modern population is around 250,000, Loy and Wright, 1972). Even accounting for the greater area, these figures are much higher than those in the southern Argolid. There is again a substantial collapse in the 'Dark Age', followed by an increase into the Hellenistic and then stability into the Roman period.

Italy

Barker (1995) suggests a gradual increase in population in the Biferno valley of east-central Italy from the Mesolithic to the early Mediaeval period, which saw a dramatic population decline, with subsequent expansions in 1000–1300 CE and fifteenth to seventeenth centuries. The nineteenth century saw a further dramatic rise, as seen in a wide range of other Mediterranean areas.

In Etruria, populations remained relatively low until the Chalcolithic, with significant expansion only taking place in the Bronze Age (Potter, 1979). A rapid phase of expansion took place in the ninth century BCE with a more rapid increase continuing through the seventh and sixth centuries BCE. Expansion continues, albeit at a slower rate in the fifth and fourth centuries BCE, despite the fact that large parts of the Etruscan landscape were conquered by the neighbouring Romans from 435 to 373 BCE. Population then continues to expand (despite local setbacks, for example, the 15,000 Faliscans killed in the war with Rome in 242 BCE) until the second century CE, where there is a slight decline. From the fourth to sixth centuries, population numbers drop off rapidly and the landscape of the early Middle Ages is relatively sparsely settled. Expansion reoccurs by the eighth or ninth centuries, with a concentration of settlement in fortified centres. Between *c*.1350 and 1550 CE, the fortified centres are replaced by nucleated settlements in the low-lying areas, although it is not clear whether there was a population decline or simply a geographical redistribution.

France

In southern France, the settlement patterns on the Montpellier Garrigue suggest a substantial increase in population from the middle Neolithic (after *c*.3500 BCE) into the late Neolithic and Chalcolithic (*c*.2200 to 1800 BCE). In the subsequent early Bronze Age, there is a significant decline (Delano Smith, 1972). Similar phenomena are recognized elsewhere in Languedoc (e.g. Guilaine, 1972; Wainwright, 1994). These population levels do not recover for a period of around 300 to 500 years. In the Valdaine, a similar pattern of slow growth in the early Neolithic, accelerating in the middle to late Neolithic and Chalcolithic followed by an early to middle Bronze Age hiatus has recently been demonstrated by Berger *et al.* (1994). A rapid increase is seen in the later Bronze Age, but the Iron Age (*c*.800 to *c*.100 BCE) is less well known although the reason for this may be the lack of detailed investigations. The Roman period saw the development of a highly structured and divided landscape, with an expansion into previously unexploited areas and the construction of a drainage system. Depopulation followed the socio-economic crisis of the Roman Empire in the late second and third centuries, with the resulting abandonment of maintenance of the drainage system and a corresponding phase of erosion and flooding. The early Mediaeval period saw lower population densities reflecting a more pastoral economy. Further south, in the Alpilles, a similar prehistoric pattern is found, albeit with a longer phase of Bronze-Age decline, which is not reversed until the middle Iron Age (Gazenbeek, 1994). Expansion starts with contact with the Etruscan and Greek world and was expanded by the Romans on very large scales, expanding onto the plains from the colluvial soils, again with the development of structured agricultural systems and drainage of marshy areas. The Haut Comtat again has evidence of Neolithic and Chalcolithic settlement, with a lacuna relating to the Bronze Age (Ballais and Meffre, 1994a). Here, there are very few traces of occupation in the later Iron Age (fourth to second century BCE), with evidence of occupation of the upland sites only in the mid-second and first half of the first century BCE. This period saw an expansion, with sites often being constructed on top of sites previously occupied in the early Iron Age. The first villa was established around 40 to 30 BCE. Expansion continued through the first and second centuries CE, and a decline in the later Roman period, starting at the end of the second century. The decline is followed by a re-expansion in the fifth and sixth centuries. More detailed studies of available evidence for the Roman period illustrate that populations were often very dynamic in relatively small areas. The Tricastin, Valdaine, Haut Comtat, Alpilles, Uzège, Beaucairois, Vaunage, Lunellois have been studied in detail for the Roman period – namely, from the establishment of the Narbonnais colony in 118 BCE, until the fourth century CE (Favory *et al.*, 1994a, b; Verhagen *et al.*, 1994; Tourneux, 1994; Favory and Girardot, 1994; Audouze *et al.*, 1994; Verhagen, 1994). Early colonization took place often in previously unexploited areas, and was very rapid, with 152 sites established in the first century BCE, and 496 in the first century CE. Only around 30 per cent of these failed in the first half century of occupation, suggesting a remarkable stability. There is typically an early population increase in the more southerly areas, possibly relating to their proximity to the important urban centres of Arles and Orange, apart from the Lunellois, which expands more rapidly in the second half of the first century CE, along with the Tricastin and Valdaine. The second century sees recession everywhere apart from the Tricastin, which continues to expand. There is general decline in the third century, although beginning later in the Tricastin and Vaunage, and the Lunellois and Vaunage in the west become depopulated to pre-Roman levels. These areas underwent some recolonization after about 350 CE. At the end of the period, the Beaucairois and Uzège have a net loss of sites because of their initially high numbers, the Alpilles, Vaunage and Lunellois have increased slightly, whereas the Haut Comtat, Tricastin and Valdaine in the north have an increase of between an eighth and a quarter. The contraction appears mainly to take place by desertion of smaller sites, with many being resettled in the fourth and fifth century. These figures can be set in the context of a maximum population of the classical Graeco-Roman world of 50 to 60 million people (Finley, 1985), and of an Italian population of around seven million during the reign of Augustus (29 BCE–14 CE: Brunt, 1971).

Spain

Similar techniques have been used to reconstruct the populations of the Vera Basin and the Aguas valley (*c.*100 km²) in southern Spain from the mid-fourth millennium BCE to the present day (Castro *et al.*, 1998). A variety of techniques were used to estimate population figures, producing upper and lower estimates, and allowing the comparison of archaeological and historical methods. The results of this study suggest four peaks in the population around

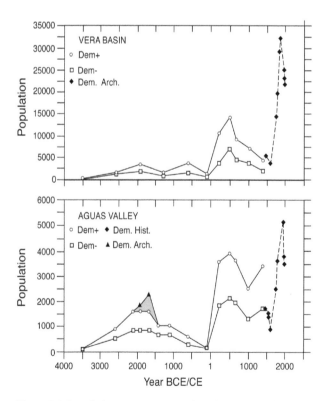

Figure 9.4 Population reconstruction based on archaeological data for the Vera Basin and the Aguas Valley, southern Spain

Source: After Castro *et al.* (1998).

*c.*1900 BCE, *c.*600 BCE, AD 500 and in the nineteenth century (Figure 9.4) in the Vera Basin as a whole, although the peak at *c.*600 BCE is missing in the Aguas valley, which also shows an additional peak in the pre-*reconquista* period. Both areas show a decline in the population after the *reconquista* due to the expulsion of the Moorish populations, as seen elsewhere in Spain (see below and McNeill, 1992). The population increase following the start of agriculture to the first peak in the second millennium BCE is of the same order, or slightly faster than that seen with the first development of agriculture in the Levant.

General patterns

McNeill (1992) has assessed the historical population data for the Taurus, Pindus, Lucanian Apennines, Alpujarras and Rif in his overview of the Mediterranean mountains. Although the specific trajectories of each of these areas is different (Figure 9.5), a number of common themes emerged between them. First, population increased dramatically in all of them in the past two hundred years, causing economic hardship in all but the Taurus. Low populations had been maintained in the Taurus until the nineteenth century because of the importance of nomadism. Emigration from the rural areas was important from the late nineteenth century, and has increased in the latter part of the twentieth century (we will return to this theme in Chapters 14 and 16).

Braudel (1975) discusses in detail the Mediterranean population in the sixteenth century. At the end of this century, he believes the population to have been around 60 million, with 38 million in the European Mediterranean and a further 22 million in the Islamic world. The former figure is considered to be much more reliable than the latter, and other authors have estimated the latter to be considerably higher at 30 to 35 million. Braudel also suggests that the Mediterranean population had probably doubled from 30 to 35 million at the start of the sixteenth century, and many areas had shown a steady growth since the aftermath of the Black Death in 1348 (which killed between a third and a half of the population of Languedoc and Provence, for example), but which continued to be important until at least 1481 and even locally 1510 (Le Roy Ladurie, 1969). However, rates of expansion slowed in the seventeenth century as the limits of growth were reached, and indeed some areas also saw a decline from around 1650 to 1700. The population remained unevenly distributed, with extensive areas of 'waste lands' (Braudel, 1975).

Summary

Human populations in the Mediterranean region first date to the period of major general cooling in both global and regional terms. These populations were very scattered and sparse, even taking into account the difficulties of obtaining reliable data that relate to this period. From the appearance of anatomically modern humans and more developed lithic technologies in the period between around 55,000 and 35,000 years ago, important differences developed. Although still scattered, population densities increased, and there were whole new ways of exploiting the landscape both in a physical and a social sense. In some cases, the physical landscape was changed in significant ways at this time, at least locally. The available evidence suggests that these phenomena were widespread.

Population increases were notable in the Levant in the period from around 20,000 years ago, and this too led to higher local concentrations of human impact on the landscape. There is also clear evidence that these concentrations also led to greater stress being imposed on the human populations. In part, these stresses were a cause of the significant changes to be discussed in the next chapter.

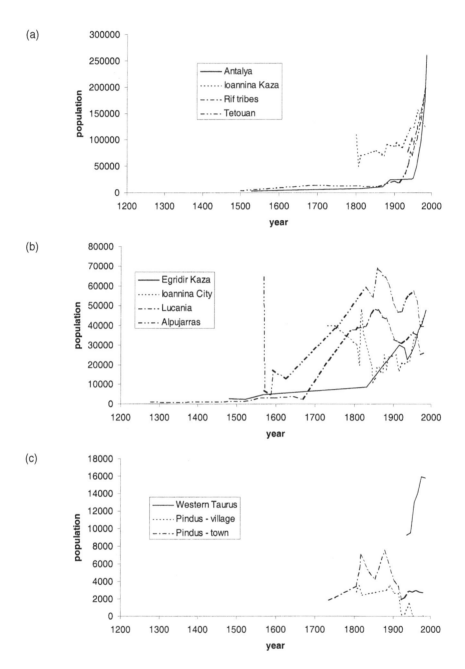

Figure 9.5 (a), (b), (c) Population data for a number of mountain regions of the Mediterranean
Source: Based on data from McNeill (1992).

Population change, when looked at from the perspective of centuries or millennia, has always been dynamic. Such dynamism is seen at local and regional levels throughout all the time periods covered. It is important, then, when trying to explain landscape change to account for this variability – not necessarily as the dominant controlling variable, but more as a constraint or boundary condition to the explanation. In Chapters 10 to 13, we will make an attempt to do so for the Mediterranean Basin.

Suggestions for further reading

Bar-Yosef (1987) provides a useful introduction to the Epipalaeolithic populations of the Levant. Bailey *et al.* (1993) demonstrate the links between late Pleistocene hunter–gatherers and their environment. Butzer (1976), Cohen (1977) and McNeill (1992) provide useful overviews of the methods and results of population estimations.

Topics for discussion

1   Why is the understanding of past population levels so important?
2   What difficulties are encountered when trying to reconstruct past populations in the absence of census or historical data?
3   In what ways do hunter–gatherer populations exploit their landscapes, and how do these relate to the environmental conditions described in the preceding chapters?

10 Traditional land-use patterns 1
Agriculture

Introduction

The transition to agriculture is probably one of the most significant stages – arguably *the* most significant – in the path of human evolution. By continuing the trend to control food supplies rather than depending on their natural availability, there were important changes in the ways human populations exploited and treated the landscapes in which they lived. As populations grew because of the greater stability of food supplies, there came the need to extract food increasingly efficiently from the environment (although some authors argue about which of these came first: see Cohen, 1977). The need to increase efficiency meant that greater areas of the landscape needed to be exploited, and ultimately these areas would need to be exploited increasingly intensively. Thus, the transition to agriculture was not only an important stage in human evolution, but also in landscape evolution. First, the landscape was modified in terms of the vegetation and, second, these vegetation changes led to changes in runoff (and therefore water resources) and erosion. Thus, to understand the different phases of agricultural development and expansion is to understand Mediterranean landscape evolution throughout much of the Holocene period. These developments and their consequences are therefore the focus of this chapter.

The origin, spread and intensification of agriculture

In the previous chapter, we noted the development of settlement patterns in the Levant until the period of the boundary with the Holocene, and noted a number of important developments including the use of specialized hunting strategies of gazelle and sometimes goat, the exploitation of very broad resource bases including the use of extensive plant foods, and probably the development of more sedentary lifestyles. Each of these were important for the subsequent development of agriculture in the region. The first sites with evidence for agriculture occur in the so-called Pre-Pottery Neolithic A (PPNA: see Box 10.1), the earliest dates for which in the Levant are found at Abu Salem (11944–11400 BCE: Kujit and Bar-Yosef, 1994). The latest PPNA dates are from Jericho (8017–7425 BCE) and Nachcharini in the Lebanon (7538–6596 BCE). PPNA Jericho has up to twenty-five building levels covering 2.5 ha with some buildings greater than 5 m in diameter and a large number of rooms interpreted as being for storage (Mellaart, 1975). The animals were still all hunted (37 per cent gazelle plus wild cattle, goat and boar) but there is the first evidence of domestication of plants, including two-row hulled barley and emmer wheat (Hopf, 1969; see Box 10.2). A number of large PPNA sites like Jericho are located in fertile alluvial terraces along the Jordan valley, are multi-phase and have thick deposits, contrasting with a number of much smaller sites which are more scattered through the region (Kujit, 1994). This pattern suggests that subsistence was carried out by fixed agricultural settlements coupled with more seasonal camps used for hunting or herding. The larger sites also contain obsidian from Anatolia, zoomorphic art, burials with ritual activities such as skull caching, and non-residential structures. The most notable of these is the 8-metre tall tower and walls surrounding the settlement of Jericho. Originally interpreted as having a defensive role (Kenyon, 1957), Bar-Yosef (1986) has more recently proposed an interesting alternative suggestion. The location of Jericho on an alluvial fan near a mountain front would make it particularly prone to flooding, but would have course made the site attractive from the point of view of having a permanent water supply. Partial burial of this wall by alluvial sediments supports this suggestion (Figure 10.1). Such flooding may have been exacerbated by the possibly slightly wetter climate at the time and the fact that vegetation near the site would have been removed due to agriculture and possibly also localized herding and the cutting of trees and shrubs for fuelwood. The tower, which seems not to be functional from a defensive purpose (nor is there any evidence that defence was required at the time), and may have been used for communal activities, possibly including rituals relating to the nearby spring. The site of Beidha which is situated on a wadi edge also seems to have a terrace wall to prevent flooding of the settlement.

The succeeding PPNB is more widespread throughout Palestine, Syria and Lebanon (dates range from 9248–8524 BCE in Syria and 8967–8089 BCE in Israel to 6164–5445 BCE: Kujit and Bar-Yosef, 1994). Although hunting continues, first domestic animals are found during this period – at Beidha, 86 per cent of the fauna is goat with twisted horn cores – and plant cultivation seems to be particularly well established, although domestication of some species is not yet complete (Mellaart, 1975). A number of plant species seem to have been imported from the northern (Zagros) zone. Hunting technology is also improved, with the use of 'desert kites' for driving and trapping game first attested in the PPNB (Bar-Yosef, 1986; 1991). Lime plaster is used in Syria during this period to make the first pottery (Mellaart,

---

Box 10.1 The 'Neolithic Revolution'

The term 'Neolithic Revolution' was coined by the archaeologist Gordon Childe in the 1950s. Because of the resolution of dating information available at the time, it often appeared that there was a sudden change to the practice of agriculture, with a relatively rapid domestication of plants and animals. Associated with this change was the development of technologies such as pottery (for storage), flint knives (for harvesting cereals) and threshing implements, querns (for processing cereals) and the use of polished stone tools rather than those made from chipped and flaked flint. These associations of objects together with evidence of domesticated plants and animals could thus be termed a Neolithic 'package' of material culture and subsistence techniques.

Today, we know this picture to be grossly oversimplified. Even in a single area, the transition to agriculture took place over a period of a thousand years or more, with different lifestyles and subsistence patterns co-existing within the landscape. Improvements in excavation techniques have also demonstrated that sites existed with only some elements of the Neolithic package. For example, sites in the Levant existed for several thousand years with evidence for domestic species before the invention of pottery (thus explaining curious culture names such as 'Pre-pottery Neolithic A'). However, because the change to agriculture is such a momentous step in human evolution, we still tend to refer to the changes of this period as a revolution. In socio-cultural and economic terms, they were certainly very much so.

Box 10.2 Evidence for domestication

One of the important developments in prehistoric archaeology has been the ability to define whether plant or animal remains found on a site belong to wild or to domestic species. These techniques depend on an understanding of how the selective breeding involved in agriculture or animal rearing causes modifications in the size, shape or physical characteristics of a species.

Plants

Cereals were the first important crops to be domesticated. As well as morphological changes which meant that the grains became increasingly fuller, domesticated cereals also develop a tough rachis. The rachis is the part of the plant that connects the seed to the stem of the plant, and is brittle in wild forms. This brittleness means that once the seed is ripe, it breaks from the stem and is scattered on the soil surface, ready to germinate and grow in the following year. Because this property meant that harvesting grains was difficult, early farmers would have tended to select for plants with the tough rachis. Processing of grains would therefore be necessary to remove the rachis – hence the need for technology such as threshing tools and querns – so cereal remains that either have the rachis still attached, or indeed rachis fragments on their own are good indicators of a domesticated crop.

Olives are found throughout the Mediterranean in their wild form in the early Holocene. However, the domesticated forms are all believed to originate from the Levant. Their domestication can be defined on morphological grounds.

Vines are again endemic in their wild form. See Zohary and Hopf (1988) for further details.

Pulses are often forgotten as an endemic staple in the Mediterranean. Their domestication is discussed by Zohary and Hopf (1975).

Animals

Ovicaprids are native to the Near East, and were introduced to the Mediterranean Basin. It is often difficult to differentiate between sheep and goat bones, hence they are often lumped together in the earlier periods.

Cattle were found in their wild form – the aurochs (*Bos primigenius*) – throughout the region. Bones are usually distinguished on the basis of size, although this leads to problems with the initial phases of domestication, and due to sexual dimorphism. The analysis of butchery patterns may be one way to resolve this problem.

Pigs occur endemically in their wild form (*Sus scrofa*) throughout the region. Similar problems occur with the distinction of wild from domesticated forms as with cattle.

Alternatively, we may be able to rely on the geographical distributions of plants, so that changes that cannot be attributed to other mechanisms may relate to deliberate human introductions. In more recent periods, we can also use historical records.

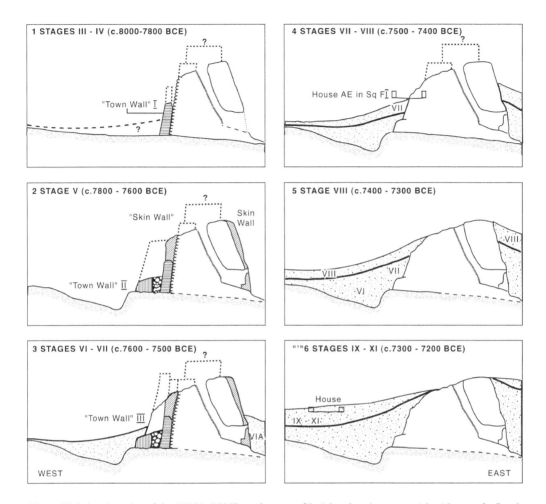

Figure 10.1 Stratigraphy of the PPNA–PPNB settlement of Jericho showing potential evidence of a flood-control structure

Source: After Bar-Yosef (1986).

1975). The area of settlement also spreads into more marginal zones such as the Negev and Sinai areas, with seasonal hunting and gathering of wild cereals attested in these areas (Avner *et al.*, 1994). Trade links were continued, including copper ore from the Arabah valley and the southern Negev, turquoise from the western Sinai, and shells from the Red Sea. In general, the pattern of large nucleated settlements and smaller groups in the desert zone continues from the PPNA, although there seems to be a break-up into a number of more local patterns of material found on PPNB sites (Bar-Yosef, 1991).

The first pottery Neolithic appears in the Syrian sites of Ras Shamra and Ramad in the second half of the eighth millennium BCE, and slightly later in the Israeli sites (Kujit and Bar-Yosef, 1994). Mellaart (1975) and others have suggested a phase of depopulation at the end of PPNB because of climatic aridification, although the summary of dates produced by Kujit and Bar-Yosef suggests a clear overlap between PPNB and PPNA, albeit with a smaller number of sites in the latter period. Some of the pottery is decorated with impressions of cockle

(*Cardium*) shells (Mellaart, 1975), which was to become a common motif throughout the Mediterranean as the technology spread. By this time, though, most of the elements of agriculture and related technologies are in place in the Levant.

A number of parallel developments were occurring in Anatolia, also leading to the adoption of agriculture. Contact with this area and the Levant had been apparent since the later Natufian period (see Chapter 9) because of the exchange of Anatolian obsidian, and thus it can also be considered that there was an exchange of technology between the Levant, Anatolia and other areas of early domestication in Mesapotamia and into the Zagros mountains to the east. In eastern Anatolia, the site of Çayönü has evidence for the use of wild emmer wheat by 9016–8271 BCE, in association with other wild species of pistachio, almond and vetch, and a hunted fauna of aurochs, boar, red and fallow deer, ovicaprids and other smaller animals (Mellaart, 1975; Yakar, 1991). By the later levels of occupation at the site, dated to 8076–7033 BCE, there are both domestic animals (ovicaprids and possibly pig) and plants (emmer and einkorn wheat, peas, lentils and vetch) as well as gathered pistachio, almond, acorn and hackberry and hunted aurochs and deer (Mellaart, 1975). The site also has stone buildings with plastered floors, some attempts to make pottery, clay bricks and figurines. Further west at the same time, the site of Cofer Höyük (first occupation before 8341–7620 BCE) has a village structure, but evidence for wild animals only (Yakar, 1991). Similarly, at Suberde (earliest date of occupation 7912–7262 BCE) a very large faunal assemblage points only to the hunting of ovicaprids and cattle as well as a wide range of other species, but in conjunction with very few faunal remains. There have been some suggestions that the ovicaprids here and at Aşıklı Höyük (occupied from at least 8327–7636 BCE) represent an early form of domestication based on the age structures (Payne, 1985), although the same pattern could equally relate to communal hunting as discussed in the previous chapter. At Can Hassan III, however, einkorn, emmer and lentils were grown from the earliest levels in the village (possibly as early as 7854–7490 BCE), with collection of wild einkorn, vetch, légumes, grasses, bulrush and blackberry (Yakar, 1991). Animal remains at the site include cattle, ovicaprids, pig, onager, red and roe deer, hare, dogs, birds, snakes, tortoise, fish and other small mammals.

On the Konya Plain, the well-known village site of Çatal Höyük (Mellaart, 1967; 1975) has extensive evidence for cereal cultivation from the earliest known levels (Kuniholm and Newton, 1996, date the occupation of Çatal Höyük to 7020 BCE ± 50 to 6500 BCE ± 100, based on radiocarbon determinations of a 570-year tree-ring chronology of juniper wood used to build the houses on the site). Plant remains include einkorn which is probably in an early stage of being domesticated, emmer, naked barley and peas, as well as collected species such as pistachio, almond and hackberry (Yakar, 1991). Domestic cattle seem to have been the principal meat source, and this is reflected in the extensive symbolism relating to bull horns and wall paintings at the site (Figure 10.2; Mellaart, 1967; 1975). The sheep from the site still appear to be wild, and goat are very rare. That hunting was important is reflected both in the faunal remains and in wall paintings of hunting scenes (Figure 10.3). The village again has extensive architecture with rooms interpreted as having communal or ritual purposes, and burials within the settlement. The combination of a large fire which destroyed the settlement at one point and waterlogging led to excellent conditions of preservation, with evidence for textiles and wooden artefacts including bowls and boxes. A recent re-evaluation of the site (see Hodder, 1996) has demonstrated that the site was founded on an alluvial plain which had only started to accumulate on the bed of the former Lake Konya at the start of the Holocene (Roberts *et al.*, 1996). This plain would have been particularly poorly drained, encouraging the upward building of the settlement tell to minimize flooding. On the other hand, as with early agricultural sites in the Levant, the proximity to a water supply – in this case the Çarşamba river – would have allowed the optimal development of agriculture. Yakar (1991)

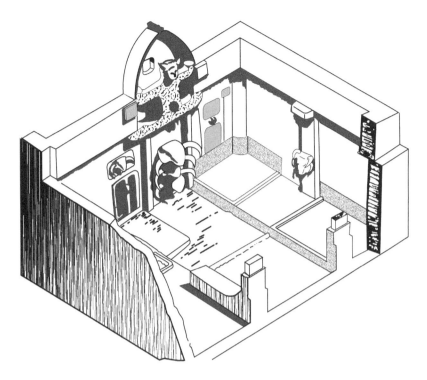

Figure 10.2 Bull horns and wall paintings relating to symbolic representations of domesticated
species at the early Neolithic settlement of Çatal Höyük

Source: After Mellaart (1975).

Figure 10.3 Wall paintings of hunting scenes at Çatal Höyük, suggesting the co-existence of hunting
with agricultural activity

Source: After Mellaart (1975).

notes that proximity to water supplies is a characteristic of a large number of early Neolithic sites in Anatolia. Roberts *et al.* (1996) also note an early case of the pollution of water-courses, with evidence for the dumping of pottery, bone and other debris in a river channel near the site during the later Chalcolithic occupation. Yet further west, the site of Hacılar again provides some evidence for early domestication. Carbonized grains from an Aceramic Neolithic level dated to 8061–7422 BCE are claimed to be of domestic forms by Mellaart (1975), although Yakar (1991) is more cautious, pointing out that the sample is very small and the grains very distorted. Agriculture was, however, definitely practised on the site before 7032–6172 BCE, with evidence for einkorn, emmer and bread wheat, hulled and naked barley, lentil, pea and vetch. Fruits including blackberry, caper and apple, as well as almonds were also collected for consumption.

Following the evidence for early occupation at Aetokremnos in Cyprus (see Chapter 9), there is an aceramic Neolithic occupation of the island, although the links between the occupants at Aetokremnos and the later inhabitants are poorly known. The earliest Neolithic dates are 8829–8022 BCE at Kalavassos Tenta (Todd, 1987) and well before 7301–6778 BCE at Shillourokambos (Guilaine *et al.*, 1998). These sites include evidence for ovicaprids, pigs and (at Shillourokambos only) cattle, as well as einkorn and emmer wheat, barley and lentils. The slightly later sites of Khirokitia (7245–6460 BCE, Le Brun, 1988) and Dhali Agridhi (7192–6600 BCE, Lehavy, 1989) have evidence of wheat, barley, lentil and pea, as well as ovicaprids and pig. Most of these sites are again located close to floodplains which may have exploited floodwater farming, although Todd (1987) also highlights the possible defensive nature of most sites. The source of the incoming populations is again unknown, and may either be from the Levant or from Anatolia (Held, 1989). Continuing direct or indirect contacts with these areas are demonstrated by the continued presence of Anatolian obsidian (e.g. Todd, 1987). A slightly later settlement at Cape Andreos Castros (7007–6373 BCE: Le Brun, 1981) on the north-eastern tip of the island suggests that there may have been several routes of migration into the island. An interesting factor in the early settlement of Cyprus is that there is an apparent gap in the early agricultural record after these aceramic Neolithic sites, which lasts for possibly as long as a millennium. No other Mediterranean island appears to have suffered a hiatus in settlement after the arrival of agricultural communities, and it remains to be seen whether the gap is real, or a result of missing data (see Cherry, 1990).

The westward spread of agriculture across the Aegean seems to occur relatively rapidly, with the first Neolithic sites in Greece being dated to around *c.*7000 BCE (van Andel and Runnels, 1995). At Franchthi cave in the Argolid, sheep and goat as well as cultivated lentils, wheat and barley are found first in levels dated to 6998–6595 BCE which show continuity with the previous settlement at the site, with pottery-making technology arriving only later (van Andel and Runnels, 1987). Early Neolithic sites seem to be sparsely scattered through Greece, apart from a concentration in Thessaly, particularly in the Larissa Basin, and unlike Franchthi are mainly in areas without a pre-existing Mesolithic population. Van Andel and Runnels (1995) suggest that this cluster is real rather than an artefact of where archaeologists have looked for sites. Sites tend to be clustered near the edges of floodplains, a fact which van Andel and Runnels interpret as being related to the use of spring floodwaters as the basis for the successful growth of crops. The need for predictable harvests was, they point out, all the more important because of the sparse distribution of Neolithic sites and because help from areas not affected by drought in years of lower rainfall would be difficult to come by. Emmer wheat may have been a preferred crop because it could tolerate the initially wet soils. Thus, as with the early agricultural sites in the Levant and Anatolia, there seems to have been a preference for locations which were naturally well irrigated. The lack of a pre-existing Mesolithic population in many of the newly settled areas in Greece leads van Andel and Runnels to suggest that people as well as the technologies were moving (although in some places such as Franchthi, it may just have been the technology). However, simple models of movement driven by population pressure (e.g. Ammerman and Cavalli-Sforza, 1984) are no longer tenable in the light of available settlement evidence.

At a similar date (7006–6230 BCE), the first settlement appeared on the island of Crete at the site of Knossos (the location of the later, Bronze Age, palace and the infamous labyrinth), with permanent structures constructed shortly afterwards (6627–6049 BCE: Cherry, 1990). Despite extensive searches, there is still no reliable evidence for a Mesolithic settlement on the island (Cherry, 1990; Strasser, 1991). The earliest, aceramic, levels at Knossos include remains of bread wheat, other cereals and légumes, ovicaprids, cattle and pig (Evans, 1968), none of which are indigenous and thus must have been introduced (Broodbank and Strasser, 1991). Broodbank and Strasser suggest that although the island is visible from Melos, which was known to have been visited in the Mesolithic to obtain obsidian (which is also found in the Neolithic levels of Knossos), it is more likely that the early colonists took a more circumspect route because of the need to water the animals being transported. The exact path is, however, unknown because of a lack of information relating to this date on the intervening islands. The location of Knossos is ideally suited for agriculture, on a fertile alluvial plain just above a permanent stream, and thus the presence of naturally watered soils (Jarman *et al.*, 1982). The site is also conveniently close to large areas of land suitable for rough grazing. Movement into the interior of northern Greece occurs slightly later, for example, at the tell of Nea Nikomedia in Macedonia, where einkorn and emmer wheat, barley, lentil, pea and bitter vetch are recorded from 6536–6180 BCE (van Zeist and Bottema, 1971), possibly on similar soils to those found around Knossos (Jarman *et al.*, 1982).

Expansion of agriculture to Sicily was seen by 7032–6544 BCE at the Grotta dell'Uzzo (Tusa, 1985; Constantini, 1989), when einkorn and emmer wheat are found with vetchling and lentil as well as cattle, ovicaprids and pig. Barley appears at the site shortly afterwards. A second series of cultivars including bread and club wheat, horse bean, bitter vetch and pea arrive at the site around 5711–5488 BCE. Thus, there seems to have been a relatively rapid dispersal of agriculture through the Aegean and into Sicily and probably southern Italy, for example, at Casa San Paolo (7039–6467 BCE: Whitehouse, 1987) and possibly Torre Sabea (Cremonesi *et al.*, 1987). Small village settlements are known from southern Calabria by 5941–5638 BCE at Piana di Curinga (Ammerman, 1985). Cultivation of einkorn and emmer wheats, barley and lentils is attested from the settlement of Torre Canne in Apulia at a similar time (5944–5595 BCE: Coppola and Costantini, 1987).

The continued westward spread of agriculture follows a coastal pattern, first to the western coast of Greece (e.g. Sidari on Corfu, dated to 6703–6216 BCE: Sordinas, 1969), then to the Tavoliere of south-eastern Italy, and along the Dalmatian coast (van Andel and Runnels, 1995; Chapman and Müller, 1990). Apart from the very early site of Coppa Nevigata (*c.*7150 BCE, although there is much debate about this date), which is a coastal site with pottery and extensive cockle shell remains, most of the Tavoliere sites are located to exploit soils ideally suited for agriculture (Jarman *et al.*, 1982). More recent re-excavation at Coppa Nevigata produced evidence for emmer and eikorn wheat as well as some barley, and provided some confirmation for a relatively early date for the site (7485–5971 BCE: Whitehouse, 1987). Early open settlements date to 6059–5669 BCE at Masseria La Quercia and 6058–5529 BCE at Rendina 2 (Cassano and Manfredini, 1993) and at 5988–5891 BCE at Masseria Giufreda and 5959–5449 BCE at Villa Communale (Guilaine *et al.*, 1981), while the earliest of the large villages on the Tavoliere occur at 5684–5258 BCE at Villagio Leopardi and 5593–5335 BCE at Madalena di Muccia (Whitehouse, 1987), and at 5359–4734 BCE at Passo di Corvo (Tinè, 1983). At Gudnja Pećina on the Dalmatian coast, domestic ovicaprids and pig are present from 6158–5870 BCE, while wheat is present at the lowland, open settlements at Pokrovnik I (6003–5632 BCE) and wheat and ovicaprids, cattle and pig at Tinj-Podlivade (6122–5527 BCE: Chapman and Müller, 1990). Pottery is found at all three of these sites.

Both Corsica and Sardinia seem to have been explored in the later Mesolithic period (Cherry, 1990), and Neolithic settlement seems to have been early. Pottery, together with domestic sheep and dog are found at Corbeddu on Sardinia possibly as early as 7269–6604 BCE. Basi on Corsica has an occupation with pottery and a fauna dominated by domestic ovicaprids and pigs at 7002–6187 BCE, while at Curacchiaghiu these appear in small numbers by 6993–6038 BCE (Lewthwaite, 1985). Filiestru cave on Sardinia has ovicaprids, cattle and pig by 5692–5444 BCE (Trump, 1984). The absence of cereal grains was tested by thorough sieving of the sediments, although the location of the site would have made it unsuitable for agriculture. Obsidian from Sardinia is found from Neolithic levels on Corsica, northern Italy and southern France, demonstrating important interactions between these areas (Hallam *et al.*, 1976; Tykot, 1998).

In southern France, the earliest dated site of Cap Rognon, now an island just off Marseille, has pottery in association with shells, similar to Coppa Nevigata in Italy (although here, too, the date of 7300–6460 BCE has been disputed, and may be nearer to the date of 6993–6038 BCE for the nearby Île Riou). Nice-Caucade has a date of 6702–6050 BCE (Binder, 1992). At Châteauneuf-les-Martigues, the transition to the Neolithic is seen at a previously occupied Mesolithic site, with levels including pottery, silos with large numbers of grains, and a fauna dominated by ovicaprids with some cattle and a range of hunted species in a level dated to

Box 10.3 Sea-level change and its impact on understanding the spread of agriculture

Although sea levels began to rise from the period of the last glacial maximum, as ice-sheets and glaciers began to melt, the process took a significant period of time to complete, in part because of the gradual nature of the warming and in part because of time delays of the changes feeding through the ocean system. Data on changing sea levels in the Mediterranean have been collected by Pirazzoli (1991). Some examples are illustrated in Figure 10.4.

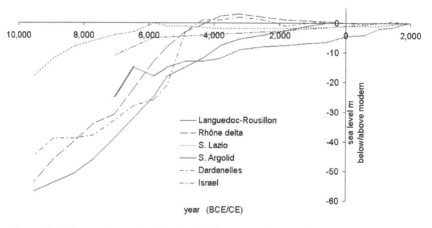

Figure 10.4 Changes in sea levels in the Mediterranean during the Holocene

The effect of sea-level rise would obviously have been a progressive drowning of the coastal zone, significantly changing the position of the coastline in places. The amount of change also depends on the topography, so a rocky coastline would tend to move inland a much shorter distance than would a coastal plain under conditions of increasing sea level (see Figure 10.5).

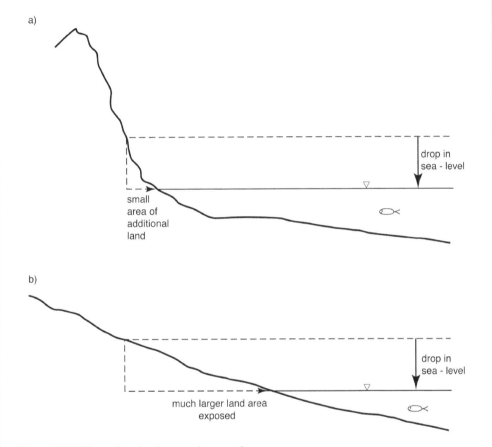

Figure 10.5 Effects of sea-level rise in the coastal zone

The sea-level rise meant that significant areas of the LGM coastal plain would have been drowned in the early Holocene, submerging important archaeological sites, relating to the first landfalls of agricultural settlers (or coastal communities in contact with farmers) as some marine excavations demonstrate. We also need to be careful in interpreting the location of sites near the coast, which may have been far distant in the early Neolithic, as for example at the important site of Franchthi in the Argolid (Lambeck, 1996).

7001–5870 BCE (Escalon de Fonton, 1980). A number of southern French sites seem to indicate the presence of ovicaprids before other domesticates and pottery, including Châteauneuf-les-Martigues, Gazel (6991–6429 BCE), the Roc de Dourgne (5941–5526 BCE: Geddes, 1983; 1985), and Gramari (7475–6424 BCE and 7042–6177 BCE, although there is some debate about the stratigraphic integrity of this site). Thus, the French early Neolithic appears to have

developed an agriculturally based economy in stages, a fact which Lewthwaite (1989) has attributed to a 'filtering' of imports via the Tyrrhenian islands. Alternative suggestions that these
animals may relate to hunted animals that escaped from adjacent Neolithic settlements (Davidson, 1988) is difficult to support at present given the lack of known sites anywhere nearby.
Part of the reason for this may be the drowning of many of the original sites of dispersal by
sea-level rise in the Early to Middle Holocene (see Box 10.3). Such a hypothesis is supported
by the discovery of the site of Leucate-Île Corrège, now submerged to a depth of 4 m below
sea level (Guilaine *et al.*, 1984). Dated to 5807–5522 BCE, the fauna is dominated by ovicaprids, cattle and pig. The presence of cereals is well attested by this time in the lowland sites
of Fontbrégoua, and again, Châteauneuf-les-Martigues.

The earliest introduction of agriculture to Spain seems to take place around the same time
as that in southern France (discounting the probably erroneous date of 7291–6455 BCE at
Verdelpino in Cuenca). At Nerja, levels with ovicaprids and pig but no cattle are dated to
7263–6383 BCE (a similar date of 7290–6375 BCE occurs at the Cueva Chica de Santiago,
further inland), while the Cueva de la Dehesilla (7499–5677 BCE) is dominated by hunted
species, with only small quantities of ovicaprid, cattle and pig (Pellicer and Acosta, 1983).
Thus, domesticates seem to have been introduced first into Andalucia and only subsequently
into Valencia (e.g. 6622–6212 BCE at Cueva Fosca) and Catalonia (Llongueras i Campana,
1987). However, the date of 5726–5523 BCE for the Cueva Chaves in Alto Aragón might
suggest that coastal sites in the north-east have been destroyed or are yet to be discovered (but
see Zilhão, 1993, for a discussion of the Andalucian evidence). Definite evidence of cereals
appears first in Andalucia (5772–4719 BCE at the Cueva del Toro and 5252–4995 BCE at the
Cueva de los Murciélagos, Zuheros: Acosta, 1987) and later in Valencia (5216–4165 BCE at
the Cova de Recambra: Capderilla, 1991). In Portugal, the earliest Neolithic dates seem to be
from the Gruta do Caldeirão (5433–5071 BCE, the dated material being a sheep bone: Zilhão,
1988, although there is possibly an earlier phase dating to 6119–5344 BCE: Zilhão, 1993).

The Balearic islands also appear to have been explored by pre-agricultural populations, and
there are now close associations with human remains and those of *Myotragus balearicus* (an
endemic species similar to antelope) suggesting that the former may have been responsible at
least in part for the extinction of the latter (Waldren, 1986; Cherry, 1990). There seems to be
no evidence for attempts to domesticate this species (Ramis and Bover, 2001). The first introduction of pottery and domesticated ovicaprids, pigs, cattle and dog appears relatively late,
around 3663–3032 BCE at So'n Matge on Mallorca.

In North Africa, the early Neolithic settlement at Les Travertins, Oran in Algeria has been
dated to 7045–6182 BCE (Guilaine, 1979), although this date has been disputed by Gilman
(1976). The site of Columnata, with a similar association, dates only to 4938–4466 BCE. Other
sites in the Maghreb with domesticates in a context with the pre-existing stone-tool industry
(the so-called Neolithic of Capsian tradition) occur as early as 5929–4928 BCE at the Grotte
Capéletti in Algeria (Roubet, 1980). The earliest site with domesticated ovicaprids and cattle
in northern Morocco is Kaf That el Ghar (5253–4694 BCE: Daugas *et al.*, 1989). Agriculture
in the Maghreb is a late establishment, typically occurring after 3800 BCE.

In Mediterranean north-east Africa, the record is more problematic. In Libya, the site of
Haua Fteah provides the first evidence of domestic caprines *c.*5650 BCE (Higgs, 1967), and
there are few other known remains apart from some surface scatters of material which appears
similar to the so-called Neolithic of Capsian tradition (Barker, 1996). Although it had been
proposed that an early centre of domestication was found at Nabta playa in the Western Desert
in southern Egypt (with domestic as well as wild barley, some cattle and ovicaprids in a dominantly hunted fauna dating to between 7486–6778 BCE and 6619–6393 BCE: Wendorf and
Hassan, 1980), a subsequent re-evaluation of the data suggested a 4038–3047 BCE date was

more appropriate, as the site stratigraphy had probably been compressed due to aeolian deflation (Wendorf and Schild, 1984). The earliest evidence for the practice of agriculture in Egypt therefore comes from Merimde Beni-Salama on the edge of the Nile delta. The site has emmer and some free-threshing wheat, barley, lentil, pea and flax in contexts dating to *c*.5000 BCE (Hassan, 1985). Further south in the Nile valley, there is possible evidence at Badari for emmer and six-row barley cultivation together with domestic cattle and ovicaprids as early as 5274–4534 BCE (Wetterstrom, 1993). In the Fayum oasis, there is the introduction of domestic ovicaprids and possibly cattle and pig to sites with an otherwise hunted subsistence base by *c*.5230 BCE, but domesticated plants, which include emmer, some barley and flax, do not arrive for several centuries after this. Wetterstrom suggests first that this pattern reflects the selective uptake of domesticates by hunter–gatherer populations, who were looking to diversify their subsistence base in years when major floods of the Nile drastically reduced available food supplies. Second, she argues that agriculture only develops following the widespread relocation of south-west Asian farming communities, possibly in response to a period of drought. By 3946–3517 BCE, the extensive cereal pollen at Nagada in the Nile valley downstream of Thebes suggests the presence of large grain fields, storage and/or processing nearby, plus pollen from large numbers of plants with medicinal properties and tree species introduced, which were not indigenous, such as cedar and lime (Emery-Barbier, 1990). Thus, once in place, the agricultural systems seem to have become rapidly well established in the Nile valley.

The spread of agriculture through the Mediterranean was, therefore, by no means a straightforward and unilinear process. Following the initial accumulation of technology in the Levant, there was an initial phase of westward movement. Probably, the new agricultural technology was accumulated slowly, and was perhaps linked to existing networks of exchange that were focused on obsidian, shells and possibly other forms of goods, which have not survived. In a relatively continuous manner, the technology spread across the Ionian Sea to Sicily and southern Italy, Sardinia and Corsica, and up into the Adriatic. The spread into the western part of the basin then continued much more slowly, with evidence for only selective uptake of agriculture, and a possible early focus on pastoral activity and hunting. Lewthwaite (1989) proposed the 'island filter model' for the early Neolithic of the western Mediterranean, suggesting that only selected techniques were passed on by communities on Corsica and Sardinia to those in southern France and Spain. The movement to North Africa was much slower, a factor that is not necessarily easily explainable in terms of an overly harsh climate at the time (see Chapter 4), and may rather reflect the networks of contacts between human populations. Indeed, as most contact seems to have been by boat, rather than the difficult hinterland, the prevailing currents of the Mediterranean may go some way to explaining this lack of contact with the south. The landscape pattern that evolves through this period is thus one of patchiness, with localized clearance for agriculture (Box 10.4) and pasture adjacent to uncleared areas – either due to the low population densities or the fact that hunter–gatherer communities co-existed for up to several thousand years with the agriculturalists. There is some evidence also to support the assertion that these may have been the same groups using the landscape in different ways at different times of the year.

Human impacts on this landscape would therefore tend to have been very localized and relatively minor because of its rather diffuse nature. Whittle (1996) emphasizes the fact that these first farmers were not necessarily the sedentary villagers that are usually assumed, and that mobility in the landscape may have been more the norm. This interpretation would be compatible with the lack of extensive landscape modification during this period, as will be seen below. Although the agriculturalists were modifying the landscape, they were yet to do so in a way that would have significant long-term impacts.

Following the initial adoption of agriculture, there is a second important development in

Box 10.4 Evidence for the use of fire in agriculture

The early history of the use of fire in the Mediterranean Basin was discussed in Chapter 7, where it was noted that the ability to make fires was known by some of the very earliest hominid populations of the region. Therefore, in this section, we will concentrate on evidence for its use for clearance in early agriculture.

The evidence for the direct use of fire for clearance for early agriculture tends to be limited, principally because of the concentration of archaeological studies on 'sites', namely settlements, caves and burials, rather than on the areas surrounding these sites. However, some recent studies are beginning to demonstrate its use. Julià *et al.* (1994) have demonstrated the presence of both pre-Neolithic wildfires and more recent fires in a series of cores from Spain by studying the frequency of charcoal fragments in the sediments. Similarly, in the Valdaine region of southern France, there is evidence for wildfires in the driest parts of the late Pleistocene both from charcoal data and vegetation assemblage, as well as in a very marked horizon in the early Neolithic, where the quantities of charcoal suggest repeated burning (Berger *et al.*, 1994). Following abandonment of some areas, there is evidence that the forest may have been burned in the Chalcolithic and in the early Iron Age. Soil studies in the Vera Basin of southern Spain have indicated the presence of a large number of charcoal fragments in levels dating to the middle and late Bronze Age (*c.*1550 to *c.*700 BCE), which probably represent repeated clearances (Courty *et al.*, 1994). Fire as a means of vegetation clearance was apparently extensively used in the Classical period (Meiggs, 1982; Hughes, 1994).

That burning and its effects were an integral part of Mediterranean life is demonstrated by the specific vocabulary developed. The French words *essart*, *issart* or *taillade* were used to reflect cultivated areas on ground recently cleared by burning (Amouric, 1992). Indeed, '*Les Essarts*' is a commonly occurring place-name in France. *Écobuage* is also used to mean a deliberate clearance of land by fire. Fires for clearance seem to have been particularly common in the figures presented for the Var *département* of southern France from the eighteenth century, according to the data presented by Amouric. Furthermore, the fact that often these clearances would get out of hand is reflected by another word in Provençale – *usclado*, which referred to a clearance fire that had become uncontrollable and damaged land and property. That this problem is not recent can be seen in the writings of Plato, who implies there may have been laws in Attica against spread of fire to trees on neighbouring property (Hughes, 1994).

the exploitation of domesticates allowing potentially more efficient use of environmental resources and thus potentially allowing further increases in population density which could be supported within a given area. This development has been called the 'secondary products revolution' (Sherratt, 1981). Rather than simply producing animals for their meat, a series of secondary products, including the drinking of milk and its conversion into dairy products, and the use of animals for transport and as sources of power – perhaps most importantly for pulling ploughs which would increase agricultural productivity – were employed. The plough seems to have first developed in the mid-third millennium BCE, from pictographic evidence on cylinder seals in Mesopotamia and Assyria and images in Egypt (Sherratt, 1981). Clay models are known from the Cypriot early Bronze Age, and there is a preserved example of an ard from the Lago di Ledro near Trento in Italy, and depictions in the rock art of Val Camonica (Anati, 1961) as well as Monte Bego in the French Alps (Masson, 1993), although Sherratt notes that there is indirect evidence of ploughing in northern Europe from the later fourth millennium. Assuming that the plough occurs at the earlier time of the first occurrence of carts in Mesopotamia, Sherratt estimates that the plough may have reached the Aegean by *c.*3400 BCE and Iberia by *c.*3200 BCE. As Gilman (1981) points out, one advantage of the plough in Mediterranean environments is that it aids water retention in the soil profile. The first appearance of circular threshing floors and threshing sledges – a technology which has been recorded into the twentieth century throughout the Mediterranean – is noted from southern Israel, again in the fourth millennium BCE (Avner, 1990; Avner *et al.*, 1994). The use of the horse for transport or traction appears to have started in the steppe areas of eastern Europe and spread into Anatolia and the Levant in the late third millennium BCE, while onager and donkeys were used for traction at a similar period in Egypt and the Levant (Sherratt, 1981). Camels also seem to have been domesticated at this time in Egypt. Milking of cattle also seems to be a Bronze Age phenomenon, and Sherratt attributes widespread changes in the style of pottery manufacture to the need for vessels to store and process milk and milk products. Milk production increases the return in terms of protein and energy by four to five times when compared to simple meat production (ibid.). Wool for textiles again seems to be a late fourth and early third millennium phenomenon, becoming important in the Aegean in the second millennium. The introduction of wheeled vehicles around 1900 to 1750 BCE is a further indicator of the importance of animals as providers of traction.

In Spain, the development of a system of 'policultivo ganadero' – selective use of animals in different locations as part of a broader economic system – has been suggested from the Bronze Age (Harrison, 1985). Prior to this time, environmental constraints seem to have been the most important factor in determining the composition of the diet and secondary uses of animals. Horses seem to have been used for draught and/or transport. While some sites show meat production or draught/transport to be the principal mode of production, there is some evidence from La Mancha that dairy herds were kept. The cattle population at Moncín in northern Spain also suggests milking (Legge, 1994), so that the use of milk herds seems to have been more common in Bronze Age than in recent historical Spain. The age structure of goats suggests an important use for the production of milk, while sheep were principally used for meat or wool – although manure is also likely to have been a further important secondary product, used to maintain the fertility of the fields. Pigs were less frequently used in the Bronze Age, and seem to have served mainly as a meat supply.

It is also possible to note the introduction of developed crops, particularly the cultivation of shrubs and trees for fruits and nuts. In particular, the Mediterranean staples of vines and olives, as well as dates and citrus fruits fall within this category. The importance of such crops is first the length of time required to establish a domesticated plant and bring it to maturity, which requires stability and planning over a period of at least five to ten years. In the case of the olive, full production is reached only after forty years (Boardman, 1976). Second, there is the need to develop special techniques of plant propagation. Plants such as the vine or olive are indigenous to the Mediterranean, and the evidence noted above shows that the wild forms were collected for foods for several millennia before their domestication. Advantages of these tree crops are that they can be interplanted with the cereals and other crops (the so-called *cultura promiscua*), having an annual schedule of work that is complementary to the other crops (Gilman, 1981). This fact not only increases the productivity of a given plot of land, it also promotes the security of the subsistence farmer because of long-term storage of the crops in the form of oil, pickled olives or wine. Together with the aspects of the secondary products revolution, Gilman suggests that the development of tree crops was an important reason for increased social complexity from the Bronze Age.

Domestication of the olive involved the selection of trees with larger, more fleshy fruits which contained more oil. Because trees produced from seed tend to be highly variable, vegetative propagation of using cuttings or grafting must be employed to produce clones which maintain the quality of the fruit (Zohary and Hopf, 1988). Olives have the unique characteristic of a biannual fluctuation in the harvest (Boardman, 1976). They are also extremely sensitive to frost. For example, following the severe winter of 1984–1985 in southern France, olive production declined from 70 tonnes in 1984 to nothing in 1985 and 2 tonnes in 1986, and was still just over half the pre-freeze production levels in 1988 (Erétéo, 1988). The first evidence of olive production was thought to come from Teleilat Ghassul, to the north of the Dead Sea, dated to *c.*4450 BCE (Zohary and Spiegel-Roy, 1975), and probably grown with the aid of artificial irrigation (Zohary and Hopf, 1988). However, underwater excavations of the submerged site of Kfar Samir, just off the Israeli coast to the south of Haifa following its exposure by winter storms in 1993 to 1994 have revealed significantly earlier evidence (Galili *et al.*, 1997). The site contains a number of pits with abundant olive stones and pulp, the earliest of which is dated to 5569–5278 BCE (Figure 10.6). The site also contains in-filled wells of up to 2 m in depth and large, flat, stone basins which have been interpreted as being used for crushing of olives. A range of other finds at the site suggest that the olives were being processed to produce oil. There seems to have been a relatively slow spread of domesticated olives, first, into Syria and thence into the Aegean, although it is entirely possible that some other early sites have been submerged. At Myrtos in eastern Crete, there is extensive evidence for the production of olive oil in the early Bronze Age (Warren, 1972). By the middle Bronze Age, olive cultivation seems to have been widespread in the Aegean (Boardman, 1976), not only for food, but also for manufacture of perfumes and unguents (Melena, 1983). Intensive cultivation of olives seems to have begun before the late Bronze Age in Cyprus (Hadjisavvas, 1992). Sherratt (1981) suggests that the widespread dispersal of ploughing technology before and during the Bronze Age may have aided the spread of tree crops by making the preparation of the ground for planting significantly easier. The spread of the domestic olive to the western Mediterranean is usually attributed to the phase of Greek colonization which took place in the seventh and sixth centuries BCE (Boardman, 1976). The Phoenician colonization phase was also significant, for example, in the introduction of the crop (and vines) around Carthage, to the extent that the Roman Senate had Carthaginian treatises on agronomy translated into Latin in the fifth century BCE (Aubet, 1993). Olive oil and wine were important components of the Phoenician trade system throughout the Mediterranean. Barker (1981) suggests that the introduction of olives and vines into central Italy in the first millennium BCE was fundamental in allowing the expansion of settlement. It is likely that the introduction of the domestic olive and vine to Spain was also a result of Phoenician colonization (Harrison, 1988). However, Gilman (1976) points to the existence of olive stones in late Neolithic and Chalcolithic sites in southern Spain, suggesting that there may have been an independent domestication from the local wild species (although there is no way of determining whether these remains were cultivated or collected from the wild). Terral and Arnold-Simard (1996) suggest an even earlier cultivation in southern Spain based on morphometric analyses of olive-wood charcoal, although the evidence in this case is somewhat ambiguous.

The vine tends to have a more northerly distribution, relating to its preference for cooler and more humid conditions than the olive (Zohary and Spiegel-Roy, 1975). Vegetative propagation is also required to stabilize the domestic plant, and also seems to have brought about self-fertile varieties (Zohary and Hopf, 1988). Domestication also seems to have taken place in the Levant, with early Bronze Age remains being found at Jericho (*c.*3950 BCE) as well as a number of other sites in Israel and Syria. In the Aegean and Cyprus, cultivation seems to have begun in the early Bronze Age, and is again well established by the late Bronze Age period. The site of Myrtos on Crete has significant evidence for wine production. Dispersion to the western Mediterranean again seems to be a later phenomenon relating to the period of colonization.

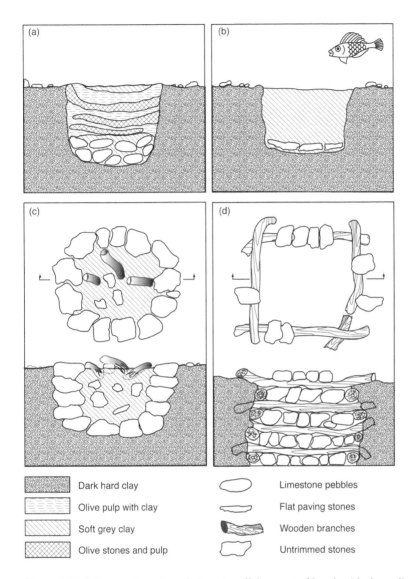

Dark hard clay

Olive pulp with clay

Soft grey clay

Olive stones and pulp

Limestone pebbles

Flat paving stones

Wooden branches

Untrimmed stones

Figure 10.6 Submerged remains of olive pits off the coast of Israel, with the earliest evidence of olive processing, which is dated to 5569–5278 BCE

Source: After Galili *et al.* (1997).

The fig and, in the southern part of the basin, the date palm, also seem to be first cultivated in the Chalcolithic or early Bronze Age (Zohary and Hopf, 1988). It is also possible that almond was first cultivated at the same time, although it is extremely difficult to distinguish between wild and cultivated forms. Walnut also first appears in Turkey and the Balkans in the mid-second millennium BCE, with pollen evidence from before 1618–1437 BCE at Beyşehir Gölü in Turkey and at 1680–1423 BCE at Edessa in northern Greece (Andrieu *et al.*, 1995). It appears to have reached Italy by the early first millennium BCE, but to have taken several hundred years to move further west. Chestnut pollen also seems to expand from the Aegean at around the same time, although there are additional early dates from Italy at Lago Riane (3992–3382 BCE) and Valle di Castiglione (1916–1676 BCE: Andrieu *et al.*, 1995). Although there are suggestions of cultivated pistachio in the late Neolithic and early Bronze Age of Greece, the evidence is questioned by Zohary and Hopf (1988) who believe that the domestication of this tree must have been later because it depends totally on grafting for propagation. In summary, though, we can suggest that the Chalcolithic and Bronze Age of the eastern Mediterranean saw important developments in the ability to produce foods by the development of tree crops. The cultivation of tree crops also implies a greater degree of sedentarism and the development of social mechanisms which allow the appreciation of longer timescales necessary for bringing the crop to full production. In many sites, this is also reflected in the development of extensive storage and production facilities, and a notably hierarchical social system (Gilman, 1981).

Stevenson and Harrison (1992) used pollen data from southern Spain to demonstrate the existence of managed woodlands over extensive periods. Open parklands or *dehesas* have historically been used to provide pasture, fuel, cork, wood and autumn fruits (acorns, olives and later chestnuts). At El Acebrón and Las Madres in Huelva, the first vegetation assemblage typical of a *dehesa* occurred between around *c.*2500 BCE and *c.*1600 BCE. The fact that such woodland could provide spring or winter pasture for sheep, goat, cattle and horses as well as acorns for pigs may suggest an intensification of forest production which ties in with other aspects of the secondary products revolution (Harrison, 1994). There followed a subsequent phase of forest destruction at the sites, probably due to overgrazing. The re-establishment of the *dehesa* landscape did not occur until *c.*500 CE (Stevenson and Harrison, 1992).

A further phase of expansion of crops from the east to the west was associated with the Moorish (Islamic) expansion, particularly into Spain from 711 CE. Although it has previously been assumed that a very wide range of crops were imported, it has been demonstrated that most of these were actually already known in the Roman world (Butzer *et al.*, 1985). Crops which were definitely involved do, however, include some which have subsequently become economically very important. These include apricot, lemon, grapefruit, bitter orange, banana and probably carob as tree crops, as well as sorghum, spinach, cauliflower, taro, safflower, sugar cane, honey melon, watermelon, aubergine, indigo, henna, jasmine and cotton. The Islamic period also seems to have seen the introduction of economically important crops such as sugar, cotton and rice into Egypt (Butzer, 1976).

The landscape implications of these subsequent agricultural developments are significant and contradictory. In some cases, more sustainable techniques were involved, such as the use of managed woodlands, tree crops or mixed planting to maintain a high ground cover. Disastrous erosion episodes could then be relatively minimized while expanding the potential areas of occupation. Such techniques would be vital in the face of expanding populations. Furthermore, the use of a variety of crops and animal species would provide stability under the naturally variable conditions of the Mediterranean environment, by means of various risk-buffering strategies (see below). However, there was also the danger that such techniques would increase the sedentary nature of populations, focusing their impacts on specific locations to a greater and greater extent. Coupled with the enhanced technology of clearing longer slopes and the creation of the furrows that could concentrate overland flows, there was a much greater likelihood that these impacts would lead to significant erosion events. This technology, and the ways of living within the landscape that it brought about, led to dramatic changes in the appearance and form of the Mediterranean landscape that is the ultimate cause of the landscape we see today. We have already discussed the general vegetation change in this respect in Chapter 4. In the following section, we go on to argue that the erosional changes – in some cases irreversible – are a response to these fundamental changes in the methods of environmental exploitation.

Erosion and agricultural exploitation

As noted in Chapters 5 and 6, one of the consequences of agriculture is the production of enhanced runoff and erosion. In the early period of clearance, the use of fire might also be expected to enhance the erosion rates (Chapter 7). We might therefore expect there to be some relationship with the development of agriculture and of increased erosion rates. Erosion as a consequence of agriculture might reasonably be expected to be a major component of landscape evolution in the Mediterranean region in the period described above, and should occur diachronically as different technologies arrived in different locations. Indeed, erosion might be likely to occur in any location where human activity and settlement have led to the clearance of vegetation. The earliest reported example of this is the slope erosion noted around the Kebaran and Geometric Kebaran (see Chapter 9) sites on the Israeli coastal plain. Schuldenrein (1986) has noted that a move from mid-slope sites was probably due to erosion following increased runoff due to locally intensive land use.

The best developed studies of landscape evolution in response to settlement have been the works of van Andel, Runnels and co-workers in various parts of Greece. In the Argolid (van Andel and Runnels, 1987), there appear to have been four major erosional phases (Figure 10.7) since the early Bronze Age (around *c*.3000 BCE). The first, early Bronze Age, phase coincides with the first rapid increase in population noted for the region (see Chapter 9). The

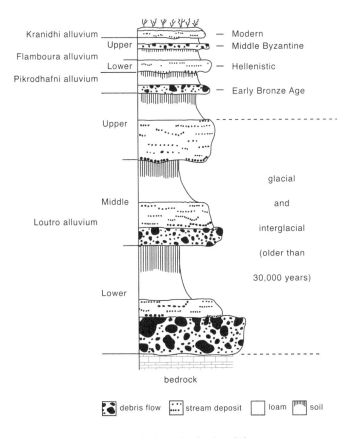

Figure 10.7 Major erosional phases in the Argolid
Source: After van Andel and Runnels (1987).

Bronze Age sites were often located near to formerly deep upland soils which have been subsequently stripped away completely to bedrock. Van Andel and Runnels suggested that a shortening of the fallow period in an attempt to produce more food for the growing population may have led to large areas of increasingly steep slopes being left freshly ploughed (presumably in association with the recently arrived ploughing technology) and vulnerable to erosion in late summer and autumn storms. Although there was an apparent drop in population after the erosional phase, van Andel *et al.* (1986) suggest that it is unreasonable to link the two as there were a number of other destabilizing socio-economic factors in the region at the time. A second erosional phase which seems to relate to the expansion of agriculture occurred in the middle Byzantine period (*c.*900 to 1200 CE), when the region which had been largely unoccupied since the late Roman period began to be reoccupied. Furthermore, there was relatively rapid clearance without reconstruction of terrace walls, again leading to erosion. As well as these two phases of erosion linked to population expansion, van Andel and Runnels (1987) also note two phases relating to population decline, occurring in the Hellenistic to early Roman period (*c.*250 to 50 BCE) and in the twentieth century. The mechanism proposed in this case is the partial abandonment of the landscape and the lack of maintenance of essential structures such as terrace walls (which may be disturbed by the trampling of passing flocks of sheep or goats) or check dams, leading to rapid gullying and erosion of the soil from behind the terraces.

In the Larissa Basin in Thessaly, van Andel *et al.* (1990) note that at least some of the earlier Neolithic settlements occur on the Aghia Sophia soil which was already eroded down to its calcic B horizon. As noted above, this area had a significant concentration of early Neolithic settlement and may have come under pressure from agricultural clearance before others in less densely settled areas. A further phase of sedimentation seems to occur between *c.*4500 and *c.*4000 BCE. On the Argive plain, the first phase of erosion takes place between *c.*5000 and 3000 BCE, with subsequent phases in the later part of the early Bronze Age and at the end of the Bronze Age. Indeed, parts of the Mycenaean city of Tiryns are buried by several metres of sediment. Erosion also seems to rapidly follow the introduction of agriculture into Macedonia, and relatively soon in Nemea and Euboea. Elsewhere in Greece, erosional phases seem to be associated with the Classical, Roman and Byzantine periods (Figure 10.8). Davidson (1980) points out that soil erosion occurs with the later Bronze Age occupation and continues into the first millennium BCE on the island of Melos, while on Santorini, soil degradation seems to have been advanced by the time of the eruption in *c.*1628 BCE (see Chapter 2). In Macedonia, there is a phase of alluviation corresponding to the occupation of the tell of Sitagroi (Neolithic to mid-Bronze Age), followed by successive events in the fifth to third centuries BCE and in the Classical period. The last few centuries BCE seem to have been particularly important in producing widespread erosional events around the Aegean, including on its eastern coast (van Andel and Zangger, 1990). In a number of cases, relatively sophisticated engineering programmes seem to have been developed to mitigate their effects. There also seems to have been a significant phase of erosion relating to the Roman settlement of Crete from the first to fifth centuries CE (Blackman and Brannigan, 1977).

In Sicily, Judson (1963) produced evidence for deposition in stream valleys at some point well before the eighth century BCE, when a phase of stream erosion ended, between the eighth century and *c.*325 BCE, and in the 'Mediaeval' period (probably fifth to fifteenth centuries CE). Two valleys in Etruria produced evidence for a different sequence, with no evidence for the earlier period. In the case of the Crescenza valley, stream deposition starts in the early third century CE, continuing until *c.*1534, whereas in the Volcetta valley the deposition is dated to some point after the second century. At Narce, aggradation phases are noted from the middle Bronze Age to *c.*250 BCE, from 150 CE to possibly as late as the eighteenth century CE, albeit

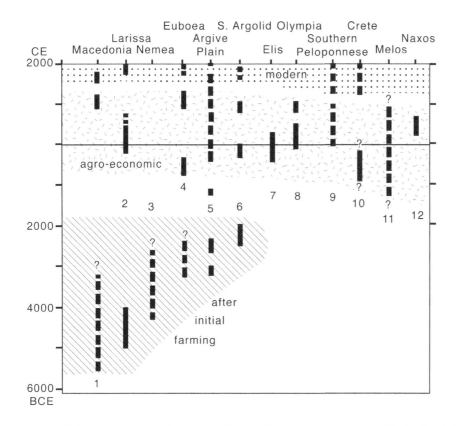

Figure 10.8 Erosional phases throughout Greece, showing an association with the Classical, Roman and Byzantine periods

Source: After van Andel *et al.* (1990).

with a break in the tenth century (Potter, 1979). The onset of both of these phases probably relates to population expansion, with the southern Etrurian landscape being particularly extensively settled by the second century CE (Figure 10.9). However, the maintenance of the second erosional phase is probably related to a number of causes, including the relative decline in the rural settlement from the fourth century onwards. It should be noted that the second phase of erosion occurs at Narce before those in the nearby valleys studied by Judson. Seven depositional phases were found in sedimentary studies of the Biferno valley system in east-central Italy (Hunt, 1995). The first of these is relatively poorly dated, but is probably Bronze Age, or slightly earlier. If the Bronze Age date is correct, then this corresponds with the first expansion of the settlement system in the valley through the altitudinal zones, rather than just in the lower reaches (Barker, 1995). These levels are separated by a phase of stability marked by soil development. There then follows a phase known from both thick alluvial and colluvial deposits, relating to the period from the fifth or fourth century BCE to no later than the second century CE, which coincides with the Samnite and then Roman large-scale clearance of the landscape. A single section contains evidence for an early Mediaeval erosion phase, dated by charcoal in the base of the silt deposits to 656–990 CE, which would correspond to a phase of declining population and localized, small-scale clearances. Other localized deposits suggest a similar pattern in the thirteenth and nineteenth centuries. There are two phases of extensive

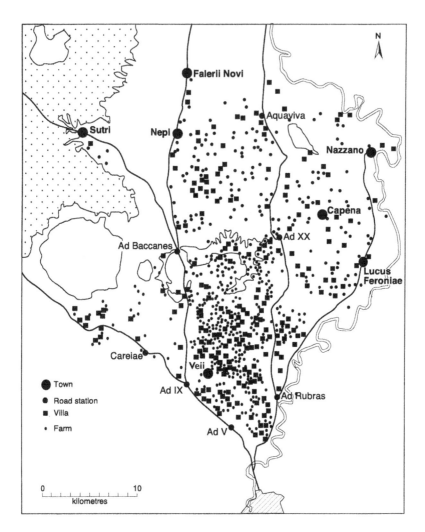

Figure 10.9 Population expansion in southern Etruria
Source: After Potter (1979).

erosion in the twentieth century, the first relating to the extensive cultivation of the inter-war years, and the second relating to the mechanization of agriculture in the last three decades.

In southern France, Ballais and Meffre (1994a) looked at the erosional history of the Haut Comtat. The first erosional phase is poorly dated, but prehistoric, topped with a palaeosol containing Iron Age pottery. Roman cultivation in the Sausses valley seems to have led to extensive colluviation, with up to two metres of sediments, and abandonment from the second century CE. This corresponds with a second phase of much wider accumulation of sediments. A further Mediaeval erosional phase post-dates 1036–1279 CE. Marked erosional phases occur in the mid to late Neolithic, the Chalcolithic to early Bronze Age, the Iron Age and during the third century CE in the Valdaine (Berger *et al.*, 1994). Limited erosion reoccurs in the twelfth century with the re-establishment of intensive agriculture, and increases more rapidly in the fourteenth century. To the south, in the Montagnette and western Alpilles, there seem to have been a number of distinct erosional phases. At Servanne, they belong to the early Iron Age and post-Roman periods (Jorda *et al.*, 1990), while at Les Barres, they belong to the early Bronze Age, mid-Iron Age and an undefined post-Roman period (Ballais *et al.*, 1993). Colluvium covers an Iron Age site (620 BCE to 580 BCE) at Mont Valence (Arcelin and Brémond, 1978). At Grande Terre, the phases are Bronze Age, around 426–973 CE, and sub-modern (Gazenbeek, 1994). Elsewhere in these regions, there seems to be extensive, multi-phase erosion following the end of the Roman occupation. In the Massif des Maures, there is evidence for erosion from colluvial deposits dating to the mid-late Neolithic (after 2845–2331 BCE) and the Gallo-Roman period, and from alluvial deposition in the mid-late Bronze Age, the early Iron Age, the tenth to thirteenth, fifteenth to seventeenth and eighteenth to twentieth centuries CE (Bernard-Allée *et al.*, 1994). The record from the Etang de Berre on the coast suggests a late Neolithic erosional phase, relating to agricultural expansion, stability from the Bronze Age to Roman period during continually increasing pressure, and a large phase of erosion in the post-Roman period (Provensal *et al.*, 1994).

On the island of Porquerolles off the southern coast of France, Provensal (1995) has noted the presence of two erosional phases. The first contains charcoal dated to 413 BCE−76 CE and 65–606 CE, while the second phase begins by 547–1218 CE. In the lower Gapeau valley of the mainland, there are two erosional phases, the first of which ends around 722–1161 CE, and the second is relatively recent, with a date of 1213–1954 CE.

At Laval de la Bretonne, near Carcassonne, Wainwright (1994; Gascó *et al.*, 1996) demonstrated using a modelling study that a phase of accelerated erosion in the early Bronze Age was probably related to the occurrence of one or more extreme rainfall events following land clearance for agriculture (Figure 10.10). The extensive removal of soil from the hillsides upslope of the site seem to have led to their unsuitability for agriculture subsequently because of the relatively slow rates of soil recovery. Other episodes of accelerated erosion have been suggested at the site of Frías de Albarracín in Teruel, Spain, based on evidence from multiple radiocarbon dates of the stratigraphic sequence (Harrison and Wainwright, 1991). Following a first occupation, the site is covered by an accumulation of 1.35 m of sediment, which was probably laid down in a period of between five and fifty-five years sometime around 2100 BCE to 2065 BCE. A second, slower erosional phase seems to have taken place from 2065 BCE to 1960 BCE, and the final destruction of the site by fire after *c*.1900 BCE seems to have set a further erosional phase of uncertain duration into motion.

Accelerated erosion in the Vera Basin in southern Spain seems to occur in the Argaric early Bronze Age (*c*.2300 to *c*.1960 BCE: Courty *et al.*, 1994), a period of rapid population expansion (Castro *et al.*, 1998). Locally there is evidence for fire disturbance, mud flows and even aeolian erosion, suggesting the extensive removal of the vegetation cover. Erosion continues, but at a slower rate in the post Argaric Bronze Age (to *c*.700 BCE) with probably a low vegetation cover maintained by repeated fires, as suggested by numerous charcoal fragments in soils. There seems to be only one erosional phase between this time and the nineteenth century CE, which is probably immediately pre-Roman. The terrace and horticultural agriculture of the Moorish period seems to have brought relative stability to this landscape, and it is only with the population expansions of the nineteenth and twentieth centuries that erosion rates accelerate again. In the Río Aguas system to the south of the Vera Basin, there are four fluvial terraces relating to accelerated erosion (Schulte, 1996a, b). The first predates *c*.5000 BCE, and is possibly related to a period of more stormy rainfall. The second is of unknown date, but the third is related to 1413–1631 CE, which probably reflects erosion relating to the abandonment of Moorish terrace and irrigation systems. McNeill (1992) suggests a similar phase of erosion in the Guadalfeo valley in Granada. The final set of deposits in the Aguas valley relates to very modern erosion, possibly as late as the 1970s, with increasingly mechanized agriculture.

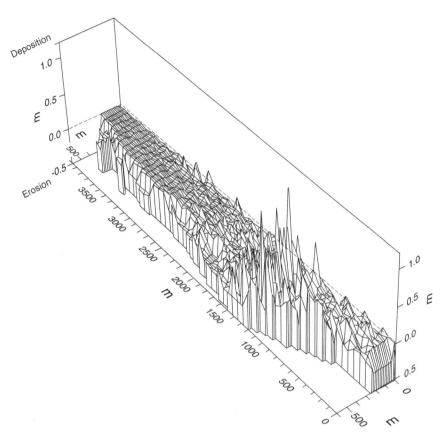

Figure 10.10 Accelerated erosion at Laval de la Bretonne, Aude, France
Source: After Wainwright (1994).

The Bronze Age and early Iron Age in the north-western Mediterranean seem to have seen a phase of repeated but localized erosion (Wainwright, 2000). In France, gullying seems to have been initiated in the Coma del Tech from the mid-seventh century BCE (Allée and Denèfle, 1989), while over 2 metres of slope deposits were laid down at St Guilhem-le-Désert in the several hundred years after *c.*500 BCE (Ambert and Gascó, 1989), and there are thick alluvial deposits in the Calvaire valley in southern Provence dating to the start of the first millennium BCE (Provensal, 1995). Erosion in the southern French Alps seems to have been triggered in various locations between *c.*2000 BCE and *c.*1500 BCE (Jorda, 1992). In Spain, there are colluvial deposits from El Estanquillo dated to *c.*2250 BCE (Borja Barrera, 1992), and in different locations in the Zaragoza region, throughout the period (Burillo Mozota *et al.*, 1984; 1986) and confined to the first half of the first millennium BCE (van Zuidam, 1975; 1976). There also seems to be the start of an erosional phase at Cerro del Mar/Toscanos in the eighth and seventh centuries BCE, relating to the Phoenician occupation of the area (Schultz, 1983). In the Algarve, deposition begins after 998–786 BCE (Chester and James, 1991). In Italy, deposition in the Po valley increases around 2911–2500 BCE, with a second depositional phase around 1516–1269 BCE (Tropeano and Olive, 1989). At Bagioletto, colluvia overlie a palaeosol dating to *c.*2200 BCE while elsewhere in northern Italy there are localized phases of erosion throughout the third to mid-first millennia BCE (Coltorti and Dal Ri, 1985; Cremaschi, 1990). In the Veneto, erosion seems to date to the middle or later part of the Bronze Age (Balista and Leonardi, 1985). In southern Italy, valley alluviation occurred in the mid to late Bronze Age until around 1100 BCE, and again in the eighth and seventh centuries BCE in the Basento and Cavonna valleys (Neboit, 1977; 1984). A first erosional phase is noted in the Bronze Age on the Dalmatian island of Brac as a result of significant population expansion (Stančic *et al.*, 1996). Although these periods do seem to represent a period of sustained erosional activity, the localized nature of the erosion and the variability of its timing suggest that human activity rather than climatic fluctuations are the dominant reason.

In Turkey, Roberts (1990) has studied the relationship between the erosional, vegetation and settlement histories in the region of Söğüt-Balboura. His results demonstrate a close link with a rapid phase of population expansion which coincides with a rapid decline in the juniper–oak–pine woodland around *c*.1050 BCE (Figure 10.9). Following the population decline after around 450 or 580 CE, a pine woodland is able to regenerate on the eroded soils until a further clearance and erosional phase in the nineteenth century.

For North Africa, the evidence is less complete. In Morocco, Lamb *et al.* (1995) suggest that the high clay content in Tigalmamine Lake since *c*.550 CE reflects soil erosion following disturbance. Rognon (1987) notes that it is difficult to distinguish climate fluctuations from vegetation after about *c*.4850 to *c*.3800 BCE because of the impacts of human clearance. Palaeosols were, however, extensively forming in the period from *c*.3200 to *c*.650 BCE, suggesting at least that some landscape stability was maintained over this period. Vita Finzi (1969) notes the onset of alluviation in Morocco and Cyrenaica in the first century CE, but not until the third or fourth centuries in Algeria, Tunisia and Tripolitania. Alluviation is a common phenomenon relating to the Romano-Libyan levels in the area covered by the Libyan valleys survey (Barker, 1996).

There appears to be a phase of sediment accumulation in the southern Dead Sea Basin associated with the early Bronze Age, which saw a phase of expanded settlement in the Levant (Gophna, 1979). Deposition in the Wadi Hasa (Copeland and Vita Finzi, 1978) was still ongoing at 2883–1980 BCE, and seems to be represented in a number of other valleys in the area (Frumkin *et al.*, 1994). The end of this phase is dated to around 2137–1689 BCE. In the Negev, Bruins

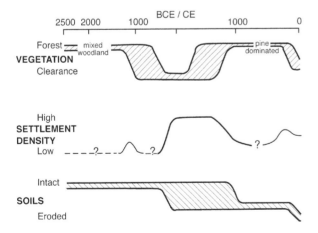

Figure 10.11 Erosional, vegetation and settlement histories in the region of Söğüt-Balboura, Turkey, demonstrating a close link with a rapid phase of population expansion which coincides with a rapid decline in the juniper–oak–pine woodland around *c*.1050 BCE

Source: After Roberts (1990).

(1986) has noted the occurrence of two phases of alluviation, the first of which is linked with agricultural expansion in the later Roman period and is dated to 62–543 CE and continues until around the seventh century CE. The second dates to *c*.1200 to *c*.1700, and is linked by Bruins to the onset of wetter conditions as suggested by tree-ring data (see Chapter 4) and previous levels of the Dead Sea. Goldberg (1986) suggested that there is a quite widespread phase of alluviation in the southern Levant, which is probably due to the expansion of settlement during this period, although he does not rule out the possibility of a climatic cause for this particular event.

Accelerated erosion in the Pindus of northern Greece was noted in the later nineteenth century, with contemporaneous accounts relating the problems to deforestation (McNeill, 1992). Similarly, the nineteenth century in Lucania saw the onset of extremely rapid erosion as a result of the deforestation and intensive agriculture which were required to meet the needs of the rapidly expanding population. The pattern of erosion seems to have accelerated further following emigration and land abandonment in the period from 1880 to 1925. In the Alpujarra of southern Spain, McNeill notes three major phases relating to soil erosion before the twentieth century. The first relates to the period immediately following the expulsion of the Moorish populations in the sixteenth century, when terrace systems fell into decline. The second followed the development of market agriculture in the late eighteenth and early nineteenth centuries, relating to expanding populations. The third was as a result of the *desamortización*, or selling off to individuals of land belonging to the state, church, villages or military orders. This process began in 1813, but accelerated in the second half of the century. Much of the land was bought up by investors trying to make a quick profit, which meant that marketable timber was felled and the land rented out to tenant farmers, sharecroppers or shepherds. The rapid deforestation to which this inevitably led (for example, in Murcia, the proportion of land under forest was 57 per cent in 1755 but only 34 per cent in 1850: López Bermúdez *et al.*, 1995), caused a sudden and very rapid increase in erosion rates. This erosion led to the expansion of the deltas at Motril and Adra. Land alienations such as this were also an important process leading to the nineteenth-century acceleration of erosion in Lucania.

Another aspect of erosion relating to agriculture which has only recently begun to be investigated is that of tillage erosion. Experiments to calculate the effects of manual tillage using hoes on steep slopes in Thailand have demonstrated that on average 8 to $18 \, \mathrm{t \, ha^{-1}}$ were lost in each tillage operation (Turkelboom *et al.*, 1997). Modern mechanical ploughing has produced measured rates of $13.9 \, \mathrm{t \, ha^{-1}}$ for contour ploughing and $28.2 \, \mathrm{t \, ha^{-1}}$ for slope-parallel ploughing (Poesen *et al.*, 1997), so the values for traditional ard or mouldboard ploughing probably lie between these values. However, care must be taking in extrapolating these values to make comparisons with other erosion rates, because the distance of movement is much smaller, and most displaced soil will remain within a particular field or terrace. There are even recorded instances of eroded soils being carried back up slopes and redeposited. However, these tillage-erosion measurements do suggest that there would have been an increase in the erosion rates following the introduction of animal-driven ards and ploughs (see above), which were in widespread use by the early Classical period (Isager and Skydsgaard, 1992; White, 1970). A further important consequence of ploughing is that it will tend to bring stones from the soil profile to the surface (Poesen *et al.*, 1997; Oostwoud Wijdenes *et al.*, 1997), making other agricultural operations more difficult and possibly also enhancing runoff and erosion rates (see Chapter 6).

An important technological means of mitigating erosion is the development of terracing. Terracing works by breaking steep slopes into a number of steps, where the soil surface is relatively flat and therefore less prone to erosion. The deeper soils in the terraces are also beneficial in that they retain more water in the soil. The introduction of terrace systems for agriculture in the Levant is suggested to have occurred in the twelfth or eleventh century BCE, at a time when water-control technology was becoming increasingly sophisticated, allowing an expansion of cultivation into increasingly marginal regions (Gophna, 1979). However, the use of retaining walls for settlement sites is known from the PPNA (Bar Yosef, 1986), and thus the system may have been used for agricultural terracing much earlier. Van Andel and Runnels (1987) suggest that terracing was known from at least the Bronze Age in Greece because of the steepness of the slopes being cultivated, and in some instances the development of soil profiles in terrace soils, which would have taken several thousand years to evolve. As they note, it may be difficult to distinguish between prehistoric and recent terracing, making the history of terracing difficult to define. Barfield (1971) has noted a series of terraces at the early Bronze Age site of Monte Tondo in northern Italy, the uppermost being used for supporting house structures, while those further down the slope were probably for cultivation. If so, this discovery probably represents an independent invention of the technology, several centuries earlier than the examples in the Aegean and Levant. Certainly, terraces seem to have been used by the late Bronze Age in central Italy (Spivey and Stoddart, 1990), and possibly also in southern France (Provensal *et al.*, 1994). As a further example of probably independent invention of the technology, Courty *et al.* (1994) suggest that a number of small stepped features may represent agricultural terraces at the late Argaric (*c.*1950 to *c.*1550 BCE) site of Gatas in southern Spain. In southern France, there are possible Iron Age examples of terraces (Lewuillon, 1991) and some examples known in the Valdaine in the fourth century BCE (Berger *et al.*, 1994).

The general phases of erosion can thus be seen to be widespread and very commonly related to changes in the human exploitation of the landscape. By comparison, the evidence for climate variability and change (Chapter 3), seems very poorly related to the phases of erosion. Moreover, the sensitivity of the erosional system is much more marked in terms of human disturbance, particularly when combined with major storm events. However, there is not necessarily a straightforward pattern in the way that human activity affects the landscape: both erosion and stability can follow phases of population expansion and contraction (Table 10.1). In the period up to the Bronze Age, population expansion and the onset of agriculture typically led to phases of accelerated erosion. The likelihood of accelerated erosion seems to increase in the later Neolithic and initial Bronze Age, probably because populations were starting to expand much more rapidly compared to the capacity of the technology on the land available. Technological improvements such as ploughing with draught animals also meant that larger areas could be opened up more easily. The coincidence of long, bare slopes, with furrows to channel water and sediment, rather than localized planting in small hollows, would have done a lot to accentuate this process. Following this period, improved technology – and probably also a greater awareness of the potential impact of agricultural activity – meant that this link was not necessarily so straightforward, and that significant expansions could take place with relatively minor impacts on the landscape. Perhaps the best example of this taking place relates to the major expansions of the Roman period across the whole Mediterranean world. However, the converse of this pattern was also true. A well-maintained, intensively used landscape needed to have this maintenance carried out constantly. Otherwise the collapse of structures may have led – though by no means in all cases (see also Chapter 16) – to rapidly accelerating erosion, often accompanied by a phase of gullying. Given the lack of general evidence for phases of significant vegetation change throughout most of the time period under consideration here (see Chapter 4), it seems most feasible to explain most of the accelerated erosional activity with human activity. The natural interannual and storm-by-storm variability of the Mediterranean landscape is able to explain these phases of erosion in conjunction with human activity. However, we must remain aware that it is often difficult to determine cause and effect, as most of the evidence is indirect or circumstantial. This problem, however, does not necessarily imply that climatic explanations are to be preferred, as the discussions relating to the 'Younger Fill' of Vita Finzi (1969, 1976) have demonstrated (e.g. Butzer, 1974; Wagstaff, 1981). There is often an inherent circularity in the climate

Table 10.1 Classification of landscape-degradation events as a function of socio-economic status of human settlement for the degradation phases noted in the text. Only broad phases are shown here for the sake of clarity. More precise datings are given, where available, in the text

Socio-economic status	Landscape status	
	Degradation	No/little degradation
Increasing pressure	Israel: Kebaran/Geom. Kebaran, Early Bronze Age, Later Mediaeval?	Argolid: Classical, late Mediaeval
	Negev: Later Roman	Epirus: Late Bronze Age-Iron Age
	Argolid: Early Bronze Age, middle Byzantine	Vera Basin: Visigothic/Byzantine, Moorish
	Argive Plain: Middle-Late Neolithic, Early Bronze Age, Late Bronze Age	Rio Aguas: Roman-*c*.15
	Melos: Late Bronze Age	Valdaine: Early Neolithic, Late Bronze Age, Early Roman
	Nemea: Early Neolithic	Haut Comtat: Early Neolithic-Chalcolithic
	Euboea: Early Neolithic	Etang de Berre: Bronze Age-Roman
	Thessaly: Middle-Late Neolithic	Etruria: Iron Age
	Macedonia: Early Neolithic, Late Neolithic, Classical	Biferno valley: Middle Bronze Age
	Epirus: *c*.19–modern	
	Santorini: Early-mid Bronze Age?	
	Crete: Roman	
	Turkey: Late Bronze Age	
	Alpujarras: *c*.18–*c*.19	
	Vera Basin: Chalcolithic, Argaric, Phoenician/Punic, Roman, *c*.18→	
	Rio Aguas: Chalcolithic, Argaric, *c*.18→	
	Valdaine: Mid/late Neolithic, Chalcolithic, Iron Age, *c*.14→	
	Alpilles: Early Neolithic-Chalcolithic, Iron Age	
	Haut Comtat: Iron Age, Early Roman, Late Middle Ages	
	South France: generally relating to *c*.17–*c*.18→ deforestation	
	Croatia: Bronze Age	
	Etruria: Late Bronze Age, Classical	
	Biferno valley: Late Neolithic/Early Bronze Age?, Samnite/Early Roman, Modern	
	Laval de la Bretonne: Early Bronze Age	
	Teruel: Early Bronze Age	
	Languedoc/Provence/Pyrenees: Iron Age	
	Étang de Berre: late Neolithic	
	Morocco: Roman, Early Mediaeval	
Decreasing pressure	Argolid: Hellenistic/Early Roman, Modern	Vera Basin: Post-Argaric
	Epirus: Post-Roman	Valdaine: Early Mediaeval
	Alpujarras: Post-*reconquista*	Haut Comtat: Bronze Age
	Vera Basin: Post-*reconquista*	Crete: Modern
	Rio Aguas: Late (Post-Argaric) Bronze Age, Post-*reconquista*	(Other periods probably fall within this category, but are not well enough documented for specific inclusion)
	Valdaine: Early Bronze Age, Late Roman	
	Alpilles: Bronze Age, Late Roman, Early Mediaeval	
	Haut Comtat: Post-Roman	
	Etang de Berre: Post-Roman	
	South France: Generally relating to post-Roman depopulation	
	Etruria: Post-Roman depopulation	
	Biferno valley: Early Mediaeval, *c*.13, *c*.19	
	Algeria, Tunisia, Libya: Late Roman	

Note: See also Wainwright (in press).

argument, as there is often assumed to be a simple causal relationship, that is implied whenever erosion rates increase. Wetter conditions do not necessarily mean more erosion, as they could also lead to more vegetation cover providing protection for the soil surface. Under natural vegetation conditions, erosion rates over most of the Mediterranean area are also rather insensitive to changes in total rainfall (see Chapter 6). We must also remember that the best evidence for erosion comes from slopes and colluvial deposits, as the links between erosion and alluviation are not necessarily direct, nor are they well understood. There may be a significant time delay between the time of soil erosion from hillslopes and the subsequent deposition of the material in floodplain and other alluvial deposits. This deposition, however, also suggests that soil erosion is not always bad – 'one person's erosion is another person's accumulation of sediment' (King and Sturdy, 1994) – particularly if it means that soil is transferred from inaccessible mountain areas to lowlands where there may be at least seasonal water supplies. However, the steepness of many Mediterranean channels and the lack of a broad continental shelf often mean that the soil will be lost to the sea, a fact that was appreciated by Plato in his *Critias*.

Climatic hazards and agriculture

[the climate is] bad in winter, sultry in summer, and good at no time.

(Hesiod, *Works and Days*)

Mediterranean agriculture must be carried out in the context of the climatic hazards described in Chapter 3. The principal problems relate to the occurrence of extreme storm events, bringing erosion and damage due to hail, rainfall variability and the recurrence of drought, the scorching of crops due to hot, dry winds or volcanic eruptions, and the loss of sensitive crops to frost. Halstead (1990) has noted that there are a number of traditional responses to these threats in rural Greek societies. Risks to the food supply may be buffered either by risk-spreading or risk-avoidance strategies. In the first category may be included the nature of extended family households, which would allow a readily available labour supply, the dispersed character of land holdings, so that the chance of catastrophic failure of all of a family's land could be minimized, and the growing of a wide variety of crops (such crop diversification is also nutritionally useful, see Zohary and Hopf, 1988). Risk prevention can include the taking on of additional labour from close kin, friends or more distant social contacts at labour-intensive times of the year such as at ploughing, sowing and harvest, so that the work needed to obtain a full crop could be carried out. In cases where the timing of the winter rains was unexpected, there are recorded instances of resowing of crops and water diversion from springs or planting in wetland areas. The principal aspect of risk prevention is the storage of surplus production from year to year, for use in periods of shortage. The underground storage of grain in airtight containers can prevent rodent and insect infestations, and occurs from the Neolithic period. Therefore, the Mediterranean farmer needs to aim for over-production, so that it is possible to break even in a bad year, because of the highly variable nature of the climate. However, such a process of repeated over-production could lead to longer-term reduction of crop yields by reducing soil and nutrient resources. Halstead notes that it was common for recent farmers to retain two years of surplus grain before they will countenance selling off any. However, direct surpluses can also be vulnerable to spoiling, so that it was also common to sell off some of the surplus and convert it to animals (flocks of fifty to a hundred sheep or goats commonly) or land, feed the surplus to livestock, or in more recent periods, by growing small areas of cash crops. Capital could also be raised by selling labour to neighbours, although this may have also been in the form of reciprocal labour exchanges. In serious crises, there may have been attempts to reduce the number of dependants, for example, by giving away a daughter in an 'unsuitable' marriage, by sending children to work for wealthier relatives, or even throwing them out to fend for themselves, and in some recorded cases, even selling them. Alternatively, land or animals may have been sold off to buy staple foods. Ultimately, rural emigration was seen as the solution in particular, repeated cases of hardship. Such rural emigration was very common in the nineteenth century (McNeill, 1992), and increased in the twentieth century (see Chapter 16). Thus, Halstead notes that there is a hierarchy of mechanisms used to deal with risks, starting with direct or indirect storage of surpluses, followed by spreading of risks and responsibilities, and ultimately leading to emigration. The system seems to have been based on 'amoral familism', or devotion to family self-interest, and envy of success of others, leading to competition between households, and social stresses which were resolved in a number of ways (ultimately through the 'blood feud' for example, in the Rif [McNeill, 1992] and the Balkans).

The impact of drought has also been suggested on a large scale in Mediterranean prehistory by several authors. The traditional hypothesis of late Bronze Age collapse of the palace economies in the Aegean was to attribute it to the invasion of tribes from the Balkans. However, the patterns of movement are wider – including the documented attacks by the 'Sea Peoples' on Egypt in *c*.1232 BCE, on Enkomi in Cyprus *c*.1200 BCE and at Ugarit on the Levantine coast (where the Sea Peoples reputedly later settled after being driven out of Egypt by Rameses III) – and this hypothesis does not provide a complete explanation, leading Carpenter (1968) to suggest that the movements could have been due to a period of prolonged drought leading to populations attempting to seek out cooler, wetter areas. Bryson *et al.* (1974) then pointed out the close similarities with the known movements and the drought experienced in the eastern Mediterranean in 1954–1955. Weiss (1982) carried out a more detailed analysis of drought patterns and demonstrated that it was possible to explain the late Bronze Age patterns by a combination of frequently recurring spatial distributions of drought. Whether this is the real cause of the late Bronze Age changes is unlikely to be known for certain, although it is likely that enhanced drought could have caused increased stress on already stressed social interactions. However, Weiss does point to the fact that the radiocarbon record suggests a period of low solar activity between *c*.1470 and 1260 BCE, which may be linked with repeated drought conditions. Further possible testing of the hypothesis may come from detailed analysis of the tree-ring records which are being built up for the Aegean (see Chapter 4).

Irrigation

The best risk-avoidance strategy against drought is of course to provide an artificial water supply. However, the principal problem with investigating the history of irrigation is with the lines of evidence which are available. For the later periods, there are available documentary, historical or literary sources, but for the formative periods of the technology, such lines are often unavailable. Archaeological data are often difficult to obtain, either because they have not been looked for, or because they are overlain or destroyed by subsequent irrigation features. Thus, the evidence described below almost certainly underestimates the age of Mediterranean irrigation technology, which would seem to be a fruitful line of further investigations.

Chapman (1978) presented an overview of the evidence for water management in the Chalcolithic and Bronze Age of southern Spain. Water conservation seems to have originated in the Chalcolithic at Los Millares, where a conduit carries water from a spring to the settlement, at Cabezo del Oficio, where there was a 10 × 8 m oval pit, 2.6 m deep surrounded by a wall and incorporating sediments indicating that it was filled with water, and at the site of Gatas which has a possible overflow system from a stream channel for filling a cistern. Possible cisterns have also been discovered at Las Anchuras (Chalcolithic) and La Bastida de Totana (Bronze Age), as well as the small cisterns at Bronze Age Marirías de Cieza in Murcia, and in the Talayotic period of the Balearics (*c*.1750 BCE: see also Waldren, 1986). The 'subterranean galleries' excavated the early Bronze Age sites at El Argar and El Castillo de Rioja may also have been used for water storage. It is also possible that deep storage pits at Tabernas were used for water storage. Evidence for water diversion is more difficult to find because the evidence is typically destroyed by later irrigation practice. However, at the Cerro de la Virgen, a 3 m wide × 2 m deep ditch on the west side of the hill contained Chalcolithic sherds and may have been used for irrigation. Certainly, there are large numbers of sites with surrounding ditches in the region by this time, so the technology of constructing ditches and an awareness of their use in diverting water can probably be assumed. Chapman also notes that there is possible evidence for floodwater farming in southern Spain.

This idea was taken further by Gilman and Thornes (1985), who carried out site catchment analyses of a number of sites in the Neolithic and Bronze Age of southern Spain. They found that there was a greater focus in the Bronze Age on the water resources close to a site in the Bronze Age when compared to the Neolithic and Chalcolithic (Figure 10.12). However, in

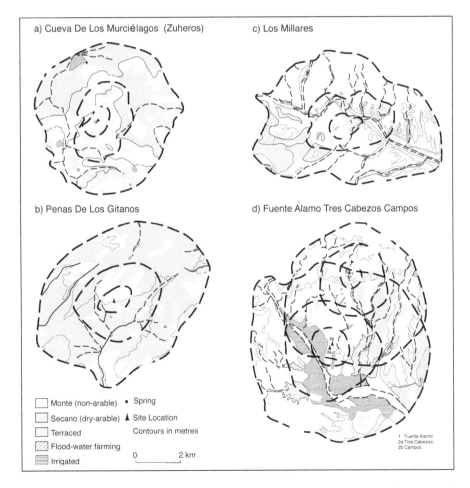

Figure 10.12 The increased emphasis on the water resources close to a site in the Bronze Age when compared to the Neolithic and Chalcolithic in southern Spain

Source: After Gilman and Thornes (1985).

the more arid zones of the region, the use of irrigation may have occurred even during the Neolithic.

As noted above, the olives found at Teleilat Ghassul have also been used to suggest that irrigation was practised by *c*.4450 BCE (Zohary and Hopf, 1988). The first irrigation in Mesopotamia probably originated by around *c*.4800 BCE (e.g. Oates and Oates, 1976), and thus the technology may have spread by contacts with the Levant.

It has been suggested that the ability to store water is an important aspect in the expansion of settlement in the Bronze Age Levant. At the start of the second millennium BCE, the Sharon area became more heavily settled and a number of limestone and dolomite regions of Israel seem to have been occupied for the first time. These changes have been linked to the development of plastering, which allowed water to be stored and transported with minimal leakage (Gophna, 1979).

The expansion of agriculture in Egypt depended on the development of water-control technologies, as discussed in detail by Butzer (1976). Although quite large areas of the Nile valley can be cultivated using floodwaters alone, the expansion of agriculture – both in terms of available area and the ability to plant a third, summer crop – requires an ability to supply water to the valley bottom during periods of low Nile flow. The earliest evidence for artificial irrigation comes from a pictorial representation on the mace head of the 'Scorpion king', who is seen ceremonially cutting an irrigation ditch around *c*.3100 BCE (Figure 10.13). There is later textual evidence suggesting that the Pharaoh Pepi I (*c*.2390 to 2360 BCE) cut a channel to place a tract of land under water. Butzer also points to indirect evidence relating to transport

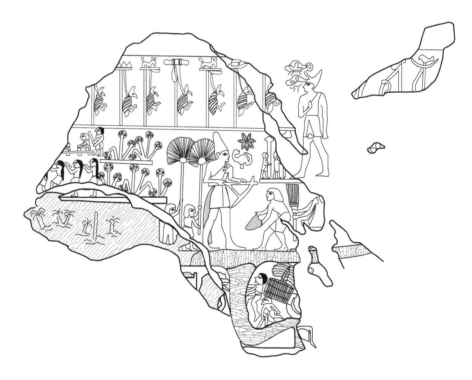

Figure 10.13 Representation on the mace head of the Scorpion king, who is seen ceremonially cutting an irrigation ditch around *c*.3100 BCE

Source: After Butzer (1975).

facilities to allow the stones to be transported to the pyramids of Khufu (*c.*2606 to 2583 BCE), Khafre (*c.*2575 to 2550 BCE), Menkaure (*c.*2548 to 2530), Unas (*c.*2430 to 2400 BCE) and Pepi II (*c.*2355 to 2261 BCE) that would have required a transverse canal to have been constructed from the Nile. Other significant technological improvements include the introduction of the shaduf for lifting water on a pivoted arm (Figure 10.14a), which is first represented in the Amarna period (*c.*1346 to 1334 BCE) and the *saqiya*, or animal-drawn waterwheel (Figure 10.14b), which did not occur until the Persian or Ptolemaic period. The lack of these technologies for lifting water seem to have restricted the expansion of agriculture until quite late, as does the lack of development of a high waterhead canalization to prevent the siltation of channels. It was only in the Ptolemaic period, for example, that extensive irrigation and land

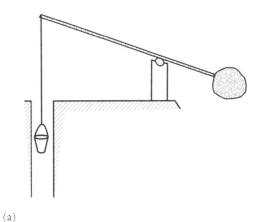

(a)

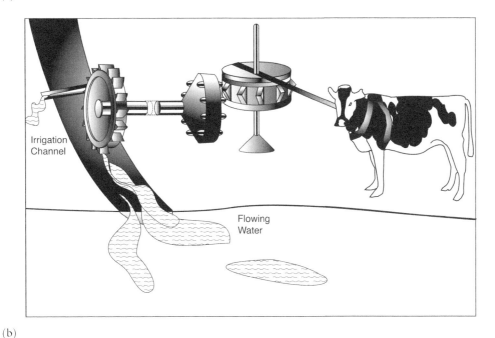

Irrigation
Channel

Flowing
Water

(b)

Figure 10.14 Technology for irrigation found in ancient Egypt: (a) the shaduf for lifting water on a pivoted arm; and (b) the saqiya, or animal-drawn waterwheel

reclamation in the Faiyum Basin allowed the tripling of the cultivated land in the basin to 1,300 km². Complex cropping, with a third annual crop per year, seems to have originated in the third century BCE, and to have allowed the support of increasingly large populations (see Chapter 9).

Balcer (1974) has noted the construction of a Mycenaean dam near Tiryns in the Argolid. The construction is of a large earth barrier supported by cyclopean masonry. The purpose of the dam was apparently to divert the course of a number of streams from the edge of the town, and prevent the modified channels from flooding onto nearby agricultural fields. He also notes the use of artificial drainage ditches around the same date to drain Lake Kopais in Boeotia, near to the Mycenaean fortress at Gla. This undertaking included the construction of a 4-mile drainage tunnel (White, 1984). Van Wersch (1972) suggested that the close correlation between the location of late Bronze Age sites in Messinia with present-day springs or small river irrigation system may indicate that irrigation agriculture was also practised then.

The abandonment of a very large number of sites in the Chalcolithic and early Bronze Age in the Tavoliere of south-eastern Italy has been linked to possible fluctuations in rainfall (Jarman *et al.*, 1982) At present one in four years is likely to suffer from crop failure due to spring drought, possibly leading to total crop failure. By comparison, this problem led to the marginalization in recent years of agriculture in the area, and has been reflected by a specific EC directive to improve irrigation in the area.

runoff farming in the Negev desert of southern Israel seems to have been practised for several thousand years. An aqueduct in the Kadesh-Barnea valley has been dated by charcoal incorporated in the mortar used to construct it to 1748–1316 BCE, although no evidence of the irrigation system to which it was connected remains (Bruins, 1986). Following a period of alluviation in the Roman period and subsequent incision, a new dam and aqueduct were constructed in the Byzantine period (537–873 CE). The valley floor has subsequently been eroded since the eighteenth century to a depth of 4.6 m below the level at which flow was abstracted from this system. More extensive runoff farming seems to have begun in the first century CE (Bruins, 1994). Poesen and Lavee (1997) have demonstrated that deliberate removal of stones from the soil surface, as suggested by the presence of clearance cairns over large areas, could have increased the runoff by as much as 250 per cent, allowing very effective control of the water supply.

The use of qanats – subterranean channels used to tap groundwater flowing beneath alluvial fans or ephemeral channels and distribute the water to lower-lying fields (Figure 10.15) – seems to have originated in the Middle East. The technology may be as old as the thirteenth century BCE, but is more likely to have originated in Persia or Armenia in the seventh or sixth centuries BCE, and subsequently transferred to Syria, Israel and Egypt shortly afterwards (Lightfoot, 1996). Most of the qanats in Syria seem to be related to Roman or Byzantine sites. In northern Italy, a similar method of water diversion called the cuniculus seems to have become very common in association with Etruscan sites in the fifth and fourth centuries BCE (Potter, 1979), although there may be earlier examples, and it is not clear whether this was an independent invention or a transfer of technology from the east. The cuniculi were used for water diversion and storage as well as for drainage, in some cases to allow the passage of roads (Figure 10.16). There are examples of quite large tunnels over 2 m wide and 40 m underground, and others which are 5,600 m in length (Judson and Kahane, 1963). There were 25 km of cuniculi near to the Etruscan city of Veii. In a number of places, the cuniculi are associated with aqueducts to allow diverted water to be transported across valley bottoms. There have been suggestions that some cuniculi may have been used to minimize surface runoff and thereby protect fields from erosion (White, 1970).

While the Greeks and Romans of the Classical period made great advances in water-

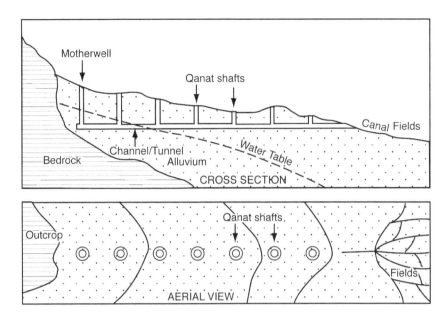

Figure 10.15 Qanats: subterranean channels used to tap groundwater flowing beneath alluvial fans or ephemeral channels and distribute the water to lower-lying fields

Source: After Lightfoot (1996).

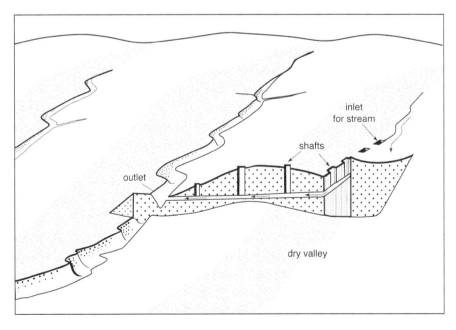

Figure 10.16 Cuniculi in Etruria, Italy

Source: After Potter (1979).

distribution technology (see Chapter 12), they seem mainly to have used irrigation agriculture in particularly marginal situations or in the drier regions of the Mediterranean such as southern Spain or North Africa. However, there are a number of references in the Classical literature to the use of irrigation for pasture as well as for tree crops and cereals (White, 1970). The earliest of these is a reference to the use of an artificial channel to irrigate trees and gardens in part of Homer's *Iliad* (Isager and Skydsgaard, 1992). Further information comes down to us from records of legal cases or systems designed to regulate the amounts of water that were available to individuals for irrigating their farms or gardens (Hughes, 1994).

In Syria, the 2-km-long dam to create the Lake of Homs was constructed in 284 CE to supply both domestic and irrigation water from the 40 km^2 reservoir that was created (White, 1984). The most extensive systems for Roman irrigation agriculture are to be found in the wadis of North Africa from Algeria to Libya. These systems involved constructing a dam across the wadi floor to trap the water from floods in the valley bottom or divert it into adjacent irrigation systems. Evidence suggests that this is an indigenous North African technology, which was expanded dramatically following the direct Roman control of the provinces in the first century CE (Vita Finzi, 1969). In the Wadi Gobbeen in Libya, terrace walls run across the wadi floors to trap floodwater, and there are a series of walls which would have directed runoff from the adjacent slopes onto the wadi-floor fields (Gilbertson *et al.*, 1994; Barker, 1996). Both tree and grain crops were grown in what was apparently a highly integrated system, probably also relying on pastured animals to provide manure to the fields (Figure 10.17). The tree crops also included olives, which may have been grown on alluvial fan surfaces near the wadis. However, some Romano-Libyan farm buildings associated with wadi agriculture seem to have concentrated on olive production (Mattingly, 1995). In some cases, highly complex sluices were built to distribute the floodwaters, probably in an attempt to minimize their erosional impact. Large cisterns, with capacities of up to 175,000 litres, were also constructed to maintain water supplies throughout the year. The wadi-farming system seems to have been successful and sustainable, with certain people making a large profit from selling on surpluses (Mattingly, 1995). Irrigation seems to have been an important part of the landscape in the hinterland of Carthage in northern Tunisia in the pre-Roman period (Charles-Picard and Charles-Picard, 1961). Large cisterns were also constructed in the city to store rainwater collected on roofs (Lancel, 1992).

Another aspect of Roman technology that was important was the implementation of large-scale drainage schemes. Attempts were made to regulate the flow of the River Velinus in the early third century BCE, although as with many flow-regulation schemes, its lack of success was demonstrated by a lawsuit to tackle unforeseen consequences in 54 BCE (White, 1970). The drainage of low-lying areas was also attempted to try to extend the areas of cultivable land, with the beneficial side-effect of reducing areas of mosquito infection. As noted above, Lake Copais was drained, and Theophrastus noted changes in local climate due to land drainage in Thessaly and Thrace (Hughes, 1994). There were unsuccessful attempts to drain the Pomptine marshes to the south of Rome as early as 312 BCE, while the Po valley was drained by a number of channels started in 109 BCE. Ballais and Meffre (1994b) indicated the presence of a number of artificial channels on the Ouvèze-Rhône terrace dating from the first century BCE to the third century CE, which probably were used for drainage and flow control. Recent archaeological work has demonstrated the widespread extent of Gallo-Roman drainage and irrigation features in the lower Rhône valley. In some places, this allowed the development of a 'mass production' form of agriculture, with some large vineyards estimated as being capable of producing up to 2,500 hl a^{-1} of wine, a figure which would not be out of place in modern agro-industrial viticulture (Favory *et al.*, 1994b). In a number of places in southern France, this system seems to have collapsed at some stage in the third century CE when population decline

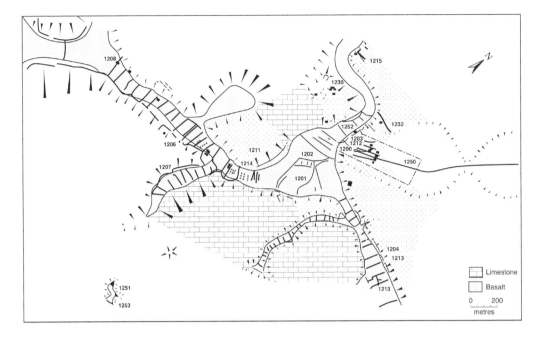

(a)

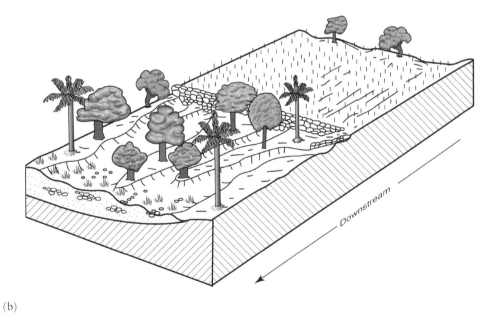

(b)

Figure 10.17 (a), (b) Integrated agricultural system used in North Africa to minimize stress in extreme
conditions. (a) Map of wall systems used for floodwater farming in the Wadi Gabbeen;
(b) Schematic plan of how floodwater farming might have been organized in this system

Source: After Gilbertson *et al.* (1994).

– possibly as a result of wider economic pressures – meant that the drainage ditches became increasingly difficult to manage. Because of the regular nature of the ditch system, the blockage of one channel could have led to large areas falling into disuse (Chartier *et al.*, 1994; Chouquer and Odiot, 1994).

The role of the spread of Islam in the Middle Ages has often been cited as being important in extending or re-establishing irrigation, particularly in the southern Mediterranean and in Spain. Watson (1983) emphasizes the importance of rebuilding of Roman irrigation systems which had fallen into disuse after the plague of the sixth century CE. Irrigation was often closely linked with the establishment of the new Islamic state, most commonly as a means of increasing the state revenue. Newly irrigated areas were also formed and new techniques introduced – perhaps most notably the qanats for tapping aquifers under alluvial fans, which spread from Iran through North Africa and even into Spain. Records speak of a qanat near Cordoba in 754 CE (Butzer *et al.*, 1985). Other new techniques introduced to a wider area involved the capture, channelling and lifting of water using dams, shadufs and water wheels. Watson also notes the importance of Islamic land-ownership and taxation practices in encouraging every possible area to be irrigated. Conversely, the right of an individual to own land which they drained also played a part in extending the agricultural area.

It has been suggested that large-scale irrigation schemes must have required centralized planning and organization to construct and run them. However, the irrigation systems of Mediaeval Valencia provide a counter-argument to this generalization, showing how an extensive system could evolve and be run at a very localized level (Glick, 1970). The *huerta* of Valencia was made up of eight main canals which were communally owned and administered. Each farmer would belong to one canal system (even if the land farmed was bounded by more than one canal), to which there would be rights to the extraction of water in return for the payment of dues and some involvement in the work required to maintain the system. Access to water seems to have been controlled in terms of the time permitted for diversions onto a particular piece of land to have been made, rather than as fixed volumes of water. Overseers of the system were employed to ensure that extractions were not made from canals to which an individual had no rights, nor for longer than the allotted time, nor at the wrong point in the cycle of turns, exacting fines from those who contravened the regulations.

Summary

The slow but sure transition to agriculture in the Mediterranean Basin over a period of about six thousand years was one that has produced the most significant single impact on the landscape. It led to major changes in the hydrological cycle and caused rates of erosion to increase by orders of magnitude. In some places, rapid erosion in the prehistoric period has been 'irreversible' – at least in the sense that the areas affected no longer support significant agricultural activity as soil-formation rates are much too slow to allow recovery. The pattern of agricultural increase meant that initial impacts were localized and minor. A significant increase in erosion from human activity came in the Bronze Age, with the onset of the secondary products revolution. More extensive modifications of the landscape could thus take place, and therefore the impacts of major storm events could be much more widespread. With these improvements in technology came techniques for minimizing erosion and storing water so that agriculture could be carried out much more efficiently, and with relatively minimal impact. Tree crops became increasingly important with a further, generally stabilizing effect on the landscape. Thus the 'Mediterranean triad' of wheat, olives and vines was born. Although erosion relating to population expansion continued to be important and occurred in a variety of places, from this point on we also see the impact of population decline on erosion processes. The lack of upkeep of structures could lead to catastrophic effects. There are a number of locations where different directions of population change have led to stability or instability in the Mediterranean landscape (see Table 10.1). If we are to be able to use past experience to suggest potential future changes in the landscape, a greater focus of research in some of these areas would prove invaluable.

Agriculture in the Mediterranean is by no means an easy subsistence option, though. The variability of the landscape and climate is such that farmers have a continual battle against the elements to secure their food supplies. The development of agriculture in the region is therefore a history of how people have come to terms with this variability, and avoid or at least minimize it. There are indeed suggestions that whole civilizations have fallen, due at least in part to stresses caused by the variability of the climate. Once irrigation and water-storage technologies were developed, it was possible to intensify further the production of food supplies. Again, an implication of this intensification was a further expansion of people into increasingly marginal areas. Too much expansion might mean that in years of extreme conditions, these populations were increasingly vulnerable to collapse, even with risk-spreading strategies over large areas.

Suggestions for further reading

Van Andel and Runnels (1995) give by far the best description of the evolution of a Mediterranean rural landscape through time under changing patterns of human settlement. There are useful introductions to the evolution of agriculture in Cowan and Watson (1992) and Gebauer and Price (1992), while Zohary and Hopf (1988) discuss the underlying mechanisms and patterns. Excellent discussions of Neolithic and Bronze Age Europe in general are given by Whittle (1996) and Harding (2000), respectively. Sherratt (1981) is an important reference for later developments of agricultural technology. Halstead (1990) provides an overview of agricultural adaptations to climate and climate variability in the region.

Topics for discussion

1   What mechanisms and technologies are fundamental precursors for the development of agriculture in the Mediterranean region?
2   Where and when did the 'Mediterranean triad' evolve, and what is its significance?
3   What impacts did the secondary products revolution have on Mediterranean landscapes?
4   Can the impacts of land-use change and climate be disentangled?
5   Given the focus on the control of water, why do erosion-control structures take so long to have developed?
6   Have the impacts of fire increased or decreased through time?

11 Traditional land-use patterns 2
Rural settlement

Introduction

As well as strictly agricultural systems, the Mediterranean world has long been the focus of important and distinctive pastoral uses of the landscape. Indeed, such systems probably evolved in response to some of the vagaries of the Mediterranean landscape. We look at these pastoral approaches and their impacts on the structure of settlement and on the environment itself in this chapter. The implications in terms of the use of water resources are also investigated. Under increasing pressure, the upland elements of these land-use systems have become increasingly unstable.

Agro-pastoral systems and vertical settlement patterns

Beyond the use of the landscape for agriculture, we have already seen how integrated systems of agro-pastoral activity developed in the Bronze Age Mediterranean. In the Upper Palaeolithic, there is ample evidence for seasonal movements within the landscape to exploit animal resources. The exploitation of the landscape in a seasonal way, taking advantage of the short distances between very different environmental zones seems to have had a long history. In historical and present times, seasonal patterns of exploiting these vertical resources have been studied in the form of patterns of transhumance. Transhumance can be defined as the seasonal movements of flocks – usually sheep or goats – to different areas of pasture by a component of the population. The fact that elements of the population remain in certain areas to tend crops differentiates transhumance from nomadism. Different forms of transhumance have been defined (e.g. Braudel, 1975). Normal transhumance involves the movements of flocks based in the lowlands to the mountains in the summer months to take advantage of the pasture and lower temperatures of the mountains and possibly also in the past to escape the malarial lowlands. Inverse transhumance is 'the frantic rush down the mountains in the winter ... to escape the cold' (ibid.: 86) and to exploit the markets of the lowlands. Braudel also notes a mixed type of transhumance, for example, in Corsica, where the base settlements are midway between the lowlands and uplands. Nomadism was also practised in parts of Turkey in the historical period (e.g. Bates, 1973), as well as through parts of the Levant and northern Africa. In modern times, movements of flocks to mountain pastures has been by truck (Barker, 1995; Musset *et al.*, 1986). However, in the past, these movements were along the extensive droveroads that formed major features in the landscape – the *cañada* of Spain, *draille* of France and *tratturi* of Italy (Figure 11.1). These roads were often protected from cultivation by the state, not least so that the movements could be controlled and tolls efficiently collected.

A further example of transhumance existing in recent times is that of the Sarakatsani in Epirus (Campbell, 1964). A detailed study of the Zagori Sarakatsani showed that of the

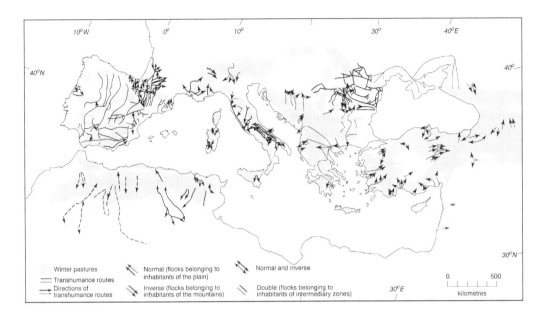

Figure 11.1 Map of transhumance routes and other droveroads in the Mediterranean region, showing the long distances often travelled by pastoralists in the course of the year

Source: Based on Braudel (1975) and other sources cited in the text.

152,000 sheep and goats in the area, 85,000 sheep and 13,000 goats were moved by the Sarakatsani in transhumant systems. (These figures reveal a further point, also made by Barker, that pastoralism also had a coexistent sedentary form.) These flocks were concentrated between May and early November in summer mountain pastures, but relatively widely dispersed in the lowland winter pastures. In the latter, the Sarakatsani had no fixed grazing rights and had to rent these rights from the local community or from individuals, often at great expense. Distinctive social groupings evolved to control the flocks, with associations of two or more related families into *stani* or companies often in competition with each other. Unity of the transhumant group usually only occurred in the case of conflict with the villagers of the lowlands. Although some have used the Sarakatsani as examples of cultural continuity from Classical Greece and earlier, Campbell points out that their control of the flocks was essentially a historical accident relating to the Ottoman occupation of Greece. Increasing populations over the nineteenth and twentieth centuries have led to increasing pressures on grazing land and a gradual decrease in flock sizes. Since Campbell's study, this decline has accelerated so that only a handful of families still practise this way of life (Green, 1994).

It has been suggested that transhumance has other impacts on the agro-pastoral landscape. Semple (1932) pointed out that it effectively meant that half of the potential manure which could be applied to lowland fields was lost. Increasing losses of soil fertility led to the development of other techniques of fertilization, including the ploughing-in of green crops from the Classical period in Greece. Other factors leading to the development of migrant pastoral systems include the avoidance of hazards, insects and disease, to reduce competition within and between groups and as a means of avoiding taxation (Dyson-Hudson and Dyson-Hudson, 1980; Mee, 1991). In the latter case, the systems of movement themselves may have been taxed, as in the Spanish Mesta system (Braudel, 1975), although this is not to say that the tolls were not avoided. For example, until the introduction of EU subsidies meant that accurate reporting of herd numbers was advantageous, there is evidence of persistent under-reporting of animal numbers in the Epirus (Green and King, 1996). Mee (1991) also suggests a further reason for expanding pastoral activities in more remote locations is that it produces products such as cheese, wool and hides which are easier to transport to market than are liquids such as olive oil or wine.

There are a number of lines of evidence to suggest that integrated agro-pastoral economies are a persistent feature of the Mediterranean landscape, and provide a relatively stable means of exploiting the environment in a relatively intensive manner. Although the ethnographic examples cited above may give some indications of how these systems may have operated, they do not necessarily provide the only model of how such systems may have operated – indeed, we know from historical parallels (e.g. Braudel, 1975) that very different systems have been in operation. In the middle Neolithic period (*c*.4800 to *c*.3200 BCE) in southern France, there seems to have been the development of a structured settlement pattern (Berger *et al.*, 1994). Major sites are located on good soils in and near the Rhône valley, while smaller villages and caves seem to have been used in the uplands by shepherds. The vegetation data support the onset of garrigue at the same time (Thiébault, 1988), and there is contemporaneous active erosion through the Drôme river system. Specialized grazing in the Camargue developed in the nineteenth century due to an excess of available water and constraints imposed by increasingly specialized neighbouring land uses (Aschan-Leygonie *et al.*, 1994).

Indications of seasonal movements of animals are given in Sophocles' *Oedipos Tyrannos* with summer grazing on Mount Kitheron and winter grazing near Corinth and Thebes, while inverse transhumance is suggested by the speech of Dio Crysostom in the mountains of Euboea (Isager and Skydsgaard, 1992). A late fifth- or early fourth-century law in Tegea to allow sacred animals en route to rest for a day and night on sacred land is also possibly related to transhumance. Alcock (1993) suggests that transhumance would have been difficult in the Classical period in Greece because of the difficulty of moving large flocks across *polis* boundaries, which may mean that most of the transhumant pastoralism suggested by the literature was over relatively short distances, for which there are also modern analogues (Hodkinson, 1988). In some cases agreements specifically allowing the movements of animals from one *polis* across the land of another (Hodkinson, 1988) were reached (and in others they escalated to war: Isager and Skydsgaard, 1992). The removal of these constraints may have led to an expansion of the activity in the Roman period, although the direct evidence for this is minimal at present (Alcock, 1993). On the other hand, Purcell (1990) also questions the limited mobility view of the *polis* on more conceptual grounds, and suggests that there may have been shepherd populations acting in a more marginal way with relation to the urban centres.

Opinions vary as to whether Albanian migrants brought transhumance to the Argolid in the Middle Ages or whether its introduction relates to early Iron Age settlers (Jameson *et al.*, 1994). In the historical period, different ethnic groups have been responsible for transhumance in the southern Argolid – Valtetsiots in eastern Arcadia and Sarakatsani in southern Korinthia. Other mechanisms of sheep and goat husbandry involve extensive herding by local inhabitants and may move (but not as a group from a settlement) up mountains on timescales of days or weeks. There is also some evidence to suggest a dispersed pastoral activity in the middle Neolithic, if not full-blown transhumance.

Frayn (1979) notes the importance of movements of large flocks in Roman Italy. She points to the presence in the Agrarian law of 111 BCE of free pasturage available to flocks on the move and the payment of tolls to pay for the upkeep of the drove roads. Some of these drove roads may have been Bronze Age in origin. An example of inverse transhumance near Naples is mentioned in one of Pliny's letters, and involves the movement of horses and cattle as well as sheep. However, there seems to have been a break between around the sixth and third centuries BCE, when the practice of large-scale herding was reintroduced (Frayn, 1984). In later periods, it also seems that the Roman state imposed taxes on pasture to benefit from the practice. Both milk and wool were important products during this period. Barker (1995) points to the evidence of an inscription at Saepinum in south-eastern Italy, registering a dispute over the collection of tolls from shepherds. The persistence of droveroads within the landscape is further suggested by the fact that the modern droveroad still runs through this location.

In central Italy, extensive historical transhumance systems were recorded with movements between summer pasture in the Apennines and winter pasture in the Tavoliere (Sprengel, 1975; Barker, 1995). These systems continued into the 1960s, and had their roots in at least the Roman period, and probably peaked in the period of the Kingdom of the Two Sicilies, when millions of animals were moved each year. Towards the end of the nineteenth century, the system developed further by the transport of flocks to railheads by train and thence of a shorter walk to the pasture. The wide droveroads employed are still highly visible within this landscape.

Camels were exploited as well as sheep and goats within a transhumant system in North Africa (Hitchner, 1994). There seems to be increasing archaeological evidence for organization of the pre-existing semi-nomadic pastoralism changing to a large-scale pastoral system in the period relating to the Roman occupation of Tunisia, fitting in with a much broader agricultural system, and continuing until the links with Rome were cut off in the early sixth century CE.

Although modern and recent analogues can be important in understanding past land uses, Halstead (1987) warns against their uncritical application. Transhumance is one of the key cases in point, in that in most cases, large-scale movements of animals were probably related to the specific historical conditions of the time. Furthermore, he points out that extensive alpine pastures in the high mountains may also be a relatively recent phenomenon, and that more extensively wooded lowlands would have hindered the pasturing of large flocks. Thus, although some form of vertical distribution of settlement and subsistence patterns appears to be a very common theme in the prehistory and history of the Mediterranean, it probably operated for the most part in a manner quite unlike the recent forms of transhumance. Most of the large-scale systems seem to have evolved under circumstances of political instability or state control of flocks. Frayn (1979) points to a passage in Virgil's *Aeneid* which suggests a vertical pattern of kitchen gardens on level ground, plough agriculture on intermediate hillslopes and pastoralism in the higher mountains.

Pastoralism and soil erosion

One further aspect of mountain soil erosion in the Mediterranean deserves explanation: the symbiotic relationship between grazing and erosion. Each promotes the other. Erosion reduces the rewards of agriculture, encouraging land abandonment, indeed encouraging shifting cultivation. To make abandoned land yield anything at all, mountain folk must loose their sheep and goats on it. As they do so, they prevent spontaneous vegetation from colonizing, thereby perpetuating conditions favourable to further soil erosion. Eventually – and it does not take long with steep slopes and friable soils – the land can support only the sparsest and hardiest vegetation; and this in turn can support only the hardiest of animals: the goat. This symbiosis, in combination with fire, has maintained landscapes in which the goat and the holly oak are the ecological dominants. This process, inaugurated by deforestation, vastly impoverished Mediterranean hill and mountain districts, sharply reducing their economic potential – to the point, in fact, where no economy more fruitful than extensive pastoralism is possible. In most places, the effect is irreversible.

(McNeill, 1992: 313)

In the Draix region of the southern French Alps, the *terres noires* have developed into badlands in the Holocene. Ballais (1996) has noted that fossilized pine trunks in their growing positions in the upper levels of what are now deeply incised surfaces show the badlands developed after 7300–6427 BCE, probably in a series of discrete time phases which probably reflect the effects of prehistoric and historic overgrazing. At Font Juvénal near Carcassonne, Brochier (1984) demonstrated a clear link for erosion relating to vegetation change which may be related to pastoral activity, dating to the early part of the fourth millennium BCE. In the Valdaine, there is extensive deposition on the alluvial fan of the River Drôme where it enters the Rhône valley, dating to *c.*4870 BCE and *c.*3780 BCE, during the period of marked population expansion (Berger *et al.*, 1994). The erosion may be linked to repeated burnings of the landscape to maintain grassland for grazing.

Métaillie (1987) notes that high erosion rates were common in the French Pyrenees at the end of the nineteenth century and produced photographic evidence to show spectacular erosional events from which the landscape has subsequently recovered (Figure 11.2). He notes that the erosion is due to the effects of human activity in geologically unstable areas which are subject to high-intensity storms. Tellingly, he provides a quotation from a governmental inspector in 1881 who noted the occurrence of a 'Tragedy of the Commons':

> The erosion has taken place under the eyes of the present generation. [The efforts of the administration] have always been broken at Montauban by a blind, ignorant and invincible resistance. Without denying the effectiveness of the works which we propose for them to undertake, the villagers have never had either the intelligence or the energy necessary for sacrificing the interests of individuals to the interests of everyone ... [There are] no controls on the areas under pasture, no calculation of the possible outcomes, no preventative actions taken, and no attempt to control the various animals which dispute the same hillslopes.

Again in the French Pyrenees, the area near Castelet seems to have undergone two phases of enhanced debris-flow activity and flooding from 1750 to 1772 and from 1850 to 1910 (Antoine, 1988). In the case of the floods of July to August 1750, crops and fields downstream at Ax were destroyed. The cause again seems to relate to extensive deforestation to support the pastures of the increasing mountain populations and their needs for wood for fuel and construction.

In the Slovenian karst, there also seems to be evidence for increasing pressure of pastoralism on the landscape in the eighteenth century (Culiberg, 1995). Although vegetation clearance has been seen since the Neolithic, catastrophic erosion leading to the widespread degradation of the karst does not seem to have occurred until this late stage. Conversely, there are reportedly successful attempts at reforestation as early as 1859.

Fire

The use of fire within a pastoral context is well attested from both modern and ancient examples. Métaillie (1981) has studied the use of fire by shepherds in the Pyrenees and its impacts on the landscape. Large areas of summer pastures – in some cases as large as 60 hectares – were burned in 1942 and 1948 according to evidence from aerial photographs. The

(a)

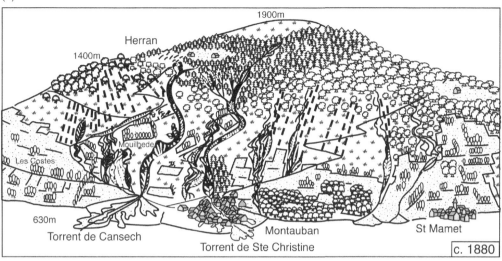

(b)

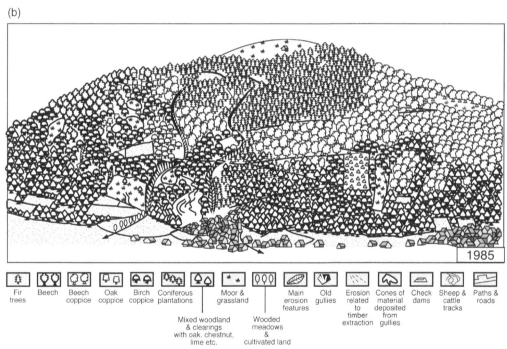

| Fir trees | Beech | Beech coppice | Oak coppice | Birch coppice | Coniferous plantations | | Moor & grassland | | Main erosion features | Old gullies | Erosion related to timber extraction | Cones of material deposited from gullies | Check dams | Sheep & cattle tracks | Paths & roads |

Mixed woodland
& clearings
with oak, chestnut,
lime etc.

Wooded
meadows
&
cultivated land

Figure 11.2 Sketches of photographic evidence showing (a) spectacular erosional events in the French Pyrenees at the end of the nineteenth century; and (b) from which the landscape has subsequently recovered

Source: After Métaillie (1986).

extent and repeated frequency of burning on some sites had significant effects on the vegetation cover. From 1957, there seems to have been something of a recovery of many of the summer pasture areas with little further burning as a result of declining herd sizes. In the nineteenth century, such burning probably reached its peak, judging from the known numbers of animals grazed at the time. The numbers increased from 50,000 cattle and 194,000 sheep and goats in 1806 to 150,000 cattle and 502,000 sheep and goats in 1867, before declining again to 115,000 and 208,000, respectively, in 1912. The impacts of the corresponding clearance is noted above. An association between fire and pastoralism is difficult to make, although the repeated clearances of the Chasséen in southern France in relatively upland areas may reflect the long history of the linkage. In the Classical period, Virgil makes reference to shepherds deliberately firing forests in *Aeneid* (and the unforeseen consequences of the intervening areas being burned accidentally: Hughes, 1994), while the imagery of the forest fire was used extensively by Homer (Meiggs, 1982). However, frequent, small fires set by shepherds may actually be beneficial in reducing fuel stocks and thus preventing catastrophic fires, as will be seen in Chapter 16.

Another significant aspect of fire in the rural landscape came during periods of warfare. For example, forests in Greece were reputedly extensively burned by the troops of Xerxes during the Persian invasion of 480 BCE (Hughes, 1994). In southern France, 300 km² burned in the Esterel forest in 1536 during the invasion of Charles V, and during the retreat of Prince Eugène of Savoy in 1707, as much land as possible was put to the torch. Other important episodes of war-related burning in Provence took place in 1524, 1590 and 1943 (Amouric, 1992).

Water resources

Although pastoral occupation of the upper slopes has been seen to have a significant impact on the erosion of slopes due to the removal of vegetation cover, the loss of this vegetation cover can also have significant impacts on the water resources. Green and King (1996) have noted the distinct differences that occur across the modern border of Greece and Albania. Where flock numbers have declined in Greece and the 'no-man's land' of the border zone, vegetation regrowth is very clear from satellite images, compared with the bare areas of the intensively used Albanian side. In turn, these changes may be related to the drying up of springs near the border and on the Greek side, as more water is taken up by the vegetation and thus does not reach the aquifers. This case study suggests that the link between vegetation and water resources is not a straightforward one (see Chapters 4 and 6), and that other local conditions are important. Similar rapid changes in vegetation have also been noted following the removal of vegetation from a very different environment in the Negev (e.g. Tsoar and Karnieli, 1996).

Butzer *et al.* (1985) looked in detail at the development of irrigation in seven mountain villages of southern Spain. This irrigation appears to have had Islamic origins. Although agriculture was an important component of subsistence in these villages, making up 26 to 67 per cent of the land area, only 2.2 to 5.8 per cent of that area was irrigated. These figures suggest the sustainability of small-scale dry farming in the Mediterranean mountain regions. Where irrigation took place, the water supply was from active springs or wells, and the water was temporarily stored in cisterns before being distributed via channels to fields or to mills. Distribution of water was controlled by a paid official, the *regador*, and commonly took the form of delivery to certain fields or villages on specific days. At least into the nineteenth century, the small areas of irrigated land were shared out between all the farmers in the community and used to grow wheat, maize, beans, cherries, almonds and apples, although by the twentieth century, changes in land holdings meant that the irrigated plots tended to be owned by the wealthier families.

Small-scale studies suggested that traditional irrigation practices were carried out on a two-year cycle to improve the redistribution of the water, and in relation to an eight-year crop cycle to maintain soil fertility.

Marginality of mountain populations

In many cases, exploitation of the Mediterranean mountains reflected the development of sustainable systems reflecting the availability of specific resources such as pasture, woodland or various specialized industries in the historical period. An example of the latter would be the production of chestnuts and of silk in the French Cévennes. A further use of the mountains was for defence, particularly on vulnerable islands such as Corsica and in the Aegean, and particularly in the turbulent social conditions of the Middle Ages. In times of hardship, a variety of more marginal resources came into use. Commonly, acorns from the Mediterranean forest were ground up to make flour and bread if wheat was unavailable. This strategy of 'balynophagy' was widespread and probably recurred through the historical and prehistoric periods (Lewthwaite, 1983). McCorriston (1994) also provides ethnographic examples of how the eating of acorns reflects times of stress.

McNeill (1992) describes in detail the implications of increasing populations in the Mediterranean mountains, particularly from the seventeenth century to the nineteenth century. In many cases, populations reached their logical limits at this time. As the various niches for exploitation filled up, populations became much more sensitive to variability in the availability of resources, particularly due to the inter-annual variability of climate. Deforestation was an important consequence of the necessary expansion. For example, in Basilicata, the forest area declined from 290 k to 180 k ha between 1800 and 1908. Rural emigration starts to become a significant process in the nineteenth century, particularly from Spain, Italy and Greece (McNeill, 1992). The emigration was commonly down into the coastal zone, but as this too became increasingly populated, there was common migration out of the region altogether, for example, to North or South America.

Summary

Transhumance and nomadism seem to have evolved at a number of points in time in the Mediterranean region in a number of guises. In some cases, the development was in relation to environmental conditions, in others to population pressure and in yet others to market forces and other socio-economic conditions. It is probably true to say that no two forms in different places and different times had the same underlying mechanisms or modes of operation, so that extreme care should be taken when using recent and historical cases as analogues for prehistoric examples. Probably the most important impact of these systems is that they allowed the spreading of population pressure across a range of altitudes and environmental niches, so that increasing human populations were more stable. This stability probably resulted from a minimization of the direct impact on the landscape – although locally, high rates of erosion did result from these systems – from an increased resilience from not being over-dependent on a narrow range of resources, and from a redistribution of important resources such as manure. Such stability resulted in the persistence of these various modes of landscape exploitation over very long time periods in some instances.

Fire was an important component in the armoury of Mediterranean herders, allowing provision of good grasslands, and the opening up of forest and shrubland areas for intensive grazing. The impact of this burning on the vegetation structure and thus landscape appearance should not be under-estimated (see Chapters 4 and 7). Indirect effects of this way of exploiting the landscape should also be noted in the way that it modified the hydrological cycle. More direct exploitation of mountain waters for irrigation in the later periods would also have had an impact on the water resources reaching lowland areas.

Ultimately, the stresses of occupying mountain areas under increasing population numbers proved too much. The stability of the system broke down, and there were important losses to the integrity of the environment, most notably through accelerated erosion. The increasing difficulty of making a living in the light of the growing apparent prosperity of the cities – or indeed of simply surviving – led to significant amounts of migration out of the upland areas. Thus the resilient system started to collapse, and increasing focus made of settlement in the coastal areas. The continuation of this process in recent times and its consequences are a theme to which we will return in Chapter 16.

Suggestions for further reading

Campbell (1964) is a classic work on a transhumant society in the Mediterranean. McNeill (1992) provides a range of case studies of life in the Mediterranean mountains, which reflect conditions in a range of different locations. The book by Barker (1995) provides a detailed case study for one area of Italy.

Topics for discussion

1   Why do different examples of transhumant activity in the Mediterranean not necessarily reflect similar underlying mechanisms?
2   What are the social and environmental advantages and disadvantages of vertically distributed settlement patterns in the region?
3   Can the loss of vegetation through grazing and through the use of fire be differentiated?
4   Why are mountain settlements more vulnerable to environmental stresses than those in the lowlands?

12 Traditional land-use patterns 3

Urban settlement

Towns are like electric transformers. They increase tension, accelerate the rhythm of exchange and constantly recharge human life. They were born of the oldest and most revolutionary division of labour: between work in the fields on the one hand and the activities described as urban on the other.... Towns, cities, are turning-points, watersheds of human history.

(Braudel, 1981: 479)

Introduction

One of the questions which has caused great controversy among geographers, archaeologists and anthropologists alike is how to define a town or city, or what constitutes an urban settlement pattern. Definitions vary from threshold population sizes (or relative population sizes) to complex models of a variety of interacting social elements (e.g. Wheatley, 1972). Childe (1950) defined three important stages in the development of urbanism. Progressive growth of the population led to the development of urban populations who were able to carry out non-subsistence economic activities, which in turn led to the specialization of labour, differential accumulation of wealth and class stratification. One problem with this model is that it fails to indicate *why* populations became urban simply as a function of growth. A major factor that needs to be explained is why the population growth was not reflected by expansion rather than nucleation. Trigger (1972) emphasized the role of increasing social and economic complexity in answering this question. When an increasingly diverse set of activities need to be carried out, individual activities are often highly localized in order to maintain efficiency in terms of accessibility and transport. He suggested that the steady increase in population associated with sedentary, agricultural lifestyles tend to produce such increases in population and types of activity (see also Chapter 10). Various authors have pointed out, though, that these sets of conditions by no means lead to the development of urban centres in all cases, and that local factors are most important when defining where and when such centres emerged. As will be seen below, their emergence varied very distinctly in space and time.

It is important to recognize the impacts of urban landscapes and the distinctive changes they imply over the very areas not limited to the boundaries of the cities themselves. As pointed out by David Clarke (1979), 'urbanization is the emergence of site *systems* with urban *elements* and we must study and classify the development process of these systems as a whole and not just their conspicuous urban nuclei'. In particular, the food supply to cities involves a complex set of relationships between rural populations, urban farmers in some cases, and urban populations who depend on the exchange of other goods or services for the provision of food. In certain examples, notably Classical Athens and Rome (Garnsey, 1998), the supply of food to major cities became very detached from the immediate hinterland and led to the modification of much more distant landscapes. In other cases, the division between rural food producers and urban consumers is very much over-simplified. Thus, Delano Smith (1979) considers the case of the 'urban farmer' and the urbanized countryside, particularly with reference to the Mediaeval period in the north-western Mediterranean. Many cities in Spain and Italy contained extensive fields, orchards and walled gardens. Thus we must be aware of the changing spatial dimensions of the move to urbanism at different points in time. Falconer (1987) suggests that the interrelationships between the urban and rural populations are complex, interdependent and usually conflicting, so that the dynamics of these relationships will tend to be very dynamic.

Development of urban settlement patterns

Egypt

Kemp (1989) suggests the first towns of Egypt were established *c*.3600 BCE, or about 700 years before the beginning of the first dynasty. The town of Nagada covered several thousand square metres, and is associated with large cemeteries and ceremonial buildings. Hierakonpolis was a further large centre in southern Egypt at this time, which at the end of the pre-dynastic period – as with Nagada – contracted in size to become a densely populated centre with protective walls. The northward spread of fortified centres was probably as a result of inter-state competition. Kemp believes that in Egypt this strong sense of competition developed initially at the individual level as a result of strong competition for resources within the restricted area of the Nile floodplain, and the strong sense of territoriality to which this led (Figure 12.1). The development of the main centres of pharaonic Egypt probably occurred *c*.3050 BCE (although firm evidence is lacking) with the establishment of Memphis, and probably around the same time, Thebes. Both cities were probably agglomerations of existing settlements, with Memphis extending for a distance of 30 km, with a densely settled core of 13×6.5 km (Brewer and Teeter, 1999). An example of the dynamism of urban centres is given by Tell el Amarna, founded as a new capital by the Pharaoh Akhenaton in the fourteenth century BCE and reaching a population of 30,000 before being abandoned. The development and abandonment of this city were related to attempts to change the religious basis of Egyptian life. The urban centres of Egypt were highly hierarchical, with a number of levels of settlement size (Butzer, 1976; Brewer and Teeter, 1999). A large number of villages provided the agricultural basis of life.

The Levant

The initial development of urban centres in the Levant was supposed to have taken place in the early Bronze Age (*c*.3000 BCE). However, Falconer (1987, 1994) doubts whether there were any real urban centres in the Levant in the early Bronze Age, and that the few significant urban forms were related to external contacts with the Middle Kingdom of Egypt. Decline occurred in the early Bronze Age IV (*c*.2350 to 2000 BCE) with only a few small centres and probably widespread nomadic pastoralism (Dever, 1995) complementing relatively small sedentary sites. Re-establishment of more widespread urban forms took place in the middle Bronze Age from *c*.2000 BCE, probably with an intensification of occupation and a spread inland from established coastal centres. As Ilan (1995) suggests, many of the new lowland sites are to be found near karst springs or locations with high water tables, implying a focus on rain-fed agriculture to support this expansion. Population expansion in the middle Bronze Age seems to have been along lines of drainage, and may have integrated both lowland and highland populations. Only later were the highlands reoccupied in a complementary way. At the same time, trade with

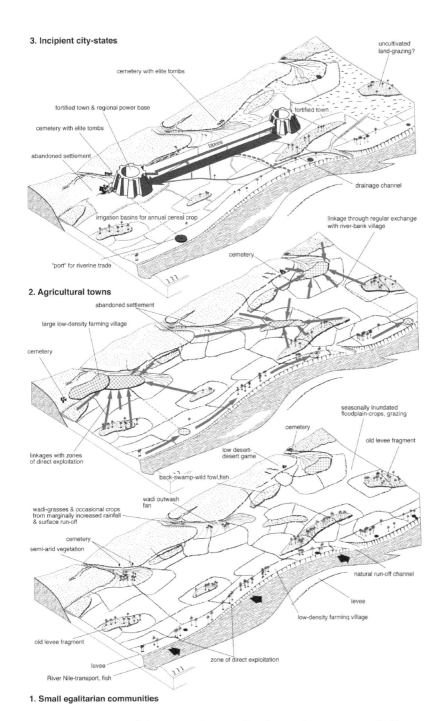

3. Incipient city-states

uncultivated
land-grazing?

cemetery with elite tombs

fortified town & regional power base

fortified town

cemetery with elite tombs

abandoned settlement

taxes

drainage channel

irrigation basins for annual cereal crop

linkage through regular exchange
with river-bank village

cemetery

"port" for riverine trade

2. Agricultural towns

abandoned settlement

large low-density farming village

cemetery

seasonally inundated
floodplain-crops, grazing

cemetery

old levee fragment

linkages with zones
of direct exploitation

low desert-
desert game

back-swamp-wild fowl,fish

wadi outwash
fan

wadi-grasses & occasional crops
from marginally increased rainfall
& surface run-off

cemetery
semi-arid vegetation

natural run-off channel

levee

low-density farming village

old levee fragment

levee

zone of direct exploitation

River Nile-transport, fish

1. Small egalitarian communities

Figure 12.1 Development of clustered and ultimately urban settlement patterns in Egypt
Source: After Kemp (1989).

Syria and Egypt led to the integration of this urban–rural system within a much broader scale of activity within the eastern Mediterranean. The urban centres again declined in the late Bronze Age following frequent Egyptian attacks (Falconer, 1994).

Greece

Nucleation of sites took place in Bronze Age Greece during the third millennium BCE, with important settlements such as Thebes (*c*.50 ha), Tiryns and Mycenae (*c*.25 ha) continuing to grow into the late Bronze Age (late second millennium BCE: Halstead, 1994). Although settlement hierarchies can clearly be defined, the size of these early urban centres is much smaller than elsewhere in the Levant. The similar palace centres of Minoan Crete also develop in the early second millennium BCE (Cadogan, 1976; Manning, 1994). Knossos reached a maximum size of around 40 hectares by the middle Bronze Age.

In the Greek cities of the Classical period, urban planning was common by the second half of the fifth century BCE, and a number of Greek cities were replanned in the Classical period (Owens, 1991). However, there are even earlier examples. Old Smyrna was destroyed by an earthquake at the end of the eighth century BCE, giving a chance to replan, which was carried out by dividing the peninsula on which the city stands into a series of building strips separated by series of parallel streets, as well as creating a religious and civic zone. Miletos near the mouth of the River Meander was also regularly laid out.

The western Mediterranean

Recognizably urban centres develop somewhat later in the western Mediterranean. The Bronze Age Argaric culture in southern Spain shows some degree of settlement organization in the period from *c*.2400–2600 BCE to 1100/900 BCE, with groupings of structures within fortified settlements. The size of settlements actually decreases slightly between the Chalcolithic and the Early Argaric, where settlements are typically one to two hectares, although this trend probably reflects a better organization of structures, often in densely packed rows or clustered cells with shared walls (Mathers, 1994). Fortifications are often present, with watch towers, and single or multiple enclosure walls, although the use of fortifications decreases with time through the Bronze Age. There is a distinct break in the final Bronze Age, with less organized dwelling spaces, and often a move to less defensive positions. At the same time, there was the development of Phoenician colonies. Nucleated settlements in northern Spain are more typical of the later Bronze Age (*c*.800 BCE), for example the 0.3 ha site of San Cristobal de Monleón with a street ending in a small plaza or pond, public space and storage areas as well as residential units (Harrison, 1994).

The Early Bronze Age in southern France is generally poorly known, although the site of Camp de Laure in the Bouches du Rhône covers an area of 1.5 ha, with external fortifications and towers (Gascó, 1994). Otherwise, large, agglomerated, fortified settlements do not occur until the Final Bronze Age, for example, at Carsac near Carcassonne (Guilaine *et al.*, 1986).

In northern Italy, agglomerated settlements occur in the Middle Bronze Age in the form of the *terremare* of the southern Po plain. The largest sites range from 12 to 20 ha in area, and although there is a hierarchy in terms of settlement size, there is little other evidence to suggest the development of specialized centres (Barfield, 1994). The *terremare* end suddenly in the twelfth century BCE and are followed by a period with more diffuse settlement patterns. In central Italy, there was little agglomeration and hierarchization of the settlement pattern until the Early Iron Age, although some sites where this agglomeration occurs are continuously occupied from the later Bronze Age (Barker and Stoddart, 1994). In southern Italy, heavily fortified settlements associated with ports developed on the coast of Apulia, with evidence of trade with the Mycenaean world and craft specialization (Malone *et al.*, 1994).

Phoenician colonies formed another phase of urban activity in the west, starting with Gadir (modern Cadiz), in southern Spain, and Utica and Lixus on the North African coast (Aubet, 1993). These developments occurred possibly in the late twelfth century BCE, providing a block of sites controlling entry into the western Mediterranean, and hence the lucrative trading routes along the Atlantic seaboard of Europe (and hence the tin supplies of Cornwall and Brittany, among other goods). Carthage in Tunisia was founded in the late ninth or early eighth centuries BCE in a second phase of development of Phoenician trading routes. The colonization of North Africa, eastern Andalusia, Sicily, Sardinia and Malta also probably took place during the eighth to seventh centuries BCE. There was also a possible Phoenician protectorate in eastern Cyprus, based in Kition from the tenth century BCE. Overall, the Phoenician cities had a coastal distribution, which reflected their trading basis, and were often located on highpoints.

The Etruscans started to develop the first true cities in northern Italy in the seventh century BCE. These centres were often developed from earlier Villanovan settlements, such as the hilltop city of Veii, which built up gradually and is unplanned. Veii formed an important regional centre, and was probably even stronger than Rome in its heyday. Etruscan expansion into northern Italy in the fifth century led to the development of a number of planned towns on grid pattern, such as Marzabotto (Owens, 1991). Rome itself developed from an Iron Age settlement, and grew gradually, in an irregular and unplanned way, even following phases of destruction. The haphazard plan probably exacerbated the consequences of the great fire in 64 CE (just as it did in London over sixteen centuries later).

It was common practice for the Greeks to plan their colonies, although the limits of the process were obviously dependent on local conditions. There is also evidence of the planning of the land surrounding a colony. For example, at Metapontum in southern Italy, regular parcels of land were divided by drainage ditches approximately 205 m apart, with transverse divisions *c.*323 m apart and further subdivisions, probably for distribution to individual colonists. These divisions probably date to the sixth century BCE. This later phase of Greek colonization reached far into the western basin, with the foundation of Massalia (modern Marseille) to exploit the wine trade with temperate Europe (Shelton, 1994), and Emporion in northern Spain. Planning became an important aspect of Roman cities from the period of expansion through Italy and into the period of the Empire. Most Roman colonies, for example, were laid out on a regular plan with regular cadastrations of the surrounding land for subdivision among the colonists, who would often be former soldiers (hence the term 'centuriation'). The town of Elche in Alicante is a clear example where the structure imposed on the landscape continued into the street plan and layout of the town itself, and a number of other examples are described by Delano Smith (1979).

Desert cities

Even in the desert conditions of the Negev, cities were established in the fourth century BCE (Negev, 1966). These heavily fortified cities were developed by the local Nabataean populations as a progression from the staging posts which provided water along trading routes through the desert. After a phase of abandonment, these cities were re-established under the rule of the Romans in the second century CE. Aqueducts and canals supplying water to these sites were important innovations (e.g. Bruins, 1986). The eastern frontier of the Roman Empire was seen as vulnerable to attack and by the second century CE, numerous fortified cities and lesser fortified centres were set up on the desert margin in a broad arc from Jordan to Iraq (Kennedy and Riley, 1990).

In Tripolitania in North Africa, important urban centres were already in place by the end of the second century BCE as part of the Numidian kingdom (Mattingly, 1995). Inscriptions from a variety of sources show that these centres were in contact with a wide geographical area in Africa. The major centres were also reinforced by a series of hillforts. By the first century CE, there were four major cities in Tripolitania, all located on the coast in order to control Roman contact with the major grain-growing areas. However, there were also a number of important towns in oases in the interior.

The Mediaeval period and beyond

In the immediate post-Roman period, cities tended to decline with the political and socio-economic changes that the decline of the empire brought. Rural populations were reduced too, and the structures that linked existing towns with their hinterland began to break down. The general depopulation and lack of maintenance of rural as well as urban structures led to a number of catastrophic failures, as we have already seen in Chapter 10. It is easy to forget, though, that these declines did not affect the entire region as strongly, or for as long a period of time as in the west. In the east, Constantinople flourished as the centre of the new eastern Roman Empire, although it was later to suffer at the hands of Rome during the period of the Crusades. The Moorish expansion of the seventh century onwards led to the development of new centres in the east and south. For example, in Egypt the old Graeco-Roman centre of Alexandria was replaced by a new capital: Cairo. Morocco, Tunisia and Egypt became some of the important centres of power in the region. In southern Spain, the Moorish presence continued up to the *reconquista* of 1492, and led to Granada, Cordoba and Seville being some of the most important cities in Europe in the early Mediaeval period.

European urban expansion did not occur on a significant scale until the eleventh century. Often, the towns were small and heavily fortified as in northern Italy (e.g. Potter, 1979), reflecting the instability of the times. Braudel (1981) also points out (for the fifteenth century, although the same argument holds for earlier periods) that the large number of small towns reflected the slowness of communications between centres. This problem would have been particularly marked for the Mediterranean world, with its multiplicity of mountains. When the new centres did develop – Genoa, Naples, Venice – they reflected a marked change in the geography of power throughout the basin. Furthermore, they often reflected the focus on the sea and international trade. The immediate hinterlands of these large cities thus became much less important, just as the hinterland of Rome had once much of her grain was shipped in from North Africa. On the other hand, the rise of the eleventh century was effectively precipitated by rapid rural developments (Braudel, 1981). Delano Smith (1979) points out that even by the late Mediaeval period and Renaissance, many Mediterranean towns would have been unrecognizable to us as such. They often contained fields, orchards and gardens used for the growing of food, most likely as a measure to safeguard supplies in times of siege, and benefiting from the rich manure of urban waste (Figure 12.2). As late as 1871, a 'big' city was defined by the Italian census as having more that 8,000 inhabitants. Overall, the Mediaeval and early modern city formed part of what Braudel (1975) has termed 'the urbanized countryside' with strong links between most small towns and the surrounding agricultural areas. This period saw a shift in urban centres on the whole away from the Mediterranean and towards the northern European centres, which tended to grow much more rapidly. From 1300 to 1500, urban population densities actually fell in many Mediterranean regions, and did not pick back up until the sixteenth century (Terlouw, 1996). From this time, though, there was a period of sustained growth in most areas that continues to the present day. We will return to this issue and its impacts in Chapter 14.

Valencia 1738   Sagunto 1806

Avignon 1635   Lucca c1640   Padua 1780

Ancona c1640   Milan c1640   Zaragoza 1769

Parma c1630   Rome 1667   Florence 1755

Barcelona 1492   Pisa 1770

Major monumental complexes
(fortifications; cathedral area at Pisa)

Open fields; walled gardens

Extent of Moorish city
at Valencia

Figure 12.2 Schematic maps showing the extent to which the walled cities of Mediterranean Europe
contained large areas of open space including enclosed and unenclosed fields, orchards
and enclosed gardens

Source: After Delano Smith (1979).

Impacts on land use

The early urban centres in Egypt provided the central structures of control of a number of agricultural settlements scattered across the Nile floodplain (Kemp, 1989). In the Levant, Falconer (1987, 1994) stressed the importance of the development of differentiated networks of production and exchange of commodities in the development of the early urban centres there. These centres grew beyond their means of independent agricultural support (Falconer estimated 35 ha as the maximum size for this in Palestine under rain-fed agriculture) and thus needed to control a supply of food from surrounding areas. The presence of 'urban' temples in small rural settlements is suggested as one means by which this supply could be controlled. However, it is important to note that simple population increases are not sufficient causative explanations for the rise of urbanism. The other mechanisms of social interaction and control needed to be put into place.

For the early palaces of Middle Minoan Crete, starting around 2050 or 2000 BCE (Manning, 1994), Renfrew (1972) stressed the role of the urban site as centres of redistribution of agricultural produce. These centres, he suggested, arose as part of a four-stage process that was initiated by the availability of the more diverse types of crop, including olives and vines, at the end of the Neolithic (see Chapter 10). As a result of this diversification of the subsistence base, it was possible to bring more land into cultivation, leading to population increases. Because farmers would therefore be increasingly interdependent on each other, Renfrew suggests that chiefs emerged to take charge of the redistribution of different agricultural products, which would require centralized storage facilities. Finally, these chiefs would encourage specialized and craft production, emphasizing the divisions between different parts of the agricultural system, and requiring further agricultural production to support those employed in making prestige goods. These four stages then provide a positive feedback to reinforce the initial change. This model of change has been criticized for being too altruistic, and others have suggested that the system arose as a result of forceful specialization from elite groups who coerced the populations into nucleated settlements (Gamble, 1979). On its own, this argument lacks a causative mechanism, which van Andel and Runnels (1988) suggested was control of the trade routes that were becoming more extensive and important at this period (see also Gilman, 1981).

The common factor between these explanations is the effect of having a central control with storage facilities for agricultural produce (Figure 12.3). Thus, the structure of the landscape would have been irrevocably changed as both a cause and consequence of the rise of urban centres. We have seen how a more diverse landscape evolved in the later Neolithic and Early Bronze Age because of the introduction of a range of new crops. The fact that these crops could exploit new niches in the landscape would have increased the rate of clearance of the natural vegetation, and thus contributed to the accelerated rates of erosion noted around these periods in time (Chapter 10). Movements of produce would have led to the development of tracks and roads to the centre, which would have led to further potential concentrations of erosion, and probably also a restructuring of the settlement pattern to reflect their locations. Such restructuring would have further concentrated the anthropic impacts into certain areas of the landscape. Future developments in landscape archaeology should help to evaluate specific histories of how this process might have occurred.

The difficulties of defining relationships between town and countryside are discussed in detail for Roman Greece by Alcock (1993). Specifically the role of land ownership and how the land was worked as absentee landlords with slave workers became significant means of exploitation are important factors in understanding how the landscape would have been maintained. The villa system of Roman times was an important consequence of hierarchical divisions within

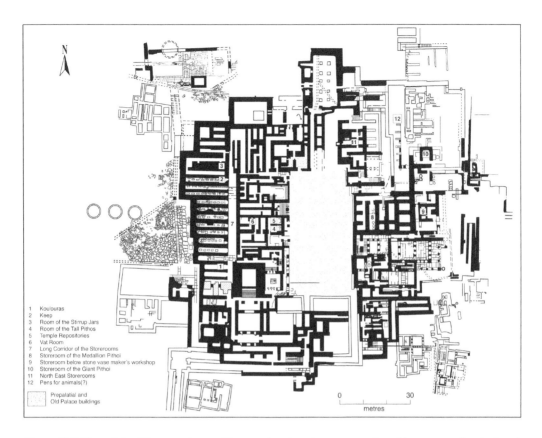

1 Koulouras
2 Keep
3 Room of the Stirrup Jars
4 Room of the Tall Pithos
5 Temple Repositories
6 Vat Room
7 Long Corridor of the Storerooms
8 Storeroom of the Medallion Pithoi
9 Storeroom below stone vase maker's workshop
10 Storeroom of the Giant Pithoi
11 North East Storerooms
12 Pens for animals(?)

Prepalatial and
Old Palace buildings

0   30
metres

Figure 12.3 Plan of the Minoan palace at Knossos showing the space employed in storage of agricultural
produce

Source: After Cadogan (1976).

the rural landscape that ultimately related to control from the urban centres. Similar issues
relating to absentee farmers and feudal or peasant farmers became significant again with the
rise of urban centres in the Mediaeval and early modern periods (Delano Smith, 1979). The
operation of such systems is examined in detail for the Valpolicella of northern Italy by
Musgrave (1992).

In the later periods, the structure of the landscape became much more formalized.
Examples of cadastration from the Classical periods have been mentioned above, and we can
still observe elements of the present landscape that essentially preserve boundaries and structures which were put in place around two thousand years ago. A clear example is the Roman
centuriation of the Rhône valley, with a highly integrated set of boundaries and drainage channels (Figure 12.4: Aschan-Leygonie *et al.*, 1994). One consequence of such systems, as we
have seen in Chapter 10, is that they require significant amounts of maintenance and thus are
very vulnerable in periods of population decline, as happened in the third century.

An important development in the Classical period was the need for long-distance contacts
to be used to supply the basic food to the major cities. In Athens, Garnsey (1998: 183) notes
'the apparent fact that ... food needs far outstripped the capacity of its home territory to
satisfy them', and points to historical evidence for significant imports from the Black Sea

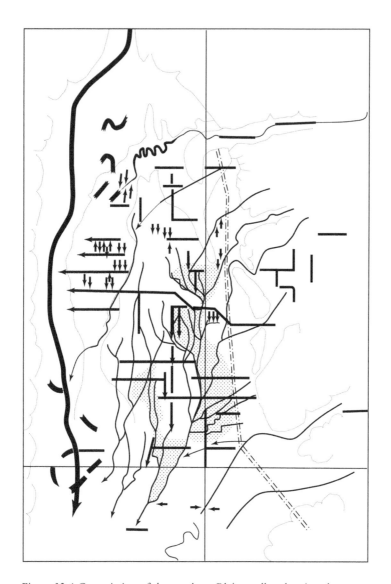

Figure 12.4 Centuriation of the southern Rhône valley showing the extent to which landscapes were modified in the Roman period as a result of structures required to support the entire hierarchy of settlement

Source: After Chartier *et al.* (1994).

region. However, the historical evidence may refer only to a particularly bad year and over-emphasize the importance of the imports. Garnsey's calculations suggest that a population of 120,000–150,000 could have been supported by local agriculture (a figure with which Sallares, 1991, is in broad concurrence), that falls far below his estimate of 200,000–300,000 for the population of Attica between 450 and 320 BCE. Thus, up to half the population at this time would have had to be supported by imported grain, suggesting the importance of trade and the political need to maintain control of the sea. The food supply of Rome became even more precarious. Finley (1985: 130) goes so far as to say 'Rome took to the seas because she had become a great city rather than the other way round. Rome was hardly typical, the complete parasite-city (though she was unique only in scale).' The 150,000 tonnes per year of grain that were required on average to support the million-strong population from the first century BCE to the third century CE were imported from central Italy, Sicily, Sardinia, North Africa, Egypt and, to a lesser extent, France and Spain. Some of the consequences of the growing of grain in these often remote parts of the empire have previously been discussed in Chapter 10. Finley estimates that 85 per cent of these imports reached Rome via the sea, explaining at least in part the importance of the development of the port of Ostia at the mouth of the Tiber. The growth of piracy from the first century BCE reveals something of the vulnerability of this system (Robinson, 1992). The distribution of food obtained centrally was, however, a central element of the structure of life in the city and elsewhere in the Empire (Veyne, 1990), but malnutrition was endemic among the poor of Rome (Garnsey, 1998).

Deforestation became an important process around the urban centres. By the fifth century BCE, the area around Athens was largely treeless (Hughes, 1994). The need for wood in the cities was not only for construction but also for fuel – Hughes notes the profits traders could make importing firewood into Athens from the surrounding countryside. Areas with suitable timber supplies then formed the basis for new urban developments at the mouth of rivers with mountainous, forested watersheds, as at Thessaloniki, Luna, Ravenna and Colchis (ibid.). The impacts of such deforestation would have been large (Chapters 5 and 6) and no doubt contributed to the magnitude of the floods that we know affected Rome during this period (see Robinson, 1992, for details). Thus, the link between urban development and rural deforestation is an important one, and one that would recur in the European Mediterranean in the centuries following the rebirth of major cities in the later Mediaeval period.

Control of the water supply

So far, we have concentrated on the control of the food supply and its impacts as being an important factor in the maintenance of urban centres. Equally importantly, the concentrations of large populations in the Mediterranean would require control of the water supply to minimize the impacts of seasonal and interannual variability (Chapter 3). There is some suggestion that the developments of the Early Bronze Age had led to significant developments in this direction (Chapman, 1978), and Crouch (1993) believes that there are strong links with an ability to control the water supply and the process of urbanism in the Mediterranean. Through time, the technology for controlling the water supply to towns and cities became increasingly more sophisticated. For example, Ortloff and Crouch (1998) describe the structure of a drainage outlet at the Hellenistic city of Priene on the Aegean coast of Anatolia. The hydraulics of the structure were designed to ensure that debris caught up in flows would not clog the outlet and lead to flooding of the town. A range of measures were taken in Rome to try to mitigate the problems of flooding, while maintaining secure water supplies (Robinson, 1992).

In the mid-sixth century BCE, a large tunnel running up to around 120 m below ground, nearly 3 m in height and width, and about 1.3 km in length was constructed on the Aegean island of Samos to carry a piped water supply from a spring to the city (White, 1984). The piped supply was carried in a second, smaller tunnel beneath the first, which is a technique seen in other tunnels at Athens, Syracuse and Akragas. Greek technology included the development of siphons for transporting water across deep valleys, for example, quite spectacularly in the case of the water supply for the city of Pergamon in Anatolia, where the siphon crosses two valleys before climbing up to the citadel at a height of only 40 m below the initial spring level, from where it passed into settling tanks before distribution to the city. Owens (1991) suggests that a number of Greek cities were deliberately located so as to include freshwater springs within city walls, for example, at Old Smyrna and Zagora on Andros (see also Crouch, 1993). In the later periods, more developed water-supply schemes were introduced. In Athens, Peistratids improved the water supply, constructing an aqueduct and fountain house, with further additions and renovations being made in the Classical and Hellenistic periods, with extensive renovations by the Roman emperor Hadrian, who had a new aqueduct and holding cistern built. At Olynthos in Macedonia, water was brought in the fifth century BCE from sources seven miles away by a pipeline of jointed terracotta sections in an underground channel (Owens, 1991). The streets here seemed to have had underground drainage. Priene in Asia Minor had paved streets with both surface and underground drainage for runoff, and storm drains have been found at Pergamon. The drainage system at Akragas was constructed by prisoners of war after the Battle of Himera. Rhodes also had a drainage system but there are records of a catastrophic flood in the spring of 316 BCE, with loss of life, large areas inundated and further catastrophe prevented only by the collapse of walls. The event was said to have occurred after the seasonal rains were thought to have finished and therefore the drains were not maintained (Owens, 1991). The Greek colonies often had water supplies brought from great distances: one of the aqueducts to Syracuse in Sicily was 17 km long.

The water supply of the Etruscan cities has already been mentioned in Chapter 10. At Veii, a cuniculus passes underneath the northern part of the city (Judson and Kahane, 1963), so that water could have been obtained without leaving it. Aqueducts were also used in conjunction with cuniculi to control the water supply, for example, to the Faliscan town of Ponte del Ponte (Potter, 1979). The planned settlements of Marzabotto and Spina also had extensive water-supply and drainage schemes.

Large aqueducts carrying water supplies to urban areas were a characteristic feature of the Roman period. The aqueducts supplying Rome itself were built progressively from 312 BCE until 52 CE, although the impressive, arcaded, overground portions were only a small element of the total system (Table 12.1: White, 1984). Most of the system was constructed in the second half of the first century BCE and in the first half of the first century CE, with progressively higher water sources used, which has been suggested as requiring a need for an increasingly secure supply in relation to population growth. Historical evidence gives further information about the supply system once it reached the city, including a series of settling tanks to remove sediment and water towers. Users would be charged on the basis of the diameter of the pipe used for their supply, and estimates of the water consumption at its height vary from 386 to 1,441 million litres per day (de Camp, 1960, cited in White, 1984).

Table 12.1 Water supplies of Rome, detailing portions running above and below ground and the date of construction

Name	Total length km	Above ground km	Underground km	Date
Aqua Appia	16.6	0.10	16.5	312 BCE
Anio Vetus	63.6	0.32	63	272/269 BCE
Aqua Marcia	91.7	10.25	81	144/140 BCE
Aqua Iulia	22.8	9.50	13	35/33 BCE
Aqua Alsietina	32.8	0.50	32.3	30/14 BCE
Aqua Virgo	20.8	1.03	19.8	21/19 BCE
A. Anio Nova	86.8	15.00	71.8	35/49 CE
Aqua Claudia	68.7	13.00	55.7	38/52 CE
Total	403.8	49.70	353.1	

Source: After White (1984); Evans (1994).

At the drier extremes of the Roman empire, dams were built to control urban water supplies artificially. The case of Homs in Syria has already been mentioned (Chapter 3), while two large dams were constructed at Mérida in Spain to control the water supply to the veteran colony of Emerita Augusta (White, 1984). The first of these seems to date from the early second century CE.

In northern Africa there were extensive aqueducts carrying water to the large Roman city of Lepcis Magna on the Tripolitanian coast (Figure 12.5: Vita Finzi, 1969). The aqueduct from the Wadi Caam and related public fountains seem to have been built in 119 or 120 CE, paid for by a local citizen, and led to the development of one of the largest provincial public baths complexes of the time (Mattingly, 1995), as a very visible means of consumption of the new supply. The Roman aqueduct supplying Carthage was 132 km long (Owens, 1991). White suggests that the choice of using a siphon or an aqueduct to transport water across a valley depended simply on the depth of the valley. The tallest known Roman aqueduct bridges are the Pont du Gard in southern France at 49 m and at Alcantara in Spain which is 50 m tall, and beyond this height, the arcaded bridge was considered unstable. The span of the Pont du Gard is around 30 m, and it carried water from a series of springs near Uzès to the Roman colony at Nîmes. It forms part of an extensive water-redistribution system of over 40 km in length (Figure 12.6; Fabre *et al.*, 1995).

The Roman water-supply system shows extensive evidence of detailed planning. One of the water commissioners of Rome, Frontinus, wrote a treatise on the water supply of Rome around 97 CE (Evans, 1994). The treatise includes a history of the development of the water system, with the recognition in 312 BCE that water from springs and the River Tiber were no longer sufficient to supply the city, leading to the construction of the Aqua Annio aqueduct. Frontinus provides details of the different supplies, the pipe diameters used, delivery and distribution figures and details of the administration and maintenance of the system. He notes, for

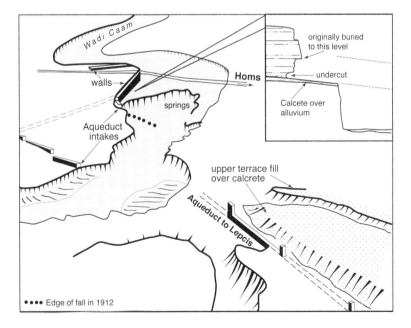

Figure 12.5 Examples of aqueduct systems used to carry water to the Roman city of Lepcis Magna in Tunisia

Source: After Vita Finzi (1969).

Figure 12.6 Map of the Roman water-distribution system bringing supplies from near Uzès to the
city of Nîmes via the Pont du Gard aqueduct

Source: After Fabre *et al.* (1995).

example, that repairs to leaking pipes should generally be carried out rapidly in spring or
autumn, and never in the summer when the water supply was most needed.

At Pompeii, water was obtained from underground sources and cisterns until construction
of an aqueduct, and the city also provides evidence of underground drains and raised walkways
(Owens, 1991). Comprehensive systems of street drains often accompanied the planning of
Roman colonies – in some cases the drained areas are greater than those eventually occupied.
Wells and cisterns were used initially until the construction of aqueducts which distributed
water to fountains and roadside basins for public use. Owens stresses the use of water to
promote the 'health and happiness' of Roman populations, notably in the provision of public
facilities, bath houses, fountains and latrines. Further Roman technologies included the use of
water wheels and pumps to provide water and power (White, 1984).

Pollution

As a final note, we should recognize that pollution is not just a recent phenomenon, although researchers have perhaps tended not to look for evidence of it as often as they should. For example, we noted above the importance of the developments as a consequence of urbanized settlement patterns. An episode of lake eutrophication is reported following the construction of the Via Cassia in *c.*171 BCE close to the Lago di Monterosi in Latium (Sallares, 1991). Hughes (1994) demonstrates that there must have been a concern for the increasing pollution of water supplies from ancient authors who indicate how water may be purified by filtration, percolation or boiling. Air pollution in Rome was also noted by Horace and other authors. A further element of urban life is the disposal of refuse, which in Rome was often simply carried out by dumping into the Tiber. However, the impacts of large amounts of decaying rubbish in the river and elsewhere in the city would have had led to significant pollution and probably the spreading of disease.

Summary

The rise of urban life in the Mediterranean was probably a consequence, at least in part, of successful exploitation of the rural landscape. Cities grew, at first in Egypt and the Levant, and then spread ultimately to the rest of the basin, albeit by a rather slow and indirect process. Although Braudel (1985: 481) considers a city to be a city forever, there are clear examples if one takes an appropriately long timescale of the decline of some centres, some to be reborn later, but others to be forgotten to all but the archaeologist.

The impacts of urbanism on the landscape as a whole were both direct and indirect. Food supplies needed to be maintained, and this required an efficient structuring of the local landscape. From the Classical period, the larger centres exerted a control on landscapes far distant whose people largely never saw the city for which they grew grain and other supplies. Tracks and roads to take supplies were also an important new development. Thus, the look of the landscape completely changed, even away from the towns and cities themselves. In the later periods, regular landscapes resulted from the need to provide structure. Indirect impacts would have largely related to the deforestation that was often concentrated around urban centres, to provide fuel and timber for houses and the ships that would be used to trade and protect. Pollution of the air and water also seems to have been a significant problem, at least locally.

Urban centres in the Mediterranean would not be possible on any scale without the technology to minimize the impact of variations in the supply of water. This technology was probably in place in at least a preliminary form by the Bronze Age, and thus provided a necessary precursor to the nucleation of settlements that took place from this time on. From the Classical period, sophisticated means of controlling the water supply and allowing drainage from the towns were available. These allowed substantial centres to develop in even the most arid parts of the region in the Negev and North Africa.

Suggestions for further reading

The edited volume *Development and Decline in the Mediterranean Bronze Age* by Mathers and Stoddart (1994) contains a number of useful papers relating to the socio-economic conditions that led to early urban developments throughout the Mediterranean. Clarke (1979) gives an invaluable insight into some of the mechanisms that may have been important in the early development of urbanization. Alcock (1993) provides an overview of the difficulties of reconstructing the operation of entire landscapes in the past, and Sallares (1991) puts the landscape–human link into an ecological perspective. Hughes (1994) brings together a large body of evidence on all aspects of environmental impacts in the Classical period.

Topics for discussion

1   Are the underlying mechanisms of the development of urbanism the same for all areas of the Mediterranean?
2   What might explain the different timing of the rise of urban centres in the east and the west of the basin?
3   What specific problems arise from an attempt to reconstruct the links between urban and rural landscapes when historical data are sparse?
4   Were the systems set in place to maintain food supplies to the large urban centres of the Classical period ultimately sustainable?
5   How might the impacts of urbanism in the prehistoric, Classical and Mediaeval periods differ from those of the modern day?

13 Traditional land-use patterns 4

Use of mineral resources

Introduction

As a final look at the impacts of land uses in the ancient world, we turn to the extractive indus-tries. Although extraction of flint would have been widespread in the Palaeolithic world, its impact would have been very localized. The same is probably the case for obsidian in the later Palaeolithic and Neolithic periods, although it could be argued that its social role had far more significant repercussions. From the later Neolithic, the extraction of metals for use in tools and other material goods started to develop. These extractive industries would go on to have significant direct and indirect impacts on the landscape of the region.

Mineral sources and their use

The earliest evidence we possess for widespread extraction, distribution and use of geological materials is that for obsidian. Obsidian is a volcanic glass, formed by the rapid cooling of lava on eruption, usually relating to contact with water. There are thus a number of sources of obsidian in the Mediterranean Basin relating to recently active, or geologically active volcanoes (Figure 13.1). Analyses of the major and minor element concentrations in obsidian have proved a reliable method for distinguishing the different sources of the material (Hallam *et al.*, 1976; Renfrew *et al.*, 1965; Tykot, 1998; Williams-Thorpe, 1995). As there are a limited number of sources – Melos, Antiparos and Giali in the Aegean, Acıgöl in Anatolia, Pantelleria, Lipari, Vulcano, Palmarola and Monte Arci in Sardinia – defining the source of obsidian used to make the artefacts found at a particular site tells us about the movements of prehistoric peoples. Indeed, we have already seen how important contacts relating to Anatolian and Melian obsidian may have been in the early spread of agriculture (Chapters 9 and 10). These movements have been explained in terms of gravity models for interaction (Hallam *et al.*, 1976) and by nomadic and/or transhumant movements (Crawford, 1978), although other mechanisms are certainly possible (Tykot, 1998).

Trace metal and isotope analysis of the semi-precious and precious metals have been par-tially successful in demonstrating trade links between different parts of the Mediterranean from the late Neolithic onwards. One of the main problems is that these metals can be smelted and mixed with metal from other sources, so that the 'signature' of the different sources became indistinct. However, there are suggestions of early sources and movements of materials over large distances (Figure 13.1). The earliest mines known are found in Serbia at Jarmovoc, Rudna Glava and Mali Šturac and were used to extract copper perhaps as early as 4000 BCE (Shepherd, 1980; Weisgerber and Pernicka, 1995); a copper mine discovered in the Sinai dates to around 3000 BCE; and works from Kozlu, 80 km south of the Black Sea in Turkey date to around 3369–3380 BCE (Shepherd, 1980). A small copper mine is known from Libiola in

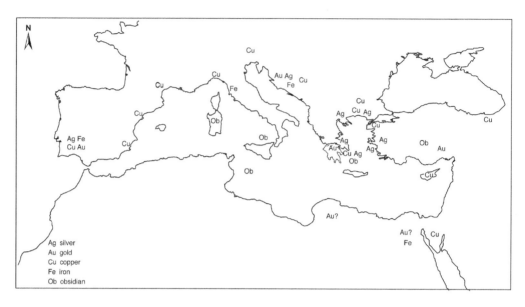

Figure 13.1 Rock and mineral resources exploited in the prehistoric and historic periods in the Mediterranean Basin

Genoa, northern Italy with a pick handle radiocarbon-dated to 3370–3100 BCE; and other small Bronze Age workings are known from Chinflon, Huelva, Spain and Cabrières near Montpellier in southern France (Weisgerber and Pernicka, 1995). Copper was probably mined and transported from Sardinia and Cyprus as early as *c*.1400 BCE, and from the source at Laurion on Kea in the Aegean from the twelfth century BCE (Dayton, 1982; Constantinou, 1981). Laurion was a source of lead from a similar time period, and silver from the fifth century BCE – used to supply Athens at the height of its power. Silver was also supplied from Siphnos in the Aegean, and from Almería from the middle to late Bronze Age, and there are important other prehistoric sources in Sardinia. Finally, some tin (used to make Bronze) was from sources in southern Spain and Portugal, although undoubtedly much of it came from further afield (Dayton, 1982). Muhly (1984) concurs with this general pattern, suggesting tin was widely transported by sea in the Bronze Age, but also points to potential sources in Egypt.

Subsequently, iron became a far more important metal for use in tools, but required more developed technologies for extracting the metal from its ores. The main development of the technology was by the Hittites in Anatolia, from around 1800 to 1500 BCE, spreading to Greece around 1100 BCE, Italy in *c*.900 BCE, and thence west and north into the rest of Europe, although some early examples are known from Mesapotamia and Egypt (Shepherd, 1980). Large-scale use of iron did not occur until the ninth century BCE in Egypt, though. Mining is known to have occurred on relatively large scales from the western Sinai, the Taurus mountains, Tuscany, Elba, Portugal and the Rio Tinto of southern Spain (Coghlan, 1988; Shepherd, 1980). Gold was also exploited more extensively from the Iron Age, with mines known from the Greek mainland and islands, Bosnia, Turkey, Libya and Egypt. Workings were so extensive that the reserves were completely worked out (Shepherd, 1980).

We must also remember the widespread use of stone as a building and construction material, with increasing use from the Bronze Age. In some areas, this use would have developed from locally available surface materials (e.g. in the Montpellier garrigue of southern France), but ultimately quite extensive technologies were developed to extract large quantities of materials. The construction of the pyramids of Egypt is perhaps the most impressive early example of this, but we should remember that megalithic constructions were also known from around the region before the pyramids, for example, in Malta (e.g. Trump, 1976). Peacock (1992) demonstrates the importance of such extractions of rock with the example of Mons Claudanius, a Roman quarrying site in the Eastern Desert 500 km to the south of Cairo. The site is exceptional in the degree of preservation of organic remains and documents written on papyrus, due to the high aridity, giving us a clear record of the sorts of activity that took place. Large items such as columns were made *in situ* and distributed from the site by wagon across the desert and then to the Nile, 120 km away. Examples of the granite quarried at this site are found in Istanbul, Split and a number of sites in Italy.

Impacts of quarrying and mining

In terms of the direct modification of the landscape, the quarrying of obsidian had a relatively minor impact. At Melos, the chipping floors where at least the primary processing of the raw obsidian took place can still be seen by the outcrops (Renfrew *et al.*, 1965), but apart from the localized removal of material, there were no other real impacts.

It is with the later Neolithic quarries that we find a more significant direct impact, albeit on a relatively small scale initially. At Rudna Glava, eleven mines have been discovered, with a maximum depth of 10 m (Weisgerber and Pernicka, 1995). Platforms were cut in the rock face, and then mine shafts sunk following the inclination of the veins in the rock. The limit to the depth reached seems to be the point where groundwater was reached. There is some evidence to suggest that fire was used as well as mechanical means for extracting the ores from the shafts. The spoil from workings would be another impact, with an area of 48 ha covered with debris and still largely unvegetated at the early copper mine at Mali Šturac (Weisgerber and Pernicka, 1995). Later mines became more extensive, for example, at Kozlu, the shaft was 3–5 m wide and extended for a depth of 300 m (Shepherd, 1980). The Rio Tinto mines led to significant reshaping of the landscape even in the later prehistoric and Roman periods (see below). Hughes (1994) points out that the silver required for minting coins in the Roman period meant removing 50 t of silver per year, or 100,000 t of rock from the mines at Laurion. Quarrying of building stone would also lead to the removal of large volumes of material, for example, at the limestone quarry at Syracuse, 112 Mt of stone were excavated with one face 27 m high and over two kilometres long (Healy, 1988).

Deforestation was perhaps an even more important direct impact of the early mines, with timber used to support shafts, to set fires to help remove ore, and to fuel furnaces to extract the metal from the ore and then to process it into finished artefacts. At Kozlu, significant use of timber from local forests is suggested as early as 3369–3380 BCE, and a late Neolithic furnace for the smelting of copper is known from the Sinai (Shepherd, 1980). Later prehistoric mines in Cyprus were also found to have made extensive use of pine, plane, alder and oak to support shafts, as well as requiring extensive timber to supply the processing furnaces (Constantinou, 1981). The onset of the use of iron would have required a significant increase in the amount of wood required for smelting. Although less wood is required than for copper in relation to the amount of metal produced, the amount of iron produced was in far higher quantities because of the more common availability of the ores. This increase would have occurred as early as the middle second millennium BCE in Anatolia (Coghlan, 1988). The use of fire-setting in mine shafts is known from some of the Etruscan mines. Meiggs (1982) points to the extensive deforestation around the Laurion mines in both prehistoric and recent times as the result of a need for fuel and pit props. In the Classical period, timber even had to be imported from Euboea to provide for the needs of the mines, causing irreversible changes to the vegetation of the island (Hughes, 1994). In the Classical period, there is evidence that timber supplies for fuel to process metals were running low. There was thus great stress for competing needs of timber, for construction, domestic fuel supplies, and for building the ships required for trade and military purposes. Hughes (1994) calculated that the 70–90 Mt of slag that were probably produced in the Graeco-Roman period would have required the cutting down of around 20–28 million ha of trees. He points to a remark by the elder Pliny that wood shortages in Campania required the use of coal to produce bronze.

Extensive use was also made of water supplies in mining. Large ore-washing tables for the repeated grinding and wet sieving of ore are known from Agrileza in Laurion from the later fifth century BCE (White, 1984). The Roman mines of the Río Tinto area in south-western Spain (see Chapter 2) employed extensive hydraulic technology to extract the ore from the overburden (Domergue and Hérail, 1978; see also Greene, 1986). Water fed from aqueducts and header tanks was led either directly across the surface of an open-cast mine to erode the lighter material directly, or down deeper, vertical shafts which would have caused slope failures and the easier extraction of materials (Figure 13.2). As previously indicated, groundwater flooding of mine shafts is also a significant problem, and there is evidence that the Romans employed chains of buckets, waterwheels and screw pumps to extract water from mines in Spain and Portugal (Coghlan, 1988). There were observations in the ancient world of the impacts of air and water pollution as by-products of mining activity. Evidence is beginning to emerge that this could have had significant downstream impacts in the Rio Tinto mining zone (Schell *et al.*, 2001). In southern Jordan, modern soil, vegetation and animal samples show

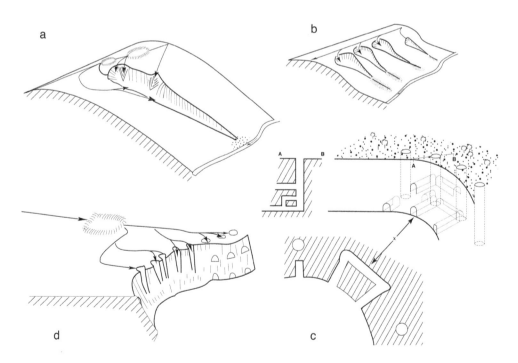

Figure 13.2 Mining technology in use at the Río Tinto mines showing the use of hydraulic technology to modify large areas of the landscape

Source: After Domergue and Hérail (1978), Greene (1986).

enhanced traces of lead and copper in areas used for mining from the Bronze Age to the Byzantine period (Pyatt *et al.*, 2000).

Summary

The mineral resources of the Mediterranean Basin (Chapter 2) have provided significant resources for prehistoric and historic civilizations alike. Their extraction was initially at very small and localized scales, resulting in minimal environmental impacts. As metal ores became the focus of extractive activities, the direct and indirect impacts became significantly more marked. Most particularly, vast amounts of wood were required – either directly or in the form of charcoal – to process the raw materials into finished products. Thus, the forests surrounding mines and processing centres were progressively removed. By Classical times the rate of deforestation was very rapid. Locally, erosion would have also increased because of the lack of vegetation cover on spoil heaps. In some areas, there is evidence for local desertification of these areas – the vegetation finds it very difficult to recover on soils with high acidity. Water erosion was also used purposefully to modify the landscape from the Roman period to remove overburden. Hydraulic techniques in mining show important cross-fertilizations with technologies originally derived for agriculture. However, it is likely that this process also led to pollution of water supplies and important off-site effects.

Suggestions for further reading

The current state of obsidian studies for the eastern and western Mediterranean are summarized by Williams-Thorpe (1995) and Tykot (1998), respectively. Weisgerber and Pernicka (1995) give a useful general introduction to early mines and metal sources in the Mediterranean. Shepherd (1980) gives a thorough overview on all aspects of prehistoric mining.

Topics for discussion

1   What were the local and regional impacts of trade in obsidian?
2   How might the relatively early use of metal resources in the region have come about?
3   When would the impacts of mining and metal extraction have become significant on the landscape?
4   What links might there be in terms of mining and other primary activities?

Part 4

The Mediterranean environment under increasing pressure

14 The present human landscape

Introduction

Although in many ways the twentieth century has followed trends started previously, there have been a number of ways in which these trends have developed to provide unique characteristics within the environment. In part, these relate to the major technological advances that have been made, partly as a response to ever-growing populations at the global scale. In this chapter, we evaluate the major population trends that have taken place within the Mediterranean Basin, their spatial patterns and the processes underlying them. Although the Mediterranean world has previously formed a core role in world systems (see Chapter 12), there have been global developments starting in the nineteenth century that mean the Mediterranean world has played a more peripheral role in the twentieth century. One upshot of this development has been the significant levels of migration out of the area, both within Europe and globally. But also, the strong north–south gradients that have been present throughout much of the historical period have, if anything, become more accentuated through the development processes of the twentieth century. Following from this, their are also strong inter-regional population movements.

The dominant forms of change have been to a more urbanized, industrialized landscape: one with very different environmental pressures and impacts from those experienced previously. These changes have again followed previous trends, but have been responsible for significant spatial restructuring within the basin, particularly within the last few decades. A much more significant development, and one with less well defined prior roots, has been the expansion of mass tourism on a basin-wide scale, albeit still concentrated in the wealthier regions of the north – these two factors, being of course, closely interlinked. But tourism has also brought with it a whole series of environmental issues, some of which threaten its very existence and long-term sustainability. Politically, there have also been new developments, most importantly the establishment of the European Community, initially comprising only two Mediterranean nations, France and Italy, but later expanding to include Greece, Portugal and Spain, and negotiating to include others. The establishment of the Community has had wide-reaching economic and social impacts, and thus by necessity has affected Mediterranean environments, both directly and indirectly, positively and negatively.

One of the positive impacts has been the setting up of research into environmental impacts, independently and in cooperation with other organizations. Under the United Nations Environment Programme, the Blue Plan was instigated to investigate the potential impacts of future growth within the basin. The final section of this chapter looks at the projections made by the Blue Plan and other organizations, and some of their likely environmental impacts.

Population dynamics in the twentieth century

Populations to 1950

As was seen at the end of Chapter 9, Mediterranean populations as a whole were on the increase, with a significant migratory move away from rural areas to the cities and even out of the basin as a whole. In many areas, these moves were reinforced by the economic catastrophe brought by the *phylloxera* virus that decimated vine production at the start of the twentieth century. The population of Italy had increased from 27 to 35 million between 1871 and 1914; in France, growth over the same period was from 37 to 40 million. These general trends continued through the first part of the twentieth century, although migration from Mediterranean Europe became more limited. In part, this reflected changing politics in the sending countries – for example, Mussolini's banning of emigration from Italy by 1929 – and in part changes in receiving countries, particularly the limitations on entry into the USA by the Quota Laws of 1924. The population of the Argolid, for example, which had risen from 7,954 in 1848 to 12,549 in 1896, dropped back to 11,255 in 1961 (Jameson *et al.*, 1994). These trends reflected general crises in rural Europe in general and the Mediterranean in particular. For example, decreasing transport costs saw the increase of cheap food imports from the USA and elsewhere, particularly in the second half of the nineteenth century (Tracy, 1989). However, populations still had a significant rural element. The French population had 42 per cent employed in agriculture in 1901, compared to 8 per cent in the UK (Tracy, 1989). To the east and south, though, change was slower to occur – 78 per cent of the Greek population were employed in agriculture in 1920, declining to 63 per cent in 1951 (Leontidou, 1990), compared to 82.3 per cent of the Turkish population in 1955 (Tachau, 1984).

Populations since 1950

The quantitative study of the population and its structure since 1950 is facilitated by more widely available statistics available through the United Nations and projects such as the Blue Plan (see below). The population of the entire Mediterranean Basin totalled around 212 million in 1950. By 1970 the figure had risen to 298 million and to 497 million by 1997 (Figure 14.1). The annual growth rate was relatively sustained, but declined from about 1.6

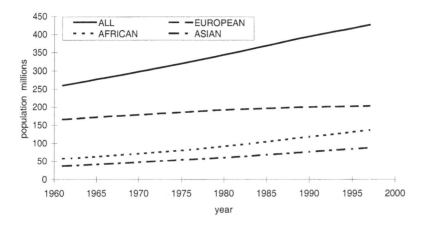

Figure 14.1 The annual total population of the Mediterranean according to region from 1960 to 1997

Source: Based on data from FAO yearbooks.

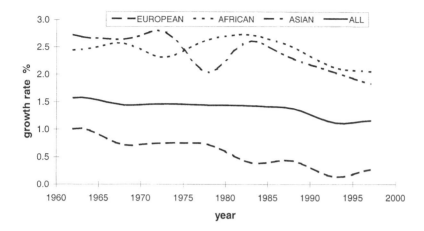

Figure 14.2 The annual growth rate of Mediterranean populations according to region from 1960 to 1997

Source: Based on data from FAO yearbooks.

per cent in 1960 to about 1.1 per cent in the later 1990s (Figure 14.2). Underlying this general trend, however, are important spatial differences. The European Mediterranean countries, which made up 64 per cent of the total Mediterranean population in 1961, have experienced much lower rates of growth, declining to about 0.3 per cent or even lower in the 1990s. The impact of this trend has been to make the European Mediterranean population the minority by 1993. The trend can be explained by declining trends in fertility – for example, Spain, France and Italy have experienced long periods where the birth rates have fallen below replacement rates (i.e. less than two children per woman) – and by internal migration within Europe, particularly in the 1960s and 1970s. Growth rates in the Asian and African countries have continued to be high, reaching almost 3 per cent in the early part of the period in question. The declines in the rates of increase in these areas to around 2 per cent per annum in the late 1990s are partly due to slight decreases in fertility, although these remain relatively high (usually above five children per woman), and to international migration as discussed below. As well as the higher birth rates, the countries of the south and east have tended to experience bigger drops in the mortality rate.

Within Europe, major differences are also apparent. Albania experienced growth rates above 2 per cent per annum into the late 1980s, while Greece and Italy have tended to grow at rates much lower than the average. Portugal, Malta and Cyprus have all experienced periods of population decline and recovery over the period. The African countries have had growth rates of around 3 per cent per annum since 1960, apart from Libya, which has averaged a 4 per cent increase. Of the Asian countries, Lebanon is markedly different with a period of net population decrease between 1976 and 1989 relating to the period of civil war. Syria has tended to grow the fastest and Turkey somewhat slower.

A further difference between the European and the other Mediterranean countries is the age structure of the population. The European Mediterranean countries show relatively uniform population pyramids (Figure 14.3), with an increasing proportion of elderly people, particularly in France, Greece and Italy. The southern and eastern countries as well as Albania have much greater proportions of children and teenagers. These differences can again be considered as resulting from the differential changes in mortality and birth rates in the different

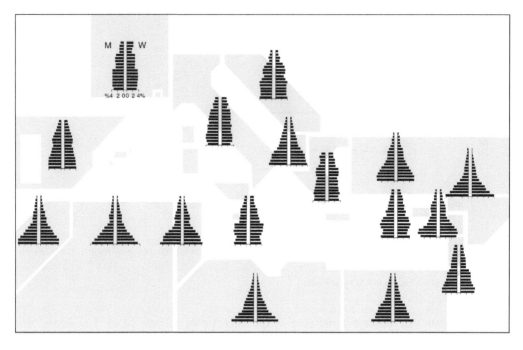

Figure 14.3 Population pyramids showing the age structure in five-year cycles of Mediterranean inhabit-
       ants by country

Source: Redrawn from Grenon and Batisse (1989), reproduced by permission of UNEP-BP/RAC and Oxford University Press.

areas. The patterns have important consequences on the numbers of economically active
people in the populations, with those in Europe growing at much slower rates than elsewhere.
On a more detailed scale, local migration has tended to lead to a concentration of more elderly
populations in more remote locations, as will be discussed further in Chapter 16.

   Another general trend of the Mediterranean population as a whole is its decreasingly rural
nature. Already by 1961, only 46 per cent of the European Mediterranean population lived in
rural areas, and this value has continued to decline to 37 per cent in the late 1990s. The cross-
over to majority urban populations had occurred by 1981 in the Asian countries, although the
figures are dominated by change in Turkey where there was strong internal and external return
migration to the cities (Tachau, 1984), and 1993 in North Africa. Only in Albania, Bosnia
Herzegovina, Egypt and Portugal are rural populations still in the majority.

   Return migration has also been an increasing factor, both to rural and urban areas, although it
has long been a significant process (Campbell, 1964; King, 1984). However, the return to rural
areas often does not result in investment in land for the returnee, so that subsequent generations
are often equally likely to migrate (Black, 1992; Hoggart *et al.*, 1995). Thus, there is a further
contribution to the increasingly elderly nature of rural populations. On the other hand, Kenna
(1993) has described the case of return migrants playing a prominent role in the development of
tourist-related enterprises in Anafi in the Cyclades. In Greece as a whole, there has been a tendency for people to migrate from rural areas to urban centres elsewhere in Europe, but to return
to Greek urban centres (Gaspar, 1984), forming an indirect move from rural to urban areas.
Although some authors suggest there is a strong process of counter-urbanization at present, for
example, in Spain (Champion, 1995), this has been challenged at a more regional level, with

rural out-migration still continuing in the case of Andalucia (Hoggart, 1997). However, this pattern too is still a simplification, with both in- and out-migration taking place in rural areas, reflecting a high population turnover overall. There is now also a trend for workers from North Africa to migrate to work in Spanish agriculture, in some cases on a seasonal basis (Hoggart and Mendoza, 1999). Leontidou (1990) points to the oil crisis of 1973–1974 as an important turning point in slowing emigration from southern Europe in general.

Gendzier (1999) has stressed the importance of migration from North Africa to the other Mediterranean countries. The role of the policies of French colonial rule is seen as fundamental in the destruction of the Algerian peasant economy and thus the exploitation of Algerian surplus labour to compensate for recurring labour shortages in France. From the Algerian perspective, this process is seen as having two important benefits. First, migrant workers sent remittances home and, second, it provided an escape valve in minimizing domestic discontent over political repression and economic inequality. After the 1973 oil-price rise, France and other European governments curbed immigration from Algeria and other countries of Maghreb, but by 1979, numbers had increased once again, so that, for example, in France alone, there were 782 k Algerian, 400 k Moroccan and 184 k Tunisian workers (Collinson, 1996). Algeria and its Maghrebian neighbours resorted again to labour exports in the 1980s to compensate for chronically inadequate development strategies (Geze, 1996). Although debt rescheduling opened the Algerian oil and gas sectors to foreign investment, the effects have been unemployment rates of 50 per cent, more than a quarter of the population under the poverty line and over half on the margins of the Algerian market economy (Tlemcani, 1999). There have been European attempts to reverse trends, but these as well as other Maghrebian initiatives of limited success, and European-led 'Mediterranean partnership' further exposed Maghrebian states to the disadvantages of trade liberalization with more powerful states. There have been similar problems with IMF-led policies in Egypt, Morocco and Tunisia (Pfeifer, 1999). In Egypt, for example, the result of neoliberal policies 'has been to concentrate public funds in different, but fewer, hands. The state has turned resources away from agriculture, industry and the underlying problems of training and unemployment. It now subsidizes financiers instead of factories, speculators instead of schools' (Mitchell, 1999). Thus, structural changes relating to basin-wide migrations have led to depopulation of the upland zone, with its consequences as seen in the north, and also to restrict the badly needed restructuring of agriculture in the lowland zones.

Whether Mediterranean countries can be treated as a whole, or whether national or even more local differences are significant is a moot issue. In their discussion of Greece, Italy, Portugal and Spain, Hoggart *et al.* (1995) point to a series of common threads such as the relatively recent development of these nations, centrality of rural issues, need for structural reforms in agriculture, and general rural–urban links due to the long history of developed urban centres. Of these factors, perhaps only the latter are less important for some areas of North Africa, but probably hold relatively true for much of the eastern part of the basin. However, the general trends of development point to different foci within the basin, reflecting north–south and east–west divisions. It could also be argued that there is a strong historical continuity of core–periphery relationships, that were particularly accentuated during the colonial period, and whose presence is still strongly felt. In part, this may have been further accentuated over the last few decades by the accession of increasing numbers of European Mediterranean countries to the EU. Even within the EU countries, Keeble (1989) notes that the Mediterranean countries tend to play a peripheral role, based on 1983 measurements of access to markets, physical proximity and a number of socio-economic indicators such as lower GDP, higher employment in the agricultural sector and more traditional manufacturing sector. Despite these generalizations, we must be aware of the continuing presence of differences at the regional and local levels, that strongly flavour the nature of Mediterranean settlement and land use.

These variations are by no means static, though. Keeble's study suggests something of a convergence between the core and periphery zones of the EU, relating to processes of deindustrialization, the development of new, small industries, and growth of the tourist sector, which is particularly marked in the Mediterranean countries, as discussed below. The development of new intra-regional differences in Greece have been evaluated, for example, by Economou (1993). The principal changes observed reflect the move of industrial focus from Athens to a more widely distributed base, the decline of old regional structures and their replacement with new territorial units and intensification and expansion (though not restructuring) of the agricultural sector. Increases in tourism have also led to the development of significant peri-urban areas and loss of rural zones, particularly as the move from agriculture to tourism is more advantageous than the switch from other activities. Specific details of these processes are given for northern Greece by Andrikopoulou (1987). Naylon (1992) discusses strong regional differences in Spain, reflecting the decline of the 'Cantabrian Corniche', the poor development in the central part of the country apart from in the immediate vicinity of Madrid, and the expansion of the Mediterranean axis, particularly from Catalonia to Murcia, and along the Ebro Basin. More recently, Andalucia has flourished as part of this general trend. The buoyancy of these areas, according to Naylon, rests on the diversity of the economic base, with the importance of tourism, networks of small and medium-sized companies, the ability of traditional industries to weather recession, the modernization of agriculture (for example in 'plasticulture' – see Chapter 15), high levels of service industries, and the presence of new sectors such as electronics and pharmaceuticals, often supported by foreign investment. Those areas that have tended to flourish depend on skills-based rather than resource-based industries. The spectacular growth of Catalonia also reflects the massive investment relating to the 1992 Olympic Games in Barcelona. A common pattern in these developments is a decentralization of development, although maintaining a general structure of regional inequalities – albeit different from the traditional ones – and often focused on maintaining existing class divisions (Hadjimichalis and Papamichos, 1990).

Overall, these general trends reflect a decline in traditional ways of life in the region, and thus their impacts on the landscape. Declining rural populations mean that there are lower populations to support the maintenance of structures such as terraces and irrigation ditches, that aim to mitigate the effects of erosion. The increase in food imports means that the diverse nature of Mediterranean agriculture has declined, and begun to concentrate on more specialized crops. The successful agricultural zones are now focused in the coastal zone, producing cash crops under intensive cultivation, often on highly modified slopes and soils. Although the move to a more structurally viable agricultural sector has economic advantages, the increasing homogenization has negative impacts both socially and environmentally, in reducing diversity and often stability. Depopulation of the uplands has also led to the decrease in pastoral ways of life and the decrease of vegetation removal due to grazing, so that areas are often increasingly more vulnerable to fires.

Urbanization

As noted above, the Mediterranean population is becoming increasingly urban in nature (Figure 14.4). Thus, the role of the city and its interaction with the surrounding landscape are also becoming an increasingly significant question. The marked increase in post-war urbanization is related to a number of general processes (Gaspar, 1984). First, those countries with more rapidly growing populations tended to experience more rapid forms of urban growth. Second, there were developments of regional imbalances, particularly in Portugal, where growth was centred around Lisbon and Oporto, and in Greece, where Athens and Thessaloniki

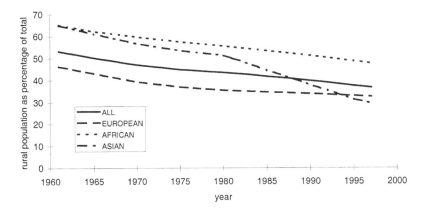

Figure 14.4 Percentages of rural Mediterranean populations by region from 1961 to 1997
Source: Based on data from FAO yearbooks.

were the major centres, although these imbalances became less marked in the 1970s. Third, there were reductions of migratory movements in the 1970s that led to the development of larger numbers of more localized, smaller urban centres. This process was accentuated by the move to 'deindustrialization', or the refocusing of industries in small, decentralized locations. Specific processes are also noted by Gaspar, for example, the internal migrations from southern to northern Italy in the 1950s' boom, and migration away from the south of Spain, at least until the 1970s.

The form of this urbanization therefore has important spatial variability (Grenon and Batisse, 1989). In the north-west, particularly in Spain, France and Italy, there are a large number of urban centres. Elsewhere, expansion tends to be concentrated in the capital, with regional centres playing a much less important role. There is clear evidence of migration from provincial centres to the capital, for example, in Morocco and Tunisia. The form of the urban expansion has been to focus on existing centres, with very few new urban centres being developed. Apart from developments around the capital, the other major form that urbanization has followed is expansion along the Mediterranean coast to form a number of major conurbations. The spatial structure of cities is changing dramatically as a result of these developments. In Europe particularly, the rapidly increasing use of cars means changes in the form of cities, with expansion of suburban areas due to apparently easier transport, but with concomitant problems due to pollution. The more rapid changes in the south and east have often led to social instability and the difficulty of providing planned centres, resulting in the development of makeshift settlements on the edge of existing cities. Grenon and Batisse (1989) distinguish four major environmentally sensitive factors which result from increased urbanization in the Mediterranean. First, there is an increase in the consumption of land that is poorly planned and regulated, so that settlement may occur in increasingly marginal locations, for example, on steep slopes which may be vulnerable to landsliding. Leontidou (1900), for example, reflects on the importance of unplanned population expansion in Mediterranean cities. In many areas, land around cities that is supposed to be for agricultural use is occupied, often by the poorer parts of the population. This pattern often reflects the tendency for Mediterranean cities to concentrate the poorer populations around their periphery. Second, the water supply becomes increasingly under pressure, and drainage and removal of waste water and sewage become increasingly more difficult, especially where development is unplanned. Issues of hygiene and public health therefore become correspondingly more important. Third, there are related issues of the production of solid waste. Fourth, air and noise pollution will increase due to increased amounts of traffic, production and consumption of energy in the home, and from the industrial expansion concentrated near the urban centres. As mentioned previously, unplanned and unregulated development may exacerbate any or all of these factors.

The trends towards increasing industrialization (see below) and urbanization go hand in hand. Industry focused population in increasingly urban centres, particularly since the growth of the 1950s and 1960s (Leontidou, 1990). In some cases, this has led to a suburbanization of the Mediterranean city, and a move away from the traditional spatial division of the city along class lines as, for example, in Athens. Particularly in the coastal zones, these suburban and peri-urban developments are also strongly influenced by the development of tourism (Economou, 1993). In Spain, there has been a strong focus in urban growth in the coastal zone (Morris, 1992). This growth is often concentrated in relatively small urban centres of 20,000–100,000 residents, and often reflects the decentralization of industry as there is a move away from a resources- to a skills-based economy. The decentralization is often a reflection of internal markets, and a move to areas that have tended to have lower labour costs. Morris also stresses the importance of infrastructural investment in this case, particularly with respect to the major motorway-building programme allowing significantly easier access and lower transport costs along the coastal zone. Similarly, in Israel, Kipnis (1997) has highlighted the development of urban centres along a concentrated coastal zone since the 1980s as a 'megalopolitan' zone, hemmed in between the mountains and sea. Population densities have continued to increase in the 1990s, not least because of the influx of large numbers of migrants from Russia, and the increased focus of capital in urban centres.

Industrialization

In the period since the Second World War, the Mediterranean region has undergone a significant process of industrialization (Grenon and Batisse, 1989: Figure 14.5). Ranked in terms of proportion of industrial output contributing to GDP (excluding oil revenues), Spain, France, Italy, Malta, Yugoslavia and Turkey had a contribution of over 20 per cent in 1985, compared to 15 to 20 per cent in Cyprus, Greece, Israel, Lebanon and Morocco, 10 to 15 per cent in Algeria, Egypt and Tunisia, and less than 10 per cent in Libya and Syria. However, there is an underlying trend of relatively slow current growth in Europe (and, indeed, decline in France and Malta), compared to much more rapid recent growth in Turkey and much of North Africa. The slow growth or decline is a function of increased competition from developing countries, both on a Mediterranean and global scale. Countrywide statistics often belie a spatial reorganization of industry which has in a number of cases become more focused on the Mediterranean. A prime example of this effect is in Spain, where decline of heavy industry on the Cantabrian coast has led to the expansion of newer industries especially in Catalonia and Valencia (e.g. Naylon, 1992). Traditional industries such as mining and iron and steel production have tended to remain static or even decline in recent years in the north-western Mediterranean, but expand elsewhere, for example, in Turkey. Cement production has increased rapidly since the 1950s, with an important concentration in the coastal zone so as to take advantage of transport and export by sea. Other important industrial developments have been the production of sulphuric acid, particularly in Italy and Tunisia, superphosphates and ammonia.

In Mediterranean Europe, the trend towards increased industrialization was spearheaded by the Italian 'economic miracle' from the 1950s, reflecting growth in the north of the country

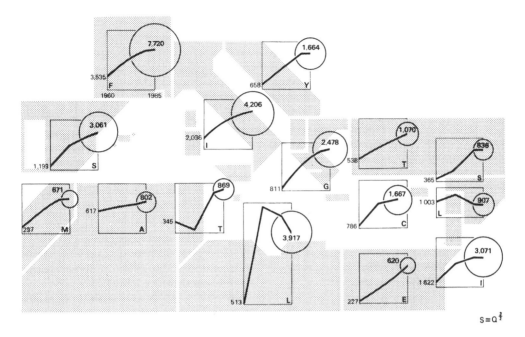

$$s = \alpha^{\frac{z}{3}}$$

Figure 14.5 Patterns of industrial growth in the Mediterranean: per capita GDP (1975 US$ equivalent)
Source: Redrawn after Grenon and Batisse (1989).

particularly, as well as in Greece and Portugal in the 1960s (Leontidou, 1990). The Italian case was spurred in particular by entry to the EEC and the free movement of foreign capital (Hudson and Lewis, 1984). For Greece, Leontidou points to the importance of foreign invest-ment and the ability of Greek workers to migrate out to other core areas. Restructuring of the industrial base was also a key factor, with decentralization and a move to more diffuse indus-trial activity being central (e.g. Kalantaridis and Labrianidis, 1999). Only in Portugal has urbanization tended to be focused on a small number of major centres – in this case, Lisbon and Oporto. Hudson and Lewis (1984) point to four major factors in the foreign investment that have accentuated Mediterranean industrialization. First, there are available natural resources, such as often small-scale mineral deposits. Second, governments have allowed activi-ties that can be seen to pollute, such as at the bauxite-processing plants in Itea, Greece. Third, there was provision of access to previously closed internal markets; and fourth, there was the availability of low-cost labour.

Rojo (1994) has highlighted a period of crisis in the industrialization of the Maghreb, despite the huge current population growth. This crisis relates to the over-emphasis on natural resources, existing pollution, often from neighbouring developed countries, and an increasing emphasis on agricultural production from unirrigated land. Furthermore, the populations tend to be underemployed and poorly educated (particularly among the older generations), there is little overseas investment and high levels of debt, and there are poor levels of technical and infrastructural support. Further development will probably depend on integration between the Maghrebi countries and the European Mediterranean, although this continues to be hindered by internal and external political structures.

The development of tourism

Probably the single most important industrial development in the region over the past fifty years has been the development of mass tourist markets. These markets have provided both financial and employment stability particularly to the coastal zones of the Mediterranean. By 1970, the Mediterranean region accounted for 36.3 per cent of international tourism, with fifty-eight million arrivals (Table 14.1). Although the proportion had dropped to 30.0 per cent by 1996, the absolute numbers have continued to increase, with 176 million arrivals. The spatial pattern of these tourists is by no means evenly distributed, with 84.4 per cent of tourists to the Mediterranean having destinations in Spain, France, Italy or Greece in 1970. Despite a significant effort to popularize other destinations, this proportion had only decreased to 75 per cent by 1996 (although, of course, this represents a significant increase in terms of total numbers). Tourism growth rates of more than 10 per cent per year in Cyprus, Egypt, Greece and Tunisia were noted from 1970 to 1986, with moderate growth of 5 to 10 per cent in Turkey, Malta, Israel, Morocco and Syria (Grenon and Batisse, 1989). Much of this tourist activity is concentrated in the coastal zone, with over 80 per cent of all tourists to Tunisia located here. In the former Yugoslavia, the equivalent figure exceeded 90 per cent for international tourists, compared to only 18 per cent in France. The financial considerations of tourism are considerable. By 1984, tourism contributed on average 6.5 per cent to GDP, increasing to about 7.0 per cent at present (Medforum, 2000). Tourism earnings in Spain were £28,447 million in 1990, and around 11.2 per cent of the workforce were directly or indirectly employed in tourism (Albert-Piñole, 1993). Portuguese earnings in the same year were £2,023 million (Edwards and Sampaio, 1993), while those in Greece were £5,626 million (Briassoulis, 1993) and those in Italy £19,850 million (Bonini, 1993). Employment in the sector is 7.2 per cent in Greece and 11.1 per cent in Italy.

The effects of tourism on the environment are classified into three major categories by the Blue Plan (Grenon and Batisse, 1989). First, it consumes land and water resources. It was estimated that in 1984, the land area occupied by tourist accommodation alone was 2.2 million m^2. A rough estimate of the infrastructure required to support this accommodation suggests this figure should be doubled. Total water-resource consumption in the same year by tourists was approximately 569 million m^3. Second, it creates waste and pollution. Tourists created 2.9 million t of solid waste and 341.4 million m^3 of waste water in 1984. A significant proportion of this waste finds its way straight into coastal waters, polluting the very resource being exploited by the tourist industry. The effects of noxious organisms in sewage can lead to diarrhoea, botulism, typhoid, cholera, dysentery, hepatitis, hookworms and tapeworms, that are hosted in fish, shellfish, or are infectious directly from faecally contaminated water (Kirkby, 1996). From the 1970s, long sewage outfalls were used supposedly to take the effluent out of the way (or at least out of sight) of the tourists, but these outfalls have subsequently come

Table 14.1 International tourist arrivals figures from 1970 to 1996. Numbers are in thousands

	1970	*1975*	*1980*	*1986*	*1996*
Spain/France/Italy/Greece	49,045	63,850	79,483	97,687	131,795
Mediterranean total	58,085	77,582	94,974	117,259	175,726
Spain/France/Italy/Greece as proportion of Mediterranean total	84.4%	82.3%	83.7%	83.3%	75.0%
Mediterranean as proportion of world total	36.3%	36.3%	34.0%	34.4%	30.0%

Source: Based on Grenon and Batisse (1989) and Medforum (2000).

under criticism for simply spreading pathogens more widely. EU legislation has subsequently aimed at improving the quality of bathing waters in general. There are also significant impacts on air pollution, particularly from exhaust fumes, as well as noise pollution particularly from air and road transport (Grenon and Batisse, 1989). Third, there are the physical and socio-cultural impacts of the tourists. Average tourist densities in 1984 for the Mediterranean coastal zone as a whole were $15.6 \, km^{-2}$, or 0.41 per metre length of coastline. The maximum density was on Malta, with 316 tourists km^{-2}, and in Spain, with 2.3 tourists per metre of coastline. By 1990, this figure had reached 5 tourists m^{-1} (Montanari, 1995), while recent estimates suggest that tourists in some parts of Italy in 2000 have as little as 15 cm of coastline each (6.7 tourists m^{-1}). There are also social imbalances relating to the ratio of numbers of tourists to local inhabitants, which averaged 0.14 in 1984 for the coastal zone as a whole, rising to 0.41 in Spain, 0.46 in France and 0.89 in the former Yugoslavia.

Unconstrained growth has long been a characteristic of Mediterranean tourism. For example, Pollard and Dominguez Rodriguez (1995) discuss the growth of tourism at Torre-molinos in southern Spain. The town of Torremolinos had a stable population of around 3,000 people in the first part of the twentieth century, in an urban area of just under 17 ha. Following the initial development of tourism, the population had reached 8,000 by 1960, and by 1990 it had grown to 26,290. The urban area has also grown rapidly, to 82.2 ha in 1957, 267.1 ha in 1971, 342.4 ha in 1976 and 440.7 ha in 1987. Although early growth was with the luxury part of the tourist market – seeing its first high-rise five-star hotel being built in 1959 – by the 1960s the focus had switched to the mass market, with the building of numerous apart-ment blocks. By 1970 there were 17,000 beds for tourists, a number that had risen to more than 27,000 by 1980. Much of this growth was uncontrolled, in part due to the lack of legis-lative bodies to follow through the municipal planning laws (and indeed the failure to develop a municipal plan for the area until 1970), and in part due to corruption leading to private developers being able to obtain land for building. Only following reform of the planning law in 1975 and the production of a second municipal plan in 1983 was there the recognition of the need to maintain the planned land zonation and to provide for environmental protection and conservation. Further developments have taken place since 1992, when the town regained its independence from the regional centre at Malaga and could develop its own coordinated, local plan.

Montanari (1995) reports on the increasing EC concern on the environmental pressures of tourism, with funding to investigate means of protecting natural resources from mass tourism. This approach follows the development in France in the 1970s of 'green tourism' to encourage responsible tourism in rural areas, and thus minimizing its impact in any one zone. Gomez (1995) suggests that the Mediterranean zone in particular will be affected by changing atti-tudes towards tourism, with more demand for individual, personalized holidays away from the mass markets. Planning for the sustainable development of Mediterranean tourism is thus seen as a key factor in future developments (Godfrey, 1995). In Catalonia, legislative efforts have been taken to improve the sustainability of tourism, for example, by controlling building in the coastal zone and designating protected areas (Priestley, 1996). There have been moves away from the mass market, towards 'green' tourism in national parks, on farms, or catering for other rural activities, for example, in France, Portugal (Cavaco, 1995) and Catalonia (Morris, 1996). It is clear that future developments in tourism must be planned assiduously with sus-tainability in mind (Priestley *et al.*, 1996). However, such plans must always take account of local conditions and needs to work effectively (Green and Lemon, 1996; Williams and Papamichael, 1995). O'Rourke (1999) has emphasized the problems of the role of attitudes of tourists to changing landscapes in Roussillon – particularly those with second homes who aim to make aesthetic changes to what they perceive as degraded landscapes, without contributing to the sustainability of indigenous rural life within them. Pridham (1999) has noted that Italy has been the most effective of the EU countries in embracing policies that work towards sustainable tourism. Environmental concerns are now fully integrated into national policy-making procedures. In contrast, in Spain and Greece, the implementation of sustainable tourism has tended to be piecemeal and at a local or regional level.

Clements and Georgiou (1998) have emphasized that problems for tourism in politically unstable areas have particularly affected certain zones of the Mediterranean. Examples include northern Africa, particularly Algeria, the former Yugoslavia, and Cyprus. Although the latter has managed to produce a thriving industry in the unoccupied zone, there are significant impacts on quality and the increase of competition with neighbouring sectors. Tourism has not always been welcomed, for example, in the case of Corsica, where it had been seen by nationalists as an intrusion by the French government (Richez, 1996).

The Mediterranean and the EU

The Mediterranean World has played an increasing role in what has become the European Union (EU), and the impact of the European Community has been significant on the Mediterranean since its inception. France and Italy were signatories of the Treaty of Rome in 1957 which set up the European Economic Community (Common Market), together with Belgium, Germany, Holland and Luxembourg. Greece joined the Community in 1981; Spain and Portugal joined in 1986. The EU itself came into being in 1994, following the signing of the Treaty of Maastricht. Since 1996, enlargement negotiations have been ongoing with a number of other states, including Cyprus, Malta and Turkey.

The major impacts of the EU in terms of the environment fall under the broad headings of social and economic impacts and of environmental policy, legislation and research. In terms of economic impacts, the most important factor since 1957 has been the ability to maintain relatively stable growth (Sapelli, 1995). This growth has obviously led to the increases in industrialization and urbanization noted above. However, its effects extend beyond the boundaries of the Community itself. For example, trade between the EC countries and the rest of the Mediterranean was worth US$88 billion in 1991 (Rojo, 1994). On the other hand, the nature of southern European industry has meant that links have tended to be concentrated with other EC and Mediterranean countries. Unless further structural changes can take place, the Mediterranean countries will fail to develop wider economic ties necessary for stability (Sapelli, 1995). However, development has been supported financially, both in terms of loans and direct aid. Blacksell (1984) notes that between 1958 and 1973, around 45 per cent of the loans made by the European Investment Bank went to projects in Italy and Mediterranean France.

The most significant socio-economic impact has been through the Common Agricultural Policy (CAP). As originally defined by Article 39 of the Treaty of Rome, the CAP had the following five main objectives:

1   to increase agricultural productivity by promoting technical progress and by ensuring the rational development of agricultural production and the optimal utilization of the factors of production, including labour;
2   to ensure a fair standard of living for the agricultural community, in particular by increasing the individual earnings of persons engaged in agriculture;
3   to stabilize markets;
4   to ensure the availability of supplies; and
5   to ensure that these supplies reach consumers at reasonable prices.

The European Agricultural Guidance and Guarantee Fund (EAGGF) administers the CAP under two parts. The Guarantee Section supports prices, so that, for example, 3,697 million ecu were allocated in the year 1982 to support the prices of 'Mediterranean' crops such as durum wheat, olive oil, oilseeds, cotton, wine, fruit and vegetables, tobacco and silkworms. The Guidance Section aimed to introduce structural improvements to the agricultural sector. These improvements included the production of farm development plans, incentives to elderly farmers to retire and training in modern techniques. Direct aid was also implemented, for example, in the 1975 directive (75/268) providing aid to farmers in 'less favoured areas'. However, the lower limit of aid to farms greater than 3 ha in area often prevented this aid from going to the traditional dispersed smallholdings of the Mediterranean. This lower limit was later relaxed to 2 ha in southern Italy and subsequently Greece. The Guidance Section also implemented a series of regionalized measures for the Mediterranean in 1978, following dissatisfaction in Italy and southern France as it was believed that the CAP favoured farmers in northern Europe. These measures included improvements to rural infrastructure and the expansion of irrigation projects. Despite these criticisms, the European Social Fund and EAGGF gave £2,622 million in grants and £3,381 million in loans between 1975 and 1981 (Blacksell, 1984). However, it has been noted that the CAP tended to favour already adjusted areas, and typically failed to address the fragmented nature of Mediterranean landholdings, which are not necessarily suited to intensive agriculture, and the generally elderly, unskilled population at work in the sector. Furthermore, the removal of existing subsidies tended to accentuate existing regional imbalances.

Widespread criticism of the CAP led to its reform in 1992 (Scheele, 1996). Reform was underpinned by both economic and environmental concerns. There was a move away from providing support for prices of goods (i.e. the Guarantee Section) towards direct payments that aimed to both control agricultural markets and lead to less intensive styles of agriculture. The changes were accompanied by an agri-environmental scheme (the so-called regulation 2078/92), which aimed to promote more environmentally sound styles of agriculture in general and a series of specific measures aimed at specific conditions in member states. The regulation also provides funds for the upkeep of abandoned land and to ensure public access for leisure activities. There are also funds available to 'set aside' agricultural land for conservation purposes. The application of these various schemes will be considered further in the next chapter. Baltas (1997) notes a number of outcomes of the CAP reform in Greece, Portugal and Spain. Set-aside by area of cereals grown in Greece and Portugal (9.39 per cent and 8.46 per cent, respectively) was much less significant than in Spain (62.88 per cent, a figure just above the whole EC average). There were also reports of reduced purchases of fertilizers and pesticides (although such trends have not necessarily been maintained in the longer term: see Figures 15.9 and 15.10). There was little impact on the milk, beef or sheep and goat sectors in these countries following the CAP reform. Other structural changes may become important, although these are more likely to relate to the retirement of more elderly farm workers than direct environmental changes such as afforestation in the Mediterranean EU countries.

Other EC and EU support for agriculture has centred around regional funds or programmes. For example, specific regulations were established to aid the focusing of production on market requirements, such as regulation 1360/78 that relates to table wines and table olives as well as to perfume products in southern France. In this way, these products have become more saleable outside the regions of production. Similarly, regulation 355/77 provided for 80 million ecu per year over the period 1978 to 1982 in aid to the development of marketing and processing of products from the Mediterranean regions. This sum was about four times over-subscribed. A 'Mediterranean package' of regionalized measures was also adopted in 1978 as a specific response to dissatisfaction in Italy and France with the alleged imbalance in the effects of the CAP. This package included: (i) the promotion of rural infrastructure (i.e. electricity and water supplies and roads) in Italy and the Mediterranean regions of France under regulation 1760/78; (ii) the development of secondary irrigation products in the Mezzogiorno under regulation 1362/78 and of livestock farming under regulation 1944/81; (iii) reconversion of viticulture in southern France under regulation 627/78, provision of irrigation in Corsica under regulation 173/79, of flood control in the Hérault *département* under regulation 174/79, and of integrated development of the Lozère *département* under regulation 1940/81; and (iv) the acceleration of agricultural development in certain less-favoured areas of Greece under regulation 1975/82, as well as the implementation of regulations previously referring only to Italy. The success of these regional measures has been somewhat variable, and has often demonstrated the need to work on an even more localized scale.

The EC and EU have also provided a body of environmental legislation and policies, although the first statement of environmental policy did not come until 1971, and the first legally binding statements did not come into being until 1975 (Krämer, 2000). The Maastricht Treaty on European Union included articles relating to environmental concern, and specifically for sustainable development in an implied environmental context. Furthermore, direct aid is provided to combat specific issues such as desertification (van Opstal, 2000) and research into the environment has been supported by five EU Framework programmes. Framework V (1998–2002) is funded at the level of 1,083 million ecu for environmental issues, compared to 914 million ecu under the previous Framework IV programme (1994–1998). These aspects will be discussed further in Chapter 21.

The Blue Plan and its projections

The Blue Plan evolved as a response to the Mediterranean Action Plan (MAP, see Chapter 21) adopted in 1975. Under the guidance of the United Nations Environment Programme (UNEP), it aimed to provide a coherent series of studies on social and economic development in an environmental context as a basis for government planning. It was hoped that optimal socio-economic development could be attained 'without causing environmental degradation' (Grenon and Batisse, 1989). Studies of recent changes, since 1950, as well as projections of potential change to 2025 have been carried out. In this section, the various scenarios by which this might be obtained are discussed and compared to other projections of socio-economic change in the region.

The Blue Plan used 1985 as a reference year, and made projections to 2000 as an intermediate time-frame reflecting already ongoing developments, and to 2025 to represent a long-term horizon. For the longer scale, it was anticipated that ecological responses to socio-economic developments could become apparent in the environment. The models used to produce projections were built up of five major components or 'dimensions':

1   The international economic context, recognizing the relations both between countries within the Mediterranean Basin, and those outside the basin.
2   The populations of the Mediterranean and their movements, particularly relating to urbanization and movements towards the coastal areas.
3   National development, reflecting the strategies for development of agriculture, industry, energy supply, tourism and transport.
4   Spatial management to observe how environmental impacts might be distributed through the landscape.
5   The extent to which environmental considerations play a part in the decision-making process.

Projections are based on two sets of scenario, distinguished as 'trend' and 'alternative'. The former reflect the continuation of existing trends whereas the latter attempt to distinguish potential differences following the adoption of more integrated and 'goal-orientated' approaches by Mediterranean governments. The three trend scenarios developed represent the continuation of current trends (T1 – the reference scenario), a worst case trend reflecting low growth and poor coordination (T2) and a moderate trend scenario (T3) in which greater international competition is thought to lead to more sustainable growth. Of the two alternative scenarios (A1 – the reference alternative scenario), the first represents an attempt towards harmonious basin-wide development, albeit with the European Union still playing a dominant role in defining north–south exchanges. The second assumes that economic cooperation will take place between integrated groups of countries such as the Maghreb and the Arab East, so that regional growth is stronger and the EU is less dominant (A2 – the integration alternative scenario).

Total Mediterranean population estimates for 2000 ranged from 435 million under scenario A1 to 448 million in scenario T2, increasing to 530 million and 560 million respectively in 2025. For the European countries, the decline in population growth continues so that projected populations actually decrease between 2000 and 2025 in scenarios T1 and T2, from 205 million to 202 million. Only under the alternative scenarios do the projected European populations increase – from 215 million to 237 million in the A2 case. Asian growth rates are greater under the T1 and T3 scenarios (from 91 million to 136 million in both cases) and least under the A1 scenario (from 88 million to 122 million). Projected population growth is greatest overall in the African Mediterranean countries, from 148 million to 225 million in the T2 scenario and even under the A2 scenario the increase is from 138 million to 193 million. Under the T1 and T2 scenarios, it is suggested that the population of the African countries will exceed those of the European countries. The continued low growth means that in general the Mediterranean population will be increasingly less concentrated in Europe, with proportions varying from 36 per cent under T2 to 43 per cent under A2.

More recent projections made by the UN Food and Agriculture Organization (FAO) suggest that the trend scenarios may be more appropriate by 2000 (Figure 14.6). For 2025,

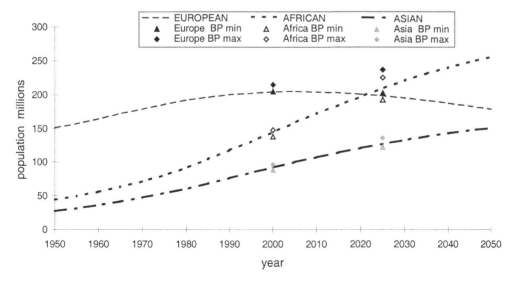

Figure 14.6 Population growth and projections according to region

Source: Based on data from FAO yearbooks and the Blue Plan (Grenon and Batisse, 1989).

the Blue Plan (BP) trend scenarios coincide with the FAO projection, but the latter produce values for the African and Asian countries which are more comparable with the Blue Plan alternative scenarios. The FAO figures also project a continued decline to 2050 of the European Mediterranean population (to 179 million) and imply that the populations of the Asian countries will also have overtaken them by around 2070. The African Mediterranean population is projected to increase continually at a slightly faster rate than that of the Asian countries, with projections of 257 million and 151 million, respectively, by 2050. FAO projections also suggest that the trend of an increasingly urban population will continue quite markedly to at least 2030, with a rural population of only 24.1 per cent in the Mediterranean Basin as a whole, with proportions of 22.4 per cent in the European, 30.8 per cent in the African and 15.7 per cent in the Asian countries. The implications of these changes for future agricultural developments will be considered in more detail in the next chapter.

Future developments in urban centres will be concentrated in the south and east of the basin, according to both FAO and Blue Plan projections. Unless deliberate policies are implemented to plan the spatial structure of development, the Blue Plan projections suggest that the rate of urbanization will continue to increase more rapidly in the coastal zones. In the worst case scenario, 41 per cent of the entire Mediterranean population will live in the coastal zone by 2025 (48 per cent of the European, 31 per cent of the Asian Mediterranean, and 39 per cent of the African Mediterranean populations). The best case scenario is for 37 per cent of the total population to be coastal by 2025 (43 per cent of the European, 28 per cent of the Asian Mediterranean, and 35 per cent of the African Mediterranean populations). Growth will probably continue to occur predominantly in existing centres, leading to the development of several megalopolizes, with consequent problems for infrastructure and planning and their associated environmental problems.

Blue Plan projections for industrial growth in the Mediterranean are focused on seven major areas. First, mining and extractive industries are likely to remain static or decline in the north and west, but expand in the south and east. Beyond the known problems with atmospheric and water pollution, the development of open-cast mining technology means that deeper mines will be constructed with important direct effects on the environment as well as aesthetic impacts on the landscape. Second, the iron and steel industry will continue to decline in Europe, but expand in several centres in the south and east, notably Turkey, Egypt and Libya. Air and water pollution will continue to be important, but decline relatively due to the development of new technologies – particularly if bio-technologies can be developed for the cold extraction of iron from ore. Third, the cement industry should continue to be important, although the European Mediterranean countries and Turkey may exhibit difficulties in maintaining production and export levels. Other eastern and southern countries are likely to expand in this area quite dramatically. Pollution, in the form of dust and fluorides emitted in the production process, will continue but could decrease with more effective controls. Fourth, petro-chemical processing to produce plastics and polymers will again remain static in the north-west, but increase in the south and east (probably also including Greece). There is a high risk of increasing pollution from this expansion, although there are moves to make more plastic products easily biodegradable. Fifth, inorganic chemical production, particularly the production of fertilizers, will continue to increase in the south and east where it is already more important than in the European Mediterranean. Chlorines will also be produced increasingly in the south and east, with corresponding increases in the risks of pollution from mercury and other heavy metals. The production of fertilizers can also lead to water pollution, for example, in the release of ammonia, and dust pollution. Sixth, the paper pulp industry will continue to have a small impact because of the lack of raw materials in the Mediterranean area. Any expansion could lead to water and air pollution from new grinding technologies. Seventh, a number of smaller industries including aluminium production and tanning may grow less slowly, but continue in many cases to have an important polluting effect. Increased use of water for industry is a further significant impact on most of these industrial developments.

Projections for tourism in the Blue Plan predict sustained growth to beyond 2025 under all scenarios. Total numbers of tourists in the Mediterranean are predicted to rise to between 379 million and 758 million, with annual growth rates from 2000 to 2025 of between 1.8 per cent and 2.5 per cent. Most of these tourists will still be concentrated in Spain, France and Italy, and to a lesser extent Greece, Malta and Cyprus. However, only the lower figures will be attained if present trends continue. The higher figures occur under the alternative scenarios because of higher predicted economic growth overall, and it is suggested that there is a need for 'goal-orientated government policies for planning (leisure time as well as space) and balanced economic development [to] allow for these flows of tourists to be absorbed without undue conflict'.

The Blue Plan can almost now be regarded as a historical document, with its mid-term predictions being testable by the fact that they have now occurred. It is notable from the discussion above that the trend scenarios have been much more realistic so far. The integration and 'goal orientation' of the alternative scenarios have failed to come about, with little obvious indication that this situation is likely to change before the longer-term Blue Plan predictions of 2025. Socio-political controls on environmental degradation will thus continue to exert pressure in the immediate future, as will be seen in the remaining chapters.

Summary

Growth rates of Mediterranean populations have been significant throughout the twentieth century, but exhibit important spatial patterns. The early growth that was concentrated in Europe into the first part of the century began to slow and become relatively negligible. By contrast, the southern and eastern parts of the Basin showed accelerating growth in the middle of the century and only start to slow in the last decades. However, the consequence of this is that the concentration of population will be highest in the southern Mediterranean by 2025, while the European Mediterranean population may even decline. This focus of population in the more extreme environmental areas within the basin is likely to have significant consequences. We will investigate these consequences further in Chapter 20. Population movements have tended to be away from upland, rural zones towards urban centres concentrated on the coast. This process, which started in the nineteenth century in Europe, is now starting to affect the eastern and southern basins significantly. There are also net fluxes from south to north. Thus, there have been greater impacts on urban infrastructures that are often poorly able to cope with the strains placed on them due to the rapidity of some of this growth. Significant issues also arise in terms of water supplies and waste relating to this increasing urbanization (see Chapters 17 and 18). Concurrent expansion in the industrial sector have added to this burden. Most notable is the rise of tourism, which has tended to be very much focused on the coastal zone, further impacting on water resources and producing large amounts of waste products that need to be dealt with. In recent years, there has been a recognition that the development of tourism as we have seen it is not sustainable, and that other models need to be found if this vital source of income into the region is to be maintained.

The EU has had significant impacts on the environment, notably through its controls on agricultural production. Many of these conflicted with traditional uses of the land in the region, and in some cases have encouraged types of exploitation that are seriously deleterious to the environment. These issues are discussed further in the following chapter. However, a number of reforms have taken place that hint at a more enlightened environmental policy, coupled with a number of direct measures which have been implemented to avoid or mitigate environmental degradation. These policies and their impacts will be addressed in more detail in Chapter 21.

Population projections into the twenty-first century suggest increasing pressures on the production base of the region. By 2050, there will be more than 2.5 times as many mouths to feed in the basin than there were in 1950, according to FAO projections. Most of this increase, as already noted, will take place in the south and the east. These areas will also see the most rapid future growth in urban centres, with many of the consequences already experienced in the north. Growth in industry is likely to lead to further air and water pollution. The doubling of the number of tourists in the region is likely to add to these problems, particularly due to the competition for scarce water resources in the coastal zone, where most of the growth will take place. Specific details relating to the impacts of these predictions will be assessed in the following chapters. It is clear, however, that important decisions will need to be made as to how to control and sustain this growth, and ensure that many of the problems that have been generated in the northern Mediterranean in the twentieth century are not repeated on a larger scale in the south in the twenty-first.

Suggestions for further reading

Although the data are a little dated, the Blue Plan publication of Grenon and Batisse (1989) is still a good first stopping place for general background and projections of future conditions. Leontidou (1990) gives an overview of urban and industrial changes in the region. A range of viewpoints on sustainable tourism is presented by Priestley *et al.* (1996).

Topics for discussion

1   How have population dynamics changed in the twentieth century, compared to previously?
2   Were existing urban centres industrialized, or has industry generated new urban centres?
3   What are the different characteristics of modern urban centres in different regions of the Mediterranean, and how might they pose quite distinctive environmental problems?
4   What difficulties are encountered when trying to predict future socio-economic scenarios that might impact on the region?
5   Overall, has the impact of the EU on the Mediterranean environment been positive or negative?
6   How effective are the predictions made by the Blue Plan? What improvements might be made to further predictive attempts, if they can be considered to be useful?

15 Agricultural changes in the twentieth century

Introduction

Agriculture continued to be the most significant factor impacting on the Mediterranean environment in the twentieth century, despite the parallel increases in industrialization and tourism that were particularly marked in the second half of the century. To support the ever growing populations at the regional scales, agriculture has seen continual intensification and reorientation towards increasingly efficient production systems, so that traditional land-use practices have fallen out of favour. This chapter describes the extent to which Mediterranean agriculture has become more intensive, and how this varies from region to region. Specific examples of how this intensification has been put into practice and its consequences are then discussed. We then assess the nature of land degradation from the changes in agricultural practice and how the continued development of agriculture into the first quarter of the twenty-first century will affect the observed trends of degradation.

Intensification, mechanization and their consequences

Mediterranean agriculture in the twentieth century has been marked by a general trend towards increased intensification. The old systems of scattered parcels of land – linked to historic patterns of inheritance – worked by hand is becoming increasingly rare, with fewer people working on the land, increased use of machinery, fertilizers, modern irrigation and other techniques for maximizing production. FAO figures demonstrate that total numbers of people employed in agriculture in the European Mediterranean has declined steadily over the period from 1961 to the present (Figure 15.1). The total number declined from 57.6 million to 17.6 million. In proportional terms, this represents a decline from 34.8 per cent to 8.7 per cent of the population. The biggest change occurred in Italy, where just under 15 million people were employed in agriculture in 1961 compared to 3.5 million in 1997. Although the total number in the North African Mediterranean countries increased slightly from 39.2 million to 45.0 million, the rapid total population increase means that this represents a proportional decrease from 68.5 per cent to 33.6 per cent of the total population. The absolute increase is dominated by changes in Egypt and to a lesser extent Morocco, with the other three countries maintaining relatively stable figures. In the countries of the eastern Mediterranean, there is again a slight total increase, from 23.8 million to 25.3 million people, but again this increase represents a proportional decrease, from 63.6 per cent to 29.0 per cent. In absolute terms, numbers in Turkey increased only into the mid-1970s, whereas Syria has experienced a slight but continuous rise.

Over the same time period, the area used in agriculture has remained relatively stable or declined slightly in the European Mediterranean countries (Figure 15.2). The total area has

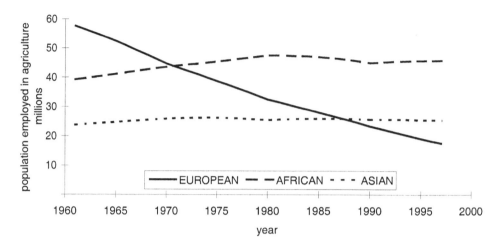

Figure 15.1 Total population in millions employed in agriculture in the European, African and Asian
        Mediterranean countries

Source: FAO yearbooks.

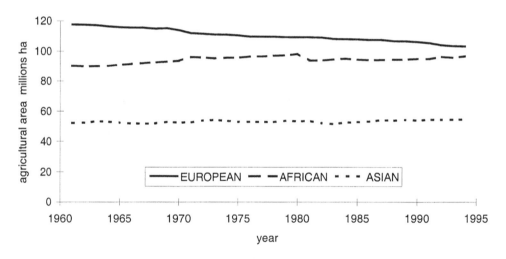

Figure 15.2 Total area used for agriculture in millions of hectares in the European, African and Asian
        Mediterranean countries

Source: FAO yearbooks.

declined from 117.6 million/ha in 1961 to 103.6 million/ha in 1994. Most notable in the
overall decreases are the countries of the EU, although the area in Spain has dropped consistently through this period, and is thus related to general trends rather than just EU policy. In
the African Mediterranean countries, the area in use has increased from 90.1 million/ha in
1961 to 96.9 million/ha in 1994. Again, there are notable spatial differences in this pattern.
Tunisia and Egypt have maintained more or less constant levels, while Morocco and Libya
have seen constant expansion. Algeria decreased slightly to 1980, but then dropped quite

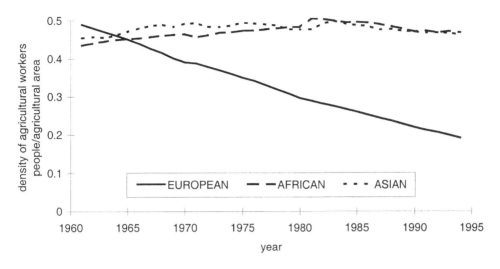

Figure 15.3 Intensity of agricultural activity measured as number of people employed in agriculture per unit area used for agriculture

Source: Based on data drawn from FAO yearbooks.

sharply, with just over 10 per cent of the total agricultural area going out of use in 1981. In the Asian Mediterranean countries, there has been a slight total increase of area, from 52.3 million/ha in 1961 to 54.9 million/ha in 1994, albeit with small oscillations over the period. This figure is dominated by the changes that have occurred in Turkey.

Combining the numbers of people employed with the area used for agriculture gives a first measure of intensification. In the European Mediterranean area, the figure has dropped steadily from 0.49 agricultural workers per hectare in 1961 to 0.19 in 1994 (Figure 15.3). The African and Asian Mediterranean countries demonstrate similar trends to each other, with lower values than in the European countries in 1961 (with values of 0.44 and 0.45, respectively), a steady increase to around 0.50 in the early 1980s, and a subsequent small decline to the present day. Coupled with this measure is the structure of land holdings (Table 15.1). Traditionally, Mediterranean land holdings have been small, outside the areas of latifundia-type farms in southern Spain and Italy, primarily due to traditional laws of inheritance that distributed land held equally to all the children, and thus repeatedly split up any block of land held. However, in a number of places, these latifundia were split up and redistributed to local peasant farmers at various points in time in the mid-nineteenth to mid-twentieth centuries. Although the presence of large microclimatic variations may mean that this approach allows specific environmental niches to be exploited, it also provides numerous disadvantages, including inefficiency due to wasted time in travelling, duplication of expenditure on infrastructure, complexity in the planning of conservation and drainage and irrigation infrastructure, and tension due to ownership disputes (see, for example, Du Boulay, 1974, for examples of the latter).

The effects of this process can be seen in the results of a land survey in Cyprus (Karouzis, 1980), with individual owners holding widely scattered plots of land (Figure 15.4). Furthermore, investigations into the time wasted by the fragmentation of plots in this study suggested that the time needed by a landholder to reach their furthest plot could vary from thirty-six minutes to four hours and eighteen minutes, and thus this could be a significant factor in

Table 15.1 Number of farms in different size categories, area covered by these farms and changes in proportion according to size categories in the EU and some other Mediterranean countries

Number of farms (thousands)

area ha	France 1987	France 1989–1990	Greece 1987	Greece 1989–1990	Italy 1987	Italy 1989–1990	Portugal 1987	Portugal 1989–1990	Spain 1987	Spain 1989–1990	Turkey 1991	Algeria 1973	Tunisia 1990	Egypt area ha	Egypt 1981–1982
<5	236.0	277.4	737.8	718.5	2,150.1	2,099.1	529.9	492.4	1,073.0	971.4	2,644.2	437.7	175.8	<0.84	2,286.2
5–20	281.9	279.6	194.2	183.9	504.3	439.5	85.3	83.5	481.5	410.4	684.9	211.8	158.7	0.84–2.1	448.5
20–50	299.1	288.0	17.5	18.0	91.6	87.7	13.0	13.6	144.6	124.9	353.2	47.8	37.7	2.1–4.2	88.1
50–100	124.4	128.6	2.9	2.5	24.7	24.7	3.5	3.9	55.6	48.8	160.7	9.8	10.3	4.2–8.4	26.3
>100	40.3	43.5	0.9	0.7	13.2	13.7	3.9	5.4	37.1	38.2	33.1	3.4	4.6	>8.4	11.6
Total	981.7	1,017.1	953.3	923.6	2,783.9	2,664.7	635.6	598.8	1,791.8	1,593.7	3,876.1	710.6	387.1	Total	2,860.7

Area covered by farms k ha

area ha	France 1987	France 1989–1990	Greece 1987	Greece 1989–1990	Italy 1987	Italy 1989–1990	Portugal 1987	Portugal 1989–1990	Spain 1987	Spain 1989–1990	Turkey 1991	Algeria 1973	Tunisia 1990	Egypt area ha	Egypt 1981–1982
<5	466.3	516.2	1,316.8	1,283.1	3,448.3	3,138.6	754.1	757.6	2,062.0	1,872.4	5,139.0	720.1	453.9	<0.84	967.1
5–20	3,436.7	3,254.1	1,660.5	1,600.2	4,616.6	4,097.2	769.3	761.3	4,617.2	3,977.0	4,478.4	2,051.3	1,753.1	0.84–2.1	496.0
20–50	9,631.7	9,346.7	493.3	517.7	2,714.8	2,637.3	383.2	407.3	4,444.6	3,844.6	4,534.3	1,373.1	1,219.5	2.1–4.2	622.2
50–100	8,434.7	8,736.8	183.9	156.4	1,685.0	1,686.3	241.2	269.2	3,805.9	3,349.4	4,300.3	642.8	733.0	4.2–8.4	220.5
>100	6,178.6	6,682.6	187.9	113.9	3,079.8	3,386.4	1,183.2	1,810.4	9,870.1	11,487.0	3,347.9	622.5	1,204.8	>8.4	479.8
Total	28,148.0	28,536.4	3,842.4	3,671.3	15,544.5	14,945.8	3,331.0	4,005.8	24,796.6	24,530.4	21,799.9	5,409.8	5,364.3	Total	2,785.6

Proportion of farms in each category relative to total number of holdings

France

year	<5ha	5–20ha	20–50ha	50–100ha	>100ha
1987	24.0	28.7	30.5	12.7	4.1
1990	27.3	27.5	28.3	12.6	4.3
1993	27.6	22.6	25.6	16.5	7.6
1995	27.3	21.5	24.1	17.4	9.6
1997	26.8	20.1	23.4	18.5	11.2

Greece

year	<5ha	5–20ha	20–50ha	50–100ha	>100ha
1987	77.4	20.4	1.8	0.6	0.6
1990	77.8	19.9	1.9	0.7	0.9
1993	75.7	21.5	2.4	0.8	1.1
1995	75.2	21.7	2.7	1.0	1.2
1997	76.3	20.7	2.6	1.0	1.3

Italy

year	<5ha	5–20ha	20–50ha	50–100ha	>100ha
1987	77.2	18.1	3.3	0.9	0.5
1990	78.8	16.5	3.3	0.9	0.5
1993	77.5	17.1	3.8	1.1	0.6
1995	78.1	16.0	4.2	1.1	0.5
1997	75.7	18.3	4.1	1.2	0.6

Portugal

year	<5ha	5–20ha	20–50ha	50–100ha	>100ha
1987	83.4	13.4	2.0	0.3	0.1
1990	82.2	13.9	2.3	0.3	0.1
1993	78.1	16.9	3	0.3	0.1
1995	76.7	17.8	3.3	0.3	0.1
1997	76.1	18.0	3.6	0.3	0.1

Spain

year	<5ha	5–20ha	20–50ha	50–100ha	>100ha
1987	59.9	26.9	8.1	3.1	2.1
1990	61.0	25.8	7.8	3.1	2.4
1993	58.1	26.9	8.3	3.6	3.1
1995	55.3	28.1	9	4.0	3.6
1997	53.6	28.7	9.5	4.3	3.9

Source: Data from EUROSTAT, 2000, and Allaya *et al.* (1995).

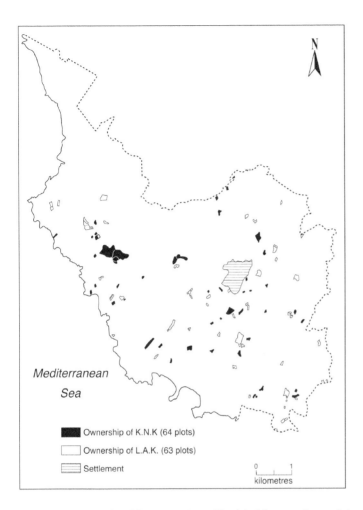

Figure 15.4 Example of fragmentation of land holdings under traditional patterns of inheritance in
     western Cyprus

Source: After Karouzis (1980).

running a smallholding, even if it was typically made up of less than five units, and provides a
natural limit to the growth of land holdings. In all countries except France, the dominant size
class of land holding was still less than 5 ha in area in the 1990s. However, there is some evidence of consolidation of holdings, with a decline in smaller land holdings and increase in larger
farms, allowing more efficient agricultural practices to take place, even if it is still a relatively
slow process. In some regions of Italy, consolidation was taking place at significant levels in the
1970s due to the activities of cooperative movements (Jones, 1984). For example, the average
size of plot in Apulia in 1970 was 2.4 ha compared to 48.4 ha in 1975, in the Molise the relative figures were 0.6 ha and 21 ha, and in Sardinia the increase was from 1.4 ha to 38.6 ha.
Land consolidation has been a planned process in Spain since 1952, with the establishment of
the Servicio de Concentración Parcelaria. By the 1960s and 1970s, between 350,000 and
400,000 ha were consolidated in this way per year (Clout *et al.*, 1994). In Cyprus, consolidation has been the policy since 1969, and is carried out at the level of each village, by swapping

fields so that single, compact units are formed (King and Burton, 1989). King and Burton note that the success of such schemes may depend on the perspective of the land owners – for example, in relation to market versus subsistence orientation, or in terms of independence relative to change imposed from outside. In this study, the benefits were felt more strongly by farmers in upland areas compared to lowland ones. Yet again, consideration of local concerns and impacts is seen to be dominant in the success of changed policies.

A second measure of intensification can be obtained by looking at yields (i.e. the amount produced per unit area of specific crop types). These too have tended to increase over the period in question, in some cases quite spectacularly (Figure 15.5). Taking wheat as a first example, yields have increased from 1,276 kg ha^{-1} in 1961 to 3,410 kg ha^{-1} in 1998 on average

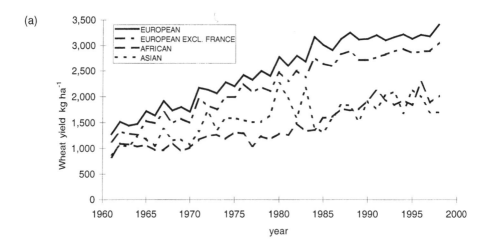

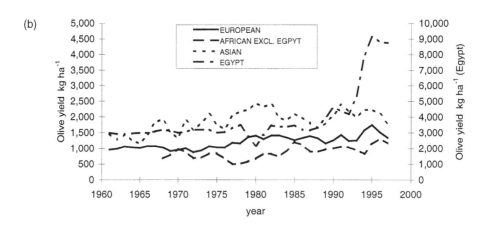

Figure 15.5 Intensity of agricultural activity measured in terms of crop yields (amount produced per unit area): (a) wheat; and (b) olives. Note that the values for olives are two-year running averages, to account for the normally occurring interannual variability in this crop, and that the curve for Egypt is measured on the scale on the right, which has a two-times vertical exaggeration compared to the other curves

Source: Based on data drawn from FAO yearbooks.

in the European Mediterranean countries. Even if we discount France from these figures, where most wheat cultivation takes place in the north of the country, the respective values are 1,136 kg ha^{-1} and 3,061 kg ha^{-1}. Large increases also occur in the African Mediterranean countries (from 832 kg ha^{-1} to 2,024 kg ha^{-1}) and the Asian Mediterranean countries (from 882 kg ha^{-1} to 1,701 kg ha^{-1}), although the values are consistently lower than in Europe. Olive yields show more variability. To a certain extent, this variability can be removed from the figures by taking two-year running averages, to account for the natural inter-annual variations in fruit production by this species. When this is done, there are still general trends of growth in olive yields – in the European countries, this growth is from 959 kg ha^{-1} in 1961–1962 to 1,343 kg ha^{-1} in 1997–1998; in the African countries (excluding Egypt) it is from 676 kg ha^{-1} in 1968–1969 to 1,171 kg ha^{-1} in 1997–1998; and in the Asian countries it is from 1,458 kg ha^{-1} in 1961–1962 to 2,091 kg ha^{-1} in 1994–1998. Yields are significantly higher in the irrigated areas of Egypt, where yields tended to be around 3,000 kg ha^{-1} in the early part of the period, but increased dramatically to up to 9,149 kg ha^{-1} in 1995–1996. The area of land used for permanent pasture has stayed relatively stable in the European Mediterranean, decreasing from 44.6 million/ha in 1961 to 38.9 million/ha in 1994 according to FAO statistics. Most of this decline occurred in France (after 1971) and Spain. In the African countries, the area increased from 67.3 million/ha in 1961 to 73.6 million/ha in 1980, before dropping back to 69 million/ha in 1994. These changes are dominated by increases in Libya and Morocco to 1980 and a sharp fall in Algeria in 1980–1981. In the Asian Mediterranean countries, the area used for permanent pasture was 20.0 million/ha in 1961 and 20.8 million/ha in 1994, but reached a minimum of 18.4 million/ha in 1971. Again, these changes are dominated by fluctuations in Turkey and Syria because of their relatively large areas. The intensity of use of this land can be assessed by looking at stocking densities (Figure 15.6). Cattle stocking rates are relatively low, and show similar trends of slight to moderate increases from 1961 to the 1980s (1984 in the European, 1989 in the African and 1982 in the Asian Mediterranean countries), followed by a decline. Only in the case of the Asian countries was the rate lower in 1994 (0.63 animals per hectare) compared to 1961 (0.66 animals per hectare). Whereas the European countries have higher rates for cattle density, the Asian countries have the highest density of sheep and goats. However, trends have been very different in the different areas. In the Asian countries, there were 3.2 sheep and goats per hectare, rising to 4.5 in 1983, before dropping back to 2.9 in 1994. In the European countries, there was a decline from 1.8 animals per hectare in 1961 to 1.5 in 1974, followed by a sustained recovery to 2.1 in 1994. In the African Mediterranean, there has been a slight tendency to rise throughout the period, from 0.6 animals per hectare in 1961 to 0.8 in 1994. These relatively high figures relate to the increased removal of vegetation by stock animals, leading to potentially important consequences relating to soil erosion and desertification. In terms of pastoral productivity also, these figures suggest the negative impacts of substantial overstocking. For example, in observations in *Quercus coccifera* shrubland in Greece, Yiakoulaki and Nastis (1995; 1996) found optimal returns for stocking densities of one goat per hectare per year.

A third measure of intensification, which is particularly appropriate to the Mediterranean zone, is that of area under irrigation (Figure 15.7). In the European and Asian countries there has been a steady rise in irrigated area from 1961 to 1996 (from 6.1 million/ha to 10.4 million/ha in the former and from 2.1 million/ha to 5.6 million/ha in the latter). The African countries have seen more variability, but still an overall increase from 3.9 million/ha in 1961 to 5.9 million/ha in 1996. Relative to the total area used for arable and permanent crops, this represents an increase over the time period from 8.2 per cent to 15.9 per cent in the European, from 17.0 per cent to 21.2 per cent in the African and from 6.4 per cent to 17.1 per cent in the Asian Mediterranean countries. Particularly high proportions are found in Egypt

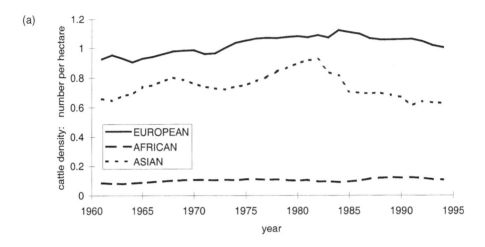

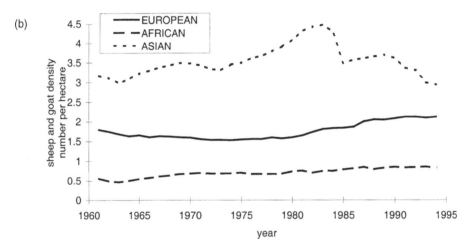

Figure 15.6 Stocking density rates in the Mediterranean region measured as number of animals per hectare of permanent pasture: (a) cattle; and (b) sheep and goats

Source: Based on data drawn from FAO yearbooks.

(100 per cent of agricultural land is irrigated), Albania (as high as 60 per cent in 1990, but dropping back subsequently), Israel and the Gaza Strip (typically around 45 per cent). Greece, Cyprus, Lebanon, Libya and Syria also saw significant increases over the period in question. An important consequence of these changes is obviously the amount of water required for this irrigation. Grenon and Batisse (1989) estimated that in 1985, approximately 72 per cent of all water drawn off from supplies in the Mediterranean – equivalent to around 110 Gm^3 – was used for irrigation. This amount represents around 62.5 per cent of sustainable water supplies in the basin (Table 15.2).

Other inputs provide a fourth measure of intensification of agriculture. The extent to which agricultural machinery is used is a key input, allowing fewer agricultural labourers and/or smaller areas to be cultivated for the same outputs. Most noticeably, mechanization has

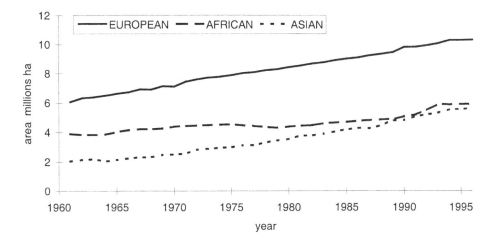

Figure 15.7 Area of agricultural land under irrigation in the different areas of the Mediterranean Basin
Source: Based on data drawn from FAO yearbooks.

Table 15.2 Water supply and demand in the Mediterranean area. The penultimate column represents the percentage of water used relative to total water supplies, while the last column represents the percentage of water used relative to sustainable water supplies

Country	Supply $Gm^3\ a^{-1}$		Demand $Gm^3\ a^{-1}$		% utilized rel. to	
	Total water resources	Stable water resources	Water drawn off	Net consumption	Total water resources	Stable water resources
Albania	31.1	7.5	13.8	11.7	1	184
Algeria	74.0	35.2	15.75	2.37	16	45
Cyprus	187.0	30.5	46.35	15.0	60	152
Egypt	0.03	0.023	0.023	0.02	98	100
France	77.5	11.5	1.5	0.28	21	13
Greece	21.3	6.5	0.2	0.036	12	3
Israel	58.6	7.7	7.0	3.65	115	91
Italy	67.0	15.6	6.7	3.27	25	43
Lebanon	0.9	0.27	0.54	0.4	15	200
Libya	4.0	2.3	0.88	0.51	229	38
Malta	4.0	2.8	0.6	0.38	77	21
Morocco	1.3	0.28	1.5	0.95	29	536
Spain	57.3	55.8	55.9	39.0	38	100
Syria	0.7	0.2	1.6	1.25	22	800
Tunisia	3.1	1.5	2.0	1.45	65	133
Turkey	10.9	2.5	1.7	1.0	10	68
Former Yugoslavia	3.8	0.9	1.1	0.57	2	122

Source: Reproduced from Grenon and Batisse (1989).

increased in the European Mediterranean countries, from 1.4 million machines in 1961 to a peak of 5.9 million in 1990, before falling back slightly to 5.4 million in 1996 (Figure 15.8). This decrease mainly relates to changes in the Former Yugoslavia, although numbers in France had also been falling slightly since the mid-1980s. Relative to the total agricultural area, this

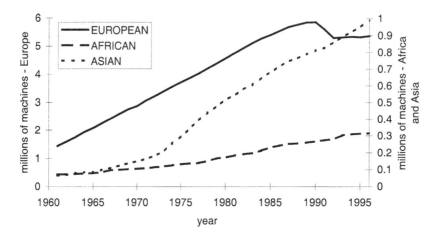

Figure 15.8 Trends in agricultural mechanization in the Mediterranean region. Note that the European Mediterranean countries are plotted on the left-hand axis, which has a six-fold exaggeration

Source: Based on data drawn from FAO yearbooks.

represents an increase from 0.012 to 0.052 machines per hectare. Although the African and Asian countries had relatively similar numbers of agricultural machines in 1961 (71 k and 65 k, respectively), it is the Asian countries that have mechanized their agriculture more rapidly – increasing to 990 k machines in 1996, compared to only 315 k in the African countries. In proportional terms, the Asian countries had 0.017 machines per hectare of agricultural land in 1994, compared to 0.003 in the African countries. In other words, the level of mechanization in the African Mediterranean in 1994 was still only a quarter of that in the European Mediterranean in 1961. Again, mechanization in North Africa was led by Egypt and Algeria, and by Turkey in the eastern basin. A second external input is the extent to which chemical fertilizers are used (Figure 15.9). In the European Mediterranean, 10.5 million t of chemical fertilizers were used in 1996, compared to the maximum level of 12.3 Mt in 1988. Throughout the Mediterranean, quantities employed have risen steadily since the 1960s, although there was a temporary decrease following the oil crisis of 1993–1994 (the fact that this is less noticeable in the African countries may relate to their oil-producing status). The Asian countries have again employed this technology more intensively than the African countries, with a use of 2.3 Mt in 1996 compared to 1.7 Mt. In terms of application rates, the European Mediterranean countries applied 159.8 kg of chemical fertilizer per hectare of land under arable and permanent crops in 1996, compared to 61.2 kg ha^{-1} in the African countries and 66.4 kg ha^{-1} in the Asian. There is a marked spatial difference in application rates, with the highest values in Egypt (365.6 kg ha^{-1} in 1996), and additionally, Cyprus, France, Israel and Slovenia all applied more than 200 kg ha^{-1} in 1996. The high use rates in Egypt balance out the low rates elsewhere in northern Africa. The third significant external input is the use of pesticides. Data available from the FAO are only in terms of the value of imports (Figure 15.10), but these show general trends of increase since 1961 in all the areas of the Mediterranean. The figures in the European zone are also likely to be underestimated, due to production in some of these countries. The figures demonstrate that the European Mediterranean countries spent $40.60 on pesticide imports per hectare of land under arable and permanent crops in 1996, compared to $6.82 in northern Africa and $5.53 in the countries of the eastern Mediterranean.

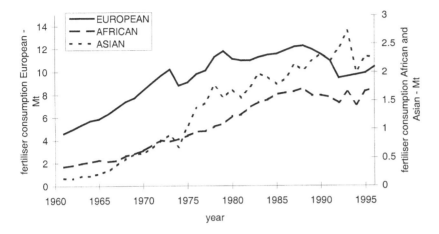

Figure 15.9 Trends in chemical fertilizer consumption in the Mediterranean region. Note that the European Mediterranean countries are plotted on the left-hand axis, which has a five-fold exaggeration

Source: Based on data drawn from FAO yearbooks.

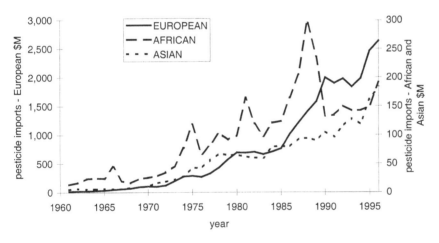

Figure 15.10 Trends in pesticide imports in the Mediterranean region. Note that the European Mediterranean countries are plotted on the left-hand axis, which has a ten-fold exaggeration

Source: Based on data drawn from FAO yearbooks.

A final measure reflecting intensification has been suggested by Jiménez-Diaz *et al.* (1987) to reflect overall effects of productivity. The index used is the ratio of gross added value from agriculture to proportion of the population employed in agriculture, and thus reflects the balance of outputs to inputs, with a higher value reflecting greater efficiency of production. Only France, Israel, Malta and Portugal show particularly high values for this index of intensification in 1989 (Table 15.3), and Turkey shows very little evidence of intensification still. Mairota (1998) has also calculated the index regionally for the five study areas of the MEDALUS project. These calculations demonstrate important regional differences, with even

Table 15.3 Index of agricultural intensification measured as the ratio of the percentage of Gross Added Value (GAV) from agriculture to the percentage of people employed in agriculture. Country-level figures are for 1989 and are calculated from data in Allaya *et al.* (1995); regional figures for 1992 for the MEDALUS-project study areas calculated by Mairota (1998)

Country	GAV/EA
Cyprus	0.17
Egypt	0.11
France	0.47
Greece	0.11
Israel	0.52
Italy	0.14
Malta	0.62
Portugal	0.61
Spain	0.28
Turkey	0.07
Region	
Languedoc-Rousillon	1.18
Murcia	0.60
Abruzzo	0.43
Sardinia	0.28
Basilicata	0.28

the less developed of these areas demonstrating higher values than the relative national levels of production.

Thus, in summary, we can say that there have been pan-Mediterranean trends of agricultural intensification, which is most distinctive in the northern and eastern parts of the basin. This intensification is reflected by lower numbers of workers, smaller areas under cultivation, higher yields and stocking rates, and increased mechanization, irrigation and use of chemical fertilizers and pesticides. However, there are still important patterns within these levels of development, both on a national scale – particularly noticeable between the northern and southern parts of the basin – and on a regional scale.

Examples of the specific nature of changes can be seen throughout the Mediterranean. In southern Spain, the expansion of cultivation under plastic greenhouses – the so-called 'plasti-culture' – has been dramatic since the 1950s (Tout, 1990). In the Campo de Dalías to the west of Almería, an area described in the 1920s as 'a stony desert without . . . a single tree or house' (Brenan, 1957, cited in Tout), development of plasticulture took place under the aegis of the National Institute for Agricultural Reform and Development (IRYDA), so that by the 1990s large areas of the 30 by 12 km coastal strip were under this form of cultivation. These developments were carried out to take advantage of groundwater supplies at up to 200 m in depth. The greenhouses are typically 25 m long by 22 m wide, 3.5 m high and orientated east–west and covered by thin polyethylene sheeting (Tout, 1990). Trickle irrigation has been used since 1971, and is now a prerequisite for loans for horticultural development. The effects of these developments on the coastal plane have been dramatic. Although the population of the area remained static in the first half of the twentieth century, it more than doubled between 1950 and 1981, with very low or even zero unemployment. By 1987, the area was exporting 315,882 t of produce, mainly peppers, cucumbers, melons and tomatoes. By 1992, Almería province made up 35 per cent of Spanish horticultural exports, despite having only 15 per cent of the horticultural land area of the country (Morris, 1992). Despite this level of intensification, the average size of farm remained small, at just over a hectare. Large levels of expenditure have been made to support the growing water supplies required for irrigation, including the construction of new dams and water pipes and tunnels, as well as to provide easier transportation of produce using the new motorway network. Other examples of intensive irrigation agriculture are provided for Spain (Wilvert, 1994), Sardinia (Pungetti, 1995), and Crete (Papanastasis, 1993). Damage to traditional agro-silvo-pastoral systems in the Alentejo has occurred as a result of perturbations due to extensification or intensification following the decline in economic value of the products of these systems (such as cork: Correira, 1993).

In some cases, even extreme Mediterranean landscape types have been converted for agricultural use. Phillips (1998) describes the case of the reclamation of badlands for agriculture in Tuscany following land reforms in the 1950s and the CAP. In the 1950s, large estates were broken up, and land shared among local farming families, leading to the reclamation of large areas of badlands. The CAP contributed further in the 1990s by providing higher subsidies for farming durum wheat that were further accentuated by devaluations of the Italian lira. Furthermore, Phillips suggests that changes in chemistry, organic matter and soil aggregation have made these areas *more* stable under agriculture – although there is no indication as to how sustainable these changes are. In Basilicata, the recovery of *calanchi* badland areas for cultivation has been carried out in recent years by creating large terraces using bulldozers (Bove *et al.*, 1995). Experiments with different cultivation methods to limit erosion suggest that while contour ploughing can help to reduce erosion from runoff in these settings, they can also increase the likelihood of slope failures. Models of the economic efficiency of the *calanchi* suggest that degradation is enhanced at present by EU price support for durum wheat cultivation, which make crop rotations and the use of fallow unprofitable. Controversy has also been sparked locally in Tuscany by the loss of the traditional badland landscapes, which were seen as a 'unique natural phenomenon' and part of the regional heritage (Phillips, 1998). Similar contrasts in attitudes can be seen in Epirus in northern Greece. The local perception of erosion is 'not so much as anomalous change in what should be a static land, but as the way the environment "is" and always had been' (Green and Lemon, 1996). Land 'degradation' is seen in the area to relate to revegetation following land abandonment as a result of recent socio-economic changes. The apparent 'untouched wilderness' as viewed through the eyes of foreign tourists appears as a collapsing, unproductive landscape to the remaining elderly populations. Similarly in Israel, Misgav (2000) found that the landscapes given the 'highest visual priority were those representing planted landscapes, forest landscapes and open forest landscapes, and those given the lowest priority were mainly grassy scrub landscapes and scrub and garrigue of the *Pistacia atlantica* and *Amygdalus korschinskii* communities'.

Irrigation has expanded dramatically in the Argolid in southern Greece since the 1930s, with an increase of irrigated area from 5,500 ha in 1945 to 12,500 ha in 1965, 16,000 ha in 1985 and 19,500 ha in 1990 (Allen *et al.*, 1994; Poulovassilis *et al.*, 1994; Alexandris *et al.*, 1994; see also Lemon *et al.*, 1994, 1995a, 1995b; Green and Lemon, 1996; da Silva *et al.*, 1997; Jeffrey and Lemon, 1996). Over the same period, the amount of irrigation water applied has risen from 45 to 145 million $m^3 a^{-1}$, although this reflects a slight decrease in the intensity of irrigation from 8,181 to 7,435 $m^3 ha^{-1} a^{-1}$. The main consequence of this has been the extension of the cultivation of citrus under irrigation. The area has also seen an in-migration to peripheral areas from the surrounding mountainous areas since the 1960s. The introduction of sprinkler irrigation systems has not led to a great increase in water-use efficiency because of the introduction of more water-intensive crops, inefficient water control, and increasing use of water to protect crops from frost. Groundwater levels have dropped by more than 300 m since the first extensive drilling of boreholes in the 1960s, and in a number of places the aquifers have undergone saltwater intrusion (Figure 15.11 – see also Box 18.3). The growth of citrus is critically dependent on the EU support for this crop based on guaranteed price support, and also follows the uprooting of the more traditional apricot in the area following attacks of the sharka virus in the 1980s. Production is organized within cooperatives, although farmers also sell without the benefit of the price support into the local market and Athens, which is only three hours away. Most farms are relatively small, with a modal size of 10–49 ha, and a modal number of parcels of five or six, which are usually, but not always located in the same village. Farms with citrus tend to be more fragmented and concentrated towards the centre of the Argolid Plain, whereas the farms on the edge of the plain tend to be larger and have a more varied crop base. As well as full-time farmers in both areas, there are a large group of part-time farmers, either because of other employment or because their land is insufficient to make a

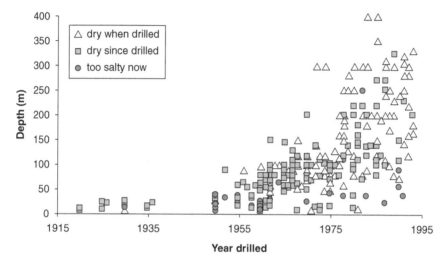

(a)

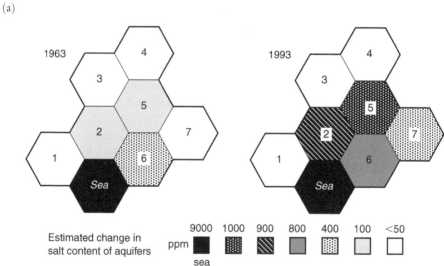

(b)

Figure 15.11 Effects of increased irrigation in agriculture on the quantity and quality of available water supplies in the Argolid, southern Greece: (a) decline in groundwater depth through time due to increased abstractions; and (b) changes of salinity of well water through time

Source: After Lemon *et al.* (1996).

living. Because contact between farmers and central agricultural services is declining, the actions of the farmers as individuals is important in determining environmental change. The perceptions – based on semi-structured and structured interviews – of the farmers towards environmental degradation includes conceptions of reductions in water quality and quantity, so that there is no ignorance of how individual actions affect the communal resource. However, there seems to be no conception of clean water as a non-renewable resource which would allow individual use to be constrained in a way as to lead to a sustainable agriculture practised by the group as a whole. Farmers within the central zone seem to be concerned about hazards from frost, leading to the increasing use of sprinklers (and thus of water) and locally air mixers in winter to prevent its occurrence, whereas farmers in the peripheral zones perceive decreasing rainfalls to be a bigger threat. There appears to be little flexibility within the current system, with many farmers suggesting that they would continue to use fertilizer, pesticide, labour and water at the same rates whatever the price they received for their crops. Price changes also apparently had little effect on the decision to change cropping patterns. This inertia may relate to the time taken for tree crops to reach maturity, but also individual attachment to areas and the crops produced from them. Farmers in the area are perceived to fall into three groups: a part-time 'inauthentic' group, with main interests elsewhere, who are less likely to reinvest profits into agriculture but are most flexible in response to changes; a part-time 'real' group, who reinvest, and are therefore more likely to feel the effects of degradation; and the full-time farmers. The first group tends to be concentrated in the centre of the plain where most water is used, so that there are important questions of equity within this division. Green and Lemon (1996) point out the contradiction between the individuality of farmers and their increasing dependence on collective action at individual, collective and large-scale levels. There is thus an increasing dependence 'upon the ability of "external" scientific and technological innovation to triumph over nature'. It may be that the response to the current decreased marginalization of the area will ultimately lead to a phase of further marginalization because the economic choices are becoming increasingly constrained within the collective structure.

It can thus be seen that agricultural intensification has brought about a mixed set of issues. Clearly, there have been significant increases in the standard of living and in the commercial value of the sector, at least locally, with a concentration on the coastal zone. This localized expansion thus sets up a tension with other developments in the coastal zone, notably tourism, both in terms of competition for increasingly scarce water resources and in terms of the aesthetic viewpoint of the visual blight of expanses of tens of square kilometres of plastic greenhouses. There are significant environmental consequences of this development, directly in relation to changes in land use, groundwater depletion, impoverishment of soil structure, salinization problems, and intensive use of fertilizers and pesticides. Indirectly, there are consequences in terms of expansion of transport infrastructure, off-site pollution from runoff carrying away fertilizer and pesticide residues, and waste polyethylene scattered around the landscape. There are also a number of significant social issues relating to the rapid expansion of certain agricultural areas, displacement of farmers to unfamiliar surroundings with the corresponding pressures of a highly competitive market in what has been likened to the boom towns of the US Far West in the nineteenth century (Tout, 1990).

Soil impoverishment and salinization

Soil impoverishment has accelerated under these conditions of intensification through the twentieth century, both in terms of quantity and quality. Chapter 6 has already highlighted the accelerated rates of erosion that occur on modern agricultural land, and the links to specific crop types. Statistics in soil erosion in the Pindus in Greece in the 1960s suggest average sediment loads of 20 million $m^3 a^{-1}$, reaching $75\,m^3 ha^{-1}$ in some watersheds, and estimated at 86 million $m^3 a^{\ 1}$ for Greece as a whole (McNeill, 1992). In most areas, 60 to 70 per cent of the land is usually considered to have been affected by serious erosion. During the 1940s and 1950s, Greek agronomists reported the loss of around 10,000 ha of cropland per year to erosion. By 1967, 1.5 million/ha of cropland in southern Italy were reputedly uneconomic to cultivate as a result of erosion. In Morocco, exceptionally high erosion rates have been reported, with values of $4,500\,t km^{-2} a^{-1}$ in the Ouerrha Basin in 1965, $4,300\,t km^{-2} a^{-1}$ in the Loukos Basin, $22,000\,t km^{-2} a^{-1}$ in the Tlata Basin, and 7,900 to $28,500\,t km^{-2} a^{-1}$ in the Nekor Basin (McNeill, 1992). Since the 1960s, it has been estimated that at least 1.5 to 2 per cent of agricultural potential has been lost to erosion every year. Accelerated erosion seems to have been ongoing since at least the 1920s, producing what McNeill has called 'a soil holocaust, the most dramatic case of overshoot in the Mediterranean in some time, and one of the most serious in the world'. The result of population and agricultural expansion, these high rates are partly due to the lack of terracing on steep slopes and the elimination of the fallow from the crop cycle. Elsewhere, abandonment of agricultural land in the higher mountain zones can be seen to have contradictory effects. In the Psilorites mountains of central Crete there is also some evidence for strong erosion on abandoned terraces at lower levels (Papanastasis and Lyrintzis, 1995). However, in eastern Crete, Grove and Rackham (1995) found little evidence for the collapse of abandoned terraces. MacDonald *et al.* (2000) indicate increased erosion due to abandonment in northern Portugal and northern and southern Italy. Part of the question as to whether erosion will take place relates to the incidence of fire (Chapter 7), which may increase following abandonment due to the greater quantities of combustible material following regeneration. Llorens *et al.* (1995) indicate that a tree cover on abandoned terraces may lead to decreased water availability (and potentially, therefore, to a greater incidence of fire), although runoff may be reduced and concentrated in former irrigation channels (Llorens *et al.*, 1992). However, the latter process may also lead to accelerated erosion if there is sufficient water to generate gully headcuts. Runoff reductions following vegetation recovery have also been noted for southern France (Rambal, 1987).

Soil-quality depletion can be defined on a number of levels. Sánchez *et al.* (1998) produced three indicators in their study of soil degradation in Castilla-La Mancha in central Spain. Biological degradation was defined as the inverse of the soil organic matter content; physical degradation as a function of the extent to which soils crust (and thus prevent water infiltration), their aggregate stability (and thus ability to minimize erosion) and water-holding capacity; and chemical degradation as a function of total salt and sodium salt contents, reflecting salinization (which is discussed further below). The two highest classes of biological degradation were found in 22 per cent of soils analysed, while 13 per cent of soils showed similar levels of physical degradation and 35 per cent of soils had this level of chemical degradation. Moderately high levels of each type of degradation were respectively found in 43 per cent, 36 per cent and 32 per cent of soils. Clearly, these figures reflect significant levels of soil degradation in this area. Cammeraat and Imeson (1998) have investigated the use of aggregate stability as an indicator of soil degradation in Murcia, Spain, and Var, France. They found that aggregate stability is increased under regeneration of natural vegetation on abandoned terraces. On bare or cultivated surfaces, there are important impacts from parent material and slope aspect, with stability tending to be lower on south-facing slopes, making them potentially more vulnerable to erosion.

Soil salinization is also becoming an increasingly important problem. If the salt content of the soil becomes sufficiently high, it reduces the ability of a plant to extract water from the soil by reducing the osmotic potential between the soil water and the plant root. Furthermore, concentrations of salts can also significantly decrease the amount of water infiltrating into soils

(Imeson and Emmer, 1992), and therefore available for extraction in the first place. The salinization process is particularly common in the southern part of the basin. For example, Goosens *et al.* (1994) noted a 4.5 times increase in the incidence of salinized soils in the western delta of the Nile between 1977 and 1989. Salt-affected lands are reported to make up 60 per cent of the agricultural land in the Lower Nile Delta (i.e. its northernmost part), 25 per cent in the Middle Delta, 20 per cent in the Upper Delta and Middle Egypt and 25 per cent in Upper Egypt (Kotb *et al.*, 2000). These levels relate both to irrigation with poor-quality water and locally to upward movement and evaporation of saline water from shallow groundwater. Extensive programmes of tile drainage have been instigated by the Egyptian Government in an aim to mitigate the latter problem, while rice cropping is being applied, particularly in the Lower Delta, in an attempt to increase leaching of salts from soils using the available, poor-quality irrigation water.

Salinization has been an important cause of land abandonment in Tunisia (Steen, 1998) and in the Souss valley of south-western Morocco (Ait Tirri, 1995, cited in Puigdefábregas and Mendizabal, 1998). Significant problems have also been noted in the Gaza Strip (Al-Agha, 1997). In the coastal zone, over-pumping of groundwater can accentuate problems of salinization of waters used for irrigation by allowing salt-water intrusion from the sea into the coastal aquifers (see Chapter 5). This problem is noted throughout the Mediterranean for Egypt (Kotb *et al.*, 2000), the Gaza Strip (Al-Agha, 1997), the Argolid (see above: Lemon *et al.*, 1996), western Crete (Grove and Rackham, 1993); Lesvos (Kosmas, 2000), Sardinia (Aru *et al.*, 1996) and Catalonia (Mas-Pla *et al.*, 1999). Morell *et al.* (1996) suggest that seawater intrusion is an important process in the salinization of groundwater in the Castellón coastal plain, but point to other sources of saline water, notably from aquifers containing older evaporites from the Mediterranean Basin (see Chapter 2). Similar processes can be seen in the groundwater of the northern Negev in Israel (Nativ *et al.*, 1997). Water-quality issues are therefore becoming an increasingly significant part of the planning of water supplies for irrigation and domestic uses (e.g. Prat and Ibañez, 1995; Comín, 1999). In these cases, abstractions can also have significant off-site impacts, for example, in wetland ecosystems of the Ebro delta.

Off-site pollution is principally caused by runoff carrying fertilizers and pesticides from agricultural land. Giupponi *et al.* (1999) illustrate the significant impacts of agriculture on the quality of both surface- and groundwater in the Venice Lagoon watershed, while Skoudilis (2000) illustrates similar problems for Greek rivers. These issues are not restricted to the drier parts of the basin, as seen from examples in Albania (Grazhdani *et al.*, 1996). Pollution issues are discussed further in Chapter 17.

Into the twenty-first century: potential impacts of future projections

Detailed Blue Plan projections have been made for Spain, Italy, Turkey, Syria, Libya, Tunisia and Morocco for the year 2025 (Grenon and Batisse, 1989). In Spain and Italy, agricultural production remains relatively static in the trend scenarios (Table 15.4), while in most of the countries of the south and east of the basin, the trend scenarios show at least a doubling of production. In all cases, the alternative scenarios show much larger increases. The proportion of irrigated land increases, although within limitations set by the availability of resources, and again more markedly out of Europe. Fertilizer consumption is projected to at least double in most countries, and commonly increases more than ten-fold in the southern Mediterranean countries. Mechanization will continue apace, both in the European countries, where levels will reach 0.110 machines per hectare (compared to 0.052 in 1990), and even more so in Africa, with projections of 0.057 in Tunisia, 0.066 in Libya, but only 0.028 in Morocco (compared to an average of 0.003 in Africa in 1990). In other words, levels of mechanization in 2025 in North Africa will be equivalent to those of the early 1990s in Europe.

Table 15.4 Blue Plan projections for agricultural productivity and other variables relating to agricultural intensification. The minimum and maximum values for 2025 give the range of projections for each variable

Variable		Spain		Italy		Turkey		Syria		Libya		Tunisia		Morocco	
		1980	2025	1980	2025	1980	2025	1980	2025	1980	2025	1980	2025	1980	2025
agricultural value added (1975 $000 M)	min	10.5	10.6	16.7	16.7	11.4	20.0	1.3	3.6	0.4	0.7	0.8	1.8	2.0	2.1
	max		35.6		33.0		49.4		9.0		1.6		4.2		9.6
productivity ($k/worker)	min	4.7	15.3	7.0	29.0	1.1	2.0	1.3	1.0	3.5	4.3	1.3	2.8	0.7	0.7
	max		28.7		45.2		5.0		2.7		10.1		5.6		2.8
irrigated land (%)	min	14.8	19.6	23.1	30.4	7.2	11.3	9.6	11.6	10.7	12.9	3.3	4.2	6.4	12.1
	max		26.1		49.0		40.3		20.1		14.3		8.9		16.7
fertilizer (kg ha^{-1})	min	77.0	136.0	175.0	300.0	47.0	151.0	23.0	187.0	35.0	250.0	15.0	200.0	28.0	250.0
	max		299.0		350.0		211.0		249.0		250.0		200.0		280.0

Source: Data from Grenon and Batisse (1989).

All these projections suggest that land and water resources will become increasingly under pressure. The patterns of river and marine pollution by fertilizers and pesticides are likely to continue, and soils and groundwater may become increasingly saline. Soil nutrient and stability depletion may further accentuate erosion problems, with their coupled off-site effects. The patterns of upland abandonment seen in the European Mediterranean may be repeated to the south and east of the basin as there is a greater need for more intensive production. The potential impacts of these changes in conjunction with projected climate changes will be considered further in Chapter 20.

Summary

The European Mediterranean zone has seen significant growth in the productivity of the agricultural sector throughout the twentieth century. This growth has been manifested by increasingly intensive exploitation, with smaller numbers employed in agriculture, rapid rises in the use of fertilizers, pesticides and agricultural machinery. Consequently, land abandonment in the mountainous zones of the region has been widespread. In recent years, this process has often been exacerbated by the policies of the EU. In certain areas this has led to industrial forms of agriculture with whole areas covered by plastic greenhouses and conversion of land which may be thought of as unsuitable for sustainable agriculture. Growth has also occurred in the south and east of the basin, but exploitation strategies still tend not to be anything like as intensive as in Europe, with extensive practices still taking place. Evidence suggests that depopulation of the uplands is only just beginning in these areas (McNeill, 1992), and that lessons can be learned from the impacts of such intensification as have been seen in the north.

These impacts have been significant and widespread. Soils have tended to become more depleted in nutrients and there is a continued reliance on artificial fertilizers to maintain yields. The soil structure has tended to become less stable, so that erosion is more active during the heavy rainstorms of the winter months. With this erosion comes off-site pollution in the form of chemicals in runoff, and silting of dams and reservoirs. Soils and groundwater are becoming more saline and groundwater levels are dropping due to over-pumping to meet the increasingly voracious needs of irrigation agriculture. Future changes in agricultural production will tend to accentuate these problems, and alternatives will need to be found to maintain sustainable food-production systems in the region.

Suggestions for further reading

Good overviews are to be found in Grenon and Batisse (1989) and Mairota *et al.* (1998). Green and Lemon (1996) provide an interesting case study of farming and environment from a range of perspectives.

Topics for discussion

1   What reasons underlie the particular changes in agriculture that occurred in the Mediterranean during the twentieth century?
2   Why is it difficult to define all the impacts of mechanized, industrialized farming?
3   Are modern agricultural systems in the region sustainable?
4   To what extent will water become the most significant limiting factor to agriculture in the twenty-first century (see also Chapter 20)?

16 The impact of depopulation

The beauty of the Mediterranean mountains is in a way a sad one. Skeletal mountains and shell villages dot the upland areas of the Mediterranean world, dominating the physical and social landscape. Rugged limestone ridges or smooth schist shoulders, bare of all but the scantiest of vegetation make the famous and apparently timeless vistas from Granada, Marrakesh, or (on a clear day) Athens. Between the ridges, usually situated so as to enjoy the Mediterranean sun, or perhaps a source of fresh water, lie quiet and moribund villages. They are shell villages, home only to the very old and sometimes the very young, but, perhaps with the brief exception of some summer weeks, home to no one in the prime of life. Both the mountains and the villages are usually picturesque. But their beauty is that of a still-life painting – *nature morte* as the French put it. They are dying villages and sterile mountains.

(McNeill, 1992: 1)

Introduction

In the previous chapter, we saw how agriculture is becoming increasingly focused in the lowland areas of the Mediterranean, and that there is a continuing move to urban and industrial centres located on the coasts. The reverse side of this story is that the upland zones, and others considered to be less productive, are becoming increasingly abandoned. This process is not new – it has occurred on numerous occasions in (pre)history, and the present phase has its roots in the nineteenth century in the northern Mediterranean – but is characterized on this occasion by its extent, and continuation through time. As these zones become less managed, there are significant consequences on the anthropic landscapes that have developed in the uplands.

Upland and 'marginal' areas

Land abandonment has reached levels of 10 to 20 per cent of previously cultivated areas in most countries, with a peak of 25 per cent in areas of Portugal (Grove and Rackham, 1995). This abandonment has accelerated since 1950, but was significant in some of the European Mediterranean uplands, for example, the Alpujarras of southern Spain, much earlier (McNeill, 1992). As noted in Chapter 14, the abandonment of upland areas has been a continuing consequence of the increasingly urban nature of Mediterranean populations, and the promise of employment in better jobs in the coastal zone, relating to the expansion of new industries and tourism. The process is by no means unilinear, but often it is the elderly part of the population that return to their original villages to retire, and thus do little to improve the economic vitality of the upland zones. The increased need to mechanize agriculture to meet modern production standards, driven largely by competition from outside the region (again a process that goes back to the latter part of the nineteenth century), has meant that the distributed nature of settlement across the landscape has declined. Narrow terraces in the high mountains are difficult if not impossible to access by tractor, and provide inefficient means of producing cereals when compared to the plains of northern Europe or North America. Pastoral activities, which spread the burden of supporting a population often by transhumant movements from the uplands to the lowlands, are declining because they are seen as less economically (and to a certain extent socially) acceptable. However, we have seen (Chapter 14) that livestock numbers have continued to increase across the basin, and this means that localized grazing is becoming excessive, with serious consequences. Thus, the upland zones are seen as being increasingly marginal areas, despite their ability to support large numbers for long time periods in traditional ways of exploitation. Either resources are insufficient for modern approaches, or the cost of accessing them and providing modern infrastructures is seen as unacceptable, and thus these areas continue to decline.

The causes and consequences of abandonment are varied spatially and temporally, as can be seen clearly in the following examples. White (1995) has studied in detail the population changes in Cilento, an isolated coastal region approximately 100 km to the south of Naples. The terrain here is highly fragmented, with low agricultural potential on the karst, and thus the area is difficult to access and has few local resources. Existing models of rural decline, based on examples in northern Europe and North America stress either the role of inter-settlement competition, so that settlements near large centres decline, or the role of residential attraction, with settlements can attract new residents if they are located near major centres to which they can commute. However, the results of White's analysis suggest different processes in depopulation over the period from 1961 to 1971. Access to large centres in fact plays a relatively minor role, and the effects of local pressure on agricultural resources, distance to the nearest settlement and other measures of accessibility proved more important in explaining the changes. The larger centres were in fact seen to decline more rapidly than elsewhere because of these factors. The presence of tourism, although poorly developed locally, can also be seen to be a key factor.

In Crete, population has decreased by 47 per cent in mountain and 36 per cent in hill communities compared to a 8.7 per cent increase in plain communities between 1951 and 1991 (Ispikoudis *et al.*, 1993). Forestry has limited importance as economic activity, although forest makes up 21 per cent of the whole area. However, the period from 1981 to 1991 has seen a 71 per cent increase in the number of livestock with a 50 per cent increase in the stocking rate. Over-stocking together with free and uncontrolled grazing and frequent pastoral wildfires have resulted in severe degradation of the landscape. Other specific case studies have been carried out within a number of projects. In western Crete, tourism and intensive agriculture have developed in the coastal zone while the uplands are being progressively abandoned (Grove *et al.*, 1993; Grove and Rackham, 1993). Despite the fact that over 300,000 visitors go to the national park of the Samaria gorge each year, leading to increased traffic on the roads (and thus pollution) and trampling (and thus a greater threat of erosion), there is little in the way of benefits to the mountain economy as very few tourists stay overnight. Agricultural change in the lowland areas has been marked by a reduction in cereals and a massive expansion of olive groves, which are often irrigated. This change has been in progress for about 350 years, but has accelerated dramatically since 1970. Olives are grown in areas previously under maquis used partly for grazing, frequently on bulldozed, unsupported terraces. These prevent a much greater erosion risk than traditional terraces, which appear to be stable on the island, even where abandoned. In part, such expansion is directly related to the availability of EU subsidies. More remote areas have been abandoned, with terraces on hard limestones with difficult access for machinery being abandoned for several decades. Many areas are becoming depopulated. Invasion of pines seems to pose a threat of increased incidence of wildfires in the area. An extreme case of abandonment is seen on the small island of Gavdos, around 40 km to the south of the main island. Only accessible by boat, the 35 km² island has an elderly winter population of fifty to one hundred. Cultivation has declined and only 1,500 goats remain. The island has experienced severe gullying on abandoned terraces, and the nature of the vegetation growth is increasing the threat of wildfires, with their consequences in terms of further erosion. In the Psilorites mountain of central Crete, emigration from the 1950s caused the first reversal of population increase (Papanastasis and Lyrintzis, 1995). Since the 1970s, however, there has been the re-establishment of a slower increase. The settlement pattern follows a bipartite division, with one series of villages established to an altitude of 700 m, which is locally the upper limit of olive growth, and a second series of seasonally occupied villages at much higher altitudes. Previously these villages were used for summer cultivation and grazing, but now have only a pastoral role. Land use has changed significantly since 1961, with a decrease in forest and agricultural land, and a corresponding increase in pasture. Sheep and goat numbers have increased from *c*.60,000 in 1961 to *c*.340,000 in 1992, largely because of the lack of controls within the communal pattern of grazing. Over the same period, irrigation and tree crops have increased, while annual crops and fallow areas have decreased. Degradation of the area seems to follow an altitudinal gradient. Higher altitudes are heavily degraded following erosion due to excessive grazing, particularly in recent years (Hill *et al.*, 1998; Kasapidis and Tsiourlis, 2000). Grazing management experiments seem to indicate that this degradation process is reversible, however, if the pressures from over-grazing are removed. A similar outcome was found by Koutsidou and Margaris (2000) on the island of Chios.

In Epirus, there are some suggestions that the badlands at Kokkinopilos developed within living memory and even that the development is due to the lack of maintenance of erosion-control structures following depopulation in the 1950s, although these claims are uncorroborated (King and Sturdy, 1994). However, to a certain extent, these badlands represent processes which have been in operation since at least the Upper Palaeolithic. As noted in the previous chapter, degradation in the area is perceived very differently by local populations as compared to outsiders (Green, 1994; Green and Lemon, 1996). Aerial photographic evidence suggests that traditional ways of life based on herding animals (see also Campbell, 1964) caused very extensive modifications of the environment within a framework of activities centred on the individual or family-group level of organization. This socio-economic organization led to the development of a number of niches, with impacts spread across the landscape. It is suggested that one problem with recent interventions in the landscape is that they are seen from a technologically-based urban viewpoint with centralized organization, and have often failed due to miscommunication, suspicion of centralized control and the historical sense of individuality. Recent thought has moved away from technological fixes as they have too great an impact on the local environment, although this viewpoint is often seen as backward and static by those with political or administrative roles. However, the role of human intervention to preserve the 'traditional' environment means that such an activity is necessary in any management strategy for the area, although the type and degree of such intervention are very much a contentious issue. At Delvinaki, a small town near the border with Albania, Green (1997a, b) demonstrated quite different issues relating to marginal areas. Although the population of the town dropped from 2,630 in 1920 to 921 in 1991 and the town is perceived as being 'passed by' by modern, rapid change, water shortages are now occurring in the summer, despite the limited requirements of the town, in part related to an apparent decline in rainfall and in part due to the installation of modern domestic appliances. Plans to divert water to improve tourism in the area thus proved highly contentious.

In the Golo valley of Corsica, the population has decreased from 25,000 inhabitants in

1891, to 6,000 in 1990 (Conventi and Dykstra, 1995). This change reflects patterns of migration towards the coastal region and an increasingly aged population. A principal components analysis indicated that important variables explaining the distribution of the loss of population were the distance of a settlement from a major road, and population structure, with indications that there is often an under-representation of females in the twenty to thirty-nine age category. Using this analysis, 60 per cent of the communes seem to be in a critical state with respect to their population, and there is a concentration of the remaining population in the lower altitude zones. These areas seem to be suffering from resource degradation at present. At higher altitudes, abandoned areas appear to be recovering, although this seems to be having major impacts on water resources. At middle altitudes, wildfires are the dominant problem, particularly when located on formerly cultivated soils on south-facing slopes, which are particularly vulnerable to erosion.

Sharecropping remained the common land use in the Abruzzo region of Italy until the 1960s, when its decline set in due to emigration and depopulation of the uplands (Niccolai *et al.*, 1995). Farm numbers and area of cultivation have decreased progressively through the 1970s and 1980s as a result of progressive urbanization and industrialization. Between 1970 and 1990 the farm numbers declined from 12,376 to 791, reflecting the decline in sharecropping leading to the consolidation of freehold properties and farms under commercial management. The type of cultivation has changed, with the area under sown crops declining by 67,006 ha at the same time as a rise in areas with olive, grape and fruit of 28,806 ha. Permanent meadow and pasture declined by 3,029 ha. Industrial farming and cultivation of new crops such as sugar beet cover large areas. Irrigation increased rapidly, with an increase of 6,692 farms using irrigation between 1982 and 1990, although the average area irrigated per farm decreased from 1.56 to 1.07 ha over the same period. Sprinkler, flood and drip irrigation are all practised and important numbers of farms belong to irrigation consortia and mechanization has increased dramatically. Within this setting, those areas where demographic and land resources have declined have been identified as the most susceptible to degradation. These areas are principally located at mid to high altitudes in the interior of the region. In part, this vulnerability is seen to be due to the ineffectiveness of economic interventions in the region to help these zones, with most of the benefits going to the more intensified lowland areas. Farming still represents the lowest income in the region, frequently leading to the development of part-time farmers, with their own particular impact on the environment.

Problems in the Basilicata region of southern Italy were in fact accentuated by the land reforms of the 1950s, when the large latifundia estates were split up and distributed to local peasant farmers (Pizzolotto and Brandmayr, 1995). The land parcels distributed were too small for subsistence agriculture (for example, the average plot size in 1970 was still only 2.1 ha: Jones, 1984), leading to abandonment even in the lowland zone, leading to migration and increased urbanization. Although populations at the regional level have continued to increase due to these local movements, the mountain zones are becoming progressively abandoned. Some areas have been recently reforested, with such areas making up 17.8 per cent of the area of the Sila plateau. Degradation occurs according to distinct spatial patterns. In the more humid upland plateau zone, human intervention is relatively recent, and soils are still well conserved. At middle altitudes, the land-use structure dates back to the early Mediaeval period, and includes terraces used for chestnut and cereals. These are now often abandoned for cultivation, are rarely used for grazing, and the corresponding regrowth of vegetation and reforestation makes these areas vulnerable to degradation as the result of wildfires.

In northern Cyprus, Makhzoumi (1997) describes the decline of olive–carob plantations, with some areas being revegetated by maquis. In part, this decline is due to the abandonment of orchards by Greek Cypriots following the invasion of 1974, in part to the unprofitability of growing the tree crops in this location, and in part to the use of some of the areas for housing development. Regrowth of natural vegetation in abandoned areas can be relatively rapid, as can be seen in several examples. Debussche *et al.* (1999) describe the case of Languedoc in southern France where Mediterranean forests have regrown in a period of tens of years under sub-humid climatic conditions. The consequences of this change have been threefold. First, the frequency of large fires is thought to be on the increase. Second, increased evapotranspiration has decreased available water resources. Third, there are important consequences on biodiversity due to the change in available habitats. Although wild boar populations are on the increase, as they prefer the forests, a number of Mediterranean bird species are decreasing as the open habitats they prefer are disappearing (Preiss *et al.*, 1997).

McNeill (1992) demonstrates clearly the spatial differences at the basin level in terms of depopulation. While areas such as the Alpujarras in southern Spain, Basilicata and Epirus have long undergone depopulation – in some cases in the mid to late nineteenth century – the process is much more recent in the Taurus of southern Turkey and the Rif of Morocco. The pull of rapidly growing urban centres is only just being felt in these locations, and population growth has been such that extreme pressure on local resources is only a recent phenomenon. In some cases, this pressure is leading to out-migration to the north of the basin and beyond, repeating the process of outmigration seen in the European Mediterranean in the later nineteenth and earlier twentieth centuries.

Margaris and Grove (1993) suggest that the effects of abandonment of agricultural land differ between areas. In Greece, Corsica, Italy and Spain, abandonment leads to desertification because vegetation colonization is suppressed by fires and grazing, whereas in southern France and Portugal similar changes are followed by recolonization. Cultivation of traditional forest products is in decline because of declining markets and this has reduced the stability of several upland zones. In Portugal, for example, demand for paper has led to the planting of eucalyptus in areas previously used for cork–oak cultivation. Grove and Rackham (1995) suggest that there is little evidence in Crete to suggest that terraces will necessarily collapse and lead to higher erosion rates as a result of abandonment. Fires seem to have increased here as a result of land abandonment leading to the development of pyrophytic vegetation and the removal of crops which would previously have formed fire breaks, but also, the planting of flammable trees such as eucalyptus or pine, and the restriction of grazing in forested areas. They suggest that the use of grazing should be employed within an integrated land-use management strategy to help reduce fires. Grazing is also seen as a fire-suppressing mechanism in the Spanish Pyrenees by removing the shrub undergrowth from trees that provides the fuel for spreading extensive fires (Valderrabano and Torrano, 2000), although it must be remembered that shepherds are traditionally cited as the cause of many fires because of their use of burning to provide new grass in pastures (Métaillie, 1986; Amouric, 1992). As seen in the previous chapter and Chapter 11, the consequences of land abandonment in terms of terrace collapse, enhanced soil erosion and runoff are equivocal. Further investigations are probably required that provide further information as to which set of processes will take place following abandonment. Certainly, the evidence to support the sweeping statement of Margaris and Grove is far from clear, and the controls on these processes probably need to be sought at the local rather than regional level.

Impacts of future projections

FAO projections suggest that rural populations will continue to decline in all areas of the Mediterranean Basin to at least the year 2030, and that the decline in relation to the total population will be even more marked (Figure 16.1). By 2030, only 22.4 per cent of the popu-

(a)

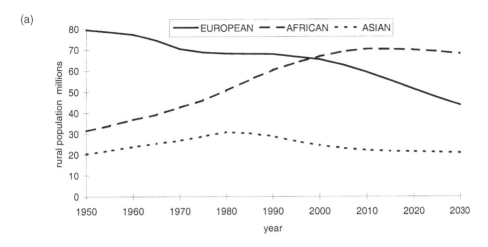

(b)

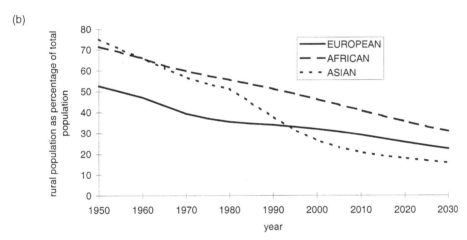

Figure 16.1 Rural populations in the Mediterranean Basin, with projections to the year 2030: (a) total rural populations; and (b) rural populations as a percentage of the total population in each region

Source: Based on data from FAO yearbooks.

lation of the European Mediterranean countries is predicted to be rural, compared to 30.8 per cent in the African countries and 15.7 per cent in the Asian countries. However, within these figures there will still be a considerable range: from 47.3 per cent in Portugal and 40.4 per cent in Albania to 17.3 per cent in France and 17.6 per cent in Spain for the European countries; from 25.2 per cent in Syria to 2.5 per cent in the Gaza Strip in the Asian countries; and from 33.2 per cent in Egypt to 6 per cent in Libya. Land abandonment of the uplands and more marginal agricultural zones is therefore likely to continue at present rates, with an increasing focus on the south and east of the basin. Locally, this will accentuate problems of fire and erosion as the local agricultural infrastructure collapses, although current evidence suggests that it will be difficult to predict which areas will be most vulnerable. Increasingly, alternative schemes for the use of these areas, such as 'green tourism' (see Chapter 21) will need to be found, to produce viable economic futures for the depopulating areas and help to minimize the effects of land degradation in them. Similarly, the continuing move towards urban centres will require the development of integrated schemes to provide sufficient, environmentally friendly infrastructures, water resources and waste-management schemes.

Summary

The consequences of land abandonment in the uplands are multiple and vary over even quite small spatial scales. The situation in these areas has flipped from one where over-population meant unsustainable conditions were produced. However, the wider landscape structure and its sustainability in the short to medium term may depend on mitigating the effects of a relatively rapid depopulation. Predicting the likely outcome of land abandonment is one of the most difficult tasks currently facing environmental managers. Vegetation regrowth is a common outcome, although in some locations this regrowth is predicted to have negative consequences, such as the increased incidence of forest fires, and a reduction in water resources due to increased evapotranspiration. Elsewhere, regrowth is insufficient to prevent accelerating erosion, and in some locations this erosion is accentuated by an increase in grazing pressure.

In some areas, active measures are being taken to try to reverse the impacts of these changes. However, measures that are simply imposed centrally are likely to fail as they do not take into account either local needs or structures. Such measures will need to become increasingly widespread in the twenty-first century, as rural depopulation continues at significant rates.

Suggestions for further reading

McNeill (1992) is by far the best general overview of the impacts of depopulation. Du Boulay (1974) and Jenkins (1979) also present very readable case studies.

Topics for discussion

1   What mechanisms have determined the marginality of upland areas in the Mediterranean?
2   Consider the factors necessary to alleviate the problem of rural depopulation and its environmental impacts.

17 Pollution

Introduction

Pollution is defined as the introduction of anything that adversely or unreasonably impairs the beneficial use of water and air, even though actual health hazard may not be involved. The main sources of pollution can be liquid or solid, organic or inorganic. Suspended solids are most abundant when water is carrying erosion products in suspension in the flow. These suspended solids can be a great problem, for example, in industrial processes involving filtering, where particles can clog up the filters and be deposited in slow-moving parts of the plant's circulation system. However, they are not usually regarded as pollutants because of their natural origin and ubiquitous occurrence.

Water pollution is inextricably coupled to the hydrological cycle in its atmospheric, hillslope and channel phases. In the atmosphere, pollutants are present as hygroscopic nuclei. In the Mediterranean, these take the form of industrial atmospheric pollution and pollution formed by fine dust from the Sahara (see Chapter 6). Both forms contribute significantly to the sediments that are accumulating on the shallow shelves surrounding parts of the Mediterranean (see Chapter 2). Their atmospheric origins can clearly be identified from the organic water proteins that can be tracked to Saharan sources or from their attached radionuclides such as caesium that can be traced to the Chernobyl incident. These provide a marker horizon in soils and sediments that has been used to estimate values of soil erosion.

Soil pollution

The main soil pollution in rural areas comes from the fertilizers that have been applied in great quantities since the 1950s, especially in irrigated systems. The most serious form of soil pollutants are those that are themselves persistent, i.e. they do not break down chemically either easily or quickly. Any pollutants that pass through the soil probably end up in groundwater and groundwater pollution is a dangerous and serious risk in Mediterranean environments, since the high rates of evaporation and heavy use of groundwater for irrigation mean that it can pass into the food chain. Pollutants that attach themselves by adsorption or absorption to soils are especially problematic because high rates of erosion transport these pollutants to wetlands and to the Mediterranean Sea itself. They can also be re-deposited as films in the dried-up beds of ephemeral channels. Fortunately the newly-introduced drip irrigation is not so wasteful of fertilizers.

The rapid growth of food processing, coupled with 'fertigation', the commercial practice of dosing irrigation water with fertilizers and pesticides, is a particular problem in modern tomato cultivation where the tomatoes are grown in massive 'grow bags' that are automatically supplied with water and nutrients. After production, the 'bags' (in reality long plastic tubes filled with sand) have to be emptied. The remains, rich in fertilizers and also toxic, can cause eutrophication if dumped into rivers, wetlands or dry channel beds (*ramblas*).

Water pollution

Ephemeral channel beds often have a deep fill of fairly porous material (sands and gravels). Waste water circulates there and this sub-alluvial aquifer is recharged at any time there is runoff across the channel bed (see Chapter 5). Some of the *ramblas* also serve as effluent recipients for industrial waste, such as the leather industry of the River Vinalopó in Alicante or the jam and fruit industries along the River Mula in Murcia. Equally, local municipalities without proper treatment plants release sewage into the *ramblas*, causing contamination widely throughout the Mediterranean.

There are extensive wetlands in the Mediterranean such as Lake Ichkeul in Tunisia, the Albufeira lagoon of Valencia, the wetlands of southern Cyprus and the Mar Menor in Murcia. These all provide a stopover for bird flocks migrating between Africa and Europe and are eco-logically important. These wetlands and their pollution are the sources of political conflicts that arise from contradictions in policy, such as urban growth, environmental conservation and the growth of the tourist industry. This problem has been most sharply focused in the Coto Doñana, a large wetland at the mouth of the Guadalquivir River in Spain. Although protected by a development plan for reconstruction of the tourist zones near to the wetlands, the plan allowed for the provision of accommodation for 10,000–15,000 new tourists in the first phase. The main debate has been about the provision of water, whether from aquifer or surface water (see Box 17.1 for more on the Coto Doñana National Park).

In heavily industrialized areas, the main sources of pollutant loads are point discharges from factories and runoff of contaminants from streets (such as petrocarbon b from traffic). The main industrial river of the Mediterranean region is the Po which drains to the Northern Adriatic, where industry causes serious pollution. Several point sources of pollution arise directly from mining of valuable metals and the toxic metals associated with them (cadmium, mercury, arsenic) at various places around the Mediterranean Sea (an example from Coto Doñana, a national park in Spain, is given in Box 17.1). There are several examples in the Mediterranean. Mine tailings produced by washing the ores are rich in toxic minerals and, if released along the coast, can cause significant pollution problems. This type of pollution has occurred at La Unión, Murcia, Spain, and downstream from Iglesias in south-west Sardinia. In both cases the local workforces have protested strongly at proposed limitations because they are fearful of losing their jobs. This is a recurrent problem. The protests against environmental damage could lead to the closure of industrial plants and, as these might may be the main employer in otherwise economically poor regions, there is a genuine fear that the loss of employment may exacerbate poverty.

Industrial development and pollution – an example from Spain

In general, the policy in southern European countries has been to encourage 'development poles' and the autonomous region of Andalucia in Spain has specifically advocated this policy since its creation. This policy has created new industrial areas mainly located on the coast. Huelva city was chosen, given the activities of the Río Tinto Company in the vicinity, on the Spanish–Portuguese frontier. The logical location for the industry, given the wind directions, would have been to the north-east or north-west of the city, but these are areas of marshland, with poor communications and hardly any infrastructure. Eventually the area of Punta del Sebo, to the south of the town, was chosen, resulting in bad atmospheric pollution for the city.

Box 17.1 Pollution of Coto Doñana National Park, Spain

Coto Doñana National Park has become an important symbol of environmental degradation and conservation both nationally and internationally. The fate of Doñana reached crisis proportions in May 1998 when a flood of toxic sludge covered 2000 ha. The disaster was caused by the breaking of the waste pond dam at the huge open-pit mine at Los Frailes near Seville, owned by the Swedish firm Bohdern. According to Mackenzie (1998), in its first year of operation, 1997, the mine produced 180,000 tons of zinc, lead, copper and silver from 4 million tons of ore. The water, crushed ore and chemicals left behind after the metals have been removed are dumped in the settling pond of another mine that was abandoned in 1996 after nineteen years of operation. An earth dam held the effluent in the pond. When the dam collapsed, 4 million cubic metres of acidic water and silt, a cocktail of dangerous pollutants swept down the Agio River heading straight for the natural park. The floodwave was diverted into the Guadalquivir and hence to the Atlantic but the water still flooded hectares of land and the pollutants will steadily enter groundwater and the river system over the years, and eventually reach the Atlantic.

A more technical treatment is given in Gómez-Parra *et al.* (2000). The high concentrations of zinc in the Guadalquivir estuary had decreased significantly within five months of the spill.

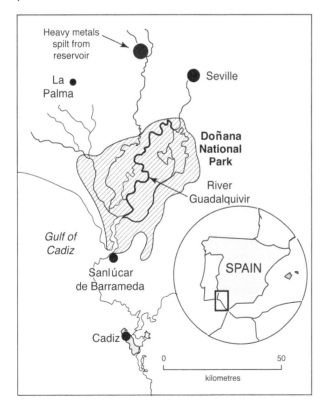

Figure 17.1 Map of the pollution episode at Coto Doñana National Park
Source: After McKenzie (1998).

Environmental and social factors were ignored in favour of the ease of urban development and water supply, the low price of the land and the ability to push the waste into the courses of the Rivers Tinto and Ochel, on the grounds that these rivers were already contaminated by the residues from existing mines in the centre and north of the province (Andalucia, 1990). The other main factor was the construction of a large power station since these industrial processes are large consumers of energy. Moreover, the importing facilities were improved, with jetties for the companies.

The industrial development pole of Huelva now has:

- a petrol refinery, providing for 83 per cent of the domestic market;
- an inorganic chemicals sector, mainly producing hydrochloric acid, sulphuric acid and methyl derivatives;
- an organic chemicals sector producing acetones, phenols and cyclohexanes;
- an agrochemical sector producing fertilizers and phosphorous derivatives;
- a paper industry based on cellulose.

This concentration gives rise to a strong atmospheric pollution as well as contamination of the land and the water (the Huelva Ria and coastal zone). Together with Bilbao and Airles, Huelva is one of the most polluted cities in Spain. The potential magnitude of the waste water is calculated to be 2.5 million m^3 day^{-1} and two to three million tons of slurry are produced in a year. Studies of the coastal zone have revealed great quantities of heavy metals in the sediments as well as in the water of the Huelva Ria (mainly lead, nickel, chrome and mercury). The accumulation of these metals provokes frequent ecological catastrophes, causing the death of thousands of fish and closure of beaches for bathing. The indices of general water quality are very low and vary considerably in time and space.

There is a plan to reduce the environmental impacts of the Huelva development pole that will enable it both to meet European standards and set an example for other areas of Andalucia, such as the Bay of Algeciras, and also other places bordering the Mediterranean. The plan attempts to bring together the different agencies and ministries to solve the problems. It emphasizes the need for a high level of participation by the principal actors and agents and a public debate. The plan is based on the 'polluter pays' principle (discussed in Chapter 21), applied in a way that would not be crippling to the industries involved, and with some financial help to the countries involved. In practice, the government of Andalucia will provide funds for environmental improvements of a more general nature and for regeneration of the physical environment affected by industrial impacts. The plan specifically schedules action by particular companies (Andalucia, 1990). A specific measure is the collector designed to prevent effluent escaping into the river. This partnership approach avoids an excessive penalty on the offending companies and thereby offsets the risk of unemployment.

Marine pollution

The Mediterranean Sea covers about 2.54 million km^2 with an average water depth of about 1.5 km. The length of the coastline totals about 46,000 km, of which 19,000 km represent islands (Milliman *et al.*, 1992). In 1985 the eighteen countries bordering the sea had a combined population of 352 million people, 37 per cent of whom lived directly in the coastal zone. Grenon and Batisse (1989) estimated that the population will reach 545 million in 2025, by which time 75 per cent is expected to be urban. The European countries have nearly stable populations with annual growth rates of less than 1 per cent. The southern countries have growth rates of 2 to 3 per cent per year.

A general overview of the environmental quality of the Mediterranean Sea was published in 1993 by de Walle *et al.* They estimated that, in 1993, 3.3 million tons of organic matter, 350 million tons of suspended solids together with heavy metals and organics were annually discharged into the sea. These can only be regarded as 'best possible estimates' because of the major difficulties in estimating the quantities, usually obtained from estimates of the concentrations of the pollutants in rivers and then multiplying by discharge to obtain the total load. Both concentrations and discharges are measured with large errors, but it is important to have some idea of the magnitude of the problem.

The pollutants include heavy metals such as mercury and cadmium. Mediterranean pelagic fish contain roughly twice the mercury concentrations as Atlantic specimens of a similar size: $1,100\,\mu g\,kg^{-1}$ near Palma de Majorca and $540\,\mu g\,kg^{-1}$ along the Slavic coast. Mercury also pollutes mussels and samples taken from the Ligurian Sea, the Gulf of Trieste and the area around Marseille exceeds international standards for strongly polluted shellfish.

Mediterranean mussels, clams and oysters were, at that time, also seriously polluted by hydrocarbons (usually oil products and by-products) and, in this respect, the mouth of the Ebro River in Spain, near to the city of Barcelona, was one of the pollution hot spots of the Mediterranean. As a result of strenuous efforts, such as the development of ship-cleaning facilities, the pollution by pelagic tar has dropped dramatically in the western Mediterranean since the 1960s, though the Rivers Rhône, Po and the Adige still contribute notable quantities of chlorinated hydrocarbons. Although the use of the pesticide DDT has been forbidden, it was still an important pollutant in 1993, conspicuously in the western Mediterranean at the Ebro delta and around Barcelona between the Llobregat and Besos rivers. Further east, it is found in important contributions in sediments near the Athens outfall in the Savonikes Gulf.

De Walle *et al.* (1993) estimated that 23 per cent of the total pollution load of the Mediterranean was discharged into the north-west Mediterranean – Spain, Italy and France. A further 35 per cent was discharged into the Adriatic, mainly from Italy. The Mediterranean received $820,000\,t\,a^{-1}$ of oil spilled, which is 17 per cent of the total pollution in the oceans, even though the surface area is only 0.7 per cent.

Of the total radioactive tritium input, 44 per cent was into the north-west Mediterranean, 51 per cent into the Adriatic and coming mainly from power plants on the major rivers.

The most recent study funded by the European Commission is the European River–Ocean System (EROS 2000) which includes the Mediterranean and especially the River Rhône. The Rhône discharges into the Gulf of Lions and the waters are carried by the Liguro–Provencal current towards the south-west across the shelf which has deep canyons. The salt front (between the fresh river water plume and the ocean water) limits the transfer of fresh water into the open sea. Nevertheless, a surface patch of nutrient-rich water extends from the Rhône outlet southwards covering half the Gulf's surface. This water layer is $10\,m$ thick near the river mouth and there are strong seasonal variations in its depth at different distances from the mouth. The river water, of which the Rhône produces $1,500\,m^3\,s^{-1}$, can be separated from the sea water by its high nitrate content with $90\,\mu mol\,l^{-1}$ in the river water and $4–6\,\mu mol\,l^{-1}$ in the sea water. The plume can be traced $18\,km$ out to sea.

Table 17.1 shows information on Mediterranean pollution by countries that are on, or discharge into, the Mediterranean. This information has been assembled by MedHYCOS and published on their website. Most data are from 1993, but some from 1980. These data show that the highest organic loadings per head of population are from Egypt, Greece, Syria, Tunisia and Turkey, all of which are one standard deviation above the mean loading. In mineral pollution, the tables give estimates of percentage breakdown by type. They indicate the main variations in economic activities of the different countries. Discharges with the highest percentages of primary metal loadings come from the Ukraine, Italy, France and Bulgaria. Discharges with

Table 17.1 Estimates of waste pollution into the Mediterranean Sea for Mediterranean Basin countries in 1993, according to main source

	Organic pollution kg day^{-1}		Mineral pollution							
			Primary metals	Paper	Chemicals	Agro-food industry	Ceramics and glass	Textiles	Timber	Other
	1980	1993	1993	1993	1993	1993	1993	1993	1993	1993
Algeria	60,290									
Bulgaria	151,016	11,310	11.5	6.2	16.2	24.1	0.3	14.8	2.0	6.9
Croatia		55,440	9.3	13.8	8.9	43.1	0.3	14.2	3.7	6.8
Egypt	169,146	198,373	11.7	7.1	9.1	50.5	0.3	17.5	0.5	3.5
France	716,285	609,940	11.9	20.7	11.0	37.0	0.2	6.7	1.8	10.8
Greece	65,304	59,701	6.0	12.1	8.0	51.8	0.3	16.6	1.5	3.8
Israel	39,113	50,030	4.1	19.3	8.4	44.3	0.2	12.3	2.1	9.3
Italy	442,712	353,906	17.0	16.1	10.5	25.8	0.3	16.1	2.1	12.1
Jordan	4,146	11,166	4.1	15.3	15.9	49.8	0.7	7.6	4.4	3.3
Lebanon	13,137									
Libya	3,532									
FYR Macedonia		29,054	16.6	8.4	6.0	37.7	0.1	24.5	2.0	4.7
Moldovia		54,263	2.1	3.2	1.5	69.0	0.3	15.0	1.7	7.1
Morocco	26,598	33,752				36.3		57.5		6.2
Portugal	105,441	77,451	6.0	8,7	0.4	44.7	0.4	31.4	3.8	4.6
Rumania	352,368	146,154		16.1	2.9	1.8		41.4	11.3	26.6
Slovenia		39,846	17.7	17.1	8.4	25.3		17.6	4.1	
Spain	376,253	318,506	10.7	15.4	9.3	45.6	0.3	8.7	2.8	9.8
Syria	36,262	23,754	2.8	1.4	7.4	65.6	0.5	16.0	5.4	7.2
Tunisia	20,294	25,610	5.6	5.6	5.1	62.7	0.8	17.4	0.7	0.9
Turkey	106,173	168,548	15.8	8.0	7.0	46.6	0.3	17.0	0.7	2.1
Ukraine		666,233	18.3	3.7	7.3	46.9	0.5	10.1	2.0	4.5
Average			10.1	11.0	8.0	39.1	0.4	19.6	2.9	7.9

Source: From MedHYCOS web page (http://medhycos.itd.fr).

very high percentages from agricultural and food industries come from Greece, Egypt, Jordan, Syria, Moldavia, Tunisia and Spain. As before, these data should be regarded as best available estimates, given the errors that surround their collection and preparation.

Protecting the Mediterranean Sea

Because the Mediterranean Sea is enclosed, there is real fear that, over the years, it will become seriously polluted, especially given the following factors:

- the wide range of activities along its shores (industrial, agricultural and tourism);
- the slow turn-over with water flowing in at the surface in the Straits of Gibraltar and out in a deep undercurrent through the same Straits;
- the large volumes of the major in-flowing streams (especially the Rhône and the Nile) and the hundreds of other smaller but more polluted rivers and streams.

It is too facile to think of the Mediterranean as a giant bathtub. Rather, because of the land surface and the submarine topography, it is broken up into a series of basins and shelves in which the circulation is controlled by the in-flowing rivers, by the winds and by the differential heating of the surface water and the steep thermal and salinity gradients that exist in this complicated body of water (see Chapter 8). Correspondingly, issues of pollution transport and deposition are rather complex, leading to great variability in the local concentrations of pollutants.

There is an international protocol for the protection of the Mediterranean Sea through the prevention, combating and control of pollution of the sea area by discharge from rivers, coastal establishments or outfalls or emanating from any other land-based sources within their territories. This protocol came into force in 1983 and involves a large number of countries ranging from the Ukraine, Georgia and Russia that surround the Black Sea to many of the countries with a Mediterranean coastline, such as Spain, France, Italy, Israel, Tunisia and Morocco.

As a consequence of the international concern, a number of international treaties, such as the 1976 Barcelona Convention, have established research programmes to monitor the changing environment. The Mediterranean Action Plan, encouraged by the United Nations, has sustained the momentum for this activity (see Box 17.2).

Box 17.2 The Mediterranean Action Plan and the Barcelona Convention for the Protection of the Mediterranean against Pollution

The Mediterranean Action Plan (MAP) was adopted in 1975 by the countries surrounding the Mediterranean and the EEC. It has the specific aim of providing information and assessments of pollution in order to allow national governments to formulate better and integrated planning and environmental policies. A progressive shift to integrated options, taking into account the many socio-economic controls on development and pollution, has occurred through its existence. An initial success of the process was the 1976 Barcelona Convention that led to the following legal agreements regarding cooperative pollution control:

- the Convention for the Protection of the Mediterranean Sea against Pollution;
- the Protocol for the Prevention of Pollution of the Mediterranean Sea by Dumping from Ships and Aircraft;
- the Protocol Concerning Cooperation in Combating Pollution of the Mediterranean Sea by Oil and Other Harmful Substances in Cases of Emergency.

The conditions leading to the development of these cooperative frameworks are covered in detail in Haas (1992).

Subsequent additions to these agreements include the 1980 Athens Protocol for the Protection of the Mediterranean Sea against Pollution from Land-Based Sources; the 1982 Geneva Protocol Concerning Mediterranean Specially Protected Areas; and the 1994 Madrid Protocol for the Protection of the Mediterranean Sea against Pollution Resulting from the Exploration and Exploitation of the Continental Shelf and the Seabed and its Subsoil.

A further Barcelona Convention in June 1995 adopted Phase II of the MAP and identified twelve priorities for action: (i) the integration of environment and development; (ii) integrated management of natural resources; (iii) integrated management of coastal areas; (iv) waste management; (v) agriculture; (vi) industry and energy; (vii) transport; (viii) tourism; (ix) urban development and the environment; (x) information; (xi) assessment, measurement and control of marine pollution; and (xii) conservation of nature, landscape and sites. There is an excellent MAP website at http://www.unepmap.org/, with extensive information and documentation relating to pollution assessments and their implementation.

In addition, a series of major research projects, funded by the European Union, have substantially enhanced knowledge of the dynamics of flow in the ocean. This activity has been strengthened by the development of remote-sensing techniques for detecting not only spatial variations in temperature, but also the plankton and diatom content of the surface waters. These species are not only important as sources of fish nutrition, but also as indicators of the state of pollution in different areas.

Without an understanding of the chemical balances derived from in-flowing water and sediments and from the deposits on the ocean floor and an understanding of the current patterns in three dimensions, it will be impossible to understand the causes of pollution and the steps necessary to reduce it. Three areas that are pollution 'hot spots' have been examined in detail in these research programmes, mainly through transects by research ships collecting data on the physical–chemical properties of the ocean waters. These are shown in Figure 17.2:

- the Gulf of Lions, where the River Rhône enters the sea and the industrial complex of the Marseille region causes high levels of pollution;
- the Northern Adriatic, with the inflow of the River Po and contributions from the urban areas of Venice and Trieste in Italy and Rijeka in Croatia;
- Attica, the coast around Athens and Piraeus in Greece.

There are other concentrations of people and industry along the coast of Israel, the Nile delta, the major cities of the Maghrebian states (Benghazi, Tripoli, Tunis, Algiers and Oran), and of the northern shore (Malaga, Alicante, Barcelona, Genoa, Livorno, Rome, Naples, Patrai, Istanbul and Izmir). The *Dobříš Assessment* of the European Environment Agency reported estimates of 500 million tonnes of raw sewage poured into the Mediterranean each year, in addition to 120,000 tonnes of mineral oils, 66,000 tonnes of detergents, 3,800–4,500 tonnes of lead, 5,000 tonnes of copper, 3,600 tonnes of phosphates and 100–120 tonnes of mercury (EEA, 1995). Data at this scale have to be highly suspect because there are simply not enough reliable measurements to give precise concentrations of the pollutants or the recurrence intervals at which such concentrations occur in rivers flowing into the sea.

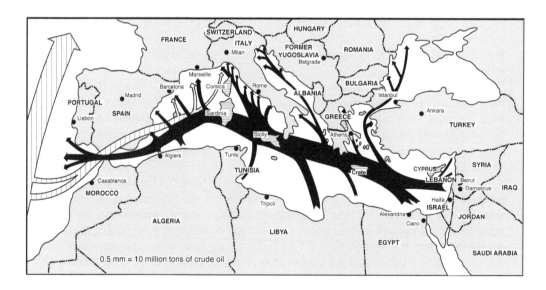

Figure 17.2 Map showing principal locations of intensive pollution in the Mediterranean Sea

The pollution problem reveals a major constraint in dealing with environmental issues in the Mediterranean, especially in terms of the paucity of data and the failure to make it available and to be open about what it shows. It is sometimes in the institutional interest (i.e. government) to hold back information on dangerous or hazardous situations for fear of public unrest. The 1997 European Union Directive on access to environmental information has broken this bureaucratic opacity and the Water Framework Directive (2000: see Box 21.1) will carry this process much further. It will also concentrate on identifying the catchment as the main avenue for water quality management.

Suggestions for further reading

McDonald and Kay (1988) is a valuable background source. De Walle *et al.* (1993) provide a masterly, but rather dated synthesis of Mediterranean pollution. The Mediterranean Hydrological Cycle Observing System (Med-HYCOS) website at http://www.medhycos.mpl.orstom.fr is a valuable source for data and background information on water quality and hydrology of the Mediterranean Sea. The Mediterranean Action Plan provides a wealth of information at http://www.unepmap.org/. Its development is discussed in detail by Haas (1990). MacKenzie (1998) gives a readable account of a threatened wetland.

Topics for discussion

1 Discuss the implications of the data shown in Table 17.1.
2 What are the main problems associated with irrigation use?
3 Describe and account for the main distribution patterns of pollution in the Mediterranean Sea.
4 Referring to Chapter 21, discuss the ways that nations and regions can control pollution.
5 What are the main changes in pollution problems in the Mediterranean Basin since 1945?

18 Water resources

Introduction

Water resources are a huge topic and a very important one, because water impinges on virtually every aspect of daily life. The topic can be broken up in many ways and the most common is into supply and demand. Another way is to attempt to account for the components of the water budget, like a balance sheet of income and out-goings. This is the most common construct when nations are trying to anticipate future water-resource problems. A further way is to think how water is used in the main economic sectors of the economy – agriculture, industry, tourism and domestic consumption.

Whichever of these approaches one takes, there are two other dimensions to water resources. The first is the scale at which we are operating. Problems that arise trans-nationally (across international borders) are quite different from those that occur within a small community. In a federal political system, such as Spain, there are two intermediate levels below the central government that have some political control over water resources: the autonomous regions and the river catchments. The autonomous regions are aggregates of the old provinces – Andalucia, for example, stretches from Almería to Huelva and Cadiz. Within these huge areas are the river basins or *Confederaciones Hidrograficas* (CH). Some of the largest of these, such as the Guadiana and the Jucar, fit within an autonomous region. Others, such as the Ebro, spread across several provinces. Yet other smaller basins are grouped together in one confederation, such as the *Aguas del Sur* (Southern Water Basins). This complex political subdivision of the surfaces into which water falls makes the development of water management a complex and difficult business. In the European Mediterranean nations, the very complex topography adds to these difficulties of defining the catchments.

This chapter deals with water resources from several of these approaches, using comparative examples from Spain, a developed European country and Tunisia, a developing North African country. Topics that are exceptionally problematic and particularly relevant in the Mediterranean environmental context are selected. We start with agriculture, usually (but not invariably) a mainstay of the economies of Mediterranean countries and from this we go on to the practices of dry farming and surface water management. These techniques are very old, but still widely practised

A brief overview of water management through time in Spain and Tunisia

It has been discovered that the economy of Bronze Age settlers in the drylands of south-east Spain were tuned to water resources in their 'catchment areas', the land that could be farmed from their settlements. There is evidence of irrigation at Antas in the province of Murcia dating to 4,000 years ago (Gilman and Thornes, 1985). Further discussion of the evolution of prehistoric irrigation systems is also discussed in the section on irrigation in Chapter 10. The history of water-resource management is one of technology brought in by colonizing powers. In Tunisia the Carthaginians introduced the 'hydraulic civilizations' to the area of Carthage and Cap Bon in about 800 BCE. With the dry farming of cereals (wheat and barley) Tunisia later became the granary of Rome. The Romans brought water-management technology to both North Africa and the Iberian peninsula, at first in the form of cisterns and a network of aqueducts that served villas, baths and small towns. In Tunisia the Roman works comprised:

*   barrages, in the centre and south of the country, at Kasserine, Sbeitla and Gabes, simple masonry structures where *wadis* emerged from the mountains;
*   reservoirs and cisterns, again in the centre and south of the country;
*   harnessing of springs, some of which still supply water today;
*   major barrages and distribution systems that made parts of the south habitable.

However, it was the later Arab invasions that transformed the use of water resources in North Africa and in eastern Andalucia, the driest part of Europe. Arabic conquests succeeded in the period 600–1500 CE in much of the southern European Mediterranean. The Arabs introduced technology such as the *norias*, large water wheels that pumped water up from deep wells, or *foggaras*, underground tunnels that capture groundwater. The engineering skills required to accomplish the latter were quite advanced, including the capacity to locate groundwater sources and then to carry out the works. In Granada province, Spain, *foggaras* were constructed under the bed of the Guadalfeo River. A tunnel 3 m high, built of stone with gaps in it, conducted water underground when flood waters infiltrated into the river bed.

In Tunisia, the next big change was brought about by French colonists after 1886. Their prime interest in irrigation was twofold: first, to improve agriculture and, second, later to stabilize and control nomadic herders, for example, in Houmt, Souk, Jerba and Gabes, in syndicates of water users. In the early days of the twentieth century, the link between irrigation and national wealth had been firmly established and the period of *la grande hydraulique* was initiated (see Box 18.1). As early as 1912, the engineer Coignet had recognized and planned the construction of some large barrages that did not come 'on tap' until the period 1930–1952. When the French left, there was a great national effort to expand irrigation agriculture. In Tunisia today, 83 per cent of active males work in agriculture and the links between irrigation and gross national product are direct and strong.

Water budgets

National planning

Most water plans start with a statement of natural resources before going on to consider how much is to be distributed and to where. It is on this kind of national budget that comparisons are made between countries. The major weakness in this approach is that it tends to hide very great regional disparities or differences. The simplest statement that can be made is the total resource, i.e. multiplying the average rainfall of an area by its area. So, for example, Spain receives a precipitation of $346 \, \text{km}^3 \, \text{a}^{-1}$, of which 68 per cent is lost by evapotranspiration, 8 per cent goes to groundwater recharge and 24 per cent to surface runoff. These data (abstracted from the official Government report, *El Libro Blanco de Espana*, Spanish Government, 1998: 55) illustrate the huge losses from evapotranspiration and imply that saving water could be better achieved by a small effect on evaporation rather than storing all the runoff in reservoirs.

Box 18.1 La grande hydraulique

In France, Italy and Spain '*la grand hydraulique*' philosophy (big dams and huge works) developed internally very early, but still on the backs of the technology of the earlier invaders. The reassertion of the control of Moorish Spain after the expulsion of the last Moors almost immediately led to the planning of large hydraulic works. The Imperial Canal of Aragón was the first to be initiated as early as 1564. Since that time, there has been a succession of water laws, major legal pronouncements to control access to, and use of, water. Water management is very advanced in Spain, as it is now in Tunisia.

Because of the central importance of irrigation to agricultural production and so to the gross national product in agriculturally based economies, the provision of water has become a major preoccupation. The first resort in water management is to provide more water storage. Water reservoirs and huge irrigation projects are prodigiously expensive and often take decades to plan and complete. They are often perceived as symbols of state virility and thought to enhance the status of the engineers who construct them. In recent years, however, the inherent fallacies of the philosophy have become apparent, around the world as well as in the Mediterranean. They may be environmentally undesirable, reducing flow in rivers downstream below ecologically desirable levels. Often they involve the loss of agricultural lands, villages, homesteads and invaluable archaeological sites.

They also can cause international tension. One example is the Ilisu Dam, whose construction threatens the River Tigris, a cradle of Mediterranean civilization, and Hasan Kyel in Turkey, one of the world's oldest cities. Iraq and Syria have claimed that they could lose 35 per cent of the flow of the Tigris (*The Times*, 19 February 2000). Some 25,000 people will be forced to move their homes from the banks of the river. Another example is the proposed Alqueva Dam on the River Guadiana in Portugal, which is almost as contentious. This would be the largest dam in the European Union, with a capacity of $4.2\,km^3$. It would flood Spanish land and affect coastal ecosystems, even though its economic feasibility has not been firmly established. When drought and water scarcity occur every few years, political pressure for large hydraulic reservoirs and water transfer schemes increases, so this philosophy is kept at the forefront of public opinion and construction of dams remains as an integral part of national water planning. Whilst most of the world is rejecting '*la grande hydraulique*', large-scale water transfer schemes are proposed in the Spanish National Water Plan (Llamas, 1997), notwithstanding the complex political, social and environmental problems that such works give rise to.

Trans-national transfers are minimal (imports of the order of 1.4 per cent, exports of the order of 0.7 per cent). The 24 per cent of surface runoff is, of course, the most accessible water for irrigation, so the exported 2.4 per cent becomes a little more significant.

The 'export' of water

The export of water by the River Guadiana to Portugal has become something of an issue between the two countries (Figure 18.1). The River Guadiana drains the great *Meseta* southeast of Madrid in the region of Castilla la Mancha and then passes into Portugal. This river sustains not only the irrigation and aquifer recharge for Castilla la Mancha, but also sustains the ornithologically important wetlands of the Tablas de Damiel near the town of Cuidad Real.

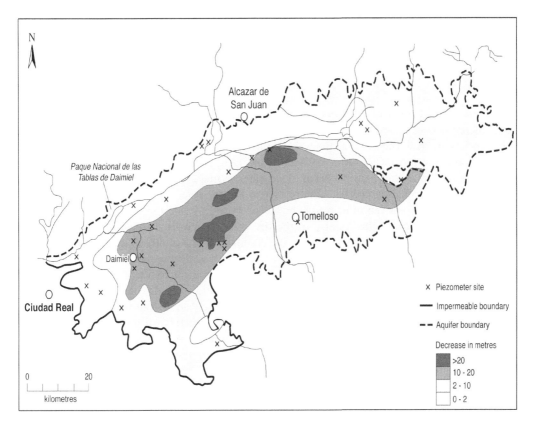

Figure 18.1 Map of Castilla la Mancha showing the Guadiana River and the Tablas de Damiel. The Guadiana flows into Portugal to irrigate areas in the lower Alentejo and the Algarve regions

The Spanish Water Law of 1993 advocates transferring water from the headwater of the River Guadiana towards the south-east and using its water to extend the irrigated lands of Murcia. This suggestion has led to serious concern in Portugal, which has built a large reservoir in the Lower Guadiana (Alqueva Dam). The Portuguese government estimated that the flow of the Guadiana would be reduced to 56 per cent of its current quantities. This fear was heightened when, in 1995, the River Guadiana almost ran dry during a severe drought. Other big rivers originating in Spain and then flowing into Portugal were also expected to suffer depletions (Duero: 20 per cent; Tagus: 15 per cent). The issue became a political football and led Llamas (1997) to conclude that 'The potential for water conflicts between Spain and Portugal have become artificially heated.' Agnew and Anderson (1992) have also noted the great potential for disruption of international security due to trans-boundary water transfers in the more arid parts of the Mediterranean region.

Regional variations

National figures also hide strong imbalances between region and river basins. Another serious problem with these general figures is that they are based on empirical rainfall records and subject to the inaccuracies of rainfall representation. This problem is especially acute at times of climate change if rainfalls are averaged over area and through time. In 1910, Spain had 150 rain gauges located mainly in urban centres in the northern wetter part of the country. By 1980 it had almost 800 in a fairly uniform distribution, so that much drier areas are now covered. Any assertions about the impact of climate change must take account of the change in rain-gauge cover, as Chappell and Agnew (2000) have demonstrated for the Sahel. This situation is probably also the case for Tunisia, where the distribution pattern of rain gauges has spread from the wetter north to the drier south.

Because of the paucity of data, regional comparisons are even more difficult to obtain, particularly where the river basins do not coincide with administrative regions, as is commonly the case. Nevertheless, in Spain the broad outlines do confirm that the areas of the country with least rainfall (the south-east provinces of Almería, Murcia and Jaen) have the most serious water shortages. This link is not so obvious as it may seem, since water shortages are produced almost as much by poor water management as by lack of rain. Herein lies the fallacy of the basic proposition that water resource problems can most readily be solved by inter-basin transfers, i.e. by taking water from wet upland parts of the country to the dry areas, by canals. It may be strategically and practically more sound to reduce demand at the destination. This approach is practically not difficult, but a great deal of money is tied up in the huge engineering schemes. The companies involved usually exert pressure on local and national government to adopt the grand schemes, especially the big dams at the centre of the new Spanish National Hydrological Plan. Greenpeace has calculated that, of the 200 new big dams projected in the Plan, at least sixty will have significant environmental effect. Aguilar (2000) argued in summary that: 'Civil engineers want to transform the remainder of our rivers and valleys into a network of waterpipes and taps. This is unacceptable for those who believe in the importance of the preservation of our nature.' Dr Aguilar is the Campaigns Director for Greenpeace, Spain.

Water transfers

Water transfers to 'balance up' national imbalances of supply and demand are always problematic. Yet it remains a favoured solution for national water-planning agencies and the new Spanish National Water Plan advocates it as a major approach to solving regional imbalances (Figure 18.2). In the same area, farmers do not like to see 'their' water going on to farmers elsewhere, especially since droughts are widespread. For example, over 400,000 people demonstrated in Zaragoza in October 2000 and a further 200,000 in Barcelona in February 2001 and 200,000 in Madrid in March 2001 against the National Hydrological Plan, which proposes the diversion of water from the River Ebro to the south-east of the country (WWF-Spain, 2001). In the receiving areas, expectations of increased productivity encourage extension of irrigation to areas that are quite unsuitable. In the case of the Tajo–Segura aqueduct, a 300 km canal takes water from the upper River Tajo to the extremely dry south-east in the provinces of Murcia and Almería, where it is used almost exclusively for irrigation of citrus trees and market garden cultivation. This canal was designed to transfer $1 \, \text{km}^3 \, \text{a}^{-1}$. It began to operate in 1980 but, until 1997, the average transfer had only been $0.3 \, \text{km}^3 \, \text{a}^{-1}$ and it is expected that this will reduce to 0.15 or $0.2 \, \text{km}^3 \, \text{a}^{-1}$ (Llamas, 1997).

The regional government of Castilla la Mancha has strongly opposed this transfer of water in law, especially for exceptionally dry years. The law regulating the Tajo–Segura aqueduct permits transfer only if there is a surplus in the upper Tajo. This law opens up great problems of what is meant by 'surplus', especially when irrigators expand their water needs (either by area or by more frequent irrigation) to match the supply. The regional government challenged the central government in the Supreme Court because of one transfer of $3.5 \, \text{km}^3$ to the Segura in the summer of 1994. Another problem is the need to sustain the ecologically acceptable

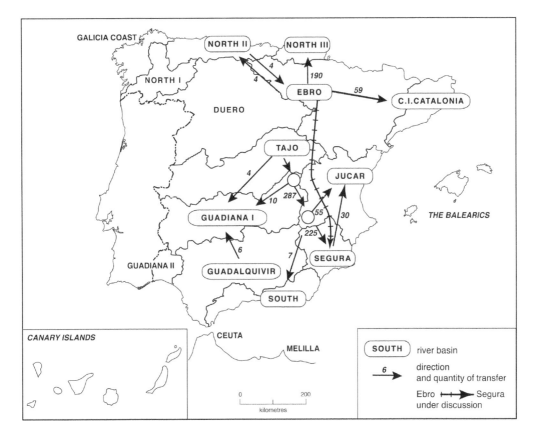

Figure 18.2 Actual and proposed water transfers in Spain according to *El Libro Blanco de Agua*. The Tajo–Segura aqueduct has been in operation since the mid-1980s

minimum flow downstream from the point of take-off. Since this water would normally go to the Tablas de Damiel wetlands and then on to Portugal, the definition of the safe ecological minimum needed to preserve the fauna and flora of the middle and lower Guadiana is a highly contentious question on which research still needs to be done and to which there are still no firm answers. As Llamas points out, the take-off has lowered the levels of two headwater reservoirs and the villages along the banks have lost their income from the recreation activities on the reservoirs. Other water-transfer schemes have been successfully employed elsewhere in the Mediterranean (Box 18.2).

Irrigation and degradation

The province of Murcia is a maze of irrigation canals and facilities, because irrigation there is as old as civilization itself. The *vegas* (irrigated gardens) spread out around the towns of Murcia, Lorca and Totana, form large areas of intensive production of oranges, lemons, tomatoes and beans, all under irrigation. As the construction of the Tajo–Segura aqueduct began and as it moved forward in the south-east as lines on planning maps, there was a boom in properties that might receive water and this was accompanied by wild speculation. About the same time, a subsidy was made available for 'land improvements'. As a consequence, large areas of the 'soft' marl bedrock were

Box 18.2 The Southern Conveyor Project, Cyprus

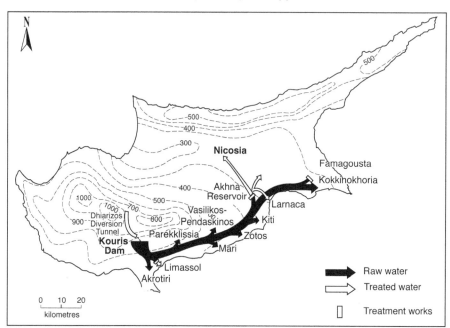

Figure 18.3 The Southern Conveyor Project for water supplies in Cyprus (Dashed lines are average annual rainfall)

Of the 4,600 million m³ of rain that fall during an average year on the island of Cyprus, only 600 million m³ are available as runoff, with around 80 per cent lost directly to evaporation (Lytras, 1993). There are strong orographic gradients on the rainfall, so that over 1,000 mm fall annually on average on the Troodos mountains, compared to 500 mm or less on the coastal areas, where population is concentrated (André and Robert, 1985). Thus, there is a significant need to develop dams and water-redistribution schemes for both domestic and agricultural water supplied in the light of significant population increases. The Southern Conveyor Project (Figure 18.3) was started in the early 1980s with the aim of increasing the irrigable area by 15,000 ha by the turn of the century, and providing domestic supplies to the cities of Limassol, Nicosia and Larnaca. The Kouris Dam to the north-west of Limassol is the ultimate supply, with a capacity of storing 115 million m³ of runoff from the south-central Troodos. The main pipeline is 110 km long and there are numerous treatment plants and irrigation distribution systems. The total cost of these constructions was around £180 million. Successful application of this type of scheme may depend on the consolidation of land holdings (King and Burton, 1989) to maximize efficiency and minimize the likelihood of disputes (see also Chapter 15).

bulldozed flat and great storage tanks were created. The land works were sometimes three to four years ahead of the arrival of the aqueduct and its irrigation waters. In many cases, this great speculative over-extension created a demand that could never hope to be filled, as the volumes transferred were fought over and progressively reduced. Without water (and without vegetation cover), the 'reclaimed land' was largely abandoned and underwent severe erosion. Freshly cleared, uncultivated, unprotected marls are among the most erodible soils in Europe and, with heavy autumn storms, they were quickly turned into some of the most spectacular badlands of Europe. The erosion rates were so high that reservoirs were quickly silted, terraces degraded beyond use and soils were lost for ever. The badlands are a landscape of greed and very little can now be done to recover the situation. As if this were not bad enough, the Guadalentín valley in Murcia has some of the maximum record rates of groundwater lowering in Europe (López Bermúdez, 1998).

Water recycling

For many years it has become increasingly obvious that recycled water can be used to offset the pressure of water shortage. Londoners, taking a glass of water from their domestic water tap accept that, on average, the same water has passed through five other individuals, such is the acuteness of the city's water supply problem. Irrigation water may also repeatedly be used. If too much is applied, it percolates to groundwater, can enter the stream system and then be drawn off for irrigation again downstream. In each application, salt is picked up and the river becomes more and more saline downstream until it may eventually become unusable for irrigation. If the excess water also drains through highly saline soils, its salt concentration can become very high. Elsewhere treated urban waste water is used to supplement irrigation applications under strict World Health Organization quality control standards (Box 18.3).

Use of computer-based water-balance models

In attempting to assess water needs in the drylands of the south-east of Spain, because of the lack of data, López Rodriguez (1991) had to use a computer-based water-balance model to estimate how soil water was distributed. This model takes account of the soils and vegetation in different areas, to produce a water budget. The results for Almería and Cordoba are shown in Figure 18.4 and, according to López Rodriguez, mainly represent the contrast in soil types between the two areas, because the model used for estimation is very sensitive to infiltration characteristics of the soils. Cordoba has deeper soils in lowland river valleys, while Almería has thin, skeletal soils, mainly on mountain slopes. Almería is generally much drier than Cordoba and the summers are exceptionally so with, on average, five to seven months per year without rain, compared to three to five months in Cordoba. Moreover, while Almería has one 'rainy' period in the year (March to April), Cordoba has two (autumn and spring). These general statistics on rainfall, evapotranspiration and so on reflect not only seasonal variability in rainfall and the vegetation cover, but also strong contrasts in rainfall from year to year. One of the strong contrasts in Figure 18.4 is the very low proportion of recharge to groundwater (VR) in Almería soils compared with Cordoba and the higher modelled proportions of the volume runoff (VE) for Almería compared with Cordoba. These differences show not only the differences in soils and seasonality, but also reflect the more intense storms that occur in Almería.

In summary, although it is generally true that drier areas have bigger (and different) water-resource problems, the main controls on the partition of water into different components are soils, land use, seasonality and inter-annual variability of rainfall. Of these the most important is land use, the factor most readily affected by human activities.

Groundwater

Another major human control is in the extraction of water for irrigation from underground and this has been a major source of contention in most Mediterranean countries. This problem

Box 18.3 Water recycling

The World Health Organization has established a system of classification for waste water that takes into account its quality, especially with respect to pathogens (bacteria harmful to humans). Many experiments have shown that fodder crops irrigated with treated sewage water of class A (the best) and then fed to cattle do not result in harm to the animals, nor are pathogens carried through to humans in the meat. Experiments on the use of waste water in Cyprus for sudax (sorghum), alfalfa and corn found that the yields of all crops were always higher when irrigated with sewage water (three times higher for fodder, corn and vegetables) compared with conventional bore-hole water. It is usually applied in drip irrigation which is safer for the operatives and better for the plants. After the plants had dried out, there was no indication of contamination. Table 18.1 shows the results obtained in another Cypriot experiment (Choukr-Allah 1996). An additional important result of this experiment is that there is no evidence of modification to the soils in the experimental plots, even after three years of irrigation. Production in grams per plot is shown in Table 18.1.

Table 18.1 Production in grammes per plot for different crop types using different water sources

Crop	Groundwater	Waste water
Tomato	2,450	3,500
Potato	3,151	3,350
Courgette	720	1,040
'Tunnel plastic' tomatoes	2,650	3,700

Source: After Choukr-Allah (1996).

A consistent complaint of Spanish farmers is that valuable water for crops is used to maintain golf courses for the tourist industry and the number of courses has grown dramatically in recent years. Choukr-Allah's work indicates that fine lawn turf can be grown with recycled sewage water or by the application of composted sewage, which has a very high nutrient content and low toxicity.

arises from the idea of water as a 'free good' that is there to be obtained by simply digging a well. Progressively, nations have legislated to make groundwater a state property.

Water that infiltrates into the soil, if it is in sufficient amounts, moves downwards to percolate into the rocks beneath at a rate that is proportional to the amount of water in the soil and the ease with which water can move through the underlying rocks. Truly impermeable rocks are those through which water cannot move, even though they may hold a lot of water in the pore spaces. Clays are generally impermeable, but porous. When rocks take in or release water, they expand or contract in a way that can cause surface subsidence. The main effects of pumping out too much water are therefore:

- The water table may fall. This is a very dramatic signal that the withdrawal is beyond the 'safe yield'. For large aquifers, the quantity of water extracted that will produce a measurable fall in the water table is enormous. Conversely, small aquifers are very sensitive. Care

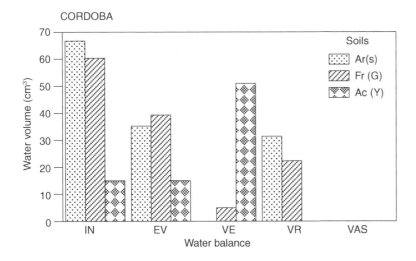

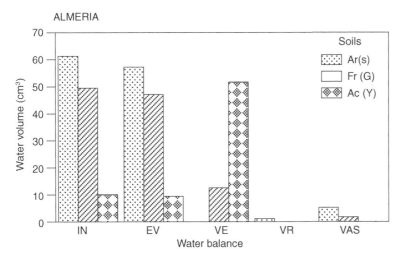

Figure 18.4 Modelled comparative water balances for Cordoba and Almería for different soil types (after López Rodriguez, 1991). The estimated water-balance components are: IN, infiltration; EV, potential evapotranspiration; VE, overland flow; VR, groundwater recharge and VAS, soil-storage volume – all measured as annual averages. Soil types are: Ar(S), Sinai sand; Fr(G), Gilat loam and Ac(Y), Yolo clay.

has therefore to be taken with figures on rates of fall (usually expressed in metres per year), otherwise it is easy to exaggerate the effects of exploitation.

• The water is obtained from greater depth. In southern Tunisia, near Kibili, water comes up under artesian pressure from 2,300 m and is so hot that it has to be cooled for application to irrigation during the day and can be used to heat greenhouses at night. In other Maghrebian states, prehistoric groundwater deposited in times when rainfall was higher than today, is being exploited. By withdrawing groundwater, lakes and rivers near the surface lose water to the bedrock. This situation happened in the Tablas de Damiel, in central Spain. This important ecological site gradually drained and became a peat bog that eventually ignited by internal combustion of the peat. There has been a prolonged debate about whether the wetlands were lost either by over-use of the rivers and groundwater through extraction for irrigation, or by changing land use (and hence the water balance) in the rush to cultivation since the 1950s, or if it was due to the falling rainfall amounts as a result of climate changes affecting the western Mediterranean (see Chapter 20). In Castilla la Mancha, in one particular aquifer (aquifer 23 in Figure 5.9), all of these causes have contributed to falls in the water level (Burke, 1998) leading to reductions in water level by 20 m in some areas since 1970. In a thorough scientific investigation, Burke (1999) showed that this fall could only be attributed to the use of the water for other purposes and that climate change could not be blamed for the decline.

- Sea water may intrude into the aquifer. Near the coast, river-valley fills and alluvial fan deposits are usually full of water. This water is replenished by infiltration either from flood flows over the river bed or from irrigation returns. The alluvial plain of Capoterra in southern Sardinia, near the town of Cagliari, is a typical example that has been studied in great detail. Sea water intrudes into the gravels as fresh water is extracted for agricultural and industrial use from about 300 wells, scattered over an area of about 60 km^2. In the central part of the plain, salinity has risen in both the confined and unconfined aquifers (Barrocou *et al.*, 1998). The deterioration of water quality renders it unusable for agriculture, industrial or domestic purposes.

Salinization

In some areas, salinization of coastal deltas occurs because upstream reservoirs are constructed and this reduces the freshwater flow that prevents saline intrusion. This situation most often occurs where streams coming from mountains have cut their beds deeply below sea level and these have been filled with sediments, often gravels or alluvial silts that form important aquifers for agriculture and domestic water supplies. The River Andarax in Almería province in Spain is a good example. Flows have been reduced, wells near the mouth have progressively turned salty and this effect is moving inland up the valley of the River Andarax.

In Greece, the complicated positive feedbacks leading to salinization have been explored in the Argolid area by Allen and his colleagues (1999), as part of the EU ARCHAEOMEDES Project, by a team led by Professor Poulovassilis at the Agricultural University of Athens, specializing in hydrology. The problems of the Argolid are not particularly driven by climate change, but rather by the actions of farmers turning increasingly to irrigated agriculture (Box 18.4).

Another form of salinization occurs when, usually as a result of too much irrigation, the water table rises to the ground surface, bringing up salt with it. As the surface water evaporates, salt is left behind. Several areas near Zaragoza, a very dry area in northern Spain, have been so affected, as in the Rambla Violada and the area between Zaragoza and Lerida, where the plateau above the River Ebro is underlain by salt-rich beds. Here, *playas* with evaporites (evaporated salts) are a major feature. As the Ebro passes through this section of its course, it is fed by salt-bearing rivers and streams and hundreds of springs bringing in very saline groundwaters. When salinity is caused by over-irrigation, either the water table has to be lowered by drainage or (as in west Australia) by increasing the vegetation cover and therefore enhancing the evaporation.

According to the European Environment Agency (EEA, 1997), salinization affects 25 per cent of the irrigated lands of the European Mediterranean (Szaboles, 1990) but they believe that climate change could lead to a further 40 to 70 per cent reduction in renewable resource. Notwithstanding the problems of predicting future Mediterranean rainfall and temperature, this can only be regarded as a 'Doomsday' forecast, which should be treated carefully.

Box 18.4 Salinization in the Argolid area of Greece

Farmers in the Argolid have turned increasingly to irrigated agriculture. This irrigation has led to the lowering of the pressures in the aquifers, the penetration of sea water and the transfer of salt to the land, reducing yields and producing a demand for an increase in the irrigated land and hence more need to pump more water for irrigation (Figure 18.5). In this way, a vicious circle is produced: the more that is pumped, the more that has to be pumped. Canals have been built to bring water from the Tripoli Mountains and more and larger canals are being constructed to allow the continued and expanding cultivation of citrus fruit trees, especially in the lower parts of the catchment. With the construction of new larger canals, the problem of their routes and the dis-

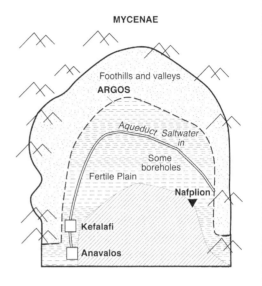

Figure 18.5 Water problems in the Argolid area, Greece

tribution of new water resources has provoked much controversy (Allen *et al.*, 1999: 428). The system of irrigation in use is also the very wasteful 'flood irrigation' method, in which water is left to stand on the surface in shallow basins in the soil, while it infiltrates (and evaporates!).

The problem now is what can be done and what policies for action should be adopted (van der Leeuw *et al.*, 1998). Allen and his group have developed a computer model to accommodate as many of the factors as possible and a set of objectives that needed to be satisfied (e.g. once the remedial project has been completed, the farming activity has to be as attractive as alternative urban activities; otherwise the farmers will move to cities to make a living).

The computer model shows how well-intentioned policies at one level of the system can have quite a negative effect at another. In this case, financial support has gradually switched from the dry farming production of olives to the irrigated production of citrus fruit and resulted in a need to pump water up from wells, bore-holes and canals. Their model considered the chain effects of pumping underground water as the area of irrigated crops increased. The model, including the farmers' decisions, showed how the system led to saltwater intrusion after 1960 and how it led to the spiral of growth, demand increase, and increased salinization as described above. Satisfied that the model is performing close to reality, they simulated change until 2020 with conditions and behaviour as they are today. The resulting outcome was the collapse of citrus cultivation, leaving soil with extremely high salt concentrations, reducing yields of the replacement crops they tried. Gradually the system returned to olives and agricultural revenue fell dramatically never to recover. They experimented with the effects of taxes on water supply and were able to estimate the tax on water required to stave off the collapse by using less water and found that it needed a tax of 150 drachmas m^{-3}.

Dry farming

The main objective of dry farming is to use natural rainfall to grow crops. This can only be achieved if the crops can tolerate very dry conditions and if measures are taken to capture and keep any rainfall in the soil itself. Some soils (e.g. clays) retain the water very well. Other soils are 'droughty' because the rain easily drains through them according to the soil texture and structure. If the surface has a seal on it, a thin crust is formed, usually made of fine elements, which reduces evaporation of the water from the soil surface and encourages runoff. Organic matter improves the water-holding capacity of soils, so sewage or kitchen waste applied to the soil may increase its capacity for producing a good yield of crop (see Box 18.2). In the Mediterranean, the traditional dry farming crops are cereals, such as the hard wheat grown in southern Italy for pasta production. Barley is even more tolerant of drought than wheat. In Portugal the government introduced a 'wheat campaign' to offset food shortage. This supported farmers by means of subsidies for the purchase of seeds, fertilizers and machinery. However, this approach had a serious impact on soil degradation that was officially recognized by the 1950s. The difficulties and hardships of this period in the Alentejo are beautifully described in *A Portuguese Rural Society* (Cutileiro, 1971).

Other major dry farming crops of the Mediterranean are tree crops such as figs, almonds, carobs and olives as well as vines. Cereals, olives and vines are all well adapted to the Mediterranean climate regime and dominate the farmer's year. Wet season cereals (wheat and barley) are usually planted after the autumn rains and need labour inputs for collecting, threshing and processing in June, July and mid-August. Then grapes are collected and later olives (Margaris *et al.*, 1998). One of the major changes in cereal production has been the abandonment of cereal cultivation on terraces. Cereals have otherwise increased greatly in importance as a result of higher yields (from an average of about $0.4\,t\,ha^{-1}$ in the 1960s to up to about $4\,t\,ha^{-1}$ today: see also Chapter 15), mainly as a result of tractors and other machinery and human endeavour. Since sunflower and maize oil (both derived from dry farming crops) have grown rapidly, the production of olive oil has declined partly because it is impossible to mechanize. In Italy, olive oil fell in its market share from 44 per cent to 33 per cent in the period 1983 to 1989. Soya cultivation is also pushing olives further into decline. The cultivation of dried fruits, figs, almonds and plums has also decreased.

A widely used ancient technology is water harvesting. The idea is to concentrate meagre rainfall to a point, such as a well or cistern, where it can be used later. This cistern may be an area of one hectare or even a hillside of several hectares. Water is discouraged from infiltrating by applying clay to the surface to create an impermeable crust, or by clearing the vegetation to encourage runoff before evaporation can create loss. The ground may be sloping towards the collection point, or small ditches cut to lead the water to the cistern. The Romans built large circular masonry cisterns and stone-lined channels leading to them, for example, in the Ebro valley at Bujaraloz near Lerida (see Chapters 10 and 12). The ancient practice of making low walls or bunds that run across hillsides oblique to overland flow can be seen in present-day Israel and as far west as the Spanish coast near Nerja, presumably introduced at the time of the Moorish invasions. Water harvesting is a small-scale activity, but is still sufficient to provide crops today where none would otherwise be possible. It is practised in southern Tunisia around Tatouine, on the desert edge.

Irrigation

As the population rises rapidly in rural areas of the Mediterranean, dry farming has been increasingly abandoned. It is too risky and the yields are generally too low. The expansion of irrigation, industry and tourism have been the linchpin of a strategy to keep emigration rates down. Today irrigation is one of the most important regional variables in the demand for water within countries. Nevertheless, there has been a dramatic drop in the agricultural sector in Spain since 1984 and this is expected to continue until 2010. The growth in the area irrigated began to level out in the 1990s and by this time national agricultural production accounted for only 5 per cent of Spain's gross national product.

By contrast, Tunisia moved into a phase of rapid irrigation after the end of the French rule, so that in the years between 1981 and 1986 it invested 1.4 times the amount invested in the previous twenty years. In Tunisia, only since 1975 has the demand for industrial water begun to increase seriously and 'explode', but so too has irrigation and hydraulics regularly absorbs an important part of agricultural investment. In 1962 to 1971 it was 29 per cent, in 1982 to 1986, it was 39 per cent. This investment is predominantly publicly financed (87 per cent – 96 per cent since 1962) mainly for large hydraulic works funded by international agencies such as the World Bank and the International Monetary Fund, but also from foreign aid from the Republic of Germany, Canada, Italy, USSR, China, Kuwait and Saudi Arabia. In the great projects, 39 per cent is for irrigation against 50 per cent for drinking water and 11 per cent for flood protection.

Consumptive use

Throughout the Mediterranean, there has been a rapid growth in domestic consumption since the 1970s, more spectacularly in North Africa than in southern Europe. In Spain, there is a decrease in the short and medium term, due to the fall in birth rates. Although tourism has also experienced rapid growth, water use for tourism is not very relevant especially when compared with the economic activity it generates. In 1996, Spain had 61.8 million tourists, 19.6 per cent in the summer months with an average stay of 12.6 days. Water is not yet a limiting factor, but it could be in the future. The main problem is sewage and water disposal. As the growth increases, the bigger impact will be on the catchments near the coast.

A case study from southern Spain

The first Water Law was in 1866 and, by 1902, the first National Plan for Irrigation was announced. The Hydrographic Confederations (CH), the main statutory bodies controlling water, were created in 1926, though each was created by separate decree. Surprisingly, the CH for Southern Basins, an aggregate of small basins draining the mountains to the coast in one of the driest parts of Spain, was not established until 1960. This CH stretches from Tarifa near Gibraltar as far as Aguilas in the east, where it abuts onto the CH of the River Segura in Murcia (see Figure 18.6). In 1978 the Autonomous Communities were formed and they took responsibility for large (hydraulic) projects, canals and irrigation works. The Water Law of 1985 included aquifers and control over groundwater for the first time, even though it was some time before the law was implemented.

The CH for Southern Basins illustrates all the problems of the Mediterranean. Rainfall and runoff are strongly seasonal, rainfall is generally low throughout, but varies enormously spatially and in time. Parts of the province of Almería have less than 200 mm per year, but the mountains in the north of the province of Murcia experience 1,000 to 1,400 mm. Mean temperatures range from 10°C in the Sierra Nevada to 19°C just south-west of Malaga. In this CH, there are 145,200 hectares of irrigated land, of which 26,000 are state irrigation projects (18 per cent). Irrigation is the main demand for water (1,210 Hm³) with urban demand in second place (220 Hm³). The industrial demand is smaller, but a greater pollution problem is generated. Only

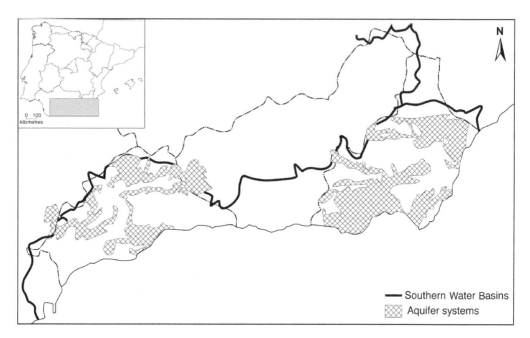

Figure 18.6 Map of the Aguas del Sur (Southern Water Basins) and the aquifer systems (shaded)

40 Hm³ a⁻¹ is needed for agriculture. These figures vary across Andalucia. In the Gibraltar region, industrial need is 37 per cent, urban 38 per cent and agriculture 25 per cent and total consumption is 39 Hm³ a⁻¹.

The most severe water deficits are in Almería province, especially in the agriculturally planned area of Dalias, where the deficit is in the order of 50–100 Hm³ a⁻¹, mainly needed for irrigation. The excesses are mainly in the extreme west. This distribution poses, together with the seasonal demand due to tourism, enormous management problems. The total capacity of the dams is 910 Hm³ in the entire area. The groundwater recharge is 1.16 Hm³ and the withdrawal is 435 Hm³ a⁻¹. Nevertheless, it is a land of great hydraulic works with huge reservoirs, Guadalhorce (82.6 Hm³), Beninar (70 Hm³) and Cuevas de Almaricera (169 Hm³) in the driest region of Europe.

Rainfall in the CH for Southern Basins is measured by radar and flow is monitored fully by automatic gauging stations. The technology is second to none in Europe, and is needed. The droughts of 1981 to 1985 had a severity that can be expected only once every 500 years. Equally, it is an area of torrential rainfalls, especially in autumn, when depressions coming from the Mediterranean with high water loads are forced up into the coastal mountains with several recorded rainfalls of 250–300 mm in a two-day period. This rain falls onto dry hillslopes where the vegetation has been dried out and the soils sealed by three months of hot dry summer. Runoff is large and swift, leading to severe flooding and mudflows in lowland towns. The flood of October 1973 devastated an area from Malaga to Benidorm – almost entirely in the CH of Southern Basins. The CH has embarked on a policy of afforestation over large areas of the basin as a methodology for flood and soil erosion protection, but it is too early yet to know the extent to which this has been successful.

Summary

Water resources are a major environmental issue in the Mediterranean, because economic growth is coupled to irrigation agriculture. However, it is incorrect to assume that the driest areas have the greatest water shortages. Water shortages are made worse by strong contrasts in the spatial excesses and deficits and poor management of the existing resources. Typically, 40 per cent of the water leaving a reservoir and intended for a plant will actually reach the plant roots. Farmers do not comply with existing legislation and water is conceived of, and treated as, a free commodity. Even where tariffs (charges for water) are applied, they are usually far less than the real cost of water. For all these reasons, water resources are not a technical supply-side question, but rather a behavioural demand-side question. Without doubt, water management is a problem of environmental management.

Suggestions for further reading

Agnew and Anderson (1992) and Beaumont (1993) both provide excellent introductions to water-resource issues in arid regions. Mairota *et al.* (1998) provide specific examples and details in the Mediterranean context.

Topics for discussion

1 Why are average water balance figures of little value in comparing water resources between countries?
2 Consider the assertion that, in Mediterranean countries, water resources are more an issue of the human rather than the natural environment.
3 Outline how water management has changed between prehistoric times and today, with reference to the main controlling factors.
4 What problems might future climate change pose for Mediterranean water resources?

19 The desertification problem

Desertification: myth or reality?

Desertification is a widely used but poorly understood word, about which there has been confusion, uncertainty and even obfuscation. This confusion has led to international difficulties in agreeing the causes, extent and remedial efforts needed to offset its impacts, so much so that, even though it is included in the World Bank's top ten global environment problems, some people question whether it actually exists as a phenomenon, especially in southern Europe.

The main sources of this difficulty arise from the following:

- Poor understanding of the phenomenon. There is still a tendency to believe in the old idea that the Sahara is advancing outwards, overwhelming adjacent areas with sand dunes. Some authors have gone so far as to suggest that the Chott el Djerid in southern Tunisia was not only protecting the north of Tunisia from the advancing desert, but also Paris and London! Algeria projected a 'green belt' of trees to protect the country from the outward spread of the Sahara, but the project was abandoned when it was found that the prevailing winds were blowing sand in the opposite direction.
- As the acceptance of desertification is coupled with the concept of international funding for remedial action, it has become a political football in the north–south power struggle. It gave the politicians the opportunity to claim that human suffering, misery and poverty in Africa were the essential outcome of colonial policy that resulted in over-reliance on, and over-stretching of, the meagre resources of the Sahelian zone. Thomas and Middleton (1994) claim too that the United Nations essentially failed in its efforts to address the crisis through a combination of exaggeration and bureaucratic ineptness and through setting unachievable targets.
- The problem is multi-disciplinary, involving the physics of weather, the economics of social science and the ecology of species. Moreover, the environment is spatially very variable, so that behaviour at one point is very different from that at another.

Desertification is an extremely complex phenomenon because:

- It is deeply rooted in history (see Chapter 10).
- It is partly brought on by natural physical causes, including climatic marginality and temporal variability, as well as human causes including rapacious use of soil and biotic resources.
- Many of the phenomena involve positive feedbacks (self-reinforcing tendencies) and thresholds that, when crossed, can lead to disaster.

Therefore, almost any definition will prove inadequate.

In 1983, Glantz and Orlovsky identified over one hundred definitions and these have been extensively discussed by Mainguet (1991). Without wishing to dismiss the complexity outlined above, we adopt the official definition of UNEP (1994) which sees desertification as 'Land degradation in arid, semi-arid and dry subhumid areas, resulting from various factors, including climatic variation and human activities.' 'Land' includes soil and local water resources, the land surface and vegetation. Recalling that soil problems include erosion, the loss of productivity for agriculture, and salinization as a result of over-irrigation practices, this statement provides a working definition against which to check claims that Mediterranean countries are indeed experiencing desertification and therefore should receive international financial support to mitigate the problems, as in the developing regions of Africa, India and Asia.

Accepting that the *impacts* of desertification include accelerated soil erosion, increased flooding and reduced agricultural productivity, as well as reservoir siltation, these phenomena can be regarded as *indicators* of desertification and help us to avoid the trap described by Mainguet (1991: 16): 'The word desertification, created four decades ago, became a trap which ambushed scientists, planners, donor countries, governments of the affected countries and the mass media.'

As Thomas and Middleton (1994) point out, the United Nations 1977 Nairobi Conference and the subsequent Action Plan:

- put desertification on the agenda of developing countries;
- gave an opportunity for the incorporation of anti-desertification measures within the overall framework of development. National plans were, and remain, one of the central catalytic activities of UNEP.

We should be under no illusion that desertification means big money and that inclusion on lists of seriously affected countries leads to the prospect of considerable international and regional financial benefits. While it is hard to argue that these outweigh the major disadvantages, they may partly account for the universal claims about the nature and real extent of the problem in countries at risk. Between 1977 and 1984, there was an increase of 3,500,000 hectares worldwide of desertification, due to redefinition of the meaning (Thomas and Middleton, 1994). Furthermore, Dregne (1987) states that the maps on which these estimates are based are founded on 'little data and a lot of opinion'.

Mediterranean desertification

The Mediterranean countries experience long dry seasons that coincide with high temperature to produce great hydrological stress. Because rainfall is highly uncertain from year to year, the conditions for plant growth and survival are generally marginal throughout, as indicated by the ratio of precipitation to potential evapotranspiration, and emphasized by runs of dry and runs of wet years (Thornes, 1993a). By contrast, desiccation is a process of long-term reduction in moisture availability, resulting from a dry period at the scale of decades. In the European Mediterranean, it is observed that:

- There has been a notable trend to less rainfall since 1960, particularly in the western part of the basin (Italy and Spain). There has been a decreasing trend for almost all regions in all seasons (Palutikof *et al.*, 1999).
- There is a marked Mediterranean oscillation in which positive anomalies in pressure in the west are matched by negative anomalies in the east, with a fulcrum line through Greece (see Chapter 3). This oscillation means that, superimposed on the longer-term trends,

there are periodic switchings of pressure that bring more or less rainfall (Conte and Giuffrida, 1991).

• The vagaries of Mediterranean climate are firmly coupled to the global circulation,
  through coupling with the North Atlantic oscillations (the fluctuations in the pressure gradient between the Azores and Iceland), according to Palutikof *et al.* (1999), so there is
  reason to couple the Mediterranean rainfall incidence more directly with the global circulation and the changes expected in it in the future.

Because vegetation is so central to desertification through its impacts on soils, runoff, flooding
and agriculture, the ratio of precipitation to potential evapotranspiration is more or less
accepted as an appropriate indicator of the desertification status. UNEP (1992) classifies
regimes according to this ratio as follows:

Hyper-arid	<0.05
Arid	0.05–<0.2
Semi-arid	0.2–<0.5
Dry sub-humid	0.5–<0.65

On this basis, nearly all of the Iberian Peninsula appears as semi-arid or dry sub-humid and virtually all the Mediterranean lands of North Africa fall into the arid class. The Israeli Levant and
Turkey are mainly semi-arid as are southern Italy and the islands of Corsica and Sardinia.

On the more localized scale, relief and proximity to the sea notably influence the outcome
of this classification. Thus for example in Spain, almost the entire country falls in categories
with a ratio of <0.5 and is therefore in the semi-arid class with only the coastal strip of Murcia

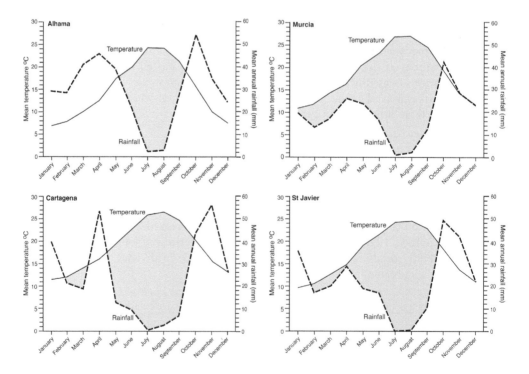

Figure 19.1 Aridity in the Iberian Peninsula

province in the arid class. At the very local scale (within the province of Murcia), it is the mountain relief that dictates the areas that are in the various categories, the most critical areas being the Guadalentín Valley in the south-west and around the Mar Menor in the east (Figure 19.1).

Notwithstanding the fact that much of the Mediterranean is classified as desertified on the basis of climatic conditions, this condition has been exacerbated by four millennia of history. Thornes and Gilman (1983) were able to show that, in south-east Spain, badland morphology was already developed by Bronze Age times (*c.*1800 BCE) and desertification could not therefore be attributed simply to deforestation and overgrazing as is usually claimed. As late as the seventeenth century, it was claimed that Iberia was so well wooded that a squirrel could cross the Peninsula by jumping from branch to branch.

Delano-Smith (personal communication 1996) concludes, after a careful and thorough review of the evidence, that the notion of a 'pristine' or 'primaeval' vegetation (unaffected by human activity) would not hold in Italy by the end of the Bronze Age (see also Chapter 4). Indeed, she questions whether a fully-developed climax woodland, unmolested by agricultural activity, ever existed in those districts where farming was established very early (see Chapter 10). She also points out that generalizations applicable to the country as a whole are likely to be misleading, as elsewhere in the Mediterranean. However, a bird's eye view of Italy at about the start of the eighth millennium BCE would have revealed a more or less continuous forest.

The three chief agencies of deforestation – fire, felling and livestock grazing – were present very early in Italy. Farming from a more or less permanent settlement site has done more to alter the world's vegetation than any other single factor. Where there is no sign of Neolithic activity, mixed oak woodland seems to have remained undisturbed until the second millennium BCE and the Bronze Age in the Italian Peninsula, much as it probably was in Spain (Gilman and Thornes, 1985). Mining, even in prehistoric times, was a cause of deforestation in many parts of the Mediterranean, particularly where charcoal was a source of fuel (see Chapter 13).

Delano-Smith documents many cases that support a thesis of a sequence of deforestation, erosion and siltation as a process and there is much geomorphological evidence for the view that vegetation and erosion are causally linked (Thornes, 1990; Wainwright, 2000; Chapters 10 and 11). However, we must be careful not to accept it as axiomatic that all evidence of erosion and sedimentation in the Holocene implies vegetation removal as the primary and necessary cause, just as we cannot always infer that climate change is the only explanation. As seen in Chapter 6, though, it is important to recognize the different sensitivities of climatic and land-use change on the erosion process. Small shifts in the mean climate conditions are likely to have a much smaller impact than either land-use change or extreme climatic events.

In areas of marginal climate and soils, hillslopes are very sensitive to changes that produce desertification. After a long period of agricultural stability in Iberia, the changes that set in during the 1950s almost certainly created the chemical, physical and biological degradation of land in the sense of the UNEP (1992) definition. Mechanization of agriculture, improved pumping systems for irrigation and the prevalence of highly production-orientated agriculture all led to desertification and eventually to desertion of rural areas in the Mediterranean. In Portugal, the sustained pressure of the wheat campaign that started in the nineteenth century, culminated in severe degradation by the mid-twentieth century as even the common lands (*baldios*) were first partitioned and then turned over to wheat. The impact of these changes, including the sharp fall in yields between 1959 and 1965, has been documented by Cutileiro (1971) in his ethnographical study of a town in Alentejo, and researched by Roxo (1993) in her doctoral thesis on the interaction between land and society in the Alentejo.

Both empirical (Kosmas *et al.*, 1997) and theoretical (Kirkby, in press) estimations have confirmed Langbein and Schumm's assessment (1958) that erosion (and therefore sediment yield) is highest in areas with a mean annual rainfall of about 250 to 400 mm. The Mediterranean regions, with their strong seasonality in climate, high rainfall intensities and poor vegetation cover under natural conditions meet these estimations. Where vegetation has been removed in the wetter Mediterranean, the effects on erosion are the same, i.e. the erosion peaks. With less rainfall, there is insufficient runoff to produce erosion. With more rainfall, the vegetation is able to grow and so offset erosion.

The processes of desertification are well known and principal among these is land degradation through soil erosion. This process is treated at length in Chapter 6. The essential control of erosion is land cover and there is a well-established 'rule of thumb' to the effect that as vegetation cover falls below 30 per cent, there is a very steep rise in the amount of erosion (relative to the bare soil value). In large parts of the Mediterranean, the cover is at or about 30 per cent. Small decreases, for example, by grazing or harvesting, will therefore produce substantial rises in erosion. This general rule applies to water erosion but it is also well known that vegetation cover has an important effect in reducing wind erosion. However, we consider wind erosion to be less important than water erosion except in the North African states of the Mediterranean, such as Tunisia, where it can assume great importance. We have discussed the controls of vegetation cover in Chapter 4. A key concern for the future is that warmer, drier conditions will lead to poorer vegetation cover and this, in turn, to higher levels of erosion. This erosion will threaten soil productivity and agricultural sustainability, as discussed in Chapter 21.

The role of the European Community

Spain and Italy both participated in the UN Nairobi Conference (1975) and, as a consequence, Spain quickly established its LUCDEME project, the battle against desertification in Mediterranean environments. In Italy, the Sardinian government assumed legislative responsibilities for land degradation in 1994, building on the law of 1766 that already recognized the need to strike a balance between 'the needs of the population and the preservation of the natural patrimony of woodland and natural pasture lands'. This statement is an early and exceptionally rare example of the concept of sustainability. The 1994 law foresaw the requirement to ascertain the existence and types of common land uses in the municipalities and the creation of a 'general inventory of unoccupied common land existing in the region. This inventory shall contain all data suitable for identifying the land and will form the base for planning measures for the use, rehabilitation and improvement of common land.' Each municipality is required to draw up measures for improving and rehabilitating common land falling into their district, aimed at social and economic development of the interested communities. This law, when fully enacted, will give the municipalities great powers of land management, especially over the common lands that are potentially very sensitive to degradation and desertification (Cau, 1994).

The European Community signed the Convention in 1994, but had already made considerable progress in contributing to the alleviation of desertification in the southern nations of Europe. The contribution is four-fold:

1   Identification and improved understanding of the problem through research (i.e. Directorate General XII). In 1986 it held a Conference in Mytilene, Greece, on Mediterranean desertification which led to the research investment in the EPOCH and Environment and Climate Programmes of 1989 and 1995. These programmes embedded a number of major research programmes on the topic, the largest of which were MEDALUS (Box 4.4), EFEDA and ARCHAEOMEDES. The outcomes of these programmes are discussed in

Burke and Thornes (*in press*) and in Peter *et al.* (1999) and in the progress and final reports of the projects held in Brussels.

2   Through the actual and potential contribution of the Structural and Cohesion Funds administered by the EU Directorate General for Regional Policy and Cohesion. These funds both require a shared effort, with the receiving nation contributing towards the cost of the projects. The Structural Funds made available 154.5 billion ecu (at 1994 prices) and the Cohesion Funds provided 15.5 billion ecu for the period 1993 to 1999. Expenditure on environmental projects varied between 6 per cent and 12 per cent of the Structural Fund. Within the Cohesion Fund, environment was much more important (around 50 per cent). In the longer term the Directorate aims 'to have an approach assisting to design an integrated strategy for combating desertification. This approach can be fully advanced by the Spatial Planning Approach which at the same time takes into account the global (thematic) and territorial (national, regional, local) perspectives' (Slavkoff, 1999).

3   The accompanying measures of the Common Agricultural Policy seek to reduce oversupply of agricultural commodities and to improve the environment by encouraging less intensive farming and afforestation and to support actions to stop soil erosion by running water, as discussed in Chapter 21.

4   The Commission is financing cooperation between regional and local authorities through the initiative called INTERREG II – C.

What can be done?

Though the desertification problem in Europe is essentially different from that in the Sahel, there are some broad similarities. In both cases, matters were brought to a head by intense droughts in the 1960s and 1970s in the Sahel and 1970s and 1980s in the Mediterranean, that led to a debate about natural versus human causes. In both cases, agriculturally productive systems were brought to a precarious instability so that even small changes in the controlling variables quickly led to substantial changes that in turn engendered the feedbacks leading to desertification. For example, in dynamically unstable pastoral systems, small changes in stocking rates can lead to extensive erosion culminating eventually in bare rock (Thornes, 1988). Long histories of unsympathetic interactions between people and nature have led to this desertified state in the Mediterranean and the Sahel.

The differences are almost equally strong since the contrast is between subsistence societies and globalized capitalistic regimes. Some caution is necessary here for, as Toulmin (1999) points out, the view of the African subsistence farmer that does not take into account that most are producing crops and livestock for both sale and home consumption is grossly oversimplified. When talking with African farmers and land managers, we have found that it is almost impossible for them to appreciate the extent to which European farmers have their livelihoods and welfare secured by an extensive system of subsidies, especially in times of agricultural crisis, as discussed in Chapter 21. In fact, this relief of risk from the market and the weather is, and probably will remain, the main difference between the African and European dimensions of desertification in spite of the intention of the 1992 GATT agreement to move in the direction of reducing the dependence on price-support mechanisms.

Notwithstanding these differences it is generally agreed that, whatever solutions are proposed to mitigate land degradation, they are unlikely to be successful unless they are developed in cooperation with the actors in communities at different levels of the decision-making hierarchy (Hudson, 1992; Thornes, 2000; Toulmin, 1999). Generally there are four modes that mitigation efforts might take: technological fixes, fiscal incentives, policy interventions and participatory approaches.

Technological fixes

This approach means attempting to stop soil losses by physical measures. These may range from simple implementation of brush barriers on slopes to the construction of terraces, bund and bank systems and check dams or even large water- and soil-retention structures (Hudson, 1971). They are the methods most likely to be adopted in projects funded by international agencies. They are essentially designed on the general proposition that the two most important controls on erosion are slope and runoff. By controlling either of these, erosion can usually be reduced, so that the physical basis of these actions is quite sound. Generally, such approaches, though often reflecting very ancient traditional water harvesting technology, have fallen into disrepute. Toulmin (1999) points out that this is a great change from the 1960s and 1970s when 'it was usually assumed that a good dosage of modern technology was the best way forward' for tackling the low yields of African agriculture.

In the Mediterranean too, the vast changes of the 1950s and 1960s were largely technology driven – fertilizers, deep-ploughing tractors and submersible pumps all contributed to increasing production and insulated the farmer financially against agricultural uncertainties. The essential problem with the technological fix approach is that it is difficult to sustain when the project has been completed and the technologists have left. A comparison of Hudson's *Soil Conservation*, first published in 1971 and his *Land Husbandry* (1992) reveals this shift in philosophy very starkly, as the titles alone indicate. Reij *et al.* (1996) also underscore the adoption of indigenous soil- and water-conservation techniques in Sub-Saharan Africa.

In addition to the sustainability limitation requiring not only resources for maintenance works on the structure, another problem is that confidence in technology encourages greater risk-taking by farmers and managers. This approach is well illustrated when land prone to flooding is re-occupied for agriculture and other forms of economic production after the threat has apparently disappeared as a result of technological fixes. The 'big dam' culture pervades even local agriculture through the construction of water and sediment-retention structures in small gullies and ephemeral channels (Thornes 1999; Shannon *et al.*, *in press*). A third problem is that such projects may have unintended side-effects, as when retention structures limit the flow of fresh water to aquifers that are therefore invaded by salt water as has occurred at several locations along the Spanish and Italian coasts (see Chapter 5).

Incentives

The economic and institutional settings within which most farmers operate provide a set of incentives which influence their behaviour. Subsidies and compensations for remedial work in soil and water conservation are ubiquitous in southern Europe and farmers adjust positively towards these, though again sometimes with unexpected outcomes. For an extended discussion of the economics of intervention, see the discussion in Chapter 21.

Incentives have frequently been used to encourage or manipulate activity. The main problem has been that resources may be directed at a different level to other needs that, at the time, seem more pressing. The chain of decision-making from Brussels down to the individual farmers varies according to the purpose and initiation of the incentives and the local structures in place. What is needed is a study of 'financial conveyance losses' analogous to water leaving a reservoir on its way to irrigate a farmer's field. Moreover, this raises important issues of the levels of information at the different decision levels, local regional and national. Both Toulmin (1999) and Hudson (1999) argue that, though incentives are provided on the basis of economic rationality, this approach can only be understood in a temporal and institutional perspective or in an atmosphere of reduced uncertainty. In Europe this is likely to be related to land tenure, but here, as in Africa, access to land is perhaps more important than tenure because many farmers are tenant farmers (Toulmin, 1999).

Francis and Thornes (1990a) advocated a system in which natural regeneration was regarded as an alternative to afforestation, but this only makes sense through a fiscal incentives arrangement if adequate assurance is given of the availability of alternative use-access, e.g. pastoralism. In some cases it has proved easier to provide imported fodder than to extensify grazing. Nor are unintended outcomes easily foreseeable. The incentives provided for replacement of old olive stocks with young plants have resulted in environmentally unfriendly tearing up of large areas of olive trees in the Greek islands, leading to transformation of the traditional landscape and encouraging soil erosion, land abandonment and even unplanned urban expansion.

Complex computer models have been developed that allow planners to consider a wide range of land-use alternatives when incentives are being considered, usually on the local scale. These 'decision-support systems' simulate the impact of changes of land use on the natural, physical and economic environment. Coupling these models to urban change is also very important because the interactions between the two, driven from both sides, are easily overlooked in urban-based economies, such as those in southern Europe. An accident on a word processor in Brussels could change the landscape of southern Europe for decades to come! Land-use inertia in Europe is such that an accidental or ill-thought-out incentive, externally imposed could have much more and a longer-term impact than anticipated changes in climate.

One major problem with regulation by incentives over and above unintended outcomes and imperfect knowledge is the uncertainty identified by Briassoulis (1999), namely the informal planning mechanism that operates in all societies. The informal sector is a potential weakness in environmental regimes that seek to achieve ambitious goals through national and supranational regulatory policies. Most policies fail to address the fact that the informal sector may play a key role in ensuring sustainable development.

Land care?

It is widely argued that participatory intervention, in which farmers, local landholders, NGOs and others are extensively involved in the decision-making process, is the strategy most likely to succeed and overcome the problems of regulatory policies designed and implemented by so-called experts. This is a major thrust of contemporary thinking about the failure of remediation projects attempted in lesser-developed countries, particularly by large national and international agencies.

According to Hudson (1992: 49):

> A basic premise of (an intervention) strategy would be that the local community would be involved at all stages. Years ago the soil conservation approach consisted of trying to sell soil conservation to an unwilling clientele. Later it progressed to trying to win the willing participation of farmers, but it was still designed and planned as an external project. Only recently have we moved to the next stage of thinking, which is that the farming community must be involved right from the very earliest stages, first to establish what are their aims, objectives and aspirations and then to plan jointly, from the beginning, a strategy to achieve *their* objectives, not those of the planners. Throughout the implementation of the programme it is essential that the collaborators fully understand and support the objectives, to avoid the old situation of reluctant participation in a programme which they did not really understand.

In Australia, this approach has been normalized and extended in the land-care concept. Government Land-Care Legislation is addressed to promote strong action against permanent damage to the land.

> Producer groups have repeatedly put forward the view that landholders can be led but not driven, which means that regulation is inappropriate for the advancement of land use policy. Today there is an increasing body of opinion which favours the development of peer pressure rather than legal pressure, to encourage the improvement of land use.
>
> (Roberts, 1992)

> Since its inception, Land Care was a popular farmer-led organization in which land holder committees were the prime action groups. These committees, called Land Care Committees (LCCs) represented agriculture, grazing, forestry, mining and public lands, reserves and local government. By 1991 more than 1000 such committees had been formed in the Commonwealth of Australia to tackle such problems as salinization of land in the Wheat Belt of Western Australia.
>
> (ibid.: 26)

As a means of focusing national attention on the work of these committees, the Federal Government declared the decade starting 1990 as 'The Decade of Land Care' and allocated a budget of A\$320 million to appropriate conservation activity to the year 2000. Specialist advice can be obtained from 'assisting departmental officers' and the local agricultural adviser can determine the success or failure for many land care committees (ibid., 1992).

With the onslaught of European Union legislation for land management, identified with Agenda 2000, it seems worth pausing to ask if lessons could be learned from the Australian example. Some of the notable failures of interventionist environmental policies in the Mediterranean may, at least in part, reflect a failure to adopt this philosophy of local empowerment in the planning process. Land- and water-resource interventions may be a good example of this (see Chapters 17 and 21).

Above all, we conclude this chapter by noting that, notwithstanding the strong constraints imposed by conditions of climate, soil and biological dynamics, the Mediterranean desertification problem, its causes, directions and remediation lie above all in the hands of the Mediterranean people who must be alerted to the undoubted difficulties that lie ahead. Desertification is not only a problem of atmospheric physics. Rather, it is caused by and exacerbated by the failure of human beings to respond institutionally to an age-old threat that is well documented in history. Others would add that it also arises from the failure of bio-physical sciences to recognize the 'human dimensions' of the problem.

Suggestions for further reading

Balabanis *et al.* (1999) provide a series of papers outlining research and policy on Mediterranean desertification. Thomas and Middleton (1994) give a very readable account of the controversy relating to the terminology and definition of desertification. The physical and human interactions of desertification are outlined by Mainguet (1991).

Topics for discussion

1 Why are there difficulties in defining desertification?
2 The title of Thomas and Midddleton's book is *Desertification: Exploding the Myth*. What is the myth to which they are referring and what evidence is there for its existence?
3 Desertification is a complicated and complex problem. What do these terms mean in the context of desertification?
4 What criteria might be used to identify areas that are experiencing desertification?
5 Examine the view that desertification is as much a product of human behaviour as of physical factors.

20 Potential climatic change and its effects

Introduction

The potential for future climate change based on evidence for ongoing anthropic effects on the environment is still hotly debated. The Intergovernmental Panel on Climate Change (IPCC) stated in its 1995 report that the 'balance of evidence suggests a discernible human influence on climate change' (Houghton *et al.*, 1995: 4) based on three sets of evidence. First, the twentieth century seems to be as warm as any since at least 1400 CE, with the data being too sparse to compare with earlier periods. Second, more studies have been able to demonstrate statistically significant trends in global mean surface air temperature over the past hundred years. Third, studies comparing model simulations with geographic, seasonal and vertical patterns of change tend increasingly to show a convergence. This convergence would be highly unlikely to occur by chance. However, the IPCC admit that there are still major uncertainties relating to disentangling the anthropic signal from the 'noise' of natural variability, particularly relating to the magnitude and frequency of the occurrence of natural changes. Despite these uncertainties, the IPCC also state that 'climate is expected to continue to change in the future' (Houghton *et al.*, 1995: 5).

Critics of this viewpoint suggest that the predictions made may be over-estimated due to a misrepresentation of basic atmospheric feedback mechanisms (e.g. Lindzen, 1994, 1997; Pearce, 1997), that mechanisms of cooling have been missed (Hansen *et al.*, 1997) or simply that there is little evidence to support the case from instrumental records (e.g. Michaels *et al.*, 1998). On the other hand, rates of change may be too small to be noticeable against background noise (Hansen *et al.*, 1998). Hansen *et al.* (2000) also suggest that rates of forcing in the twenty-first century may also slow due to the reduction of non-CO_2 greenhouse gas emissions.

Probably the most logical outcome of this debate is a realization that the global climate change scenarios published in the 1980s and early 1990s significantly over-estimated the problem. It may be that the most reasonable estimate of change for 2050 is an increase of around 1.5°C to 3°C. This range of values is at the lower to middle part of the IPCC assessment, which estimates a maximum global change of around 4.5°C by the same date. However, Forest *et al.* (2000) have challenged this range in that it only represents about 80 per cent of the true set of possible changes according to the predictions of the scenarios used, so that the actual value could easily be lower or higher.

In this chapter, we will look at the implications of some of these predictions for the Mediterranean Basin. To be able to do so, however, first requires an assessment of potential change at the appropriate regional scale because the General Circulation Model (GCM) estimates are only made at a very coarse spatial resolution (at best 4° latitude × 5° longitude and at worst 7.83° latitude × 10° longitude: Palutikof and Wigley, 1996). We will then look in turn at the potential effects of these changes on the hydrological cycle, soil erosion and fertility, sea-level rise and consequent impacts on the coastal zone, and finally on issues of human health.

Estimating potential future climatic change

The most useful estimates of potential climate change in the Mediterranean Basin are those from the downscaling studies of Jean Palutikof and her colleagues at the Climate Research Unit at the University of East Anglia. Downscaling is the process of taking data at a coarse spatial resolution as produced by the GCM studies, and applying statistical models to provide more detailed spatial information about the changes. Typically, this approach involves taking a set of historical data for the climate parameter of interest – usually temperature or precipitation – and building a statistical model using GCM predictions of actual conditions (Palutikof *et al.*, 1992, 1996; Palutikof and Wigley, 1996). For present purposes, we will assume that the GCM predictions used in this approach are the best estimates available, although the problems discussed above mean that future estimates may differ quite radically from those put forward – certainly there is no firm consensus at the moment even between different GCMs (Palutikof *et al.*, 1996; Palutikof and Wigley, 1996). Further information on model reliability can be found in Gates *et al.* (1995) and Trenberth (1992).

Estimated changes in temperature relative to predicted global temperature changes show quite significant seasonal differences and spatial patterns (Figure 20.1). In winter, and to a lesser extent in spring, temperatures are predicted to increase more rapidly than global temperatures throughout much of southern Europe, Turkey and parts of the Maghreb. Elsewhere in the south and east, as well as in southern Spain, southern Italy, the Balkans and Greece, temperatures are still predicted to increase, albeit less rapidly than average global changes. Summer-temperature increases are predicted to be most marked in Morocco, southern Spain, Greece and Turkey. In autumn, the more rapid changes are predicted for central and northern Spain as well as Morocco. Indeed, Morocco is one of the few countries that is predicted to have almost uniformly more rapid increases than the global mean change.

Predicted precipitation changes range from a decrease of about 22 to 23 per cent to an increase of 24 to 30 per cent, according to the season and location (Figure 20.2). Winter precipitation is predicted to increase throughout most of the northern part of the basin, as well as in Algeria, Tunisia and parts of northern Egypt. Some of the existing drier parts of Spain and France, Israel and the rest of North Africa will see a decline, with the most extreme cases occurring in the areas that are already more arid. There is a general north–south gradient on the predictions for spring rainfall, with most of the north of the basin seeing an increase, except for southern and central Spain, Sardinia, Sicily and Calabria, which are predicted to have significant decreases. The southern basin is generally predicted to see decreases. Predictions for the summer are more difficult, largely because of the relatively small amount of rainfall that has occurred in the instrumental period. The pattern that results is also more patchy, with very localized highpoints and lowpoints. It must also be remembered that these values are in terms of percentages of current rainfall, so that increases in summer rainfall will still only represent a relatively small amount of extra water. Autumn increases are most marked in Tunisia, Libya and parts of Greece, with smaller increases from eastern Italy into the eastern part of the basin. The western part of the basin is most likely to see slight decreases, apart from Algeria, where the rainfall is predicted to decrease significantly.

Palutikof *et al.* (1996) also produced scenarios for changes in potential evapotranspiration throughout the basin, so that the impacts on the water balance might be better estimated (Figure 20.3). The results of this analysis suggests that any increase in rainfall will be more than offset by the increase in temperature and other factors. Although there are some

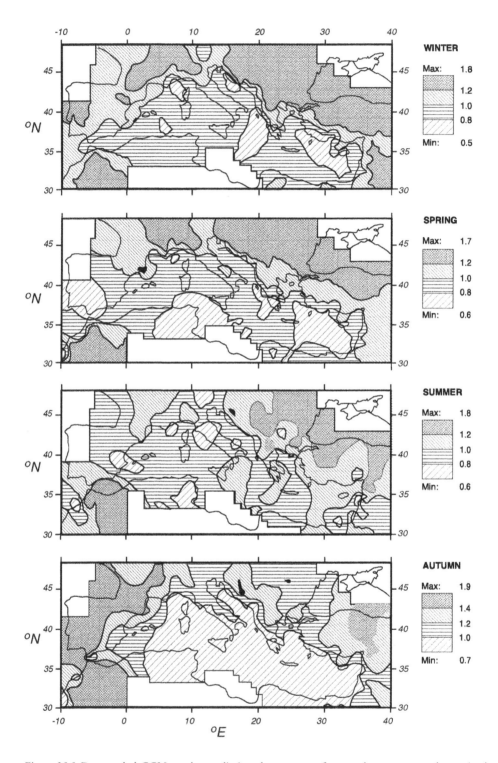

Figure 20.1 Downscaled GCM results predicting the amount of seasonal temperature change in the Mediterranean region in °C per 1°C change in global mean temperature

Source: Redrawn from Palutikof *et al.* (1996).

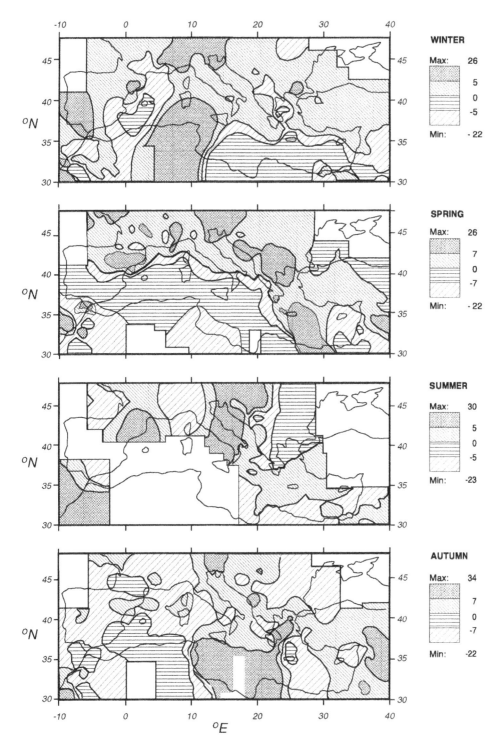

Figure 20.2 Downscaled GCM results predicting the seasonal percentage precipitation change in the Mediterranean region in per cent per 1°C change in global mean temperature

Source: Redrawn from Palutikof *et al.* (1996).

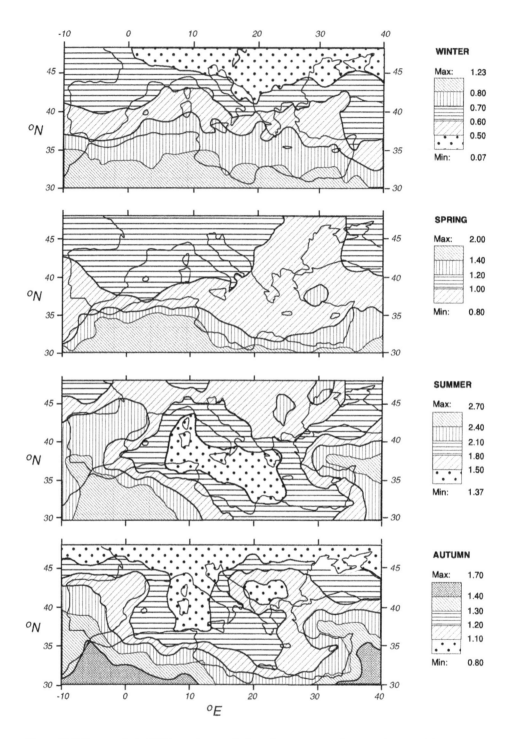

Figure 20.3 Downscaled GCM results predicting the amount of seasonal potential evaporation change in the Mediterranean region in mm day^{-1} per 1°C change in global mean temperature

Source: Redrawn from Palutikof *et al.* (1996).

irregularities in the pattern of these estimates, there is a general trend with the lowest increases in the north of the basin, and the highest in the south, with maximum increases of $2.0\,\text{mm}\,\text{day}^{-1}$ in the spring and $2.7\,\text{mm}\,\text{day}^{-1}$ in the summer.

Overall, these results suggest that future climate warming will act to make the climates of the region generally harsher, but in some respects mitigate the more severe aspects of the climate in the European Mediterranean countries. There are some data to support ongoing increases in temperature in the basin. Price *et al.* (1999) have analysed temperatures on Cyprus and found an increase of about 1°C over the last one hundred years. Diurnal changes in temperature have also decreased as minimum temperatures have increased more rapidly than maximum ones. Similar changes to less variable temperatures have been found in other global datasets by Michaels *et al.* (1998). Pasquale *et al.* (2000) have estimated increases in surface temperatures from groundwater temperatures in northern Italy of the order of 0.8–1.0°C since the 1980s, which are consistent with instrumental records at Genoa. However, some studies have also demonstrated more complex patterns, for example, including periods of cooling, particularly from the 1920s or 1940s to the mid-1970s (Giles and Balafoutis, 1990; Metaxas *et al.*, 1991). Other comprehensive studies suggest general warming, but the timing and spatial pattern of its occurrence are rather complex, and again there are localized patterns of cooling also present (Maheras and Kutiel, 1999). There is some evidence to suggest that annual precipitation has decreased slightly throughout the western Mediterranean in the period 1950–1989 (Palutikof *et al.*, 1996). There is a suggestion that the forcing mechanism may be autovariation within the Mediterranean climate system (see Chapter 3) rather than the enhanced greenhouse gas effect, although any potential links between these factors on the global scale are thus far unknown. In some places, though, this pattern contradicts the predictions made by the GCMs. At Barcelona, there was again a decrease in precipitation over this time period, but this compares to an increase in the period from the 1870s to the 1920s, which itself followed a sharp decline in precipitation (Rodriguez *et al.*, 1999). These data could either suggest that evidence for climate change due to the enhanced greenhouse gas effect is equivocal, or that any change that does occur will be spatially complex and non-linear.

Effects of predicted climatic change on the hydrological regime

Estimates of the basic changes in the hydrological regime can be made by assuming that over long time periods and averaged over large areas, changes in runoff can be simply related to changes in precipitation and evapotranspiration (Lindh, 1992). Using this method, estimates for the northern part of the basin suggest an average increase of 33 per cent of runoff, for an 11 per cent increase in precipitation and 6 per cent increase in evapotranspiration. These figures compare favourably with the direct estimates of potential evapotranspiration mentioned above. Realistically, though, estimates of the change in runoff need to take into account any change in the intensity of rainfall and its seasonality. For example, Palutikof *et al.* (1996) suggest increased storminess from detailed analysis of rainfall data from the Alentejo in Portugal. More intensive rainfall will tend to lead to more runoff as infiltration capacities of the soil are more readily exceeded (see Chapter 5). Further attempts at downscaling GCM results for small areas also suggest that in southern Spain there will be important future changes in rainfall intensity (Goodess and Palutikof, 1998). These results suggest the occurrence of larger storms in the spring, affecting agriculture and groundwater recharge, and more days with rain in the summer, increasing the amount of loss to evapotranspiration. This result supports the suggestion that an increase in temperature in the region will tend to lead to an increase in convective storms because of an increased supply of warm, moist air to the atmosphere, and stronger land–ocean thermal gradients. The effects of these changes on groundwater recharge in south-western France have been estimated to be a reduction of groundwater levels of between 1 m and 3 m, assuming no increase of abstractions (Chabert *et al.*, 1996). The critical period for groundwater recharge is from November to April throughout the Mediterranean (ibid.), so that any potential changes in precipitation and temperature regimes during these months are likely to have a significant effect on subsurface water availability.

The concentration of storms also means that there will be longer time periods between the availability of water, and the combination of this factor with the increase in potential evapotranspiration means that drought conditions may become increasingly more likely. Indeed, drought may become an increasingly common occurrence in western Europe by 2050 (Arnell, 1999), as well as in the south of the basin. Extension of the dry period in summer is also supported by the analyses of Scora *et al.* (1999) for northern Greece according to various scenarios up to the year 2100. These authors also suggest an increase in runoff in the winter months. Avila *et al.* (1996) also suggest that warmer and wetter conditions will lead to an increase in the alkalinity of streamwater, with implications for both soil and water quality. They predicted no distinct changes for warmer and drier conditions. However, their results were made with an estimated change in temperature of 4°C, so the results may be over-estimated.

Effects of predicted climatic change on vegetation

Le Houérou (1992) carried out an extensive investigation of the potential changes on equilibrium vegetation in the Mediterranean, assuming a temperature increase of $3 \pm 1.5°C$ by 2050. He predicted that a temperature increase of 3°C would shift the altitudinal ranges of vegetation upwards by 545 m, although the corresponding value for a 1.5°C increase would be 273 m. This prediction implies a shift of up to an entire vegetation zone (Figure 20.4). Because of the more rapid increase in potential evapotranspiration compared to rainfall, vegetation productivity is also predicted to fall by up to 10 per cent in the semi-arid to sub-humid zones of the region, although it may increase at higher altitudes and latitudes with more favourable conditions for growth. In the south of the basin, the prediction is for steppe conditions to be replaced by desert conditions, although whether the impact from potential climate change will be as significant as that from overgrazing is debatable. In terms of agriculture, Le Houérou suggests the potential for an increase in citrus cultivation in the north of the basin, and tropical species in the south. Cereal cultivation will be severely affected where it is carried out under rain-fed conditions, due to the estimated worsening of the water balance, particularly over large areas with thin soils in Spain, Italy and Greece. Crop returns will become increasingly poor in the semi-arid zones of the south. Kosmas and Danalatos (1994) demonstrated using a rainfall-exclusion study on wheat grown in Greece, that the timing of rainfall changes is critical in patterns of crop growth. The growth rate was found to be sensitive even to small reductions in available rainfall in the months immediately before harvest (March–May), although much larger reductions in yield were observed if rainfall was reduced in the entire growing season from November.

In the arid and semi-arid areas (or those that become so), increased CO_2 is predicted to be unlikely to improve crop yields significantly because of the increase in moisture stress. Areas with annual rainfall of more than 800 mm might benefit from an increase in productivity of up to 10 per cent from this mechanism. Similar benefits may also be achievable for irrigated agriculture. More detailed modelling studies have also demonstrated that the increase in productivity due to increased atmospheric CO_2 is likely to be offset by increased moisture stress (Osborne *et al.*, 2000). Artificial enrichment of CO_2 in a holm-oak *macchia* in Italy has been shown to favour the growth of the holm oak, although this study did not account for potential changes in moisture regime, and has only been carried out over a three-year period (de Angelis

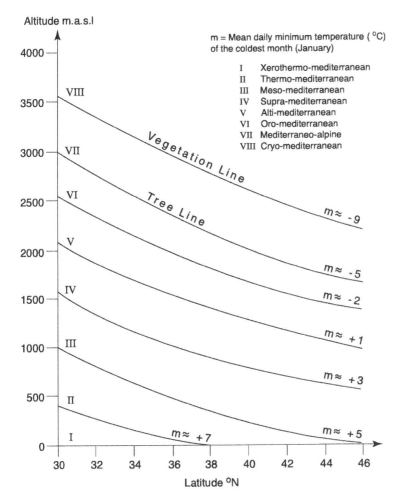

Figure 20.4 Equilibrium natural vegetation zones as a function of latitude, altitude and mean minimum temperature of the coldest month. Predicted changes in climate may lead to an upward shift of between a half and a complete vegetation zone

Source: After Le Houérou (1992), (Reproduced by permission of Hodder Arnold).

and Scarascia-Mugnozza, 1998). However, the use of Mediterranean forests as a carbon sink will be relatively minor when compared to more temperate regions (Liski *et al.*, 2000).

Kutiel *et al.* (2000) used transect studies from Mediterranean to extremely arid conditions in Israel to try to predict potential vegetation changes. They suggest that even under rather extreme scenarios for precipitation change, the proportions of different life-forms (annual, perennial or woody) would be unlikely to change significantly. In contrast, the diversity of life-forms was predicted to decrease dramatically under much drier conditions. Under more realistic scenarios, Mulligan (1996) has predicted that Mediterranean systems will continue to exhibit similar inter-annual variability to that observed at present, with any impact of climate change being minimal. Diversity of Mediterranean ecosystems at the regional scale is considered to be moderately high due to habitat and environmental diversity, but may be significantly under threat in the southern part of the basin (Cowling *et al.*, 1996). In their global review, Sala *et al.* (2000) suggest that climate is likely to be second only to land-use change as a mechanism for reducing biodiversity. In Mediterranean environments, however, they estimate the impact to be more indirect than due to localized changes. Any such decrease in diversity, however, may significantly affect the stability of the ecosystems.

Indirect changes of vegetation productivity due to climate change have also been suggested by Avila and Peñuelas (1999). An increase in dust production from the Sahara and arriving in northern Spain has been observed in the period from 1984 to 1997, compared to the period from 1944–1974. Combined with this increase is an increase in rainwater pH and nutrient input from the dust. These effects may be important in countering the effects of acid rain from pollution and increasing soil fertility. Increasing aridity and decreasing vegetation cover in northern Africa are likely to increase these nutrient transfers to soils across the basin in the future. Under certain scenarios, the change in weathering process releasing more nutrients into the soil and soil water may also benefit plant growth (Avila *et al.*, 1996).

A further indirect impact of the predicted climatic changes may be an increased incidence of forest fires in some areas (e.g. Imeson and Emmer, 1992; Pinol *et al.*, 1998). This increase would be largely due to the extension of the dry period in conjunction with the period when most fires are already started. Stronger winds relating to higher thermal gradients would tend to reinforce this effect. Any increase in the rate of biomass growth due to enhanced CO_2 would also increase the rate, as the fuel supply would be more readily available during a more rapid burn–regrowth cycle. There may, however, also be a negative feedback, in that aerosols produced during burning can cause cooling, and form cloud-condensation nuclei and thus enhanced precipitation. Depending on the temperature reached, a further negative feedback could be the enhanced volatilization of organic matter and nutrients (see Chapter 7), making the soil less able to support regrowth.

Effects of predicted climatic change on soils and soil erosion

Imeson and Emmer (1992) predict that three soil processes will be most strongly affected in Mediterranean regions over the next half century. First, the balance and composition of salts in the soil will change, leading to changes, most notably in the swelling and dispersion of clay particles. A reduction in effective rainfall, either due to an increase in potential evapotranspiration or increased incidence of crusting, will tend to lead to greater concentrations of salts in soils. High percentages of exchangeable sodium will tend to lead to more water retention and swelling in soils containing the clay montmorillonite, whereas those soils with the clay illite will become more dispersive. High salt contents have also been linked to a decrease in infiltration, with a further feedback to vegetation growth and reduction in organic matter production, which will tend to reinforce this effect. The presence of high salt concentrations will also tend to reduce the ability of plants to take up water and nutrients, by directly affecting their osmotic potential and by a number of other feedback mechanisms (Várallay, 1994). Várallay points out that the relative timing in the year of a temperature increase will tend to exacerbate these effects if it is concentrated in the summer, but lessen them if in the winter. An increase in effective precipitation may lessen the impacts of salinization, although Imeson and Emmer point to examples in Australia and South Africa where the increased precipitation has simply led to the formation of perched water tables, and thus concentrated the effects in the rooting zone. Second, areas with 100–500 mm of rain per year will tend to increase the precipitation of calcium/magnesium carbonates in the soil profile. These carbonate layers can form very rapidly, and often act to restrict the water-storage capacity of the soil by the formation of an impeding horizon. Third, there will be changes to the generation and breakdown of organic matter to the soil. The high lignin content of many Mediterranean species tends to limit the rate of decomposition, although in other similar regions, higher rainfall and higher evapotranspiration have been demonstrated to be linked to higher rates of production of organic carbon and its rate of decomposition, respectively (although the former will be also limited by moisture stress: see the discussion above). However, Imeson and Emmer point out that the critical links are the vegetation cover left standing and the activity of the soil biota, both of which are difficult to predict.

Any change in vegetation cover will tend to affect changes in erosion potential significantly (see Chapter 6). However, as we have seen, these changes are difficult to predict, and may involve increases or decreases in cover, or indeed a continuation of existing oscillations around mean conditions. Increased cover, for example, due to enhanced productivity in response to enhanced atmospheric CO_2, will tend to reduce potential erosion rates. Decreased cover due to enhanced moisture stress or more frequent burning will have the opposite effect. More severe burning may lead to the loss of organic matter and a reduction in the infiltration rate and moisture-holding capacity of the soil, and thus lead to continued erosion problems.

In terms of the effects on erosivity, the results of Goodess and Palutikof (1998), mentioned above, suggest increased rain during the summer months in the Guadalentín Basin of southern Spain, with the potential to increase rates of erosion in a period as ground cover is typically low during this period. Any increased storminess will tend to enhance erosion by producing more rapid runoff (e.g. Wainwright, 1996c), but a concentration in summer or autumn would also tend to accentuate the high rates of erosion that have been observed when large events follow fires, as seen in recent years in southern France (Ballais *et al.*, 1992) and Greece (e.g. Vafeidis *et al.*, 1997). Otherwise, an increase in rainfall may lead to a decrease in erosion rates, because of the increase in vegetation cover (e.g. Imeson and Lavee, 1998; Lavee *et al.*, 1998; Kosmas *et al.*, 2000), although areas moving from arid to semi-arid conditions may experience an increase based on equilibrium data (see discussion in Wainwright *et al.*, 1999a). Because of the difficulties in combining these various factors, it may be that the best estimates of potential change in erosion will be achieved via continuing modelling studies that can account for these various interactions (e.g. Kirkby *et al.*, 1996; Thornes *et al.*, 1996; Lukey *et al.*, 2000).

Even less predictable are the impacts of climate change on other water-erosion processes, not least because of the problems in understanding their formative mechanisms (see Chapter 6). The incidence of piping may be enhanced due to the increase of salt concentrations in soils (see, for example, Imeson *et al.*, 1982). These pipes may, in turn, accelerate the growth of gullies. Whether gullying increases under climate change will depend on the specific interactions between rainfall intensities and vegetation. This area should be one where more research effort should be focused in the future, due to the predominance of concentrated forms of erosion in total soil-loss estimates. There will also be impacts elsewhere in the catchment system, due to changes in sediment loads from the uplands, with the possibility of increased rates of arroyo formation (e.g. Imeson and Emmer, 1992).

Climatic changes will also lead to changes in the potential for mass movements. In the southern French Alps, Flageollet *et al.* (1999) found most landslides were triggered by high levels of cumulative rainfall in the month prior to failure, or due to periods of heavy rainfall (see also Wainwright, 1996c). Similar results have been found for the eastern Pyrenees (Corominas and Moya, 1999), northern Spain (Domínguez Cuesta *et al.*, 1999) and Portugal (Zêzere *et al.*, 1999). Clearly, both of these conditions may be expected to increase given the climate scenarios discussed above, although the relevance of the former may be limited by increased evapotranspiration. Using a more detailed downscaling study, Dehn and Buma (1999) suggest that under a variety of GCM scenarios, the activity of one particular landslide in the French Alps was predicted to both increase and decrease over the next century, showing the level of uncertainty in predicting changes in this erosion process. The studies discussed above that suggest enhanced weathering rates or mineral inputs to soil will also lead to greater slope loading and thus increased instability, although this process will obviously only be important over much longer timescales.

Impacts of potential sea-level rise in the coastal zone

> Few global problems have been studied in as much detail with as few reliable data as sea level.
>
> (Milliman, 1992: 45)

Although satellite altimetry is now beginning to play a more important role in estimating absolute sea levels, it by no means provides a long enough data set with which to estimate changes in sea level reliably. Estimates therefore must rely on local tide-gauge data and historical and archaeological data. Although many authors agree that there has been a gradual rise of sea level of between 1 and $2\,\mathrm{mm\,a^{-1}}$ over the last century (Milliman, 1992), this viewpoint has been disputed by Pirazzoli (1989), who believes that the figures are based on biased data sets and probably represent relatively stable sea levels since about 1930. If the former figures are correct, then they can probably be explained by thermal expansion of the oceans and from the melting of alpine glaciers. Flemming (1992) suggests that there is some evidence for an acceleration by comparing tide-gauge with archaeological records, but warns that the short-term picture can sometimes be severely misleading. Zerbini *et al.* (1996) also stress this point, and provide detailed supporting evidence for recent sea-level increases of about $1\,\mathrm{mm\,a^{-1}}$. The Mediterranean Action Plan predicted an accelerated rate of increase, possibly to as much as 2.2 or $4.4\,\mathrm{mm\,a^{-1}}$ by the year 2025, which is at the lower end of the range predicted globally (Milliman, 1992). About half this figure is accounted for by thermal expansion, but the biggest unknown in predicting future rates is the fate of the Greenland and Antarctic icecaps. Opinions differ as to whether massive melting of these ice bodies will lead to much more rapid increases in sea level, or whether increased precipitation in the polar regions (due to a warmer, more moist atmosphere and stronger poleward air currents) may offset other mechanisms of sea-level rise.

As the Mediterranean region is tectonically very active (Chapter 2), the *actual* amount of sea-level change that is experienced at any one place will also be a factor of relative land movements. Estimates of vertical crustal movement over the historical period range from subsidence of $2.28\,\mathrm{mm\,a^{-1}}$ in the Adriatic to an uplift of $1.01\,\mathrm{mm\,a^{-1}}$ in southern Greece (Flemming, 1992: Figure 20.5). Locally, the rates may exceed these average values, particularly over very short time periods. For example, the Bay of Naples area is very dynamic due to volcanic activity (Zerbini *et al.*, 1996). There is also important spatial variability because of the nature of crustal movements, so that locally measured values should not be extrapolated over distances of more than 10 to 20 km (ibid.). Some areas are now subsiding more rapidly, due to groundwater removal, as is the case for Venice, or sediment loading near large deltas, for example, in the region of the Gulf of Lions. In these areas with tectonic and other forms of subsidence, the effective rate of sea-level rise will be greater than the average, while in some uplifting areas, the effective rate will be slower, and in some cases will be counter-balanced so that the sea level is relatively stable or even falling slowly. A summary of these effective rates assuming the continuation of present rates of change, as well as the MAP scenarios for change, is illustrated in Figure 20.6.

Nicholls and Hoozemans (1996) define five major impacts of sea-level rise. First, there will be an increase in rates of coastal erosion. This factor may be exacerbated by increased storm surges if the storminess of the climate increases, as is suggested above. There, may, thus be

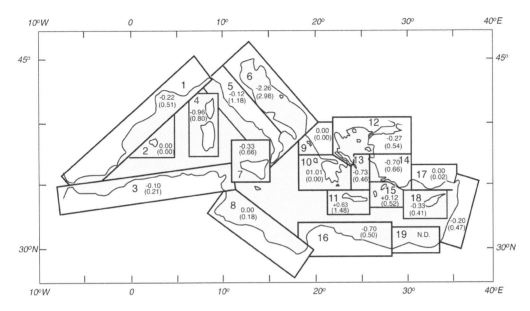

Figure 20.5 Mean rates of vertical displacement around the Mediterranean region, based on historical and archaeological data by Flemming (1992). The rates are given in m ka^{-1} (which are directly equivalent to mm a^{-1}), with negative values reflecting subsidence of the land relative to the sea and standard deviations in brackets. Although there is no attempt to remove eustatic sea-level change, over the time period in question, this can be assumed to be relatively minor (<0.15 mm a^{-1}), reproduced by permission of Hodder Arnold

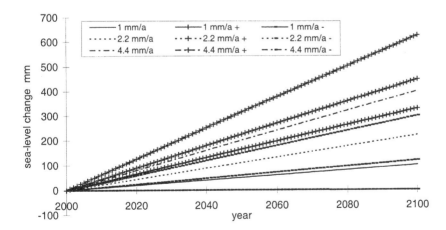

Figure 20.6 Graph showing the range of variability in effective sea-level rise to the year 2100. Three sets of condition are illustrated: a 1 mm a^{-1} rate of rise, reflecting the continuation of observed rates over the last century; a 2.2 mm a^{-1} rate; and a 4.4 mm a^{-1} rate, reflecting the lower and upper bounds of the MAP estimates (Milliman, 1992). Curves marked with a '+' show the maximal rates of effective sea-level rise, due to local crustal subsidence, and those with a '−' show minimal effective rise, due to uplift, based on the data of Flemming (1992). These rates, however, do not include any potential effect of accelerated increase from melting of the Greenland or Antarctic ice sheets

significant impacts on the beaches of the Mediterranean (e.g. Attard *et al.*, 1996; Randić *et al.*, 1996). Second, there will be inundation of coastal areas. The impact of this will be most marked on the southern coast of the Mediterranean (Nicholls *et al.*, 1999). Third, there are increased risks of flooding and of impeded drainage. Fourth, salt water will intrude into aquifers used for drinking and irrigation. Fifth, higher water tables may affect the stability of some building foundations. Some of the most sensitive areas to such changes will be the large deltas, where rates of accretion, sedimentation, soil formation and plant and animal habitats will be affected. Potential impacts have been studied on the Ebro (Sánchez-Arcilla *et al.*, 1996), Nile (Sestini, 1992b; El-Raey *et al.*, 1997), Po (Sestini, 1992a) and Rhône (Corre *et al.*, 1992). Attard *et al.* (1996) suggest that there may also be important impacts on fisheries, particularly relating to the migration patterns of various species, and the distribution of habitats. Increasing temperatures may also increase the productivity of fisheries, but may also have a negative impact by encouraging the growth of algal blooms (Randić *et al.*, 1996).

Other potential impacts

Although not particularly extensive due to the prevalent climatic conditions, there are a significant number of wetlands located throughout the region. Nicholls and Hoozemans (1996) suggest that 200 km² of wetlands in the northern Mediterranean and 2,600 km² in the south will potentially be threatened by a 1-m sea-level rise. According to most recent GCM scenarios, most coastal wetlands around the Mediterranean could be lost (Nicholls *et al.*, 1999), although these scenarios ignore the potential for local variability in sea-level rise. In north-eastern Malta, for example, the Ghadira Nature Reserve may be threatened, affecting an important refuge for migrating birds, and a number of threatened animal and plant species (Attard *et al.*, 1996).

At the other end of the spectrum in the Mediterranean, mountain environments may also be particularly affected by climate change. Seasonal changes in the availability of resources can significantly affect the agro-pastoral systems developed in the high mountains, even if at first they prove resilient to change as in the Moroccan Atlas (Kaufman *et al.*, 1999). In the Alps, simulations based on GCM scenarios suggested forests may be highly sensitive to climate change (Fischlin and Gyalistras, 1997). Some forest areas were predicted to disappear completely, while elsewhere significantly different responses were observed within short ranges of one another.

Effects of predicted climatic change on human health

Increasing temperatures, with the possibility of increased occurrence of temperature extremes, can have important impacts on human health. The main reasons for this are increased cardio-vascular stress under high temperatures, and increased susceptibility to pollution, particularly where it is trapped by summer temperature inversions. For example, Alberdi *et al.* (1998) found a direct correlation between daily mortality rates in Madrid with high temperatures and high summer humidities. On the other hand, they also found a link with cold-temperature extremes, so that winter mortality might be expected to decline under predicted climate-change conditions. The results of Saez *et al.* (1995) also support the potential effect of increased temperatures. Their study in Barcelona found a link between cardiovascular mortality and periods with three or more days of elevated temperatures. Katsouyanni *et al.* (1993) have also found a significant interaction between high temperatures ($\geq 30°C$) and mortality, and found that the effect increased in conjunction with high levels of SO_2 pollution. Deaths from respiratory failure relating to levels of high pollution in Milan were also found to be linked more commonly to periods of high temperatures (Vigotti *et al.*, 1996). A general link between pollution and high temperatures in southern Europe has been suggested in the review by Katsouyanni (1995) of work on the topic.

Indirect impacts of climate change on health may relate to increased mobilization of pollutants adsorbed onto eroded sediments. The effect of changes in water quality and salinity may lead to decreased tolerance of disease in more marginal areas, particularly those with insufficient access to sanitary conditions. This change, in turn, may lead to increased incidence of food poisoning. Disease vectors themselves may be favoured by change. These may include increased incidence of malaria, leishmaniasis and dengue fever in southern Europe and Turkey (Beniston and Tol, 1998), where some outbreaks are already being observed, as well as in North Africa (Zinyowera *et al.*, 1998). Beyond the direct impacts of flooding should the number of severe weather events increase, more floods will give a further opportunity for disease vectors to spread. Changes in sea level and processes in the coastal zone may cause health problems, especially where effluent continues to be pumped into the Mediterranean. Even currently effective long outfall systems may cause problems if patterns of sea currents are modified.

Linking potential impacts of climate change with socio-economic changes

Many of the arguments for the impacts of potential change noted above have been presented assuming no change in other conditions. However, as noted in the preceding chapters, there are significant socio-economic changes ongoing around the Mediterranean Basin. These changes will in a number of instances exacerbate the effects of potential climate change. Most notable is the increase in population, which will continue to grow most rapidly in the south and to a lesser extent in the eastern parts of the region (Chapter 14). It is obvious from the discussion above that these are the very same regions where most of the potential impacts will be more severely felt. Furthermore, many of the areas that may experience increasing moisture stress, declining groundwater levels and saltwater intrusion into aquifers are those that are most dependent on reliable, good quality water supplies for supporting the tourist industry. Unless the continually growing tourist numbers that are predicted to come to the Mediterranean over the next decades are encouraged to diversify away from the traditional beach holiday, these numbers will quickly become unsustainable. On the other hand, some authors suggest a possible positive impact of climate change on tourism, in that temperatures may increase the length of the season (Perissoratis *et al.*, 1996; Randić *et al.*, 1996). However, a longer season during a longer period of moisture stress will ultimately be deleterious, as it simply means increasing the pressure on the scarce water resource. Beyond this issue, the concentration of urban centres in the coastal zone of both the north and the south will also accentuate any of these problems.

Continued pumping of groundwater in the coastal zone for drinking and irrigation will tend to accentuate any potential local effects of sea-level rise, by increasing subsidence as seen dramatically in Venice (Sestini, 1992a). Although it is difficult to predict how much this effect may be offset by sea-water intrusion, this is obviously not a desirable solution. Nicholls and Hoozemans (1996) suggest that a 50-cm sea-level rise between 1990 and 2020 will put an additional 1.5 million people at risk from the 1,000-year storm surge in the northern Mediterranean. In the south, an additional 6.6 million people will be at risk in these conditions. For the case of a 1 m rise in sea level, they put the protection cost at approximately 2 per cent of the GNP of the northern countries, compared to 7 per cent for those in the south.

There are possible positive benefits of increased water in the mountain regions. This water could be used for hydroelectric schemes (Scora *et al.*, 1999), and be piped to the drier coastal regions for irrigation agriculture or drinking water. Sustainable development of resources may mean a realization of the interconnectedness of different elements of the Mediterranean system, but this will take political and social reorganization on regional and often international scales.

Summary

It can be seen from this chapter that all elements of trying to predict future climate changes and their impacts provide equivocal results. There are uncertainties about the amount and often direction of global climate change, how these changes will be converted into regional and local changes, and how these local impacts will affect a variety of environmental factors. In a number of cases, both positive and negative benefits can be envisaged, but it may take significant planning to allow the increasing populations of the basin to take full advantage of the benefits and mitigate the negative impacts. What is most likely is that the most significant effects will be felt due to the rapidly increasing population and its spatial pattern of growth, focused very clearly in the more marginal areas of the south and east of the basin. These patterns will tend to reinforce and increase existing inequalities within the region.

In terms of understanding potential impacts, our understanding of natural and human environments are still imperfect and are often based on equilibrium conditions. Although modelling approaches continue to allow us to make some improvement on this situation, we are still unable to make testable predictions that take into account transient conditions. System resilience, adaptation, delayed response times and unforeseen feedbacks are only four of the factors that can significantly change the impacts we predict. This problem is made worse with the difficulties encountered with non-linear systems that may be sensitive to measurement of initial conditions, or exhibit properties of deterministic chaos.

Whatever the impacts and our ability to predict them, the human outcomes are not necessarily straightforward. Thompson and Rayner (1998) discuss in detail the issues of how humans react to impending climate change, and how this may or may not be modified by prior knowledge of changes that are predicted to occur. There is often an unwillingness to accept predictions made by scientists and supported by local or central government, particularly if there is an awareness of the level of uncertainty involved (see Scherm, 2000, on predicting the change in pests following climate change, as well as the examples above). Indeed, in many instances, local information may be much more useful in making predictions, and is absolutely necessary for the success of policy, as we shall see in the next chapter. It is also important to appreciate the ability of human populations to adapt to change (Smithers and Smit, 1997), a factor that has been amply demonstrated at numerous times of crisis in prehistory and history. However, this argument should not be taken to mean that we should simply sit back and let the changes happen.

Suggestions for further reading

Palutikof *et al.* (1996) give the best current overview of potential climate changes in the Mediterranean. Pearce (1997) is a very readable overview of the debate on predicting climate change. Numerous implications and case studies are reviewed in the volumes edited by Jeftić *et al.* (1992; 1996).

Topics for discussion

1   Why is future climate change so unpredictable?
2   If climate change *is* so unpredictable, why should we bother trying to estimate what its potential impacts might be?
3   Try to build a simple model of the hydrological cycle in various parts of the Mediterranean Basin – to what extent do you think reliable projections of the impacts of climate change could be made based on this model, and what data would you require to carry this out?
4   Following some of the linkages with the hydrological cycle described elsewhere in the book, what other outcomes might there be from these changes?
5   Are population changes likely to be more important than climate changes over the next hundred years?
6   What are the main environmental and human interactions in potential climate change?

21 Key issues for mitigation

Temporal variability

There is a common misperception of the Mediterranean as a Garden of Eden, with almost ubiquitous high levels of sunshine and winter rainfall and where almost anything can grow. Certainly the translucent quality of the light and the often azure blue of the sea give an almost iridescent beauty to the coastline that is rarely matched in the world. Yet this is a tough environment for people and nature. The strong variations in rainfall throughout the year, and from year to year, lead to great uncertainty in food production. Annual rainfall totals vary from one to two standard deviations between successive years and runs of drought years produce dramatic reductions in the support for farmers and their families. This uncertainty is passed on to natural growth that is both low in biomass and tenuous in its hold in the environment. Throughout the centuries this has resulted in the traditional basis of wheat, treecrops and live-stock. Land management has encouraged the spread and growth of shrubland, and clearance has led to localized and often catastrophic erosion since the Bronze Age. From the same time period, water conservation and irrigation have become fundamental techniques for reducing uncertainty in food production.

As commercial agriculture progressed, the marginal conditions became even more haz-ardous. Concentration on monocultures, such as olives and vines and irrigated horticulture, exposed the farmer even more to the intermittency of rainfall and other disasters. The ravages of the vine disease, *phylloxera*, at the beginning of the twentieth century led to massive and desperate failure of viticulture in many parts of the Mediterranean. The relief came through overseas migration and desertion of rural areas, especially in mountain lands. Some areas never recovered from this abandonment. The almost universal soil degradation, though spatially highly variable in its intensity, has added to the ruggedness of the Mediterranean environment. The huge transformation of agriculture through fertilization, fertigation and mechanization that took place from the middle of the last century, exposed agriculture and farmers more and more to the global market place and its vicissitudes. Land degradation cannot be properly understood without reference to the history of land use.

Spatial variability

Besides the huge inter-annual variability and uncertainty, the other great theme underpinning environmental issues in the Mediterranean is the great spatial variability. This variability arises initially from the topographic variability, a function principally of the geological and erosional history. Climate, soils and nutrients reflect the topography quite closely and, since Neolithic and Bronze Age times, they have controlled the patchwork character of agricultural potential, settlement and commercial activity. Indeed, the mineral wealth of the Mediterranean played a key role in the Punic development of commercial activity, in contrast to the Roman emphasis on agriculture and the consequent early urbanization phenomena. Just as the agricultural opportunities and possibilities vary greatly over very short distances, so too do the commercial activities, the population densities and the appropriate approaches to environmental management. There is no standard problem and certainly no standard solution to the problems of environmental degradation. Rather, a basket of options is required that address the spatial variability of the natural environment and the hugely complex variations that arise from the national histories. It is for this reason that we have emphasized the importance of their historical legacy and knowledge of it in tackling environmental issues, such as land degradation and water.

Historical antecedents

Soil degradation can be accounted for by localized agricultural traditions and practices such as the imported irrigation traditions of the Moorish invaders of the Kingdom of Granada in the areas where the Christian successors arrived 200 years behind the general re-population of south-east Spain. These historical impacts can have great longevity. It has been shown that the Bronze Age settlements of eastern Andalucia were not only favourably endowed by natural resources, but also that the concentrations of peoples led to local intensification of erosion by positive feedback (Gilman and Thornes, 1985). The shift from prehistoric broad-band impact to historical narrow-band impact was tracked in detail by the EU ARCHAEOMEDES project in the Vera Basin on the Levant of Spain, as reported by van der Leeuw (1998). These historical influences highlight the questionable quality of defining soil erosion potential by over-simplified empirical models, such as the Universal Soil Loss Equation (Wischmeir, 1976) and its variants (Hudson, 1995), notwithstanding the sound physical realism underlying such approaches. Equally it brings into question what can be derived for resource management from satellite-borne remote sensing at low spatial resolutions. What is clear is that the same landscape can be occupied and exploited successfully using relatively diffuse settlement patterns, but can be over-exploited leading to significant land degradation that may take millennia to be reversed. Chapters 10 to 13 give numerous other examples where these same phenomena have occurred.

Complicated and complex impacts

It is now broadly accepted that climate change is already occurring in a broad swathe across the Mediterranean, with important differences between the eastern and western basins, and between the north and south. These differences are an inherent function of the characteristics of the Mediterranean climatic system. Although there are variations between the outputs of the various global change scenarios and of the models that purport to show their impacts at the sub-regional scale, there is broad acceptance that the future climates will yield more difficulties as the overall rainfall and river flows decline. Working out the impacts are both complicated and complex.

They are complicated in the sense that the vegetation-cover variability is very great over short distances, mainly as a result of sharp variations in relief. Climate-change impacts are often mainly propagated through the vegetation cover. The vertical ensemble of vegetation in mountains is not easily generalized and the meteorology of topographic gradients cannot readily be incorporated into the general circulation models. They are complex in the sense that small changes in unstable ecosystems may produce far-reaching consequences. It is not true, especially in the Mediterranean, that big consequences have big causes. One of the main management issues for environmental systems in the Mediterranean is to accept the complexity of responses in changes to these unstable ecosystems. Thornes (*in press*) illustrates the problems of understanding complexity of response in the problem of land degradation in Mediterranean lands. First, land managers have to accept that, in seeking the causes of desertification, historical records may in fact be a rather poor guide, especially in relation to climate change. It was commonplace in the middle of the last century to account for most hydrological and erosional changes by climate change and, even today, this outmoded conceptualization still dominates the interpretations of environmental change as reflected by soil and alluvial stratigraphies, sometimes at the expense of socio-economic causes and their non-linear impacts. Simple data records may fail to reveal the special features of time series induced by chaotic behaviour in non-linear dynamical systems (Malamud and Turcotte, 1999) that abound in Mediterranean environments.

Management, scale and sustainability

As time passed, the environmental management problems shifted in scale from an essentially local to regional, to national and ultimately trans-national level of administration, with a progressive uncoupling from the land. First, the accession of Mediterranean states to the European Community and, more recently, the globalization of the regional economy, have led to direct and indirect environmental impacts of human activity through external control. Southern Europe adapted quickly to the productionist under-pinning of the Common Agricultural Policy so that southern farmers, like their northern counterparts, had nearly half their incomes derived from subsidies paid for by the European tax-payers (as discussed later in this chapter).

The emphasis on production led to further intensification of farming methods and to demonstrable negative impacts on the environment in the form of increased soil erosion and reductions in water quality. The increased soil erosion in turn has led to lower tolerances for the further intensification of production. In dry conditions in the Aegean, soils less than 25 cm deep are unable to sustain crop growth. Erosion produces more erosion as the capacity to store rainfall reduces and overland flow increases in volume and speed. Badlands intensify the rate of runoff, producing higher runoff peaks and more flood damage. Extremely dry soils are less absorbent of rainfall, so that 'drying-out' of the Mediterranean over the past half century, combined with soil loss, resulting from agricultural intensification, has led to a powerful combination that is unsustainable. Desertification is a serious problem in the Mediterranean, leading to rural desertion and falling farm incomes. We reiterate the role of seasonality in the Mediterranean climate as a prime factor in environmental problems. Although centuries of adjustment to this seasonality have offset some of the main difficulties it presents for agricultural sustainability, the apparently obvious solution, irrigation, brings on new problems.

Water

The sustainability of water supplies is a critical issue, especially for groundwater. The Mediterranean is not alone in this. In virtually all semi-arid areas of the world, irrigation has been seen as the way out of national under-production of agriculture in the face of growing populations. Irrigation not only meets the problems of seasonal and inter-annual drought that devastates production in dry farming systems, it also creates jobs and basic infrastructure, thus developing food industry, handicrafts and services. It is estimated that 1 Hm³ of water can give rise to twenty-two labour units and have a multiplying effect of up to forty times in the case of forced crops. It allows the population to concentrate spatially and hence develop better and greater social services. These, together with the availability in time and space of agricultural raw materials, can allow industry and services to develop (Carcelen, 2000). The expansion of the area under irrigation has been the favoured method of expanding agricultural production in Egypt, Cyprus and Spain even though, especially in the last two decades, agricultural products are playing a declining role in the GNP of these states. But irrigation is not without cost and damage to the environment, though these costs are often dismissed by economists as externalities and the true price of water is rarely properly accounted for in cost–benefit exercises. Ignoring these facts will ultimately lead to environmental degradation at regional scales and the likelihood of political instability relating to the unequal access to water resources.

Agenda 21

Recognizing the inadequacy of national efforts to practise sustainable agriculture, the UN Rio Conference of 1992 not only identified desertification as a major problem in the world and promoted the International Convention on Combating Desertification, but it also pressed nations to recognize the need to adopt sustainable agriculture at all levels of government, from the municipalities to the nation states. Agenda 21 requires participating states to prepare plans for sustainable economic activities. The European Union had already adopted this approach in its efforts to revise the Common Agricultural Policy in the form of the new Agenda 2000. In this, the Commission not only promoted sustainability, but shifted the emphasis of agriculture away from increased production towards environmental protection through a threat to reduce the level of agricultural subsidies and through the provision of incentives for environmental measures. The signs are that the 'transition to sustainability' (O'Riordan and Voisey, 1998) is taking place only very slowly, at least in Greece, but where it has occurred (as in Portugal) it is profound. Sustainability goes well beyond the conservation of natural resources and requires a more radical shift from centralized management of agriculture to local empowerment, a shift that is politically and socially difficult to promulgate. However, it is a shift that is absolutely necessary for the future environmental and political security of the Mediterranean region.

Canons of Mediterranean environmental management

In this book we have examined a selection of the major environmental issues of the Mediterranean regions. From this examination, a number of canons of Mediterranean environments emerge.

1   The *historical dimension* is crucial in understanding environmental problems, their causes, main controlling agents and the formulation of sustainable solutions. This factor is nowhere better illustrated than in the context of land degradation (in particular soil erosion) and desertification. The changes of land ownership are especially crucial in this problem. A consequence of this canon is that it is unlikely that the problems and their causes can be identified, even less solved, at the very large scale. There is no substitute for local knowledge and this is almost invariably in the hands of local people (Thornes, 1998).

2   *Complexity* is a major aspect of many problems. Small interventions can have large and unexpected effects because of non-linearities in the environmental systems. This is illustrated in attempts to mitigate river flooding. Small changes in vegetation cover, in the channel or catchment, can lead to the onset of erosion that may produce catastrophic losses of soil through badland development.

3   *Climate* is a key to many environmental issues because of seasonality and inter-annual variability. Already the Mediterranean climate is getting drier and things will get worse before

they get better. Rainfall and river flow could fall by as much as 17–25 per cent per year. More extreme rainfalls, deeper droughts and more frequent crop failures are on the future agenda (Conte *et al.*, *in press*).

4   *Diversity* is enormous. Spatial variations in geology, topography, soils and vegetation cover are so great that universal solutions do not generally apply. There is great cultural, political and demographic as well as biophysical diversity.

5   *Sustainability* does not only mean conservation. It is a complicated concept involving the empowerment of local populations in environmental decision-making (O'Riordan and Voisey, 1998). The prophetic vision of Dovers and Handmer (1992), quoted in the Preface, is nowhere more apposite than in the contemporary situation of Mediterranean countries.

6   Without local involvement, *technical fixes* are likely to prove unsuccessful. This is especially the case where land use changes are involved. The patterns of Mediterranean land use and land tenure involve complementarity. Chain-like reactions almost invariably follow changes.

7   *Hierarchical decision systems* exist and this usually means hierarchically structured management plans. There must be discourse between, as well as within, levels of the hierarchy.

8   The *human dimensions* of virtually all environmental issues are stronger and closer than in most other regions because of the evolution of a symbiosis between humans and nature in the Mediterranean. At the same time a preoccupation with the needs of end-users should not lead to a neglect of the fundamental physical controls in environmental problems. The interplay between people and nature is a deep and recurrent theme in this unique and complicated environment.

9   *Uncertainty and risk* are major aspects of the Mediterranean environment, especially in relation to lack or excess of water. The Mediterranean is a hazardous place and environmental managers must engage in these issues. Unfortunately risk and uncertainty will always exist, whatever data-gathering and analysis exercises can be accomplished. The provision of a numerical estimate of the risk (e.g. in flooding) is not an adequate approach to the problem. Nevertheless, the preoccupation with indicators certainly stresses a greater faith in risk evaluation than is justified. The concentration should be on learning to live with risk and change and providing the capacity to deal with them, rather than continually attempting to estimate them. As Dovers and Handmer (1992) recognized: '*Like natural systems, human systems need to be flexible enough to cope with uncertainty and unanticipated shocks.*'

In the Mediterranean, environmental managers must be capable of responding positively to problems such as drought, flooding, catastrophic land degradation and disastrous marine pollution. Fundamentally, the uncertainty and risk are not externally imposed, but rather they are generated as a result of human activity and misunderstanding. Groundwater over-exploitation and the associated marine incursion and groundwater pollution exemplify this point only too well.

Frameworks for action

At each stage in the hierarchy (canon 7), the participants are different and so are the decision-making rules. This process can lead to delays and obstructions in the processes of implementing the rules for action. The framework at different levels is considered in this section of the chapter. Without an appreciation of this framework, the prospects for real and timely action to implement changes are unlikely to emerge. We discuss how the nations are approaching the problems posed by the main environmental issues that have been identified and discussed in this book. What are the instruments (tools), financial and political, that can be used for mitigation and what emerging strategies might be available in the coming century?

With regard to water resources, the main questions are the quality and the quantity of resources available, in ground as well as surface waters, at different spatial and timescales: international and intra-national, seasonal and inter-annual. There is also the contrast between 'hard' and 'soft' engineering approaches to environmental conflict resolution that is a recurrent theme in Mediterranean countries.

The issues in land degradation range from local intensive erosion to large-scale rural depopulation problems. Degradation is, above all, a socio-cultural as well as a physical and technical issue.

In relation to agricultural intensification, all Mediterranean states have gone through, or are still embarking on, feverish attempts to modernize and intensify production, with the consequent environmental impacts of increased soil loss and reduced yields, environmental pollution and rural instability through demographic change.

All states are pushing the growth of tourism, mainly in a loosely controlled or unplanned fashion that reflects the importance of this activity to national economic growth, even in the most advanced economies in the region. Coupled with urbanization and rural desertion, it engenders, or is involved with, the other problems listed above.

All these issues have a number of common characteristics that constrain the lines of action that can be taken. These issues are outlined in the canons of Mediterranean environmental management in the previous section.

Technical fixes versus creative solutions

Van der Leeuw (1998) separates the 'technical fix' from the 'creative solution'. In the former, a problem is identified, a range of technical solutions are analysed and one of them is implemented. In the latter, the problem is discussed or negotiated with the local community which is experiencing environmental disasters, and then implemented by common consent in a politically agreed fashion. There are two general approaches to environmental problem solving: regulatory, where a directive or legal act is imposed from above (politically speaking); and fiscal instruments, such as positive financial incentives (e.g. price support) or negative product taxation (e.g. on a pollution load in water or soil). These are discussed in the following section.

In Chapter 15 we showed that Mediterranean agriculture underwent rapid change in the 1960s and 1970s. Intensification and mechanization were pursued in order to increase the yield of the crops already grown and different varieties were chosen to give higher yields. The sale and application of fertilizers went up dramatically in all the European countries. This increase was a response to two main factors. First, northern Mediterranean countries were entering into the second stage of the Rostow growth cycle after a long period of relatively limited growth in the previous two centuries. Second, they were responding to European Union and state intervention in the form of subsidies on production. We dealt earlier with the Portuguese wheat campaign (Chapter 19), a clear example of state intervention and its impact.

The intensification of activities produced what economists call negative *externalities on the environment*, i.e. unwanted effects that are the products of a particular policy. Among these are pollution, erosion, flooding, reservoir siltation and other deleterious environmental impacts, such as salinization. If farmers had somehow to pay for this damage to the environment, then their revenue/income would be smaller and their quality of life might suffer. As it is, over the years the policy-makers have failed to come up with a good mechanism for charging the farmers for this damage, which is not an easy task and for which it is difficult to carry out practical experiments. Because the farmers were subsidized heavily to increase production, they are not so sensitive to revenue variations and so do not reckon the externalities into their budgets. Why should they, if someone else (the tax-payer) is covering these costs? The farmers' prime objective, encouraged by existing legislation, is to increase the yield of a product, especially grain, and obtain as much subsidy as possible, by reacting quickly and effectively to policy interventions.

For almost forty years, the European non-agricultural citizens accepted this position, partially through ignorance, partially from the belief that the farming sector was essential, first in producing the nations' essential food stuff and, second, because the agricultural sector growth was equated with economic development. The farmers were regarded as an integral and essential part of the countryside culture. In Britain and even more in France, especially at the time of the Second World War, farmers had a key strategic role in the war effort. There was also a strong sentimental attachment to the countryside and a fear that the newly-favoured prime component of development, industrialization, would result in the loss of the rural ethic, its population and skills. As we showed in Chapter 16, rural depopulation (the French call it desertification – and as we have seen, this sort of desertification can often lead to the other) did indeed occur throughout Europe on a massive scale. This process had all sorts of consequences and is a downward spiral. Once depopulation starts, it is a strongly re-enforcing tendency. As people leave, services decline (e.g. loss of post offices and village shops) and there is less incentive for people to stay in the countryside.

The growth of environmental awareness over the second half of the last century highlighted the dilemma. More production in agriculture was accompanied by more and more environmental impacts, ranging from river and groundwater pollution to soil loss (and soil productivity loss) by erosion and reduction in biodiversity. These consequences have led to a disenchantment with farming and less acceptance by the non-farming public of the externalities imposed by the farming communities.

In response to public demand, economists and politicians have tried to force the farming communities either to reduce over-production or to meet the costs of the externalities from their own revenue (the polluter pays principle) by a series of legal requirements (policy instruments) at different levels – national, international and European. Many of these instruments had been explored elsewhere, especially in the United States of America.

Two recent books, *Against the Grain* (Potter, 1998) and *Rural Planning from an Environmental Systems Perspective* (Golley and Bellot, 1999), have examined these policy instruments and their impacts in detail and are the source for much of what follows. Rural planning is an attempt to change land use to produce an improvement in the quality of life of rural people. It is often criticized as being too formal (analytical) in its approach, when, in fact, 'muddling through' is probably a better description of how things happen. The European Union's Common Agricultural Policy is rural planning on a staggering scale, fortified by legal provisions to enforce compliance with the plans on millions of people in the north Mediterranean and beyond. It is an awesome responsibility for the planners in Brussels. A wrong step and the face of Europe could be changed for centuries. Thus, for example, a subsidy designed to reduce over-production of olive oil by paying owners to uproot olive trees has already had a dramatic effect on landscapes from the Atlantic to the Aegean.

Fortunately there are some factors that slow down the processes and mitigate their impacts. One of these is imperfect knowledge. Managing through the market assumes that all participants have good information on the market. Another is conservatism – the risks and disadvantages of getting it wrong are too high, especially in economically less-favoured regions, though the risks in the lesser developed states is even higher. Third, if European policy requires inputs from the states (often on a cost-sharing basis), the nations may fail to press a particular Directive on its own farming communities, especially if there are strong local pressure groups or non-governmental organizations (NGOs) who oppose them. This problem is why Briassoulis (1999) describes planning as 'muddling through'. The end product is often an imperfect outcome, sometimes far away from what was intended. Farming systems are also complex. This complexity means a small change can have a large, sometimes quite unexpected effect. This effect can be partly offset by negotiations with the communities involved to reach the outcome desired. The approach has to be creative rather than analytical, cooperative rather than technical, according to van der Leeuw (1998).

Basic economic concepts

Manipulation of policy to stem the impacts of market failures arising from externalities was advocated by the economists Sidgwick and Pigou (see, for example, Pigou, 1920) and is now called Pigouvian economics.

There is a spectrum of control methods, classified by Baumol and Oates (1988) from command-and-control to research-and-education. In command-and-control, the authority mandates the behaviour in law, in order to achieve a socially desirable objective. This is the centralized approach to the externality problem and may be applied either to the process or the product. The tax-or-charge approach attempts to restore efficiency by manipulating, directly or indirectly, the price of polluting goods. Subsidies also aim to change the cost–benefit ratios of polluting (as in the case of over-production by intensification) in order to achieve a reduction in residues or emissions.

Monitoring and control are of major importance. If monitoring is ineffective by either sensitivity (how good the measurement is) or frequency (how often the measurements are made), then the authority cannot obtain compliance.

If a single irrigated farm allows its irrigation water to return to a river or an irrigation canal, the return water may be salty as a result of picking up salt from the soil. A downstream farmer, on using the water, suffers damage that rises with the amount of salt. The first farmer could abate the effect by reducing the solute output (either the load or the concentration) by recycling treatment, different irrigation applications or dilution. This process would lead to a rise in the abatement costs of the first farmer and should reduce the damage costs of the second farmer. Without government intervention, the emitter has no incentive to take into account the negative environmental externalities arising from his action. Policy intervention has to reach a position where both farmers benefit as much as possible – the optimal solution. For the first farmer this is the lowest possible abatement costs and for the second the lowest possible damage costs.

For a detailed explanation, we follow the arguments of Bonnieux and Guyomard (1999). It turns out that:

- The optimal level of pollution is not zero. Rather, the socially optimal level of emissions (water pollution) corresponds to the point where marginal damage and marginal abatement costs are equal. Higher emissions expose society to greater costs stemming from increased environmental damage; lower emissions require society to pay more abatement costs.
- Mechanisms for achieving these changes towards the optimum are:
  - to tax (the Pigouvian tax). The first farmer (polluter) pays for the efforts needed to reduce the salt levels until the socially optimal level is reached.
  - to set an emission standard which fixes the salt level that the first farmer can release into the river.

The first is price control; the second is quantity control by a command-and-control strategy. The implementation of either of these two strategies requires government intervention and public regulation, usually relying on a specific body such as an environmental agency, that holds the right to collect environmental taxes and to enforce standards. Under the tax strategy, the agency places a negative price on emissions which, if it works, results in a decentralized reduction in emission levels. Under the regulation strategy, the agency sets the amount of emissions that are permitted without penalty. Only in the case of a single emitter are both strategies the same as optimal and lead to the same amount of abatement for the same resource cost.

- With several polluters, each has its own marginal abatement cost, so there is a different optimal emission and different output from a Pigouvian tax. It appears that the company with the lower marginal abatement cost will reduce its emission level more. These variations mean it is questionable as to whether there is a uniform tax or a standard that can be applied where there are several or many emitters, as is usually the case.

- The tendency is for each source to reduce its emission until its marginal abatement cost equals the tax. Marginal abatement costs are then reduced across all the sources, resulting in minimum overall abatement costs.

- *The tax strategy is generally preferred*, because it provides a continuing incentive for improved abatement performance, while there is no incentive for abatement beyond that required by the standard set in the regulatory approach.

- It appears that agencies are generally against penalties. Standards are more often imposed than taxes, because the polluters can earn rents from standards, especially where these are in the form of permits to emit, which can be sold on. Bonnieux and Guyomard (1998) suggest four advantages of market-based strategies:

  - they are more likely to achieve a more efficient allocation of environmental protection than a regulation policy;
  - they provide a clear incentive for research into the development of less-polluting activities;
  - they mean less government interference and greater flexibility. These lead to less administrative costs; and
  - they can provide a source of government revenue.

Bonnieux and Guyomard conclude that emission charges or taxes are good in theory, but are often set at too low a level, so they cannot have any steering effect.

European policy and environmental management

These material efforts pale into insignificance when compared with the manipulations at the European level. In industrialized countries, the main objectives of agricultural policies are to support farm income and ensure a stable and reasonably-priced food supply. The Treaty of Rome embodied the Common Agricultural Policy (CAP) of the European Union. This was not revised until 1992 and essentially determined the pattern of production and land use over Europe, including the Mediterranean states. Under the productionist pre-1992 CAP, the most favoured mechanism of control was market price support for products and CAP has traditionally provided the bulk of its assistance in this form (see Chapter 14).

By 1986 it had been realized that manipulation at international (European) level distorts the market at global level and the Uruguayan round of GATT (General Agreement on Tariff and Trade) called for a common purpose in price control in the major economic groups. In 1992 the major review of CAP led to the MacSharry Reform that shifted the burden of support from consumers to tax payers, by cutting the product price support (subsidies) that had dominated farmers' behaviour and revenue for thirty-five years. At the same time, the CAP reforms sought to combine new controls with environmental protection, habitat protection and scenic landscape production, all of which are positive externalities but, as Bonnieux and Guyomard state (1998: 274): 'Every income support mechanism has a unique, secondary and unintentional effect on environmental quality.'

The pre-reform CAP produced two kinds of changes: more land was put into production; or there were changes in the intensity of use. It resulted in excessive use of fertilizers and pesticides and a socially inefficient use of chemicals. Production also extended on to marginal lands. In the Mediterranean this has meant that it was pushed beyond the limits of sensible production on to stony grounds with thin soils and little or no capability for irrigation.

In 1992, joint pressures from the GATT and the member states, combined with preparations for the Single European Act, led to the reform of CAP. The new (so-called post-productionist) CAP reduces the guaranteed price of the products at the market and the ratio of output to input, and should induce farmers to reduce variable input at the intensive margin. It also attempted to provide some protection for the environment. Potter (1998) points out that expectation of complete success in this regard might be offset by three factors:

- cutting the output price at institutional level does not mean an equivalent decrease in market price;
- the chemical fertilizer companies, faced with a potential decrease in their sales, may cut their prices in order to sell more fertilizers;
- price elasticities of yields and input-use may still be around the pre-reform equilibrium point.

He concludes, therefore, that the effects of price changes are likely to be limited. Certainly, their effects will depend on the enthusiasm with which the governments of the member states embrace them. Also, the farmers do not have complete flexibility, because they have to produce certain crops, such as cereals and oil seeds. The genetically modified food debate will, moreover, change the pattern of pieces on the European agricultural chess board. On the extensive margin of production, a support price cut should induce farmers to convert low-quality land, previously cultivated at high intensity, back into more extensive production. Much of the higher production in Mediterranean lands has been at this extensive margin and so this seems likely to be a significant impact with, perhaps, further reduction of effort in the marginal land. It seems as if the reform of product support prices (subsidies) began to bite quite soon. Support prices for grain went down from 155 green ECU/tonne in 1992/93 to 100 green ECU/tonne by 1995/96. These effects should be felt mostly at the intensive margin and intensive farmers might be expected to reduce input and adopt less intensive production techniques (such as low-input farming or no-tillage farming), which should both reduce growth rates in the medium term and reduce environmental damage externalities in the longer term.

CAP is not the only legislation that farmers have to contend with or that is directed to the environment. For example, the Nitrate Directive (91/676) restricts leakage of nitrates in vulnerable zones. Roughly this means that manure may not be applied at rates exceeding $170 \, kg \, N \, ha^{-1}$. The vulnerable zones (groundwater recharge zones) were to be defined before 1999, to include three to four zones around the well heads, with different permitted activities in each zone. The EU directives are legally binding documents. Potter (1998) discusses two alternative measures: incentive-based water effluent charges at point sources, based on water quality monitoring, and emission taxes. The latter are more difficult than the former to implement because diffuse sources are much more difficult to identify than point sources. In Box 21.1 we outline the main features of the Water Framework Directive of December 2000, potentially one of the most far-reaching environmental directives and one of enormous importance for the Mediterranean.

Erosion, a major problem in the Mediterranean countries, appears to have increased as a result of agricultural intensification in general, and specific directives in particular. Two examples are given in Box 21.2. The first illustrates the impact of price support on land use (and ultimately on erosion). The second illustrates that it may be unfair to assign all environmental disasters to European policy control and that each case deserves careful investigation.

Cena (1998: 242) concludes that:

> Agricultural development is today no longer seen as simply a matter of increased production. The political agenda also includes considerations such as efficient land use, efficient management of natural resources and protection of the environment. These aspects are vital to sustainable development and therefore have to be taken into account in designing agricultural development programmes.

Sustainability and environment

We have painted agriculture as a key issue in environmental problems. This sector also illustrates the key connection between sustainability and environment.

Sustainability has become a 'buzz' word since the United Nations Conference on Environment and Development, held in Rio de Janeiro in 1992, which agreed Agenda 21, the framework for implementing sustainable development. O'Riordan and Voisey (1998) avoid the

---

Box 21.1 EU Water Framework Directive, 2000

In December 2000, the European Commission issued a new Water Framework Directive, whose main objective is to provide surface and groundwater of good status and to avoid further deterioration of water quality. Good status means chemically and ecologically satisfactory on the basis of eleven indicator variables. The EU hopes that the directive will be fully implemented by 2015. It should lead to improved raw water quality, the protection and enhancement of aquatic wildlife, better targeting of protection, improvement of water resources and more transparency and accountability. A major emphasis of the directive is the integrated approach to river-basin management. This is a recognition that water quality and quantity is largely controlled by the condition of the catchment. In particular the changing land use, the increased application of fertilizers and the mechanization of farming have led to a need to reduce diffuse pollution, especially from agriculture and to improve river habitats. River rehabilitation and restoration will now have to start at the watershed. Another aspect of the directive, that is especially important in the drylands, is the alleviation of low flows that can lead to serious water pollution of ephemeral channel beds, which are used as waste repositories. A timetable for the programme is suggested, indicating that proper pricing policies should be in place ten years from the enactment of the directive. When enacted and carried through, this directive will have a major impact on water resources and water usage. We consider this to be the major environmental issue in Mediterranean countries at present.

Box 21.2 Two examples of EU policy impacts

Sardinia

A well-documented case comes from Sardinia, where it has been shown that the impacts of agro-pastoral activities on land degradation can be direct (animal hoof pressure) and indirect (bad land-use practices associated with grazing, such as burning to create artificial pastures on unsuitable areas). Sardinia has 3.3 million sheep, a remarkable increase in the last thirty years, favoured mainly by the rise in sheep milk prices in the 1970s and 1980s. The regional government of Sardinia adopted policies to increase forage production through price support. This resulted in a severe impact on the environment, wide areas of Mediterranean *maquis* were cleared to create artificial pastures by using fire, deep mechanical tillage on slopes that are too steep and seeding forage. The Desertification Research group of the University of Sassari carried out a wide range of detailed investigations in municipalities on the eastern side of the island using GIS, modelling and in-depth field studies (Enne, 1999). They discussed with the farmers and breeders what actions could and should be taken. From this they identified areas that could be suitable and eligible for mitigation actions. The results showed that, in the period 1955–1996, there were marked changes in land use in some parishes in eastern Sardinia that are the result of a general marked intensification of anthropogenic activities in the area. The 1996 pattern reflects a shift from extensive to semi-extensive production systems. Whereas the coastal areas experienced more tourist development, the hilly areas experienced a marked intensification of agro-pastoral systems. Between 1955 and 1996, there was a large increase in pasture lands (+26 per cent), a considerable reduction in woodlands (−36 per cent), a marked expansion of agricultural areas (+20 per cent) and a remarkable increase in urban areas (+61 per cent). Nearly all these changes reflect the increase in agro-pastoral activities and the policy efforts to increase forage. These increases have occurred despite a small reduction in arable lands.

Not all bad

In Lesvos, the area of olive groves fell by about 10 per cent between 1970 and 1996. This decrease had been attributed to support prices for sunflower oil. With the abandonment of olive groves, terrace maintenance was reduced, terraces collapsed and erosion has followed. There has been some debate about the impact of EU policy but MEDALUS research shows olive oil production fluctuated around its usual mean and there is no significant relationship between olive oil production and subsidy per kilogram in the 1991–1994 period. In other words, reaction was slow or non-existent to a policy instrument application. Current trends in the area of olive groves show that no changes occur, despite price support and other support measures for farmers. Moreover, if the regulation for the support of terraces is exploited, improvement of existing mountainous olive groves may occur. Also local support for the protection of olive groves may contribute to their maintenance and prevent conversion to other more profitable uses (e.g. tourism as second homes).

problem of defining sustainability by specifying the indicators of its achievement and looking to see if nations are moving towards it. These indicators are:

- continuity, durability and reliability of economic performance;
- stewardship, trusteeship and a duty of care towards vulnerable ecosystems and people and to future generations;
- localism, democratic innovation and greater self-reliance in communities in the face of environmental, economic and social insecurities as a measure of collective defence against threat.

The connection with the first of these is that, if existing practices cause so much environmental damage that economic performance is impaired, then the practices are not sustainable. With the second it is that unsustainable practices may increase susceptibility to catastrophe and undermine in-built resilience and so make it difficult or impossible for future generations to subsist. The third expresses the concept of empowerment, the idea that the 'owners' of the environment have both a responsibility for, and a right to be involved in, its management, because it is crucial to their futures (and usually to their present) quality of life.

O'Riordan and Voisey attempt to assess the extent to which the fine words of Rio have been worked out in practice in the European states and in the European Commission itself. The participating nations have signed the agreement, as has the European Union. This legally requires them to meet the articles of the Rio agreement. In Europe, these have been slightly reformulated in the Fifth Environment Action Plan. In the Treaty of Rome (1957), neither environment nor sustainability were formal components of the Treaty. Even in the Single European Act (1987–1991), the words 'sustainable development' were not included (Wilkinson, 1998). Implicitly, however, it was contained in the Maastricht Treaty on European Union (1993). This treaty also required that the European Union should integrate environmental policy in with the policies of the other sectors that are formulated and monitored by other Directorates-General, such as DG-VI (agriculture), DG-III (industry) and DG-XVII (energy). Each signatory state is also required to integrate environmental concerns into sectoral policies. In other words, the states have to come to terms with the requirements of Agenda 21 in their policies.

In a review of Portugal's progress towards sustainability, Ribeiro and Rodrigues (1998) found that, by the late 1990s, progress had been frankly slow, as did Fousekis and Lekakis (1998) reporting on Greece. Some reasons for this slow progress are:

- Newly emerging democracies do not have mature institutional arrangements. In both cases, the Ministries of Environment were newly formed, with wide-ranging and fragmented responsibilities.
- There was a low level of public participation because people were not accustomed to being invited to participate.
- There was insufficient expertise and technical competence in the Departments to develop the national sustainability plans.
- Emphasis was on major projects, such as the new bridge across the River Tejo at Lisbon or the new international airport at Spata near Athens. Both are said to have experienced low public involvement.
- The average citizen has only a very poor (or no) concept of the ideas involved in sustainable development.

But the progress is different and more encouraging, at least in the case of Portugal, where the public seems more aware of the issues. In a survey in Portugal, 25 per cent of the respondents claimed that the environment was one of the most serious problems in the country. Some 85 per cent of the respondents believed that the health of the next generation would be substantially affected by the degradation of the environment. About 60 per cent said that they would be willing to pay higher prices to protect it. The Portuguese government enacted the National Environment Policy Plan in 1995 and established public participation through the later Environmental Protection Associations Act, which not only established the Environmental Protection Associations (EPA), but also gave them the right to use public TV and radio channels to broadcast their messages and ideas, and to participate in defining environmental policy and its major legislative outcomes. The EPAs are of different sizes at different political levels and they present projects to the Environmental Ministry. In 1995, seventy-four were approved, out of the 178 submitted. The EPAs sought 100 million Portuguese Escudos (about 0.5 million ecu) and received 40 per cent of it. Ribeiro and Rodrigues (1998) suggest that it would be more effective if some areas were earmarked for funding (guaranteed support). The quality of the work done by each EPA varies according to its technical abilities, resources and the competency of the staff in conducting independent studies on issues concerning sustainable development. At the moment, lack of independent information seems to be the main problem at the national level. Ministries are jealous of their power and, because environment cuts across so many, it invariably faces difficulties.

Another form of national intervention for environmental protection comes from the national committees set up by the International Conventions, such as that for desertification (see Chapter 19). Here, however, the control tends to be 'top-down', with government experts telling local people how to manage their affairs.

These changes are encouraging, despite the slow rate of adoption. They bring Europeans closer to the southern continents of Africa and Australia in their recognition of the need for Land Care.

The Land Care approach in Australia and New Zealand, in which the government supports Land Care Committees to address environmental issues, such as soil salinization, appears to have been more oriented towards a 'bottom-up' approach than the European model. The approach involves end-users (farmers) and local community activists, as well as technical expertise. The Portuguese EPAs offer a similar 'bottom-up' approach that has been widely advocated in recent years. If successful, it could provide a good example for other Mediterranean states.

In concluding this section, we emphasize that much of what has been said refers to the developed countries of the Mediterranean in particular and to Europe in general. In developing countries, over the longer timescale, environmental reforms have mainly coincided with major socio-political changes as new regimes seek to boost production in industry and agriculture, which is generally the largest economic sector. The most common aims of agricultural reform are to increase output and stabilize farm prices. This in turn improves the capacity to increase foreign currency, generates revenue for the government through agricultural taxes and reduces the cost of foodstuffs to urban populations as well as providing raw materials for food processing and textile industry at below free market prices.

By contrast, government intervention in industrialized countries seeks to ensure efficiency, equity and sustainability in the agricultural sector and to facilitate agricultural reform to meet these aims. Naturally the EU policy is closely aligned to that of industrialized countries. These contrasts throw into doubt the wisdom of identifying a 'Mediterranean environmental policy' when the within-region economic contrasts are so great.

From this discussion, we must conclude that agricultural production, though largely constrained through physical factors, is a complex and delayed outcome of economic and social forces operating at several different scales, from local to global, that vary through time as political regimes come and go. Even more importantly, the contrasts between the northern and southern Mediterranean states make it difficult, if not impossible, to speak of Mediterranean agriculture and environmental issues in terms that imply there are common economic solutions to the problems of the environment across the region.

The future

The results of the first estimates of the Blue Plan (see Chapter 14) are now already testable. In a sense, they make chilling reading in that they show very little has changed in the ways that socio-economic change is driving environmental change since the early 1980s when the predictions were made. The trend scenarios for the year 2000 fit much more clearly with what has happened compared to the alternative scenarios, in which much more social and political co-operation would have led to conditions in which pressure on the environment was much less marked. This outcome is in spite of the changes that have seen integrated policies being developed via the Barcelona Conventions and the Mediterranean Action Plan (Chapter 17). At the same time, the sorts of predictions that are being made for potential climate change show significant impacts will take place (Chapter 20), often in areas that are already marginal and experiencing the greatest rates of population growth (Chapter 14). Using the shared experiences of the characteristics of the Mediterranean together with the localized approaches discussed above, it is clear that the kind of cooperation envisaged by the Blue Plan is absolutely fundamental in providing the means by which environmental sustainability can be maintained. Regional security and stability will become increasingly entwined in the twenty-first century with environmental issues in the Mediterranean.

Suggestions for further reading

Golley and Bellot (1998) is a valuable resource, written from a Mediterranean perspective. The contents cover almost all the issues covered in this chapter from a practical point of view. It is also useful as a complement to other chapters in this book. It contains the valuable reference to Bonnieux and Guyomard that is a beginner's guide to environmental economics and to Cena, who shows how they work out in practice. Dovers and Handmer (1992) provide a stimulating and prophetic discussion including the meaning of sustainability. O'Riordan and Voisey (1998) also discuss the meaning at length. A comparison of the two shows how the concept was developed in the intervening years.

Topics for discussion

1   What are externalities and how are they relevant to the policy for the European environment?
2   Comment on the quotation from Dovers and Handmer (1992: 262). How meaningful is it for today's Mediterranean environments?
3   Discuss the canons of Mediterranean environmental management listed in this chapter. Illustrate each one by example and suggest other canons. How universal are they and how specific to the Mediterranean?
4   Discuss the differences between productionist and post-productionist aspects of the EU Common Agricultural Policy.
5   Outline the main connections between environmental degradation and agricultural policy at different spatial scales.
6   Discuss how and why the physical environment should enter into policy planning at the municipal level.
7   Outline the main differences in agricultural policy in developed and lesser-developed countries. How does this relate to Rostow's model of economic development?
8   Why was 1992 a 'good year' for European environments?

Bibliography

Abrahams, A.D., A.J. Parsons and S.-H. Luk (1991) 'The effect of spatial variability in overland flow on the downslope pattern of soil loss on a semi-arid hillslope, southern Arizona', *Catena* 18, 255–270.

Acosta, P. (1987) 'El neolitico antiguo en el Suroeste español: la cueva de la Dehesilla (Cadiz)', in J. Guilaine, J. Courtin, J.-L. Roudil and J.-L. Vernet (eds) *Premières Communautés Paysannes en Méditerranée Occidentale*, 653–659, Éditions du CNRS, Paris.

Ager, D.V. (1980) *The Geology of Europe*. McGraw-Hill, London.

Agnew, C. and E. Anderson (1992) *Water Resources in the Arid Realm*. Routledge, London.

Agnew, C. and A. Chappell (2000) 'Hydrological response of desert margins to climatic change: the effect of changing surface properties', in S. McLaren and D. Kniveton (eds) *Linking Climate Change to Land Surface Change*, 27–48, Kluwer Academic, The Hague.

Aguilar, R. (2000) 'The view from a conservation NGO: the ethics of Spanish water policy', in R. Llamas (ed.) *Water and Ethics*, 9–15, Papeles del Proyecto, Series A, No 5, Aguas subterraneas, Fundación Marceline Botín, Madrid.

Aitkin, M.J., H.N. Michael, P.P. Betancourt and P.M. Warren (1988) 'The Thera eruption – continuing discussion of the dating', *Archaeometry* 30, 165–182.

Al-Agha, M.R. (1997) 'Environmental management in the Gaza Strip', *Environmental Impact Assessment Reviews* 17, 65–76.

Albaiges, J., M. Aubert and J. Aubert (1984) 'The footprints of life and man', in R. Margalef (ed.) *Key Environments: Western Mediterranean*, 317–352, Pergamon, Oxford.

Albaladejo, J., V. Castillo and A. Roldán (1991) 'Analysis, evaluation and control of soil erosion processes in a semiarid environment: S.E. Spain', in M. Sala, J.L. Rubio and J.M. García-Ruiz (eds) *Soil Erosion Studies in Spain*, 9–26, Geoforma Ediciones, Logroño.

Albaladejo Montoro, J. and S. Díaz Martinez (1983) *Planificación y Medio Ambiente de la Región de Murcia*, Editoria Regional de Murcia, Murcia.

Alberdi, J.C., J. Diaz and J.C. Montero (1998) 'Daily mortality in Madrid community 1986–1992: relationship with meteorological variables', *European Journal of Epidemiology* 14(6), 571–578.

Albert-Piñole, I. (1993) 'Tourism in Spain', in W. Pompl and P. Lavery (eds) *Tourism in Europe: Structures and Developments*, 242–261, CAB International, Wallingford.

Alcock, S.E. (1993) *Graecia Capta: The Landscapes of Roman Greece*. Cambridge University Press, Cambridge.

Alexander, D. (1982) 'Difference between "calanchi" and "biancane" badlands in Italy', in R.B. Bryan and A. Yair (eds) *Badland Geomorphology and Piping*, 71–85, GeoBooks, Norwich.

Alexandre, P. (1987) *Le climat en Europe au Moyen-Age*. CNRS, Paris.

Alexandris, S., P.M. Allen, I. Black, C. Blatsou, N. Calamaras, P. Giannopoulos, M. Lemon, T. Mimides, A. Poulovassilis, N. Psyhouyou and R.A.F. Seaton (1994) 'Agricultural production and water quality in the Argolid valley, Greece: a policy-relevant study in integrated method', in S.E. van der Leeuw (ed.) *Understanding the Natural and Anthropogenic Causes of Soil Degradation and Desertification in the Mediterranean Basin*. Volume 6: *Synthesis*, 281–326, Final Report on Contract EV5V-CT91-0021, EU, Brussels.

Allard, P., J. Carbonnelle, D. Dajleric, J. Le Bronec, P. Morel, M.C. Robe, J.M. Maurenas, R. Faivre-Pierrot,

D. Martin, J.C. Sabroux and P. Zettwoeg (1991) 'Eruptive and diffuse emissions of CO_2 from Mount Etna', *Nature* 351, 387–391.

Allaya, M. *et al.* (1995) *MEDAGRI: Annuaire des Économies Agricoles et Agro-Alimentaires des Pays Méditerranéens et Arabes*, CIHEAM-IAM, Montpellier.

Allée, P. and M. Denèfle (1989) 'La Coma del Tech: un exemple de ravinement protohistorique dans les Pyrénées Orientales', *Bulletin de l'Association de Géographes Français* 1989-1, 57–72.

Allen, H.D. (2001) *Mediterranean Ecogeography*. Prentice-Hall, London.

Allen, P.M., I. Black, M. Lemon, R.A.F. Seaton, C. Blatsou and N. Calamaras (1994) 'Agricultural production and water quality in the Argolid valley, Greece: a policy-relevant study in integrated method', in S.E. van der Leeuw (ed.) *Understanding the Natural and Anthropogenic Causes of Soil Degradation and Desertification in the Mediterranean Basin.* Volume 5: *Agricultural Production and Water Quality in the Argolid, Greece*, 3–166, Final Report on Contract EV5V-CT91-0021, EU, Brussels.

Allen, P.M., M. Lemon and R. Seaton (1999) 'Sustainable water use in the Argolid: a sub-project of Archaeomedes', in P. Balabanis, D. Peter, A. Ghazi and M. Tsogas (eds) *Mediterranean Desertification: Research Results and Policy Implications*, 423–429, EC DG XII, EUR 19303, Brussels.

Almeida-Teixera, M.E., R. Fantechi, R. Oliveira and A. Gomez Coelho (eds) (1991) *Natural Hazards and Engineering Geology. Prevention and Control of Landslides and Other Mass Movements.* EUR 12918, European Commission, Brussels.

Aloïsi, J.-C., A. Monaco, N. Planchais, J. and Y. Thommeret (1978) 'The Holocene transgression in the Golfe du Lion, Southwestern France: palaeogeographic and palaeobotanical evolution', *Géographie Physique du Quaternaire* 32, 145–162.

Alonso Sarria, F. and F. López Bermúdez (1994) 'Rainfall time and space variability during short storms in south-east Spain', *Geoökodynamik* XV, 261–278.

Alvarez Vazquez, J.A. (1986) 'Drought and rainy periods in the province of Zamora in the 17th, 18th and 19th centuries', in F. López-Vera (ed.) *Quaternary Climate in the Western Mediterranean: Proceedings of the Symposium on Climatic Fluctuations during the Quaternary in the Western Mediterranean Regions*, 221–235, Universidad Autónoma de Madrid, Madrid.

Ambert, P. (1974) 'Les dépressions nivéo-éoliennes de Basse Provence', *Comptes Rendus de l'Académie des Sciences de Paris D* 279, 729.

Ambert, P. and J. Gascó (1989) 'Les tufs de Saint-Guilhem-le-Désert: évolution holocène et pression anthropique sur le milieu karstique', *Bulletin du Musée d'Anthropologie Préhistorique de Monaco* 32, 63–85.

Ambraseys, N.N. and C.F. Finkel (1987) 'Seismicity of Turkey and neighbouring regions, 1899–1915', *Annales Geophysicae* 5B, 701–726.

Ambraseys, N.N. and C.F. Finkel (1988) 'The Anatolian earthquake of 17 August 1668', in W.H.K. Lee, H. Meyers, and K. Shimazaki (eds) *Historical Seismograms and Earthquakes of the World*, 173–180, Academic Press, London.

Ambraseys, N.N. and C.P. Melville (1988) 'An analysis of the eastern Mediterranean earthquake of 20 May 1202', in W.H.K. Lee, H. Meyers, and K. Shimazaki (eds) *Historical Seismograms and Earthquakes of the World*, 181–200, Academic Press, London.

Ammerman, A.J. (1985) *The Acconia Survey: Neolithic Settlement and the Obsidian Trade.* Institute of Archaeology Occasional Paper no. 10, London.

Ammerman, A.J. and L.L. Cavalli-Sforza (1984) *The Neolithic Transition and the Genetics of Populations in Europe.* Princeton University Press, Princeton, NJ.

Amouric, H. (1992) *Le Feu à l'Épreuve du Temps.* Narration, Aix-en-Provence.

Anati, E. (1961) *Camonica Valley.* Jonathon Cape, London.

Andalucia, Junta de (1990) *El Medio Ambiente en Andalucia en 1989.* Informe 89. Consejeria de Cultura y Medio Ambiente, Sevilla.

André, F. and A. Robert (1985) 'Maîtrise de l'eau et développement agricole dans le sud Chypriote', *Méditerranée* 56(4), 3–11.

André, J., J.-L. Brochier, J. Guilaine, G. Jalut, H. de Lumley, R. Letolle and J.-L. Vernet (1979) 'Éléments pour une approche paléoclimatique et paléoécologique du Plateau de Lacamp à l'atlantique', in

J. Guilaine, J. Gascó and M. Barbaza (eds) *L'Abri Jean Cros*, 244–248, Centre d'Anthropologie des Sociétés Rurales, Toulouse.

Andreu, V., J. Forteza, J.L. Rubio and R. Cerni (1994) 'Nutrient losses in relation to vegetation cover on automated field plots', in R.J. Rickson (ed.) *Conserving Soil Resources: European Perspectives*, 116–126, CAB International, Wallingford.

Andrieu, V., E. Brugiapaglia, J.-L. de Beaulieu and M. Reille (1995) 'Enregistrement pollinique des modalités et de la chronologie de l'anthropisation des écosystèmes méditerraneéens d'Europe et du Proche-Orient: Bilan des connaissances', in M. Dubost (ed.) *MEDIMONT Second and Final Scientific Report. Volume II/III Ecological Regional Approaches.* Final Report on Contract EV5V-CT91-0045, EU, Brussels.

Andrikopoulou, E. (1987) 'Regional policy and local development prospects in a Greek peripheral region: the case of Thraki', *Antipode* 19, 7–24.

Angelier, J., N. Liberis, X. Le Pichon, E. Barrier and P. Huchon (1982) 'The tectonic development of the Hellenic arc and the sea of Crete: a synthesis', *Tectonophysics* 86, 139–196.

Angelier, J. and X. Le Pichon (1978) 'L'arc hellénique, clé de l'évolution cinématique de la Méditerranée orientale depuis 13 Ma', *Comptes Rendus de l'Académie de Sciences de la France* 287D, 1325–1328.

Angelier, J. and X. Le Pichon (1980) 'Néotectonique horizontale et verticale de l'Egée: subduction et expansion', in J. Aubouin, J. Debelmas and M. Latreille (eds) *Géologie des Chaînes Alpines Issues de la Téthys*, 249–260, Mémoire du BRGM no. 115, Editions du BRGM, Orléans.

Antoine, J.-M. (1988) 'Un torrent oublié mais catastrophique en Haute-Ariège', *Revue Géographique des Pyrénées et du Sud-Ouest*, 59(1), 73–88.

Arcelin, P. and J. Brémond (1978) 'Le gisement protohistorique du Mont-Valence, commune de Fontvielle (Bouches-du-Rhône)', *Cypsela* 2, 161–172.

Arenson, S. (1990) *The Encircled Sea: The Mediterranean Maritime Civilization.* Constable, London.

Arianoutsou-Faraggitaki, M. (1984) 'Post-fire successional recovery of a phryganic (East Mediterranean) ecosystem', *Acta Oecologica – Oecologia Plantarum* 59, 387–394.

Ariztegui, D., C. Chondrogianni, G. Wolff, A. Asioli, J. Teranes, S. Bernasconi and J.A. McKenzie (1996) 'Paleotemperature and paleosalinity history of the Meso Adriatic Depression (MAD) during the Late Quaternary: a stable isotopes and alkenones study', in F. Oldfield and P. Guilizzoni (eds) *Palaeoenvironmental Analysis of Italian Crater Lake and Adriatic Sediments: Memorie dell'Istituto Italiano di Idrobiologia*, 55, 219–230.

Arnell, N.W. (1999) 'The effect of climate change on hydrological regimes in Europe: a continental perspective', *Global Environmental Change* 9(1), 5–23.

Aru, A., P. Baldaccini, M.R. Lai, R. Puddu, D. Tomasi and A. Vacca (1996) 'Santa Lucia Field site, Sardinia, Italy', in P. Mairota, J.B. Thornes and N. Geeson (eds) *Atlas of Mediterranean Environments in Europe: The Desertification Context*, 116–120, John Wiley and Sons, Chichester.

Aru, A. and G. Baroccu (1993) 'Field site investigations: Rio Santa Lucia, Sardinia', in J.B. Thornes (ed.) *MEDALUS I Final Report*, 534–559, Final Report on Contract EPOC-CT90-0014-(SMA), EU, Brussels.

Aschan-Leygonie, Chr., F. Durand-Dastès, H. Mathian, D. Pumain and L. Sanders (1994) 'Degradation, desertion and urban impact in the Lower Rhône valley', in S.E. van der Leeuw (ed.) *Understanding the Natural and Anthropogenic Causes of Soil Degradation and Desertification in the Mediterranean Basin. Volume 4: Degradation, Desertion and Urban Impact in the Lower Rhône Valley 1800–1990*, 3–159, Final Report on Contract EV5V-CT91-0021, EU, Brussels.

Atherden, M., J. Hall and J.C. Wright (1993) 'A pollen diagram from the northeast Peleponnese, Greece: implications for vegetation history and archaeology', *The Holocene* 3, 351–356.

Atkinson, B.W. (1981) *Meso-Scale Atmospheric Circulations.* Academic Press, London.

Attard, D.J., V. Axiak, S. Borg, S.F. Borg, J. Cachia, G. De Bono, E. Lanfranco, R.E. Micallef and J. Mifsud (1996) 'Implications of expected climatic changes for Malta', in L. Jeftić, S. Kečkeš and J.C. Pernetta (eds) *Climate Change and the Mediterranean.* Volume 2, 322–430, Arnold, London.

Aubet, M.E. (1993) *The Phoenicians and the West: Politics, Colonies and Trade.* Cambridge University Press, Cambridge.

Aubreville, A. (1949) *Climats, Forêts et Désertification de l'Afrique Tropicale*. Société d'Éditions Géographiques Maritimes et Coloniales, Paris.

Audouze, F., J. Argant, J.-L. Ballais, A. Beeching, V. Bel, J.-F. Berger, M. Bois, J.-L. Brochier, G. Chouquer, F. Favory, J.-L. Fiches, M. Gazenbeek, J.-J. Girardot, C. Jung, F. Magnin, J.-Cl. Meffre, T. Odiot, C. Raynaud, S. Thiébault, F.-P. Tourneux, X. Tschanz, Ph. Verhagen, M.-P. Zannier and S.E. van der Leeuw (1994) 'Land use, settlement pattern and degradation in the Ancient Rhône valley', in S.E. van der Leeuw (ed.) *Understanding the Natural and Anthropogenic Causes of Soil Degradation and Desertification in the Mediterranean Basin*. Volume 6: *Synthesis*, 175–221, Final Report on Contract EV5V-CT91-0021, EU, Brussels.

Avila, A., C. Neal and J. Terradas (1996) 'Climate change implications for streamflow and streamwater chemistry in a Mediterranean catchment', *Journal of Hydrology* 177, 99–116.

Avila, A. and J. Peñuelas (1999) 'Increasing frequency of Saharan rains over northeastern Spain and its ecological consequences', *The Science of the Total Environment* 228, 153–156.

Avner, U. (1990) 'Ancient agricultural settlement and religion in the Uvda Valley in southern Israel', *Biblical Archaeologist* 53, 125–141.

Avner, U., I. Carmi and D. Segal (1994) 'Neolithic to Bronze Age settlement of the Negev and Sinai in light of radiocarbon dating: a view from the southern Negev', in O. Bar-Yosef and R.S. Kra (eds) *Late Quaternary Chronology and Paleoclimates of the Eastern Mediterranean*, 265–300, Radiocarbon, Tucson, AZ.

Axelrod, D.I. (1993) 'History of the Mediterranean ecosystem in California', in F. di Castri and H.F. Mooney (eds) *Mediterranean Type Ecosystems*, 225–277, Springer Verlag, Berlin.

Baggioni, M., J.-P. Suc and J.-L. Vernet (1981) 'Le Plio-Pléistocène de Camerota (Italie Méridionale): géomorphologie et paléoflores', *Géobios* 14, 229–237.

Bailey, G.N. and I. Davidson (1983) 'Site exploitation territories and topography: two case studies from Palaeolithic Spain', *Journal of Archaeological Science* 10, 87–115.

Bailey, G.N., G.P.C. King and D. Sturdy (1993) 'Active tectonics and land-use strategies: a palaeolithic case study from north-west Greece', *Antiquity* 67, 292–312.

Baillie, M.G.L. and M.A.R. Munro (1988) 'Irish tree rings, Santorini and volcanic dust veils', *Nature* 332, 344–346.

Baker, V.R. (1998) 'Palaeohydrology and the hydrological sciences', in G. Benito, V.R. Baker and K.J. Gregory (eds) *Palaeohydrology and Environmental Change*, 1–12, John Wiley and Sons, Chichester.

Balabanis, P., D. Peter, A. Ghazi and M. Tsogas (eds) (1999) *Mediterranean Desertification: Research Results and Policy Implications*, EC DG XII, EUR 19303, Brussels.

Balcer, J.M. (1974) 'The Mycenaean dam at Tiryns', *American Journal of Archaeology* 78(2), 141–149.

Balista, C. and G. Leonardi (1985) 'Hill slope evolution: pre- and protohistoric occupation in the Veneto', in C. Malone and S. Stoddart (eds) *Papers in Italian Archaeology IV. Part i: The Human Landscape*, 135–152, BAR International Series 243, Oxford.

Ballais, J.-L. (1996) 'L'âge du modelé de roubines dans les Préalpes du Sud: l'exemple de la région de Digne', *Géomorphologie: Relief, Processus, Environnement* 1996(4), 61–68.

Ballais, J.-L., M.-C. Bosc and A. Sandoz (1992) 'La morphogenèse sur la montagne Sainte-Victoire après l'incendie: l'exemple du ruissellement (1989–1992)', *Méditerranée* 75(1–2), 43–52.

Ballais, J.-L. and A. Crambes (1992) 'Morphogenèse holocène, géosystème et anthropisation sur la montagne Sainte-Victoire', *Méditerranée* 75(1–2), 29–41.

Ballais, J.-L., M. Jorda, M. Provensal and J. Covo (1993) 'Morphogénèse holocène sur le périmètre des Alpilles', in P. Leveau and M. Provensal (eds) *Archéologie et Environnement de la Sainte-Victoire aux Alpilles*, 65–74, Université de Provence, Aix-en-Provence.

Ballais, J.-L. and J.-Cl. Meffre (1994a) 'Vaison et ses compagnes dans l'Antiquité et le Haut Moyen Age (Haut Comtat Venaissin, Vaucluse): archéologie de l'espace rural', in S.E. van der Leeuw (ed.) *Understanding the Natural and Anthropogenic Causes of Soil Degradation and Desertification in the Mediterranean Basin*. Volume 3: *Dégradation et Impact Humain dans la Moyenne et Basse Vallée du Rhône dans l'Antiquité (Part I)*, 37–53, Final Report on Contract EV5V-CT91-0021, EU, Brussels.

Ballais, J.-L. and J.-Cl. Meffre (1994b) 'La terrasse du Plan de Dieu: contraintes, occupation du sol, aménagement', in *L'Homme et la Dégradation de l'Environnement: XVe Rencontres Internationales d'Archéologie et d'Histoire d'Antibes*, 231–244, Éditions APDCA, Juan-les-Pins.

Baltas, N.C. (1997) 'The restructured CAP and the periphery of the EU', *Food Policy* 22, 329–343.

Barberi, F. and L. Villari (1994) 'Volcano monitoring and civil protection problems during the 1991–1993 Etna eruption', *Acta Vulcanologica* 4, 157–165.

Barbero, M., G. Bonin, R. Loisel and P. Quézel (1990) 'Changes and disturbances of forest ecosystems caused by human activities in the western part of the Mediterranean basin', *Vegetatio* 87, 151–173.

Barbero, M., R. Loisel and P. Quézel (1992) 'Biogeography, ecology and history of Mediterranean *Quercus ilex* ecosystems', *Vegetatio* 99–100, 19–34.

Bard, E., M. Arnold, R.G. Fairbanks and B. Hamelin (1993) 'Th^{230}-U^{234} and C^{14} ages obtained by mass-spectrometry on corals', *Radiocarbon* 35, 191–199.

Barfield, L. (1971) *Northern Italy Before Rome*. Thames and Hudson, London.

Barfield, L. (1994) 'The Bronze Age of northern Italy: recent work and social interpretation', in C. Mathers and S. Stoddart (eds) *Development and Decline in the Mediterranean Bronze Age*, 129–144, J.R. Collis Publications, Sheffield.

Barker, G.W.W. (1981) *Landscape and Society: Prehistoric Central Italy*. Academic Press, London.

Barker, G.W.W. (1995) *A Mediterranean Valley: Landscape Archaeology and Annales History in the Biferno Valley*. Leicester University Press, London.

Barker, G.W.W. (ed.) (1996) *Farming the Desert: The UNESCO Libyan Valleys Archaeological Survey*. Volume 1: *Synthesis*. UNESCO/Department of Antiquities (Tripoli)/Society for Libyan Studies, London.

Barker, G.W.W. and S. Stoddart (1994) 'The Bronze Age of central Italy: *c*.2000–900 BC', in C. Mathers and S. Stoddart (eds) *Development and Decline in the Mediterranean Bronze Age*, 145–166, J.R. Collis Publications, Sheffield.

Bar-Matthews, M., A. Ayalon and A. Kaufman (1997) 'Late Quaternary palaeoclimate in the eastern Mediterranean region from stable isotope analysis of spelaeothems at Soreq Cave, Israel', *Quaternary Research* 47, 155–168.

Barrocou, G., J.J. Collin and C. Mouvet (1998) 'The increasing demand for water', in P. Mairota, J.B. Thornes and N. Geeson (eds) *Atlas of Mediterranean Environments in Europe: The Desertification Context*, 98–104, John Wiley and Sons, Chichester.

Barrocou, G., M.G. Sciabia and C. Panisconi (1994) 'Three-dimensional model of salt water intrusion in the Capoterra coastal aquifer system (Sardinia)', *Proceedings of 13th Salt Water Intrusion Meeting, Cagliari*, Vilasmius-Cagliari.

Barry, R.G. and R.J. Chorley (1992) *Atmosphere, Weather and Climate*. Routledge, London.

Baruch, U. (1986) 'The late Holocene vegetational history of Lake Kinneret (Sea of Galilee), Israel', *Paléorient* 12(2), 37–48.

Baruch, U. (1994) 'The Late Quaternary pollen record of the Near East', in O. Bar-Yosef and R.S. Kra (eds) *Late Quaternary Chronology and Paleoclimates of the Eastern Mediterranean*, 103–119, Radiocarbon, Tucson, AZ.

Baruch, U. and S. Bottema (1991) 'Palynological evidence for climatic changes in the Levant *ca.* 17000–9000 B.P.', in O. Bar-Yosef and F.R. Valla (eds) *The Natufian Culture in the Levant*, 11–20, International Monographs in Prehistory, Ann Arbor, MI.

Bar-Yosef, O. (1986) 'The walls of Jericho: an alternative interpretation', *Current Anthropology* 27(2), 157–162.

Bar-Yosef, O. (1987) 'Prehistory of the Jordan Rift', *Israel Journal of Earth Sciences* 36, 107–119.

Bar-Yosef, O. (1991) 'Stone tools and social context in Levantine prehistory', in G.A. Clark (ed.) *Perspectives on the Past: Theoretical Biases in Mediterranean Hunter-Gatherer Research*, 371–395, University of Pennsylvania Press, Philadelphia, PA.

Bates, D.G. (1973) *Nomads and Farmers: A Study of the Yörük of Southeastern Turkey*. Anthropology Papers of the Museum of Anthropology no. 52, University of Michigan, Ann Arbor, MI.

Baumol, W.J. and W.E. Oates (1988) *The Theory of Environmental Policy*. Cambridge University Press, Cambridge.

Beaumont, P. (1993) *Drylands: Environmental Management and Development.* 2nd edn. Routledge, London.

Bebber, A.E. (1990) 'Una cronologia del Larice (*Larix decidua* Mill.) delle Alpi Orientali Italiane', *Dendrochronologia* 8, 119–139.

Beckinsale, M. and R. Beckinsale (1975) *Southern Europe: The Mediterranean and Alpine Lands.* University of London Press, London.

Benech, B., H. Brunet, V. Jacq, M. Payen, J.-C. Rivrain and P. Santurette (1993) 'La catastrophe de Vaison-la-Romaine et les violentes precipitations de septembre 1992: aspects météorologiques', *La Météorologie* 8(1), 72–90.

Beniston, M. and R.S.J. Tol (eds) (1998) 'Europe', in R.T. Watson, M.C. Zinyowera, R.H. Moss and D.J. Dokken (eds) *The Regional Impacts of Climate Change: An Assessment of Vulnerability*, 149–185, IPCC/Cambridge University Press, Cambridge.

Benito, G., M. Gutiérrez and C. Sancho (1991) 'Erosion patterns in rill and interrill areas in badland zones of the middle Ebro Basin (NE Spain)', in M. Sala, J.L. Rubio and J.M. García-Ruiz (eds) *Soil Erosion Studies in Spain*, 41–54, Geoforma Ediciones, Logroño.

Benito, G., M. Gutiérrez and C. Sancho (1992) 'Erosion rate in badland areas of the Central Ebro basin (NE-Spain)', *Catena* 19, 269–286.

Benson, R.H. and K. Rakic-El Bied (1991) 'Biodynamics, saline giants and Late Miocene catastrophism', *Carbonates and Evaporites* 6(2), 127–168.

Ben Tiba, B. and M. Reille (1982) 'Recherches pollenanalytiques dans les montagnes de Kroumirie (Tunisie septentrionale): premiers résultats', *Ecologia Mediterranea* 8(4), 75–86.

Bergadà, M.M., J.M. Fullola, D. Serrat, J. Montserrat and J.M. Vilaplana (1992) 'Aproximación a la evolución paleoecologica del periodo tardiglaciar y postglaciar del Pirineo central (Ribagorza y Noguera)', *Cuaternario y Geomorfología* 6, 45–57.

Berger, A.L. (1978) 'Long-term variations of caloric insolation resulting from the Earth's orbital elements', *Quaternary Research* 9, 139–167.

Berger, J.-F., with contributions from A. Beeching, J.-L. Brochier and J. Vital (1994) 'Le Bassin de la Valdaine', in S.E. van der Leeuw (ed.) *Understanding the Natural and Anthropogenic Causes of Soil Degradation and Desertification in the Mediterranean Basin. Volume 3: Dégradation et Impact Humain dans la Moyenne et Basse Vallée du Rhône dans l'Antiquité (Part I)*, 159–256, Final Report on Contract EV5V-CT91-0021, EU, Brussels.

Bernard, J. and M. Reille (1987) 'Nouvelles analyses polliniques dans l'Atlas de Marrakech, Maroc', *Pollen et Spores* XXIX, 225–240.

Bernard-Allée, Ph., C. Martin, J.L. Guendon and A. Delgiovine (1994) 'Approche historique de l'érosion méchanique des sols', in *Actes du Workshop des 1 et 2 Octobre 1994, Barcelone*, Report on Contract EV5V-CT91-0039, EU, Brussels.

Bertoldi, R., D. Rio and R. Thunell (1989) 'Pliocene–Pleistocene vegetational and climatic evolution of the south-central Mediterranean', *Palaeogeography, Palaeoclimatology, Palaeoecology* 72, 263–275.

Betancourt, P.P (1987) 'Dating the Aegean Late Bronze Age with radiocarbon', *Archaeometry* 29, 45–49.

Betancourt, P.P. and H.N. Michael (1987) 'Dating the Aegean Late Bronze Age with radiocarbon – addendum', *Archaeometry* 29, 212–213.

Bethoux, J.P. (1980) 'Mean water fluxes across sections in the Mediterranean Sea, evaluated on the basis of water and salt budgets and of observed salinities', *Oceanologica Acta* 3, 79–88.

Bethoux, J.P. and B. Gentili (1994) 'The Mediterranean Sea, a test area for marine and climatic interactions', in P. Malanotte-Rizzoli and A.R. Robinson (eds) *Ocean Processes in Climate Dynamics: Global and Mediterranean Examples*, 239–254, Kluwer Academic, Dordrecht.

Beucher, F. (1979) 'Étude palynologique', in C. Roubet (ed.) *Économie Pastorale Préagricole en Algérie Orientale: Le Néolithique de Tradition Capsienne. Exemple: L'Aurès*, 419–426, Éditions du CNRS, Paris.

Beug, H.J. (1967a) 'Contributions to the postglacial vegetational history of northern Turkey', in E.J. Cushing and H.E. Wright Jr. (eds) *Quaternary Paleoecology*, 349–356, Yale University Press, New Haven, CT.

Beug, H.J. (1967b) 'On the forest history of the Dalmatian coast', *Review of Palaeobotany and Palynology* 2, 271–279.

Beug, H.J. (1977) 'Vegetationsgeschtiche Untersuchungen in Küstenbereich von Istrien (Jugoslawien)', *Flora* 166, 357–381.

Bigg, G.R. (1994) 'An ocean general circulation model view of the glacial Mediterranean thermohaline circulation', *Paleoceanography* 9, 705–722.

Bigg, G.R. (1995) 'Aridity of the Mediterranean Sea at the last glacial maximum: a reinterpretation of the $\delta^{18}O$ record', *Paleoceanography* 10, 283–290.

Biju-Duval, B., J. Dercourt and X. Le Pichon (1977) 'From the Tethys Ocean to the Mediterranean Seas: a plate tectonic model of the evolution of the western Alpine system', in B. Biju-Duval and L. Montadert (eds) *Structural History of the Mediterranean Basins*, 143–164, Éditions Technip, Paris.

Binder, D. (1992) *Le Néolithique Ancien Provençal: Typologie et Technique des Outillages Lithiques.* Éditions du CNRS, Paris.

Biondi, F. (1992) 'Development of a tree-ring network for the Italian peninsula', *Tree-Ring Bulletin* 52, 15–29.

Black, R. (1992) *Crisis and Change in Rural Europe: Agricultural Development in the Portuguese Mountains.* Avebury Press, Aldershot.

Blackman, D.J. and K. Brannigan (1977) 'An archaeological survey of the lower catchment of the Ayiofarango valley', *Annals of the British School at Athens* 72, 13–84.

Blacksell, M. (1984) 'The European Community and the Mediterranean region: two steps forward, one back', in A. Williams (ed.) *Southern Europe*, 269–288, Harper and Row, London.

Blamey, M. and C. Grey-Wilson (1993) *Mediterranean Wild Flowers.* HarperCollins Publishers, London.

Blumler, M.A. (1993) 'Successional pattern and landscape sensitivity in Mediterranean and Near East', in D.S.G. Thomas and R.J. Allison (eds) *Landscape Sensitivity*, 287–308, John Wiley and Sons, Chichester.

Bluth, G.J.S., C.C. Schnetzler, A.J. Krueger and L.S. Walter (1993) 'The contribution of explosive volcanism to global atmospheric sulphur dioxide concentrations', *Nature* 366, 327–329.

Boardman, J. (1976) 'The olive in the Mediterranean: its culture and use', *Philosophical Transactions of the Royal Society of London* B275, 187–196.

Bocco, G. (1991) 'Gully erosion: processes and models', *Progress in Physical Geography* 15(4), 392–406.

Boero, V. and U. Schwertmann (1989) 'Occurrence and transformations of iron and manganese in a colluvial terra rossa toposequence of northern Italy', *Catena* 14(6), 519–531.

Bolle, H.J. (1995) 'Climate and desertification', in R. Fantechi, D. Peter, P. Balabanis and J.L. Rubio (eds) *Desertification in a European Context: Physical and Socio-Economic Aspects*, 15–34, EC, Luxembourg.

Bonacci, O. (1987) *Karst Hydrology with Special Reference to the Dinaric Karst.* Springer-Verlag, Berlin.

Bonini, A. (1993) 'Tourism in Italy', in W. Pompl and P. Lavery (eds) *Tourism in Europe: Structures and Developments*, 302–323, CAB International, Wallingford.

Bonnieux, F. and H. Guyomard (1999) 'Public policies, markets and externalities', in F.B. Golley and J. Bellot (eds) *Rural Planning from an Environmental Systems Perspective*, 267–288, Springer-Verlag, Berlin.

Bono, P. and C. Boni (1996a) 'Water supply of Rome in antiquity and today', *Environmental Geology* 27, 126–134.

Bono, P. and C. Boni (1996b) 'Mineral waters of Italy', *Environmental Geology* 27, 135–142.

Bordes, F. (1976) *Leçons sur le Paléolithique.* Vol. II: *Le Paléolithique en Europe.* Cahiers du Quaternaire No. 7, Éditions du CNRS, Paris.

Bordes, F. (1984) *Leçons sur le Paléolithique.* Vol. II: *Le Paléolithique en Europe.* Cahiers du Quaternaire No. 7, Éditions du CNRS, Paris.

Borja Barrera, F. (1992) 'Cuaternario Reciente, Holoceno y Periodos Historicos del SW de Andalucia. Paleogeografia de Medios Litorales y Fluvio-Litorales de los Ultimos 30000 Años', unpublished PhD thesis, University of Seville.

Bottema, S. (1974) 'Implications of a pollen diagram from the Adriatic Sea', *Geologie en Mijnbouw* 53, 401–405.

Bottema, S. (1979) 'Pollen analytical investigations in Thessaly (Greece)', *Palaeohistoria* 21, 19–40.

Bottema, S. (1980) 'Palynological investigations on Crete', *Review of Palaeobotany and Palynology* 31, 193–217.

Bottema, S. (1982) 'Palynological investigations in Greece with special reference to pollen as an indicator of human activity', *Palaeohistoria* 24, 257–289.

Bottema, S. and H. Woldring (1984) 'Late Quaternary vegetation and climate of southwestern Turkey, part II', *Palaeohistoria* 26, 123–149.

Bove, E., G. Quaranta and M. Dubost (1995) 'Desertification in southern Italy – the case of clay-hill areas in Basilicata region', in M. Dubost (ed.) *MEDIMONT: A Multinational, Multidisciplinary Research Programme on the Rôle and Place of the Mountains in the Desertification of the Mediterranean Mountain Regions. Second and Final Scientific Report.* Final Report on Contract EV5V-CT91-0045, EU, Brussels.

Bradley, R.S. (1999) *Paleoclimatology: Reconstructing Climates of the Quaternary.* 2nd edn. Academic Press, London.

Brande, A. (1973) 'Untersuchungen zur postglazialen Vegetationsgeschichte im Gebeit der Neretva-Niederungen (Dalmatien, Herzogowina)', *Flora* 162, 1–44.

Brandt, C.J. (1989) 'The size distribution of throughfall drops under vegetation canopies', *Catena* 16, 507–524.

Brandt, C.J. (1990) 'Simulation of the size distribution and erosivity of raindrops and throughfall drops', *Earth Surface Processes and Landforms* 15(8), 687–698.

Brandt, C.J. and J.B. Thornes (eds) (1996) *Mediterranean Desertification and Land Use.* John Wiley and Sons, Chichester.

Braudel, F. (1975) *The Mediterranean and the Mediterranean World in the Age of Philip II.* 2 vols. Collins, London.

Braudel, F. (1981) *Civilization and Capitalism 15th–18th Century,* Volume 1: *The Structures of Everyday Life, The Limits of the Possible.* Collins, London.

Braun-Blanquet, J. (1932) *Plant Sociology* (translated, revised and edited by G.D. Fuller and H.S. Conrad). McGraw and Hill, New York.

Brewer, D.J. and E. Teeter (1999) *Egypt and the Egyptians.* Cambridge University Press, Cambridge.

Briassoulis, H. (1993) 'Tourism in Greece', in W. Pompl and P. Lavery (eds) *Tourism in Europe: Structures and Developments,* 282–301, CAB International, Wallingford.

Briassoulis, H. (1999) 'Sustainable development and the informal sector: an uneasy relationship', *Journal of Environment and Development* 8(9), 211–213.

Brochier, J-E. (1984) 'Chênes à feuillage caduc, chênes verts et stabilité des versants', in *Influences Méridionales dans l'Est et le Centre-Est de La France au Néolithique: Le Rôle du Massif Central. Actes du 8ᵉ Colloque Interrégional sur le Néolithique. Le Puy 1981,* 321–327, CREPA, Clérmont Ferrand.

Brochier, J.-L., P. Mandier, J. Argent and P. Petiot (1991) 'La cône détritique de la Drôme: une contribution à la connaisance de l'Holocène du sud-est de la France', *Quaternaire* 2, 83–99.

Broecker, W.S., G. Bond and M. Klas (1990) 'A salt oscillator in the glacial North Atlantic? 1. The concept', *Paleoceanography* 5, 469–477.

Broecker, W.S. and G.H. Denton (1990) 'The role of ocean-atmosphere reorganisations in glacial cycles', *Quaternary Science Reviews* 9, 305–341.

Broecker, W.S., J.P. Kennett, B.P. Flower, J.T. Teller, S. Trumboe, G. Bonani and W. Wölfli (1989) 'Routing of meltwater from the Laurentide ice sheet during the Younger Dryas cold episode', *Nature* 341, 318–321.

Broodbank, C. and T.F. Straesser (1991) 'Migrant farmers and the Neolithic colonization of Crete', *Antiquity* 65, 233–245.

Brosche, K.-U., H.-G. Molle and G. Schulz (1976) 'Geomorphologische Untersuchungen im östlichen Kroumirbergland (Nordtunisien, Gebiet östlich von Tabarka)', *Eiszeitalter und Gegenwart* 27, 143–158.

Bruins, H.J. (1986) 'Desert Environment and Agriculture in the central Negev and Kadesh-Barnea during Historical Times', doctoral thesis, The Agricultural University of Wageningen.

Bruins, H.J. (1994) 'Comparative chronology of climate and human history in the southern Levant from the Late Chalcolithic to the Early Arab period', in O. Bar-Yosef and R.S. Kra (eds) *Late Quaternary Chronology and Paleoclimates of the Eastern Mediterranean*, 301–314, Radiocarbon, Tucson, AZ.

Brun, A. (1979) 'Recherches palynologiques sur les sédiments marins du golfe de Gabès: résultats prélimi-naires', *Géologie Méditerranéene* 6, 247–264.

Brun, A. (1983) 'Étude palynologique des sédiments marins holocènes de 5000 b.p. à l'actuel dans le Golfe de Gabès (Mer Pélagienne)', *Pollen et Spores* XXV, 437–460.

Brun, A. (1985) 'La couverture steppique en Tunisie au Quaternaire supérieur', *Comptes Rendus de l'Académie des Sciences Série II* 301, 1085–1090.

Brun, A. (1987) 'Étude palynologique des limons organiques du site de l'oued el Akarit (sud Tunisien)', *Bulletin de l'Association Française pour l'Étude du Quaternaire* 29, 19–25.

Brun, A. (1989) 'Microflores et paléovégétations en Afrique du Nord depuis 30 000 ans', *Bulletin de la Société Géologique de la France* V, 25–33.

Bruno, N., T. Caltabiano, M.F. Grasso, M. Porto and R. Romano (1994) 'SO_2 flux from Mt. Etna volcano during the 1991–1993 eruption: correlations and considerations', *Acta Vulcanologica* 4, 143–148.

Brunt, P.A. (1971) *Italian Manpower 225 BC–AD 14*. Clarendon Press, Oxford.

Bryson, R.A., H.H. Lamb and D.L. Donley (1974) 'Drought and the decline of Mycenae', *Antiquity* XLVIII, 46–50.

Bücher, A. (1989) 'Fallout of Saharan dust in the Northwestern Mediterranean region', in M. Leinen and M. Sarnthein (eds) *Palaeoclimatology and Palaeometeorology: Modern and Past Patterns of Global Atmospheric Transport*, 565–584, Kluwer Academic Publishers, Dordrecht.

Bufalo, M., C. Oliveros and R.E. Quélennec (1989) 'L'érosion des Terres Noires dans la région du Buëch (Hautes-Alpes): contribution à l'étude des processus érosifs sur le bassin versant représentatif (BVRE) de Saint-Genis', *La Houille Blanche* 1989(3/4), 193–195.

Bunte, K. and J.W.A. Poesen (1993) 'Effects of rock fragment covers on erosion and transport of non-cohesive sediment by shallow overland flow', *Water Resources Research* 29, 1415–1424.

Burillo Mozota, F., M. Gutiérrez Elorza and C.S. Marcón (1986) 'Geomorphological processes as indi-cators of Climatic Changes during the Holocene in North-East Spain', in F. López-Vera (ed.) *Quaternary Climate in the Western Mediterranean: Proceedings of the Symposium on Climatic Fluctuations during the Quaternary in the Western Mediterranean Regions*, 31–43, Universidad Autónoma de Madrid, Madrid.

Burillo Mozota, F., M. Gutiérrez Elorza and J.L. Peña Monné (1984) 'Acumulaciones holocenas y su datación Arqueológica en mediana de Aragón (Zaragoza)', *Cuadernos de Investigación Geográfica* XI(1–2), 193–207.

Burke, S.M. (1998) 'Groundwater over-exploitation: a case study in Castilla la Mancha, Spain', in P. Mairota, J.B. Thornes and N. Geeson (eds) *Atlas of Mediterranean Environments in Europe: The Desertification Context*, 82–84, John Wiley and Sons, Chichester.

Burke, S.M. (1999) 'Modelling groundwater and recharge under variable Mediterranean conditions', in G. Ghazi, Ch. Zanolla and D. Peter (eds) *Desertification in Europe: Mitigation, Strategies and Land-Use Planning*, 79–86, Directorate General for Research, Environment and Climate Programme, EUR19390, European Union, Brussels.

Burke, S.M. and J.B. Thornes (eds) (in press) *Desertification in the Mediteranean*. European Union, Brussels.

Butcher, G.C. and J.B. Thornes (1979) 'Spatial variability in runoff processes in an ephemeral channel', *Zeitschrift für Geomorphologie Supplementband* 29, 83–92.

Butzer, K.W. (1974) 'Accelerated soil erosion: a problem of man–land relationships', in I.R. Manners and M.W. Mikesell (eds) *Perspectives on Environment*, 57–77, Association of American Geographers, Washington, DC.

Butzer, K.W. (1976) *Early Hydraulic Civilization in Egypt: A Study in Cultural Ecology*. University of Chicago Press, Chicago.

Butzer, K.W., J.F. Mateu, E.K. Butzer and P. Kraus (1985) 'Irrigation agrosystems in eastern Spain: Roman or Islamic Origin?', *Annals of the Association of American Geographers* 75(4), 479–509.

Buxó i Capderilla, R. (1991) 'Recent studies on the plant remains found in the Cova de Recambra in

Valencia, Spain', in J.M. Renfrew (ed.) *New Light on Early Farming*, 237–245, Edinburgh University Press, Edinburgh.

Byrd, B.F. (1994) 'Late Quaternary hunter–gatherer complexes in the Levant between 20,000 and 10,000 BP', in O. Bar-Yosef and R.S. Kra (eds) *Late Quaternary Chronology and Paleoclimates of the Eastern Mediterranean*, 205–226, Radiocarbon, Tucson, AZ.

Cadogan, G. (1976) *Palaces of Minoan Crete*. Methuen, London.

Cadogan, G. (1987) 'Unsteady date of a big bang', *Nature* 328, 473.

Cadogan, G. (1988) 'Dating of the Santorini eruption – reply', *Nature* 333, 401–402.

Caldeira, K. and M.R. Rampino (1992) 'Mount Etna CO_2 may affect climate', *Nature* 355, 401–402.

Calvari, S., M. Coltelli, M. Neri, M. Pompilio and V. Scribano (1994) 'The 1991–1993 Etna eruption: chronology and lava flow-field evolution', *Acta Vulcanologica* 4, 1–14.

Cammeraat L.H. and A.C. Imeson (1998) 'Deriving indicators of soil degradation from soil aggregation studies in southeastern Spain and southern France', *Geomorphology* 23, 307–321.

Campana, D.V. and P.J. Crabtree (1990) 'Communal hunting in the Natufian of the Southern Levant: the social and economic implications', *Journal of Mediterranean Archaeology* 3, 223–246.

Campbell, I.A. (1997) 'Badlands and badland gullies', in D.S.G. Thomas (ed.) *Arid Zone Geomorphology: Process, Form and Change in Drylands*. 2nd edn. 261–291, John Wiley and Sons, Chichester.

Campbell, J.K (1964) *Honour, Family and Patronage: A Study of Institutions and Moral Values in a Greek Mountain Community*. Clarendon Press, Oxford.

Camuffo, D. (1984) 'Analysis of the series of precipitation at Padova, Italy', *Climate Change* 6, 57–77.

Camuffo, D. (1987) 'Freezing of the Venetian lagoon since the 9th century AD in comparison to the climate of Western Europe and England', *Climatic Change* 10, 43–66.

Camuffo, D. (1992) 'Acid rain and deterioration of monuments: how old is the phenomenon?', *Atmospheric Environment* 26B, 241–247.

Camuffo, D. (1993) 'Reconstructing the climate and the air pollution of Rome during the life of the Trajan column', *The Science of the Total Environment* 128, 205–226.

Camuffo, D. and S. Enzi (1992) 'Reconstructing the climate of northern Italy from archive sources', in R.S. Bradley and P.D. Jones (eds) *Climate Since A.D. 1500*, 143–154, Routledge, London.

Camuffo, D. and S. Enzi (1994) 'Chronology of dry fogs in Italy, 1374–1891', *Theoretical and Applied Climatology* 50, 31–33.

Camuffo, D. and S. Enzi (1995) 'Impact of the clouds of volcanic aerosols in Italy during the last 7 centuries', *Natural Hazards* 11, 135–161.

Cantù, V. (1977) 'The climate of Italy', in C.C. Wallén (ed.) *World Survey of Climatology*, Vol. 6: *Central and Southern Europe*, Elsevier, Amsterdam.

Cantù, V. and P. Narducci (1967) 'Lunghe serie di osservazioni meteoroligiche', *Revista di Meteorologica Aeronautica* XXVII, 71–79.

Capecchi, F. and P. Focardi (1988) 'Rainfall and landslides: research into a critical precipitation coefficient in an area of Italy', in C. Bonnard (ed.) *Landslides*, 1131–1136, A.A. Balkema, Rotterdam.

Carbonell, E., J.M. Bermúdez de Castro, J.L. Arsuaga, J.C. Díez, A. Rosas, G. Cuenca-Bescós, R. Sala, M. Mosquera and X.P. Rodriguez (1995) 'Lower Pleistocene hominids and artefacts from Atapuerca-TD6 (Spain)', *Science* 269, 826–829.

Carcelen, V. (2000) 'The view from the Ministry of Agriculture', in R. Llamas (ed.) *Ethical Issues in Spain's Water Management*, 22–30, *Water and Ethics*, special issue. Fundación Marceline Botín, Madrid.

Carlson, T.N. and F.H. Ludlam (1968) 'Conditions for the occurrence of severe local storms', *Tellus* 20, 203–226.

Carpenter, R. (1968) *Discontinuity in Greek Civilization*. Cambridge University Press, Cambridge.

Carraro, F. with contributions from 35 others (1996) 'Revisione del Villafranchiano nell'area-tipo di Villafranca d'Asti', *Il Quaternario* 9, 5–120.

Carson, M.A. and M.J. Kirkby (1972) *Hillslope Form and Process*. Cambridge University Press, Cambridge.

Cassano, S.M. and A. Manfredini (1993) *Studi sul Neolitico del Tavoliere della Puglia: Indagine Territoriale in un'Area-Campione*. BAR, Oxford.

Castri, F. di and H.A. Mooney (1973) *Mediterranean-type Ecosystems: Origin and Structure*. Springer-Verlag, Berlin.

Castri, F. di, D.W. Goodall and R.L. Specht (eds) (1981) *Mediterranean-Type Shrublands*. Elsevier Scientific, Amsterdam.

Castro, P.V., R.W. Chapman, S. Gili, V. Lull, R. Micó, C. Rihuete, R. Risch and M.E. Sanahuja Yll (eds) (1998) *Aguas Project: Palaeoclimatic Reconstruction and the Dynamics of Human Settlement and Land-Use in the Area of the Middle Aguas (Almería), in the South-East of the Iberian Peninsula*. EUR 18036, European Commission, Brussels.

Cau, L. (1994) 'Common land and land degradation in Sardinia', in G. Enne, G. Pulina and A. Aru (eds) *Land Use and Soil Degradation: MEDALUS in Sardinia*, 291–301, University of Sassari, Sassari.

Cavaco, C. (1995) 'Tourism in Portugal: diversity, diffusion, and regional and local development', *Tijdschrift voor Economische en Sociale Geographie* 86(1), 64–71.

Cavazza, S. (1961) 'Precipitazioni brevi e intense in Lucania e Calabria', *L'Energia Elettrica* 8, 746–748.

Cena, F. (1998) 'The farm and rural community as economic systems', in F.B. Golley and J. Bellot (eds) *Rural Planning from an Environmental Systems Perspective*, 229–247, Springer-Verlag, Berlin.

Chabert, M., J.J. Collin and J.P. Marchal (1996) 'Modelling short-term water resource trends in the context of possible "desertification" of Southern Europe,' in C.J. Brandt and J.B. Thornes (eds) *Mediterranean Desertification and Land Use*, 389–429, John Wiley and Sons, Chichester.

Champion, A.G. (1995) 'Internal migration, counterurbanization and changing population distributions', in R. Hall and P.E. White (eds) *Europe's Population: Towards the Next Century*, 99–129, UCL Press, London.

Champion, T., C. Gamble, S. Shennan and A. Whittle (1984) *Prehistoric Europe*. Academic Press, London.

Chapman, J. and J. Müller (1990) 'Early farmers in the Mediterranean Basin: the Dalmatian evidence', *Antiquity* 64, 127–134.

Chapman, R.W. (1978) 'The evidence for prehistoric water control in south-east Spain', *Journal of Arid Environments* 1, 261–274.

Chappell, A. & C.T. Agnew (2000) 'Desiccation in the Sahel', in S. Maclaren (ed.) *Linking Climate Change to Land Surface Change: Proceedings of RGS-IBG conference, Leicester. January 1999*. Kluwer, Hague.

Charles-Picard, G. and C. Charles-Picard (1961) *Daily Life in Carthage at the Time of Hannibal*. Allen and Unwin, London.

Chartier, A., G. Chouquer, C. Jung and X. Tschanz (1994) 'La gestion de l'eau', in S.E. van der Leeuw (ed.) *Understanding the Natural and Anthropogenic Causes of Soil Degradation and Desertification in the Mediterranean Basin*. Volume 3: *Dégradation et Impact Humain dans la Moyenne et Basse Vallée du Rhône dans l'Antiquité (Part II)*, 283–296, Final Report on Contract EV5V-CT91-0021, EU, Brussels.

Cheddadi, R. and M. Rossignol-Strick (1995) 'Eastern Mediterranean Quaternary paleoclimates from pollen and isotope records of marine cores in the Nile cone area', *Paleoceanography* 10, 291–300.

Cheddadi, R., M. Rossignol-Strick and M. Fontugne (1991) 'Eastern Mediterranean palaeoclimates from 26 to 5 ka B.P. documented by pollen and isotopic analysis of a core in the anoxic Bannock Basin', *Marine Geology* 100, 53–66.

Cherry, J. (1990) 'The first colonization of the Mediterranean islands: a review of recent research', *Journal of Mediterranean Archaeology* 3, 145–222.

Chester, D.K., A.M. Duncan, J.E. Guest and C.R.J. Kilburn (1985) *Mount Etna: The Anatomy of a Volcano*. Chapman and Hall, London.

Chester, D.K. and P.A. James (1991) 'Holocene alluviation in the Algarve, Southern Portugal: the case for an anthropogenic cause', *Journal of Archaeological Science* 18, 73–87.

Childe, V.G. (1950) 'The urban revolution', *Town Planning Review* 21, 3–17.

Chisci, G., C. Zanchi and G. d'Egidio (1981) 'Erosion investigations – plots: temporary disconnected experiments on crop and land management', in Proceedings, IAHS Symposium, Florence, 40–45, IAHS Press, Wallingford.

Chisholm, A.J. and J.H. Renick (1972) *The Kinematics of Multicell and Supercell Alberta Hailstorms: Alberta Hail Studies*. Report 72-2, Research Council of Alberta Hail Studies, Alberta, ON.

Chorley, R.J. (1968) *Water, Earth and Man*. Methuen, London.

Choukr-Allah, R. (1996) 'Unconventional water resource management', in A. Hamdy (ed.) *International Conference on Land and Water Resources Management in the Mediterranean Region, 4–8 September 1994*, volume 1, 403–417, IAM, Bari.

Chouquer, G. and Th. Odiot (1994) 'La dynamique spatio-temporelle du front colonial en vallée du Rhône', in S.E. van der Leeuw (ed.) *Understanding the Natural and Anthropogenic Causes of Soil Degradation and Desertification in the Mediterranean Basin*. Volume 3: *Dégradation et Impact Humain dans la Moyenne et Basse Vallée du Rhône dans l'Antiquité (Part II)*, 267–282, Final Report on Contract EV5V-CT91-0021, EU, Brussels.

Cirugeda, J. (1973) *Informe Relativo a las Crecidas de Octubre de 1973 en el Sureste: Estudio de Caudales*. Centro de Estudios Hidrograficos, Madrid.

Cita, M.B. (1982) 'The Messinian salinity crisis in the Mediterranean: a review', in H. Berckhemer and K.J. Hsü (eds) *Alpine-Mediterranean Geodynamics*, 113–140, American Geophysical Union, Washington, DC.

Cita, M.B., G.J. Delange and E. Olausson (1991) 'Anoxic basins and sapropel deposition in the eastern Mediterranean – past and present – introduction', *Marine Geology* 100, 1–4.

Clark, G.A. and L. Strauss (1983) 'Late Pleistocene hunter–gatherer adaptations in Cantabrian Spain', in G.N. Bailey (ed.) *Hunter–Gatherer Economy in Prehistory: A European Perspective*, 131–148, Cambridge University Press, Cambridge.

Clark, S.C. (1996) 'Mediterranean ecology and an ecological synthesis of the field sites', in C.J. Brandt and J.B. Thornes (eds) *Mediterranean Desertification and Land Use*, 271–302, John Wiley and Sons, Chichester.

Clark, S.C., J. Puigdefábregas and I. Woodward (1998) 'Aspects of the ecology of the shrub-winter annual communities of the Mediterranean Basin', in P. Mairota, J.B. Thornes and N. Geeson (eds) *Atlas of Mediterranean Environments in Europe: The Desertification Context*, 44–48, John Wiley and Sons, Chichester.

Clarke, D.L. (1979) 'Towns in the development of early civilisation', in N. Hammond *et al.* (eds) *Analytical Archaeologist: The Collected Papers of D.L. Clarke*, 435–443, Academic Press, London.

Clauzon, G., J.-P. Suc, F. Gautier, A. Berger and M.-F. Loutre (1996) 'Alternative interpretation of the Messinian salinity crisis: controversy resolved?', *Geology* 24, 363–366.

Clauzon, G. and J. Vaudour (1971) 'Ruissellement, transport solides et transport en solution sur un versant aux environs d'Aix-en-Provence', *Revue de Géographie et de Géologie Dynamique* XIII(5), 489–504.

Clemens, S.C. and R. Tiedemann, (1997) 'Eccentricity forcing of Pliocene-Early Pleistocene climate revealed in a marine oxygen-isotope record', *Nature* 385, 801–804.

Clements, M.A. and A. Georgiou (1998) 'The impact of political instability on a fragile tourism product', *Tourism Management* 19(3), 283–288.

Clottes, J., A. Beltrán, J. Courtin and H. Cosquer (1992) 'The Cosquer cave on Cape Morgiou, Marseille', *Antiquity* 66, 583–598.

Clout, H., M. Blacksell, R. King and D.A. Pinder (1994) *Western Europe: Geographical Perspectives*. Longman, Harlow.

Coghlan, H.H. (1988) *Notes on Prehistoric and Early Iron in the Old World*. Pitt Rivers Museum, Oxford.

Cohen, M.N. (1977) *The Food Crisis in Prehistory*. Yale University Press, New Haven, CT.

COHMAP members (1988) 'Climatic changes of the last 18,000 years: observations and model simulations', *Science* 241, 1043–1052.

Colacino, M. and R. Purini (1986) 'A study on the precipitation in Rome from 1782 to 1978', *Theoretical and Applied Climatology* 37, 90–96.

Colacino, M. and A. Rovelli (1983) 'The yearly averaged air-temperature in Rome from 1782 to 1975', *Tellus Series A – Dynamic Meteorology and Oceanography* 35, 389–397.

Collin, J.J., Movet, C. and Barrocu, G. (1998) 'The increasing demand for water', in P. Mairota, J.B. Thornes and N. Geeson (eds) *Atlas of Mediterranean environments in Europe: The Desertification Context*, 100–101, John Wiley and Sons, Chichester.

Collinson, S. (1996) *Shore to Shore: The Politics of Migration in Euro–Maghreb Relations*. Royal Institute of International Affairs, London.

Collison, A.J.C. (1996) 'Unsaturated strength and preferential flow as controls on gully-head development', in M.G. Anderson and S.M. Brooks (eds) *Advances in Hillslope Processes*. Volume 2, 753–769, John Wiley and Sons, Chichester.

Coltorti, M., M. Cremaschi, M.C. Delitala, D. Esu, M. Fornaseri, A. McPherron, M. Nicoletti, R. Van Otterloo, C. Peretto, B. Sala, V. Schmidt and J. Sevink (1982) 'Reversed magnetic polarity at Isernia La Pineta, a new lower Palaeolithic site in Central Italy', *Nature* 300, 173–176.

Coltorti, M. and L. Dal Ri (1985) 'The human impact on the landscape: some examples from the Adige valley', in C. Malone and S. Stoddart (eds) *Papers in Italian Archaeology IV*. Part i: *The Human Landscape*, 105–134, BAR International Series 243, Oxford.

Comani, S. (1987) 'The historical temperature series of Bologna (Italy) – 1716–1774', *Climatic Change* 11, 375–390.

Combourieu-Nebout, N. (1993) 'Vegetation response to Upper Pliocene glacial/interglacial cyclicity in the Central Mediterranean', *Quaternary Research* 40, 228–236.

Combourieu-Nebout, N. and C. Vergnaud Grazzini (1991) 'Late Pliocene Northern Hemisphere glaciations: the continental and marine responses in the central Mediterranean', *Quaternary Science Reviews* 10, 319–334.

Comín, F.A. (1999) 'Management of the Ebro river basin: past, present and future', *Water Science and Technology* 40, 161–168.

Conacher, A.J. and M. Sala (eds) (1998) *Land Degradation in the Mediterranean Environments of the World: Nature and Extent, Causes and Solutions*. John Wiley and Sons, Chichester.

Confederación Hidrográfica de Aguas del Sur (1987) http://www.chse.es/.

Console, R. and P. Favali (1988) 'Historical seismograms in Italy', in W.H.K. Lee, H. Meyers and K. Shimazaki (eds) *Historical Seismograms and Earthquakes of the World*, 447–450, Academic Press, London.

Constantini, L. (1989) 'Plant exploitation at Grotta dell'Uzzo, Sicily: new evidence for the transition from Mesolithic to Neolithic subsistence in southern Europe', in D.R. Harris and G.C. Hillman (eds) *Foraging and Farming: The Evolution of Plant Exploitation*, Unwin Hyman, London.

Constantinou, G. (1981) 'Geological features and ancient exploitation of the cupriferous sulphide ore-bodies of Cyprus', in *Early Metallurgy in Cyprus, 4,000–500 BC, Acts of the International Archaeological Symposium*, 13–24, Pierides Foundation, Larnaca.

Conte, M. and M. Colacino (1995) 'Climate', in R. Fantechi, D. Peter, P. Balabanis and J.L. Rubio (eds) *Desertification in a European Context: Physical and Socio-Economic Aspects*, 79–109, EC, Luxembourg.

Conte, M. and A. Giuffrida (1991) 'L'Oscillazione Mediterranea', *Memorie della Società Geografica Italiana* 46, 115–124.

Conte, M., R. Sorani and E. Previtali (2002) 'Extreme climate events over the Mediterranean', in N.A. Geeson, C.J. Brandt and J.B. Thornes (eds) *Mediterranean Desertification: A Mosaic of Processes and Responses*. John Wiley and Sons, Chichester.

Conventi, Y. and G. Dykstra (1995) 'MEDIMONT – Zone pilote de Corse. Rapport scientifique final', in M. Dubost (ed.) *MEDIMONT: A Multinational, Multidisciplinary Research Programme on the Rôle and Place of the Mountains in the Desertification of the Mediterranean Mountain Regions. Second and Final Scientific Report*, Final Report on Contract EV5V-CT91-0045, EU, Brussels.

Cooke, R.U., A. Warren and A.S. Goudie (1993) *Desert Geomorphology*. UCL Press, London.

Cope, C. (1991) 'Gazelle hunting strategies in the Southern Levant', in O. Bar-Yosef and F. Valla (eds) *The Natufian Culture in the Levant*, 341–358, International Monographs in Prehistory, Ann Arbor, MI.

Copeland, L. and C. Vita Finzi (1978) 'Archaeological dating of geological deposits in Jordan', *Levant* 10, 10–25.

Coppola, D. and L. Costantini (1987) 'Le néolithique ancien littoral et la diffusion des céréales dans les Pouilles durant le VIᵉ millénaire: les sites de Fontanelle, Torre Canne et Le Macchie', in J. Guilaine, J. Courtin, J.-L. Roudil and J.-L. Vernet (eds) *Premières Communautés Paysannes en Méditerranée Occidentale*, 249–256, Éditions du CNRS, Paris.

Cornu, S., J. Pätzold, E. Bard, J. Meco, and J. Cuerda-Barcelo, (1993) 'Paleotemperature of the last interglacial period based on $\delta^{18}O$ of *Strombus bubonius* from the western Mediterranean Sea', *Palaeogeography, Palaeoclimatology, Palaeoecology*, 103, 1–20.

Corominas, J. and J. Moya (1999) 'Reconstructing recent landslide activity in relation to rainfall in the Llobregat River basin, Eastern Pyrenees, Spain', *Geomorphology* 30(1–2), 79–94.

Corre, J.-J., A. Berger, J.-P. Béthoux and 22 others (1992) 'Implications des changements climatiques. Etude de cas: le golfe du Lion', in L. Jeftić, J.D. Milliman and G. Sestini (eds) *Climate Change and the Mediterranean*, Volume 1, 328–427, Arnold, London.

Correggiari, A., S. Guerzoni, R. Lenaz, G. Quarantotto and G. Rampazzo (1989) 'Dust deposition in the central Mediterranean (Tyrrhenian and Adriatic Seas): relationships with marine sediments and riverine input', *Terra Nova* 1, 549–558.

Correia, T.P. (1993) 'Threatened landscape in Alentejo, Portugal – the Montado and other agro-silvo-pastoral systems', *Landscape and Urban Planning* 24, 43–48.

Cosandey, C., A. Billard and T. Muxart (1986) 'Present day evolution of gullies formed in historical times in the Montagne du Lingas, Southern Cévennes, France', in V. Gardiner (ed.) *International Geomorphology*, Volume II, 523–531, John Wiley and Sons, Chichester.

Courty, M.A., N. Federoff, M.K. Jones, P. Castro and J. McGlade (1994) 'Environmental dynamics', in S.E. van der Leeuw (ed.) *Understanding the Natural and Anthropogenic Causes of Soil Degradation and Desertification in the Mediterranean Basin*. Volume 2: *Temporalities and Desertification in the Vera Basin*, 19–84, Final Report on Contract EV5V-CT91-0021, EU, Brussels.

Coutinho, M.A. and P.P. Tomás (1994) 'Comparison of Fournier with Wischmeier rainfall erosivity indices', in R.J. Rickson (ed.) *Conserving Soil Resources: European Perspectives*, 192–200, CAB International, Wallingford.

Cowan, C.W. and P.J. Watson (eds) (1992) *The Origins of Agriculture: An International Perspective*, 71–100, Smithsonian Institution Press, Washington, DC.

Cowling, R.M., P.W. Rundel, B.B. Lamont, M.K. Arroyo and M. Arianoutsou (1996) 'Plant diversity in Mediterranean-climate regions', *Trends in Ecology and Evolution* 11(9), 362–366.

Crawford, H. (1978) 'The mechanics of the obsidian trade: a suggestion', *Antiquity* LII, 129–132.

Creer, K.M. and A. Morris (1996) 'Proxy-climate and geomagnetic palaeointensity records extending back ca. 75,000 bp derived from sediments cored from Lago Grande di Monticchio, southern Italy', *Quaternary Science Reviews* 15, 167–188.

Cremaschi, M. (1990) 'Pedogenese medio olocenica ed uso dei suoli durante il neolitico in Italia settentrionale', in P. Biagi (ed.) *The Neolithisation of the Alpine Region*, 71–89, Monografie di 'Natura Bresciana', 13, Museo Civico di Scienze Naturali di Brescia, Brescia.

Cremonesi, G., J. Guilaine, M. Barbaza, J. Coularou, O. Fonto, R. Grifoni and J. Vaquer (1987) 'L'habitat de Torre Sabea (Gallipoli, Puglia) dans le cadre du Néolithique Ancien de l'Italie du sud-est', in J. Guilaine, J. Courtin, J.-L. Roudil and J.-L. Vernet (eds) *Premières Communautés Paysannes en Méditerranée Occidentale*, 377–385, Éditions du CNRS, Paris.

Creus Novau, J. (1991–1992) 'Tendencia secular de la temperatura de Mayo en el Pirineo Oriental', *Notes de Geografia Física* 20–21, 41–49.

Creus Novau, J. and J. Puigdefábregas (1976) 'Climatología histórica y dendrocronología de *Pinus uncinata* R.', *Cuadernos de Investigación* 2(2), 17–30.

Creus Novau, J. and J. Puigdefábregas Tomás (1983) 'Climatología histórica y dendrochronología de *Pinus Nigra* Arnold', in *Avances Sobre la Investigación en Bioclimatologia, VIII Reunión de Bioclimatología, Estación Experimental Aula Dei de Zaragoza, May 1983*, 121–128.

Cristina Peñalba, M. (1994) 'The history of the Holocene vegetation in northern Spain from pollen analysis', *Journal of Ecology* 82, 815–832.

Crouch, D.P. (1993) *Water Management in Ancient Greek Cities*. Oxford University Press, Oxford.

Crozier, M. (1986) *Landslides: Causes, Consequences and Environment*. Croom Helm, London.

Cruise, G.M. (1990a) 'Holocene peat initiation in the Ligurian Apennines, northern Italy', *Review of Palaeobotany and Palynology* 63, 173–182.

Cruise, G.M. (1990b) 'Pollen stratigraphy of two Holocene peat sites in the Ligurian Apennines, northern Italy', *Review of Palaeobotany and Palynology* 63, 299–313.

Culiberg, M. (1995) 'Dezertifikacija in reforestacija Slovenskega Krasa (Desertification and reforestation of the Karst in Slovenia).' *Porocilo o raziskovanju paleolitika, neolotika in eneolitika v Sloveniji XXII*, 201–217, Ljubljana.

Culiberg, M. and A. Šercelj (1996) 'Slovenia', in B.E. Berglund, H.J.B. Birks, M. Ralska-Jasiewiczowa and H.E. Wright (eds) *Palaeoecological Events during the Last 15,000 Years: Regional Syntheses of Palaeoecological Studies of Lakes and Mires in Europe*, 687–700, John Wiley and Sons, Chichester.

Cutileiro, J. (1971) *A Portuguese Rural Society*. Clarendon Press, Oxford.

Daget, P. (1980) 'Un élément actuel de la caractérisation du monde méditerranéen: le climat', in *Colloque de la Fondation L. Emberger sur) 'La Mise en Place, L'Évolution et La Caractérisation de la Flore et de la Végétation Circumméditerranéennes.' Naturalia Monspelliensia, N° Hors Série*, 101–126.

Dall'Olio, L., M. Ghirotti, E. Semenza and M.C. Tunrrini (1988) 'The Tessina landslide (eastern Pre-Alps, Italy): evolution and possible intervention methods', in C. Bonnard (ed.) *Landslides*, 1317–1322, A.A. Balkema, Rotterdam.

Darnajoux, H. (1976) *Caractéristiques Climatiques des Saisons Froides en France Jusqu'à la Fin du Xe Siècle*. Bibliographie Signalétique Hebdomadiare Sélectionnée, Supplément n° 8, Secrétariat d'État aux Transports, Direction de la Météorologie, Paris.

Da Silva, L.M., M. Lemon and J. Park (1997) 'A systems approach to land-use planning in irrigated areas', in P.S. Teng *et al.* (eds) *Applications of Systems Approaches at the Farm and Regional Levels*, 357–366, Kluwer Academic Publishers, Dordrecht.

Daugas, J.P., J.P. Raynal, A. Ballouche, S. Occhietti, P. Pichet, J. Elvin, J.P. Texier and A. Debenath (1989) 'Neolithic of north-Atlantic Morroco – 1st radiocarbon chronology attempt', *Comptes Rendus de l'Académie Des Sciences Série II* 308, 681–687.

Davidson, D.A. (1980) 'Erosion in Greece during the first and second millennia BC', in R.A. Cullingford, D.A. Davidson and J. Lewin (eds) *Timescales in Geomorphology*, 143–159, John Wiley and Sons, Chichester.

Davidson, I. (1983) 'Site variability and prehistoric economy in Levante', in G.N. Bailey (ed.) *Hunter-Gatherer Economy in Prehistory: A European Perspective*, 79–95, Cambridge University Press, Cambridge.

Davidson, I. (1988) 'Escaped domestic animals and the introduction of agriculture to Spain', in J. Clutton-Brock (ed.) *The Walking Larder*, 92–111, George Allen and Unwin, London.

Davidson, I. (1991) 'A great thick cloud of dust: naming and dating in the interpretation of behaviour in the Late Palaeolithic of Spain', in G.A. Clark (ed.) *Perspectives on the Past: Theoretical Biases in Mediterranean Hunter-Gatherer Research*, 194–203, University of Pennsylvania Press, Philadelphia.

Davies, F.B.M. and G. Notcutt (1988) 'Accumulation of fluoride by lichens in the vicinity of Etna volcano', *Water, Air and Soil Pollution* 42, 365–371.

Davies, P. and B. Gibbons (1993) *Field Guide to the Wild Flowers of Southern Europe*. The Crowood Press, Marlborough.

Davy, L. (1989) 'Une catastrophe naturelle: l'averse nîmoise du 3 octobre 1988 et ses conséquences hydrologiques', *Hydrologie Continentale*, 4, 75–92.

Dayan, U. and J. Miller (1989) *Meteorological and Climatological Data from Surface and Upper Air Measurements for the Assessment of Atmospheric Transport and Deposition of Pollutants in the Mediterranean Basin: A Review*. MAP Technical Reports Series no. 30, UNEP, Athens.

Dayton, J.E. (1982) 'Geology, archaeology and trade', in J.G.P. Best and N.M.W. de Vries (eds) *Interaction and Acculturation in the Mediterranean*, 153–168, B.R. Grüner Publishing Co., Amsterdam.

De Angelis, P. and G.E. Scarascia-Mugnozza (1998) 'Long-term CO_2 enrichment in a Mediterranean natural forest: an application of large open top chambers', *Chemosphere* 36(4–5), 763–770.

Dearing, J., I. Livingstone, and L.P. Zhou (1996) 'A late Quaternary magnetic record of Tunisian loess and its climatic significance', *Geophysical Research Letters*, 23, 189–192.

De Beaulieu, J.-L. and M. Reille (1984) 'A long Upper Pleistocene pollen record from Les Echets, near Lyon, France', *Boreas* 133, 111–132.

De Beaulieu, J.-L. and M. Reille (1992) 'The last climatic cycle at La Grande Pile (Vosges, France): a new pollen profile', *Quaternary Science Reviews* 11, 431–438.

Debussche, M., J. Lepart and A. Dervieux (1999) 'Mediterranean landscape changes: evidence from old postcards', *Global Ecology and Biogeography* 8, 3–15.

Degg, M.R. and J.C. Doornkamp (1989) *Earthquake Hazard Atlas*. Reinsurance Offices Association, London.

Dehn, M. and J. Buma (1999) 'Modelling future landslide activity based on general circulation models', *Geomorphology* 30(1–2), 175–188.

Delano Smith, C. (1972) 'Late Neolithic settlement, land use and garrigue in the Montpellier region, France', *Man 7*, 397–407.

Delano Smith, C. (1979) *Western Mediterranean Europe: A Historical Geography of Italy, Spain and Southern France since the Neolithic*. Academic Press, London.

Delibasis, N., J. Drakopoulos and G. Stavrakakis (1987) 'The Kalamata (Southern Greece) earthquake of 13 September 1986', *Annales Geophysicae Series B – Terrestrial and Planetary Physics* 5, 731–733.

De Lima, M.I.L.P. and J.L.M.P. de Lima (1990) 'Water erosion of soils containing rock fragments', in U. Shamir and C. Jiaqi (eds) *The Hydrological Basis for Water Management (Proceedings of the Beijing Symposium, October 1990)*, 141–147, IAHS Publication No. 197, IAHS Press, Wallingford.

Dennell, R. and W. Roebroeks (1996) 'The earliest colonization of Europe: the short chronology revisited', *Antiquity 70*, 535–542.

Dercourt, J.L.P. Zonenshain, L.-E. Ricou, V.G. Kazmin, X. Le Pichon, A.L. Knipper, C. Grandjacquet, I.M. Sbortshikov, J. Geyssant, C. Lepvrier, D.H. Pechersky, J. Boulin, J.-C. Sibuet, L.A. Savostin, O. Sorokhtin, M. Westphal, M.L. Bazhenov, J.P. Lauer and B. Biju-Duval (1986) 'Geological evolution of the Tethys belt from the Atlantic to the Pamirs since the Lias', *Tectonophysics* 123, 241–315.

Dever, W.G. (1995) 'Social structure in the early Bronze IV period in Palestine', in T.E. Levy (ed.) *The Archaeology of Society in the Holy Land*, 282–296, Leicester University Press, Leicester.

De Walle, E., M. Nikolopoulou-Taruvakli and J. Heinen (eds) (1993) *The Environmental Condition of the Mediterranean Sea*. Kluwer Academic Publishers, Dordrecht.

Dewey, J.F., M.L. Helman, E. Turco, D.H.W. Hutton and S.D. Knott (1989) 'Kinematics of the western Mediterranean', in M.P. Coward, D. Dietrich and R.G. Park (eds) *Alpine Tectonics*, 265–283, Geological Society Special Publication No. 45, London.

Dewey, J.F., W.C. Pitman III, W.B.F. Ryan and J. Bonnin (1973) 'Plate tectonics and the evolution of the Alpine system', *Geological Society of America Bulletin* 84, 3137–3180.

Diamantopoulos, J. (1993) 'Field site investigations: Petralona, Thessaloniki, Greece', in J.B. Thornes (ed.) *MEDALUS I Final Report*, 560–580, Final Report on Contract EPOC-CT90-0014-(SMA), EU, Brussels.

Díaz-Fierros, F., E. Benito and B. Soto (1994) 'Action of forest fires on vegetation cover and soil erodibility', in M. Sala and J.L. Rubio (eds) *Soil Erosion and Degradation as a Consequence of Forest Fires*, 163–176, Geoforma Ediciones, Logroño.

Díaz-Fierros, F., E. Benito, J.A. Vega, A. Castelao, B. Soto, R. Pérez and T. Taboada (1990) 'Solute loss and soil erosion in burnt soil from Galicia (NW Spain)', in J.G. Goldammer and M.J. Jenkins (eds) *Fires in Ecosystem Dynamics: Proceedings of the Third International Symposium on Fire Ecology, Freiburg, FRG, May 1989*, 103–116, SPB Academic Publishing, The Hague.

Díaz-Fierros, F., E. Benito Rueda and R. Pérez Moreira (1987) 'Evaluation of the USLE for the prediction of erosion in burnt forest areas in Galicia (NW Spain)', *Catena* 14, 189–199.

Di Castri, F., D.W. Goodall and R.L. Specht. (eds) (1981) *Mediterranean-type Shrublands*. Elsevier Scientific, Amsterdam.

Di Castri, F. and H.A. Mooney (1973) *Mediterranean-type Ecosystems: Origin and Structure*. Springer-Verlag, Berlin.

Diniz, F. (1995) *Palynological Study on Desertification in South-Western Europe: Timing, Natural Trends and Human Impact. Portugal*. Final Report on Contract EV5V-CT91-0027, EU, Brussels.

Di Palma, S., F. Drago, E. Galanti and V. Pennisi (1994) 'Earthen barriers and explosion tests to delay the lava advance: the 1992 Mt. Etna experience', *Acta Vulcanologica* 4, 167–172.

Dissmeyer, G.E. and G.R. Foster (1981) 'Estimating the cover-management factor (C) in the universal soil loss equation for forest conditions', *Journal of Soil and Water Conservation* 36, 235–240.

Domergue, C. and G. Hérail (1978) *Mines d'Or Romaines d'Espagne*. Publications de l'Université de Toulouse-Le Mirail, Toulouse.

Domínguez Cuesta, M.J., M. Jiménez Sánchez and A. Rodríguez García (1999) 'Press archives as temporal records of landslides in the north of Spain: relationships between rainfall and instability slope events', *Geomorphology* 30(1–2), 125–132.

Douguedroit, A. (1980) 'Le sécheresse estivale dans la région Provence-Alpes-Côte d'Azur', *Méditerran-née* 39(2/3), 13–21.

Douguedroit, A. (1988) 'The recent variability of precipitation in north-western Africa' in S. Gregory (ed.) *Recent Climatic Changes*, 130–137, Bellhaven Press, London.

Doumas, C.G. (1983) *Thera: Pompeii of the Ancient Aegean*. Thames and Hudson, London.

Dovers, S.R. and J.W. Handmere (1992) 'Uncertainty, sustainability and change', *Global Environmental Change* 2, 260–276.

Downey, W.S. and D.H. Tarling (1984) 'Archaeomagnetic dating of Santorini volcanic eruptions and fired destruction levels of late Minoan civilization', *Nature* 309, 519–523.

Downey, W.S. and D.H. Tarling (1985) 'Archaeomagnetism, Santorini volcanic eruptions and fired destruction levels on Crete – reply', *Nature* 313, 75–76.

Dregne, H.E. (1987) 'Reflections of the PACD', *Desertification Control Bulletin* 15, 8–11.

Du Boulay, J. (1974) *Portrait of a Greek Mountain Village*. Clarendon Press, Oxford.

Durrell, L. (1969) *Justine*. Faber and Faber, London.

Dyson-Hudson, R. and N. Dyson-Hudson (1980) 'Nomadic pastoralism', *Annual Review of Anthropology* 9, 15–61.

Eagleson, P.S. (1970) *Dynamic Hydrology*. McGraw and Hill, New York.

Economou, D. (1993) 'New forms of geographical inequalities and spatial problems in Greece', *Environment and Planning D – Society and Space* 11(5), 583–598.

Edwards, J. and F. Sampaio (1993) 'Tourism in Portugal', in W. Pompl and P. Lavery (eds) *Tourism in Europe: Structures and Developments*, 262–282, CAB International, Wallingford.

Edwards, P.C. (1989a) 'Revising the broad spectrum revolution', *Antiquity* 63, 225–246.

Edwards, P.C. (1989b) 'Problems of recognizing early sedentism: the Natufian example', *Journal of Mediterranean Archaeology* 2, 5–48.

EEA (1995) *Europe's Environment: Statistical Compendium for the Dobříš Assessment*. Office for Official Publications of the European Communities, Luxembourg.

Elias Castello, F. and L. Ruiz Beltran (1977) *Agroclimatología de España*. Instituto Nacional de Investigaciones Agrarias, Madrid.

El-Raey, M., Y. Fouda and S. Nasr (1997) 'GIS assessment of the vulnerability of the Rosetta area, Egypt, to impacts of sea rise', *Oceanographic Literature Review* 44(11), 1376.

Elwell, H.A. and M.A. Stocking (1976) 'Vegetal cover to estimate soil-erosion hazard in Rhodesia', *Geoderma* 15, 61–70.

Emery-Barbier, A. (1990) 'L'homme et l'environnement en Egypte durant la période prédynastique', in S. Bottema, G. Entjes-Nieborg and W. van Zeist (eds) *Man's Role in the Shaping of the East Mediterranean Landscape*. A.A. Balkema, Rotterdam.

Enne, G. (1999) *Sardinian Field Guide to the Advanced Study Course on Desertification in Europe: Mitigation Strategies, Land Use and Planning*. Alghero, 1–9 June 1999. Università di Sassari, Nucleo Ricerca Desertifazione.

Epema, G.F. and H.Th. Riezebos (1983) 'Fall velocity of waterdrops at different heights as a factor influencing erosivity of simulated rain', in J. de Ploey (ed.) *Rainfall Simulation, Runoff and Soil Erosion*, 1–17, *Catena* Supplement 4, Braunschweig.

Erétéo, F. (1988) *L'Olivier: Plantation, Taille, Entretien, Récolte, le Gel de 1985*. Solar Nature, Paris.

Ergenzinger, P. (1992) 'A conceptual geomorphological model for the development of a Mediterranean river basin under neotectonic stress (Buonamico basin, Calabria, Italy)', in D.E. Walling *et al.* (eds) *Erosion, Debris Flows and Environment in Mountain Regions*, 51–60, IAHS Publication no. 209, IAHS Press, Wallingford.

EROS (2000) 'EROS 2000 – The European Rivers and Oceans System', http://www.marine.ie/datacentre/projects/eros2000/.

Escalon de Fonton, M. (1980) 'Circonscription de Provence – Côté d'Azure', *Gallia Préhistoire* 23, 525–548.

Esteve Chueca (1972) *Vegetacíon y Flora de las Regiones Central y Meridional de la Provincia de Murcia*, Murcia, BELMAR.

Estienne, P. and A. Godard (1970) *Climatologie*. Armand Colin, Paris.

European Environmental Agency (1997) *Water Resources Problems in Southern Europe.* Topic Report 15, Inland Waters. EEA Office for Official Publications of the European Communities, Luxembourg.

EUROSTAT (2000) 'The EU Statistical Office', http://europa.eu.int/comm/eurostat/.

Evans, H.B. (1994) *Water Distribution in Ancient Rome. The Evidence of Frontinus.* The University of Michigan Press, Ann Arbor, MI.

Evans, J.D. (1968) 'Knossos Neolithic part II: summary and conclusions', *Annual of the British School at Athens* 63, 267–276.

Fabre, G. (1989) 'Les karsts du Languedoc méditerranéen (S.E. de la France)', *Zeitschrift für Geomorphologie Supplementband* 75, 49–81.

Fabre, G., J.-L. Fiches, P. Laveau and J.-L. Paillet (1995) *The Pont du Gard: Water and the Roman Town.* CNRS, Paris.

Falconer, S.E. (1987) *Heartland of Villages: Reconsidering Early Urbanism in the Southern Levant.* UMI, Ann Arbor, MI.

Falconer, S.E. (1994) 'The development and decline of Bronze Age civilisation in the Southern Levant: a reassessment of urbanism and ruralism', in C. Mathers and S. Stoddart (eds) *Development and Decline in the Mediterranean Bronze Age*, 305–333, Sheffield Archaeological Monographs 8, J.R. Collis Publications, Sheffield.

Fanning, D.S. and M.C.B. Fanning (1989) *Soil Morphology, Genesis and Classification.* John Wiley and Sons, Chichester.

FAO (1974) *Soil Map of the World, 1:5,000,000.* UNESCO, Paris.

Farres, P., J.W.A. Poesen and S. Wood (1993) 'Soil erosion landscapes', *Geography Review* 6, 38–41.

Favory, F., J.-L. Fiches, J.-J. Girardot and Cl. Raynaud (1994a) 'Analyse statistique de l'habitat rural', in S.E. van der Leeuw (ed.) *Understanding the Natural and Anthropogenic Causes of Soil Degradation and Desertification in the Mediterranean Basin.* Volume 3: *Dégradation et Impact Humain dans la Moyenne et Basse Vallée du Rhône dans l'Antiquité (Part II)*, 3–72, Final Report on Contract EV5V-CT91-0021, EU, Brussels.

Favory, F. and J.-J. Girardot (1994) 'Habitat et milieu: Analyse statistique générale croisant les données archéologiques et les données relatives au milieu environnant les sites gallo-romains', in S.E. van der Leeuw (ed.) *Understanding the Natural and Anthropogenic Causes of Soil Degradation and Desertification in the Mediterranean Basin.* Volume 3: *Dégradation et Impact Humain dans la Moyenne et Basse Vallée du Rhône dans l'Antiquité (Part II)*, 169–198, Final Report on Contract EV5V-CT91-0021, EU, Brussels.

Favory, F., J.-J. Girardot, S. van der Leeuw, F.P. Tourneux and P. Verhagen (1994b) 'L'habitat rural romain en basse vallée du Rhône', *Les Nouvelles de l'Archéologie* 57, 46–50.

Feingold, G. and Z. Levin (1986) 'The lognormal fit to raindrop spectra from frontal convective clouds in Israel', *Journal of Climate and Applied Meteorology* 25, 1346–1363.

Fernández Cancio, M.A. Gaertner, C. Gallardo and M. Castro (1995) 'Simulation of a long-lived meso-β scale convective system over the Mediterranean coast of Spain. Part I: Numerical predictability', *Meteorology and Atmospheric Physics* 56, 157–179.

Fernández Cancio, A., M. Genova Fuster, J. Creus Novau and E. Gutiérrez (1994) 'Dendroclimatological investigation for the last 300 years in Central Spain', in *International Conference on Tree-Ring. Environment and Humanity: Relationships and Processes.* Tucson, AZ.

Fernández Cancio, A., E. Manrique Menéndez, M. Génova Fuster and J. Creus Novau (1993) 'Estudio fitoclimatico de la Serrania de Cuenca en los ultimos 300 años', in F.J. Silva-Pando and G. Vega Alonso (eds) *Congreso Forestal Español, Lourizán 1993.* Volume I: *Mesa Temática I. El Medio Forestal*, 93–98, Grafol, Madrid.

Fernández Mills, G.F., X. Lana and C. Serra (1994) 'Catalonian precipitation patterns – principal component analysis and automated regionalization', *Theoretical and Applied Climatology* 49, 201–212.

Ferreira Bicho, N. (1993) 'Late Glacial prehistory of central and southern Portugal', *Antiquity* 67, 761–775.

Ferrucci, F., R. Rasa, G. Gaudiosi, R. Azzaro and S. Imposa (1993) 'Mt Etna – a model for the 1989 eruption', *Journal of Volcanology and Geothermal Research* 56, 35–56.

Figueiredo, T. and A.G. Ferreira (1993) 'Erosão dos solos em vinha de encosta na região do Douro, Portugal', in *Actas do XII Congesso Latinoamericano da Ciência do Solo*, 79–88.

Finley, M.I. (1985) *The Ancient Economy*. 2nd edn. Hogarth, London.

Fischer, J. (1930) 'Les Inondations du Bassin de l'Adour en Mars 1930', *Études Rhodaniennes* 6, 149–168.

Fischlin, A. and D. Gyalistras (1997) 'Assessing impacts of climatic change on forests in the Alps', *Global Ecology and Biogeography Letters* 6, 19–37.

Flageollet, J.-C., O. Maquaire, B. Martin and D. Weber (1999) 'Landslides and climatic conditions in the Barcelonnette and Var basins (Southern French Alps, France)', *Geomorphology* 30(1–2), 65–78.

Flannery, K.V. (1969) 'Origins and ecological effects of early domestication in Iran and the Near East', in P.J. Ucko and G.W. Dimbleby (eds) *The Domestication and Exploitation of Plants and Animals*, 73–98, Aldine Publishing Company, Chicago.

Flemming, N.C. (1992) 'Predictions of relative coastal sea-level change in the Mediterranean based on archaeological, historical and tide-gauge data', in L. Jeftić, J.D. Milliman and G. Sestini (eds) *Climate Change and the Mediterranean*, Volume 1, 247–281, Arnold, London.

Flocas, A.A. and B.D. Giles (1984) 'Air-temperature variations in Greece: 2. Spectral analysis', *Journal of Climatology* 4(5), 541–546.

Flohn, H. (1981) 'Sahel droughts: recent climate fluctuations in north Africa and the Mediterranean', in A. Berger (ed.) *Climatic Variations and Variability: Facts and Theories*, 399–408, Reidel, Dordrecht.

Flohn, H. (1985) 'A critical assessment of proxy data for climatic reconstruction', in M.J. Tooley and G.M. Sheail (eds) *The Climatic Scene*, 93–103, Allen and Unwin, London.

Florschütz, F., J. Menéndez Amor and T.A. Wijmstra (1971) 'Palynology of a thick Quaternary succession in southern Spain', *Palaeogeography, Palaeoclimatology, Palaeoecology* 10, 233–264.

Follieri, M., D. Magri and L. Sadori (1988) '250,000-year pollen record from Valle di Castiglione (Roma)', *Pollen et Spores* XXX, 329–356.

Follieri, M., D. Magri, L. Sadori, B. Narcisi, M. Giardini, L. Ciuffarella, C. Mulder and A. Celant (1995) *Palynological Study on Desertification in South-Western Europe: Timing, Natural Trends and Human Impact. Italy*. Final Report on Contract EV5V-CT91-0027, EU, Brussels.

Font Tullot, I. (1988) *Historia del Clima de España. Cambios Climáticos y Sus Causas*. Instituto Nacional de Meteorología, Madrid.

Fontes, J.Ch., F. Gasse, Y. Callot, J.-C. Plaziat, P. Carbonell, P.A. Dupeuble and I. Kaczmarska (1985) 'Freshwater and marine-like environments from Holocene lakes in northern Sahara', *Nature* 317, 608–610.

Fontugne, M.R. and S.E. Calvert (1992) 'Late Pleistocene variability of the carbon isotopic composition of organic matter in the eastern Mediterranean: monitor of changes in carbon sources and atmospheric concentrations', *Palaeoceanography* 7, 1–20.

Ford, D. and P. Williams (1989) *Karst Geomorphology and Hydrology*. Unwin Hyman, London.

Forest, C.E., M.R. Allen, P.H. Stone and A.P. Sokolov (2000) 'Constraining uncertainties in climate models using climate change detection techniques', *Geophysical Research Letters* 27(4), 569–572.

Fousekis, P. and J. Lekakis (1998) 'Adjusting to a changing reality: the Greek response', in T. O'Riordan and H. Voisey (eds) *The Transition to Sustainability: The Politics of Agenda 21 in Europe*, 214–228, Earthscan, London.

Francis, C.F. and J.B. Thornes (1990a) 'Matorral: erosion and reclamation', in J. Abaledejo, M.A. Stocking and E. Diaz (eds) *Soil Degradation and Rehabilitation under Mediterranean Environmental Conditions*, 87–117, Consejo Superior de Investigaciones Científicas, Madrid.

Francis, C.F. and J.B. Thornes (1990b) 'Runoff hydrographs from three Mediterranean vegetation cover types', in J.B. Thornes (ed.) *Vegetation and Erosion*, 363–384, John Wiley and Sons, Chichester.

Francus, P., S. Leroy, I. Mergeai, G. Seret and G. Wansard (1993) 'A multidisciplinary study of the Vico maar sequence (Latium, Italy): part of the last cycle in the Mediterranean area, preliminary results', in J.F.W. Negendank and B. Zolitschka (eds) *Palaeolimnology of European Maar Lakes*, 289–304, Springer-Verlag, Berlin.

Frank, A.H.E. (1969) 'Pollen stratigraphy of the Lake of Vico (central Italy)', *Palaeogeography, Palaeoclimatology, Palaeoecology* 6, 67–85.

Frayn, J.M. (1979) *Subsistence Farming in Roman Italy*. Centaur Press, Fontwell.

Frayn, J.M. (1984) *Sheep-Rearing and the Wool Trade in Italy during the Roman Period*. Cairns, Liverpool.

Freeze, R.A. and J.A. Cherry (1979) *Groundwater.* Prentice Hall, New York.

Friedrich, W.L., H. Pichler and S. Kussmaul (1977) 'Quaternary pyroclastics from Santorini/Greece and their significance for the Mediterranean palaeoclimate', *Meddelelser fra Dansk Geologisk Forening*, 26, 27–39.

Frumkin, A., I. Carmi, I. Zak and M. Magaritz (1994) 'Middle Holocene environmental change determined from the salt caves of Mount Sedom, Israel', in O. Bar-Yosef and R. Kra (eds) *Late Quaternary Chronology and Paleoclimates of the Eastern Mediterranean*, 315–322, Radiocarbon, Tucson, AZ.

Fryberger, S.G. and G. Dean (1979) 'Dune forms and wind régimes', in E.D. McKee (ed.) *A Study of Global Sand Seas*, 137–140, USGS Professional Paper 1052.

Fulton, A.R.G., D.K.C. Jones and S. Lazzari (1987) 'The role of geomorphology in post-disaster reconstruction: the case of Basilicata, southern Italy', in V. Gardiner (ed.) *International Geomorphology*, 241–262, John Wiley and Sons, Chichester.

Gajić-Čapka, M. (1994) 'Periodicity of annual precipitation in different climate regions of Croatia', *Theoretical and Applied Climatology* 49, 213–216.

Galili, E., D.J. Stanley, J. Sharvit and M. Weinstein-Evron (1997) 'Evidence for earliest olive-oil production in submerged settlements off the Carmel coast, Israel', *Journal of Archaeological Science* 24, 1141–1150.

Gallart, F. and N. Clotet-Perarnau (1988) 'Some aspects of the geomorphic processes triggered by an extreme rainfall event: the November 1982 flood in the eastern Pyrenees', *Catena Supplement*, 13, 79–85.

Gamble, C. (1979) 'Surplus and self-sufficiency in the Cycladic subsistence economy', in J.L. Davis and J.F. Cherry (eds) *Papers in Cycladic Prehistory*, 122–134, UCLA Institute of Archaeology, Monograph XIV, Los Angeles, CA.

Gamble, C. (1986) *The Palaeolithic Settlement of Europe.* Cambridge University Press, Cambridge.

García-Ruiz, J.M., J. Arnaez, L. Ortigosa and A. Gomez Villar (1988) 'Debris flows subsequent to a forest fire in the Najerilla River valley (Iberian System, Spain)', *Pirineos* 131, 3–24.

García-Ruiz, J.M., T. Lasanta, L. Ortigosa, P. Ruiz-Flaño, C. Martí and C. González (1995) 'Sediment yield under different land uses in the Spanish Pyrenees', *Mountain Research and Development* 15(3), 229–240.

García-Ruiz, J.M., S.M. White, C. Marti, C. Blas Valero, M. Paz Errea and A. Gomez Villar (1998) *La Catastrofe del Barranco de Aras (Biescas, Pirineo Aragones) y su Contexto Espacio Temporal.* CSIC, Instituto Pirinaico, Zaragoza.

Garnier, M. (1974a) *Longues Séries de Mesures de Précipitations en France. Zone 2 (Ouest et Sud-Ouest).* Mémorial de la Météorologie Nationale 53 Fasc. no. 2, Paris.

Garnier, M. (1974b) *Longues Séries de Mesures de Précipitations en France. Zone 4 (Méditerranéenne).* Mémorial de la Météorologie Nationale 53 Fasc. no. 4, Paris.

Garnsey, P. (1998) *Cities, Peasants and Food in Classical Antiquity.* Cambridge University Press, Cambridge.

Gascó, J. (1994) 'Development and decline in the Bronze Age of southern France', in C. Mathers and S. Stoddart (eds) *Development and Decline in the Mediterranean Bronze Age*, 99–198, J.R. Collis Publications, Sheffield.

Gascó, J., L. Carozza and J. Wainwright (1996) 'Un Petit Habitat Agricole de l'Age du Bronze Ancien en Languedoc Occidental: Laval de la Bretonne (Monze, Aude). Hypothèses et Conséquences d'un Enfouissement sur la "Courte Durée" de l'Occupation Humaine', in *Fondements Culturels, Techniques, Économiques et Sociaux des Débuts de l'Âge du Bronze, 117ᵉ Congrès national des Sociétés savantes, Comité des Travaux Historiques et Scientifiques*, 373–385.

Gaspar, J. (1984) 'Urbanization: growth, problems and policies', in A. Williams (ed.) *Southern Europe*, 208–235, Harper and Row, London.

Gasse, F., J.C. Fontes, J.C. Plaziat, P. Carbonel, I. Kaczmarska, P. De Deckker, I. Soulié-Marsche, Y. Callot and P.A. Dupeuble (1987) 'Biological remains, geochemistry and stable isotopes for the reconstruction of environmental and hydrological changes in the Holocene lakes from north Sahara', *Palaeogeography, Palaeoclimatology, Palaeoecology* 60, 1–46.

Gates, W.L., A. Henderson-Sellers, G.J. Boer, C.K. Folland, A. Kitoh, B.J. McAvaney, F. Semazzi, N.

Smith, A.J. Weaver and Q.-C. Zeng (1995) 'Climate models – evaluation', in J.T. Houghton, L.G. Meira Filho, B.A. Callander, N. Harris, A. Kattenberg and K. Maskell (eds) *Climate Change 1995. The Science of Climate Change*, 229–284, IPCC/Cambridge University Press, Cambridge.

Gazenbeek, M. (1994) 'Occupation du sol et évolution environnementale durant la seconde moitié de l'Holocène dans la Montagnette et la partie occidentale des Alpilles'. in S.E. van der Leeuw (ed.) *Understanding the Natural and Anthropogenic Causes of Soil Degradation and Desertification in the Mediterranean Basin*. Volume 3: *Dégradation et Impact Humain dans la Moyenne et Basse Vallée du Rhône dans l'Antiquité (Part I)*, 55–158, Final Report on Contract EV5V-CT91-0021, EU, Brussels.

Gebauer A.B. and T.D. Price (eds) (1992) *Transitions to Agriculture in Prehistory*, Prehistory Press, Madison, WI.

Geddes, D. (1983) 'Neolithic transhumance in the Mediterranean Pyrenees', *World Archaeology* 15, 51–66.

Geddes, D. (1985) 'Mesolithic deomestic sheep in West Mediterranean Europe', *Journal of Archaeological Science* 12, 25–48.

Geipel, R. (1982) *Disaster and Reconstruction: The Friuli (Italy) Earthquakes of 1976*. George Allen and Unwin, London.

Gendzier, I.L. (1999) 'Labour exodus: market forces and mass migration', *Global Dialogue* 1, 89–101.

Génova, R. (1986) 'Dendroclimatology of mountain pine (*Pinus uncinata* Ram.) in the central plain of Spain', *Tree-Ring Bulletin* 46, 3–12.

Geze, F. (1996) 'Le grand gaspillage de l'economie de rente', in *Le Drame Algèrien*. Reporters sans Frontières, Paris.

Giammanco, S., M. Valenza, S. Pignato and G. Giammanco (1996) 'Mg, Mn, Fe and V concentrations in the ground waters of Mount Etna (Sicily)', *Water Research* 30, 378–386.

Gilbertson, D.D., C.O. Hunt, N.R.J. Fieller and G.W.W. Barker (1994) 'The environmental consequences and context of ancient floodwater farming in the Tripolitanian Desert', in A.C. Millington and K. Pye (eds) *Environmental Change in Drylands: Biogeographical and Geomorphological Perspectives*, 229–252, John Wiley and Sons, Chichester.

Giles, B.D. (1984) 'Water need in Provence, Languedoc-Rousillon (Southern France)', *Journal of Climatology* 4, 53–69.

Giles, B.D. and C.J. Balafoutis (1990) 'The Greek heatwaves of 1987 and 1988', *International Journal of Climatology* 10, 505–517.

Giles, B.D. and A.A. Flocas (1984) 'Air-temperature variations in Greece: 1. Persistence, trend, and fluctuations', *Journal of Climatology* 4(5): 531–539.

Gillespie, R. (1984) *Radiocarbon User's Guide*. Oxford University Committee for Archaeology, Monograph Number 3, Oxford.

Gilman, A. (1976) *A Later Prehistory of Tangier, Morocco*. American School of Prehistoric Research, Bulletin 29. Peabody Museum, Harvard University, Cambridge, MA.

Gilman, A. (1981) 'The development of social stratification in Bronze Age Europe', *Current Anthropology* 22, 1–8.

Gilman, A. and J.B. Thornes (1985) *Land Use and Prehistory in South East Spain*. George Allen and Unwin, London.

Giovannini, G. and S. Lucchesi (1983) 'Effect of fire on hydrophobic and cementing substances of soil aggregates', *Soil Science* 136(4), 231–236.

Giovannini, G., S. Lucchesi and M. Giachetti (1987) 'The natural evolution of a burned soil: a three-year investigation', *Soil Science* 143, 220–226.

Giovannini, G., S. Lucchesi and M. Giachetti (1988) 'Effect of heating on some physical and chemical parameters related to soil aggregation and erodibility', *Soil Science* 146(4), 255–261.

Giovannini, G., S. Lucchesi and M. Giachetti (1990) 'Beneficial and detrimental effects of heating on soil quality', in J.G. Goldammer and M.J. Jenkins (eds) *Fires in Ecosystem Dynamics. Proceedings of the Third International Symposium on Fire Ecology, Freiburg, FRG, May 1989*, 95–102, SPB Academic Publishing, The Hague.

Giraudi, C. (1989) 'Lake levels and climate for the last 30,000 years in the Fucino area (Abruzzo-Central Italy) – a review', *Palaeogeography, Palaeoclimatology, Palaeoecology* 70, 249–260.

Giupponi, C., B. Eiselt and P.F. Ghetti (1999) 'A multicriteria approach for mapping risks of agricultural pollution for water resources: the Venice Lagoon watershed case study', *Journal of Environmental Management* 56, 259–269.

Glantz, M.H. and N. Orlovsky (1983) 'Desertification: a review of the concept', *Desertification Control Bulletin* 9, 15–22.

Glick, T.F. (1970) *Irrigation and Society in Medieval Valencia*. The Belknap Press of Harvard University Press, Cambridge MA.

Godfrey, K.B. (1995) 'Planning and sustainable tourism development in the Mediterranean', *Tourism Management* 16(3), 243–245.

Goldberg, P. (1986) 'Late Quaternary environmental history of the southern Levant', *Geoarchaeology* 1(3), 225–244.

Goldreich, Y. (1987) 'Advertent/inadvertent changes in the spatial-distribution of rainfall in the central coastal-plain of Israel', *Climatic Change* 11, 361–373.

Golley, F.B. and J. Bellot (eds) (1999) *Rural Planning from an Environmental Systems Perspective*. Springer-Verlag, Berlin.

Gomez, M.J.M. (1995) 'New tourism trends and the future of Mediterranean Europe', *Tijdschrift voor Economische en Sociale Geographie* 86(1), 21–31.

Gomez-Parra, A., J.M. Forja, T.A. Devalls, I. Saenz and I. Riba (2000) 'Early contamination by heavy metals of the Guadalquivir estuary after the Aznalcollar mining spill (SW Spain)', *Marine Pollution Bulletin* 40(12), 1115–1123.

Goodess, C.M. and J.P. Palutikof (1998) 'Development of daily rainfall scenarios for southeast Spain using a circulation-type approach to downscaling', *International Journal of Climatology* 10, 1051–1083.

Goodess, C.M., J.P. Palutikof and T.D. Davies (1992) *The Nature and Causes of Climatic Change*. Bellhaven Press, London.

Goosens, C. (1985) 'Principal component analysis of Mediterranean rainfall', *Journal of Climatology* 5, 379–388.

Goosens, R., T.K. Ghabour, T. Ongena and A. Gad (1994) 'Waterlogging and soil salinity in newly reclaimed areas of the Western Nile area of Egypt', in A.C. Millington and K. Pye (eds) *Environmental Change in Drylands: Biogeographical and Geomorphological Perspectives*, 365–377, John Wiley and Sons, Chichester.

Gophna, R. (1979) 'Post-neolithic settlement patterns', in A. Horowitz (ed.) *The Quaternary of Israel*, 319–321, Academic Press, London.

Goudie, A.S. (1990) 'Desert degradation', in A.S. Goudie (ed.) *Techniques for Desert Reclamation*, 1–33, John Wiley and Sons, Chichester.

Govers, G., W. Everaert, J. Poesen, G. Rauws, J. De Ploey and J.P. Lautridou (1990) 'A long flume study of the dynamic factors affecting the resistance of a loamy soil to concentrated flow erosion', *Earth Surface Processes and Landforms* 15, 313–328.

Grant, M. (1969) *The Ancient Mediterranean*. Weidenfeld and Nicolson, London.

Grazhdani, S., F. Jacquin and S. Sulce (1996) 'Effect of subsurface drainage on nutrient pollution of surface waters in south eastern Albania', *The Science of the Total Environment* 191(1), 15–21.

Green, S. (1994) 'Contemporary change in use and perception of the landscape in Epirus: an ethnographic case study', in S.E. van der Leeuw (ed.) *Understanding the Natural and Anthropogenic Causes of Soil Degradation and Desertification in the Mediterranean Basin*. Volume 1: *Land Degradation in Epirus*, 171–324, Final Report on Contract EV5V-CT91-0021, EU, Brussels.

Green, S. (1997a) 'Pogoni, Epirus (Greece)', in N. Winder and S.E. van der Leeuw (eds) *Environmental Perception and Policy Making. Cultural and Natural Heritage and the Preservation of Degradation-Sensitive Environments in Southern Europe*. Volume 1: *Perception, Policy and Unforeseen Consequences: An Interdisciplinary Synthesis*, 45–66, Final Report on Contract EV5V-CT9-486, EU, Brussels.

Green, S. (1997b) 'Notes on the making and nature of margins in Epirus', in N. Winder and S.E. van der Leeuw (eds) *Environmental Perception and Policy Making. Cultural and Natural Heritage and the Preservation of Degradation-Sensitive Environments in Southern Europe*. Volume 3, Final Report on Contract EV5V-CT94-0486, EU, Brussels.

Green, S. and G. King (1996) 'The importance of goats to a natural environment: a case study from Epirus (Greece) and southern Albania', *Terra Nova* 8, 655–658.

Green, S. and M. Lemon (1996) 'Perceptual landscapes in agrarian systems: degradation processes in north-western Epirus and the Argolid valley, Greece', *Ecumene* 3, 181–199.

Greene, K. (1986) *The Archaeology of the Roman Economy*. Batsford, London.

Gregori, G.P., R. Santoleri, M.P. Pavese, M. Colacino, E. Fiorentino and G. de Franceschi (1988) 'The analysis of point-like historical data series', in W. Schröder (ed.) *Past, Present and Future Trends in Geophysical Research*, 146–211, IAGA, Bremen-Roennebeck.

Gregori, G.P., R. Santoleri, M.P. Pavese and G. de Franceschi (1988) 'The analysis of point-like historical data series', in W. Schroder (ed.) *Past, Present and Future Trends in Geophysical Research*, 146–211, Roennebeck, Bremen.

Grenon, M. and M. Batisse (eds) (1989) *Futures for the Mediterranean Basin: The Blue Plan*. Oxford University Press, Oxford.

Griffiths, J.F. (1972) 'Mediterranean zone', in J.F. Griffiths (ed.) *Climates of Africa*, Elsevier, Amsterdam.

Grove, A.T., J. Ispikoudis, A. Kazaklis, J.A. Moody, V. Papanastasis and O. Rackham (1993) *Threatened Mediterranean Landscapes: West Crete*. Final Report on Contract EV4C-CT90-0112, EU, Brussels.

Grove, A.T. and O. Rackham (1993) 'Threatened landscapes in the Mediterranean: examples from Crete', *Landscape and Urban Planning* 24, 279–292.

Grove, A.T. and O. Rackham (1995) 'Physical, biological and human aspects of environmental change', in J.B. Thornes (ed.) *MEDALUS II: Managing Desertification*, 39–64, Final Report on Contract EV5V-CT92-0165, EU, Brussels.

Grove, J. (1988) *The Little Ice Age*. Methuen, London.

Grüger, E. (1977) 'Pollenanalytische Untersuchungen zur würmzeitlichen Vegetationsgeschichte von Kalabrien (Süditalien)', *Flora* 166, 475–489.

Guibal, F. (1985) 'Dendroclimatologie du cèdre de l'Atlas (*Cedrus atlantica* Manetti) dans le sud-est de la France', *Ecologia Mediterranea* XI, 87–103.

Guichard, F., S. Carey, M.A. Arthur, H. Sigurdsson and M. Arnold (1993) 'Tephra from the Minoan eruption of Santorini in sediments of the Black Sea', *Nature* 363, 610–612.

Guilaine, J. (1972) *L'Age du Bronze en Languedoc Occidental, Rousillon, Ariège*. Klincksieck, Paris.

Guilaine, J. (1979) 'The earliest neolithic in the western Mediterranean: a new appraisal', *Antiquity* LIII, 22–30.

Guilaine, J., M. Barbaza, D. Geddes, J.-L. Vernet, M. Llongueras and M. Hopf (1982) 'Prehistoric human adaptations in Catalonia (Spain)', *Journal of Field Archaeology* 9, 408–416.

Guilaine, J., F. Briois, J. Coularou, J.-D. Vigne and I. Carrère (1998) 'Les débuts du néolithique à Chypre', *L'Archéologue* 33, 35–40.

Guilaine, J., J. Coularou, A. Freises and R. Montjardin (eds) (1984) *Leucate-Corrège: Habitat Noyé du Néolitique Cardial*. Centre d'Anthropologie des Sociétés Rurales and Musée Paul Valéry, Toulouse and Sète.

Guilaine, J., G. Rancoule, J. Vaquer, M. Passelac and J.D. Vigne (eds) (1986) *Carsac: Une Agglomération Protohistorique en Languedoc*. Centre d'Anthropologie des Sociétés Rurales, Toulouse.

Guilaine, J., L. Simone, J. Thommeret and Y. Thommeret (1981) 'Datations C14 pour le Néolithique du Tavoliere (Italie)', *Bulletin de la Société Préhistorique Française* 78(5), 154–159.

Guiot, J. (1985) 'The extrapolation of recent climatological series with spectral canonical regression', *Journal of Climatology* 5, 325–335.

Guiot, J. (1987) 'Late Quaternary climate change in France estimated from multivariate pollen time series', *Quaternary Research* 28, 100–118.

Guiot, J. (1992) 'The combination of historical documents and biological data in the reconstruction of climate variations in space and time', *Palaeoclimatic Research* 7, 93–104.

Guiot, J., A.L. Berger, A.V. Munaut and C. Till (1982) 'Some new mathematical procedures in dendroclimatology, with examples from Switzerland and Morocco', *Tree-Ring Bulletin* 42, 33–48.

Guiot, J., S.P. Harrison and I.C. Prentice (1993) 'Reconstruction of Holocene precipitation patterns in Europe using pollen and lake-level data', *Quaternary Research* 40, 139–149.

Guiot, J., A. Pons, J.L. de Beaulieu and M. Reille (1989) 'A 140,000-year continental climatic reconstruction from two European pollen records', *Nature* 338, 309–313.

Gutiérrez, E. (1989) 'Dendroclimatological study of *Pinus sylvestris* L. in southern Catalonia (Spain)', *Tree-Ring Bulletin* 49, 1–9.

Haas, P.M. (1990) *Saving the Mediterranean: The Politics of International Environmental Cooperation.* Columbia University Press, New York.

Hadjimichalis, C. and N. Papamichos (1990) ' "Local" development in southern Europe: towards a new mythology', *Antipode* 23, 181–210.

Hadjisavvas, S. (1992) *Olive Oil Processing in Cyprus from the Bronze Age to the Byzantine Period.* Studies in Mediterranean Archaeology XCIX, Paul Åströms Förlag, Nicosia.

Haigh, M.J. (1990) 'Evolution of an anthropogenic desert gully system', in D.E. Walling, A. Yair and S. Berkowicz (eds) *Erosion, Transport and Deposition Processes (Proceedings of the Jerusalem Workshop, March–April 1987)*, 65–77, IAHS Publication, no. 189, IAHS Press, Wallingford.

Hallam, B.R., S.E. Warren and C. Renfrew (1976) 'Obsidian in the Western Mediterranean: characterisation by neutron activation analysis and optical emission spectroscopy', *Proceedings of the Prehistoric Society* 42, 85–110.

Halstead, P. (1987) 'Traditional and ancient rural economy in Mediterranean Europe: plus ça change?', *Journal of Hellenic Studies* 107, 77–87.

Halstead, P. (1990) 'Waste not, want not: traditional responses to crop failure in Greece', *Rural History: Economy, Society, Culture*, 1, 147–164.

Halstead, P. (1994) 'The North–South divide: regional paths to complexity in prehistoric Greece', in C. Mathers and S. Stoddart (eds) *Development and Decline in the Mediterranean Bronze Age*, 195–220, J.R. Collis Publications, Sheffield.

Hammer, C.U., H.B. Clausen, W.L. Friedrich and H. Tauber (1987) 'The Minoan eruption of Santorini in Greece dated to 1645 BC?', *Nature* 328, 517–519.

Hammer, C.U., H.B. Clausen, W.L. Friedrich and H. Tauber (1988) 'Dating of the Santorini eruption – reply', *Nature* 333, 401.

Hanley, A. (1998) 'Tree-felling blamed for fatal Italian mudslide', *The Independent* 7 May, 15.

Hansen, J., M. Sato, J. Glascoe and R. Ruedy (1998) 'A common-sense climate index: is climate changing noticeably?' *Proceedings of the National Academy of Sciences of the United States of America* 95(8), 4113–4120.

Hansen, J., M. Sato, A. Lacis and R. Ruedy (1997) 'The missing climate forcing', *Philosophical Transactions of the Royal Society of London Series B* 352, 231–240.

Hansen, J., M. Sato, R. Ruedy, A. Lacis and V. Oinas (2000) 'Global warming in the twenty-first century: an alternative scenario', *Proceedings of the National Academy of Sciences of the United States of America* 97(18), 9875–9880.

Harding, A.F. (2000) *European Societies in the Bronze Age.* Cambridge University Press, Cambridge.

Harrell, J.A. and V.M. Brown (1992) 'The world's oldest surviving geological map: the 1150 B.C. Turin Papyrus from Egypt', *Journal of Geology* 100, 3–18.

Harrison, R.J. (1985) 'The "Policultivo Ganadero", or the secondary products revolution in Spanish agriculture, 5000–1000 bc', *Proceedings of the Prehistoric Society* 51, 75–102.

Harrison, R.J. (1988) *Spain and the Dawn of History: Iberians, Phoenicians and Greeks.* Thames and Hudson, London.

Harrison, R.J. (1994) 'The Bronze Age in northern and northeastern Spain, 2000–800 bc', in C. Mathers and S. Stoddart (eds) *Development and Decline in the Mediterranean Bronze Age*, 73–97, J.R. Collis Publications, Sheffield.

Harrison, R.J. and J. Wainwright (1991) 'Dating the Bronze Age in Spain. A refined chronology for the high-altitude settlement of El Castillo (Frías de Albarracín, Prov. Teruel)', *Oxford Journal of Archaeology* 10(3), 261–268.

Harvey, A.M. (1982) 'The role of piping in the development of badlands and gully systems in south-east Spain', in R.B. Bryan and A. Yair (eds) *Badland Geomorphology and Piping*, 317–335, Geobooks, Norwich.

Harvey, A.M. (1984) 'Geomorphological response to an extreme flood: a case from southeast Spain', *Earth Surface Processes and Landforms*, 9, 267–279.

Hassan, F.A. (1985) 'Fluvial systems and geoarchaeology in arid lands: with examples from North Africa, the Near East and the American Southwest', in J.K. Stein and W.R. Farrand (eds) *Archaeological Sediments in Context*, 53–68, University of Maine, Orono.

Hatch, F.H., A.K. Wells and M.K. Wells (1972) *Petrology of the Igneous Rocks*, 13th edn. Thomas Murby and Co., London.

Healy, J.F. (1988) 'Mines and quarries', in M. Grant and R. Kitzinger (eds) *Civilization of the Ancient Mediterranean: Greece and Rome*, Volume 2, 779–794, Scribner's, New York.

Held, S.O. (1989) 'Colonization cycles on Cyprus 1: the biogeographic and palaeontological foundations of early prehistoric settlement', *Report of the Department of Antiquities of Cyprus, 1989*, 7–28, Nicosia.

Heller, F., W. Lowrie and A.M. Hirt (1989) 'A review of palaeomagnetic and magnetic anisotropy results from the Alps', in M.P. Coward, D. Dietrich and R.G. Park (eds) *Alpine Tectonics*, 399–420, Geological Society Special Publication No. 45, London.

Henbest, N. (1986) 'Dust from Santorini and darkness over Egypt', *New Scientist* 110, 35.

Henry, D.O. (1987) 'Climatic change, settlement mobility and technological evolution during the paleolithic', *Israel Journal of Earth Sciences* 36, 121–129.

Henry, D.O. (1989) *From Foraging to Agriculture: The Levant at the End of the Ice Age*. University of Pennsylvania Press, Philadelphia, PA.

Henry, D.O. (1991) 'Foraging, sedentism and adaptive vigour in the Natufian: rethinking the linkages', in G.A. Clark (ed.) *Perspectives on the Past: Theoretical Issues in Mediterranean Hunter-Gatherer Research*, 353–370, University of Pennsylvania Press, Philadelphia, PA.

Herrero, J., R. Aragüés and E. Amezketa (1993) 'Salt-affected soils and agriculture in the Ebro basin', in *Second Intensive Course on Applied Geomorphology: Arid Regions, Zaragoza*, 139–150.

Heusch, B. and A. Millies-Lacroix (1971) 'Une méthode pour estimer l'écoulement et l'érosion dans un bassin', *Mines et Géologie* 33.

Higgs, E. (1967) 'Environment and chronology', in C.M. McBurney (ed.) *The Hauah Fteah (Cyrenaica) and the Stone Age of the South-East Mediterranean*, 16–74, Cambridge University Press, Cambridge.

Hill, J., P. Hostert, G. Tsiourlis, P. Kasapidis, Th. Udelhoven and C. Diemer (1998) 'Monitoring 20 years of increased grazing impact on the Greek island of Crete with earth observation satellites', *Journal of Arid Environments* 39, 165–178.

Hillaire-Marcel, C., C. Gariépy, B. Ghaleb, J.-L. Goy, C. Zazo and J. Cuerda Barcelo (1996) 'U-series measurements in Tyrrhenian deposits from Mallorca – further evidence for two last-interglacial high sea levels in the Balearic islands', *Quaternary Science Reviews* 15, 53–62.

Hindson, R.A., C. Andrade and A.G. Dawson (1996) 'Sedimentary processes associated with the 1755 Lisbon earthquake on the Algarve coast, Portugal', *Physics and Chemistry of the Earth* 21, 57–63.

Hitchner, R.B. (1994) 'Image and reality: the changing face of pastoralism in the Tunisian high steppe', in J. Carlsen, P. Ørsted and J.E. Skydsgaard (eds) *Landuse in the Roman Empire*, 27–43, Analecta Romana Instituti Danici Supplementum XXII, 'L'Erma' di Bretschneider, Rome.

Hodder, I. (ed.) (1996) *On the Surface: Çatalhöyük 1993–95*. McDonald Institute, Cambridge and British Institute of Archaeology at Ankara.

Hodges, W.K. and R.B. Bryan (1982) 'The influence of material behaviour on runoff initiation in the Dinosaur Badlands, Canada', in R.B. Bryan and A. Yair (eds) *Badland Geomorphology and Piping*, 13–46, Geobooks, Norwich.

Hodkinson, S. (1988) 'Animal husbandry in the Greek polis', in C.R. Whittaker (ed.) *Pastoral Economies in Classical Antiquity*, 35–74, Cambridge Philological Society, Supplementary vol. 14, Cambridge.

Hoggart, K. (1997) 'Rural migration and counterurbanization in the European periphery: the case of Andalucia', *Sociologia Ruralis* 37, 134–153.

Hoggart, K., H. Buller and R. Black (1995) *Rural Europe: Identity and Change*. Arnold, London.

Hoggart, K. and C. Mendoza (1999) 'African immigrant workers in Spanish agriculture', *Sociologia Ruralis* 39(4), 538–562.

Hollis, G.E., C.T. Agnew, F. Ayache, M. Grundwell, R.C. Fisher, A. Millington, K. Selmi, M. Smart, A.C. Stevenson and A. Warren (1992) 'Implications of climatic changes in the Mediterranean Basin, Gararet, El Ich Keul and Lac de Bizerte, Tunisia', in L. Jeftić, J.D. Milliman and G. Sestini (eds) *Climate Change and the Mediterranean*, Volume 1, 602–664, Arnold, London.

Holmes, A. (1965) *Principles of Physical Geology*. Nelson, London.

Homewood, P. and C. Caron (1982) 'Flysch of the Western Alps', in K.J. Hsü (ed.) *Mountain Building Processes*, 157–168, Academic Press, London.

Hooper, J. (1997) 'Italian earthquakes destroy priceless medieval art treasures', *The Guardian*, 27 September, 1.

Hooper, J. (1998a) 'Tides of mud sweep 33 to their deaths', *The Guardian*, 7 May, 2.

Hooper, J. (1998b) 'Italy's unnatural disaster', *The Guardian*, 8 May, 19.

Hopf, M. (1969) 'Plant remains and early farming at Jericho', in P. Ucko and G.W. Dimbleby (eds) *The Domestication and Exploitation of Plants and Animals*, Aldine Publishing Company, Chicago.

Hopkins, T.S. (1999) 'The thermohaline forcing of the Gibraltar exchange', *Journal of Marine Systems* 20, 1–31.

Horden, P. and N. Purcell (2000) *The Corrupting Sea: A Study of Mediterranean History*. Blackwell Publishers, Oxford.

Hornberger, G.M., J.P. Raffensperger, P.L. Wiberg and K.N. Eshleman (1998) *Elements of Physical Hydrology*. Johns Hopkins University Press, Baltimore, MD.

Horowitz, A. (1979) *The Quaternary of Israel*. Academic Press, London.

Horowitz, A. (1987) 'Subsurface palynostratigraphy and paleoclimates of the Quaternary Jordan Rift Valley fill, Israel', *Israel Journal of Earth Sciences* 36, 31–44.

Horowitz, A. (1989) 'Continuous pollen diagrams for the last 3.5 m.y. from Israel: vegetation, climate and correlation with the oxygen isotope record', *Palaeogeography, Palaeoclimatology, Palaeoecology* 72, 63–78.

Horowitz, A. (1992) *Palynology of Arid Lands*. Elsevier, Amsterdam.

Horowitz, A. and M. Horowitz (1985) 'Subsurface late Cenozoic palynostratigraphy of the Hula Basin, Israel', *Pollen et Spores* XXVII, 365–390.

Horowitz, A. and M. Weinstein-Evron (1986) 'The late Pleistocene climate of Israel', *Bulletin de l'Association Française pour l'Étude du Quaternaire*, 25–26, 84–90.

Horvath, F. and H. Berckhemer (1982) 'Mediterranean backarc basins', in H. Berckhemer and K.J. Hsü (eds) *Alpine-Mediterranean Geodynamics*, 141–173, American Geophysical Union, Washington, DC.

Houghton, J.T., L.G. Meira Filho, B.A. Callander, N. Harris, A. Kattenberg and K. Maskell (eds) (1995) *Climate Change 1995: The Science of Climate Change*. IPCC/Cambridge University Press, Cambridge.

Houston, J.M. (1964) *The Western Mediterranean World*. Longman, London.

Houze, R.A. and P.V. Hobbs (1982) 'Organization and structure of precipitating cloud systems', *Advances in Geophysics*, 24, 225–315.

Howard, A.D. (1994) 'Badlands', in A.D. Abrahams and A.J. Parsons (eds) *Geomorphology of Desert Environments*, 213–242, Chapman and Hall, London.

Hsü, K.J. (1978a) 'Post-Miocene depositional patterns and structural displacement in the Mediterranean', in A.E.M. Nairn, W.H. Kanes and F.G. Stehli (eds) *The Ocean Basins and Margins*. Volume 4A: *The Eastern Mediterranean*, 77–150, Plenum Press, New York.

Hsü, K.J. (1978b) 'Tectonic evolution of the Mediterranean Basins', in A.E.M. Nairn, W.H. Kanes, and F.G. Stehli (eds) *The Ocean Basins and Margins*. Volume 4A: *The Eastern Mediterranean*, 29–75, Plenum Press, New York.

Hsü, K.J. (1983) *The Mediterranean Was a Desert. A Voyage of the Glomar Challenger*. Princeton University Press, Princeton, NJ.

Hsü, K.J. (1989) 'Time and place in Alpine orogenesis – the Fermor Lecture', in M.P. Coward, D. Dietrich and R.G. Park (eds) *Alpine Tectonics*, 421–443, Geological Society Special Publication No. 45, London.

Hudson, N. (1971) *Soil Conservation*. 1st edn. B.T. Batsford, London.

Hudson, N. (1992) *Land Husbandry*. B.T. Batsford, London.

Hudson, N. (1995) *Soil Conservation*. 3rd edn. B.T. Batsford, London.

Hudson, R. (1999) 'Putting policy into practice: policy implementation problems with special reference to the European Mediterranean', in P. Balabanis, D. Peter, A. Ghazi and M. Tsoges (eds) *Mediterranean Desertification. Research Results and Policy Implications*, 243–254, European Commission, EUR 19303, Brussels.

Hudson, R. and J.R. Lewis (1984) 'Capital accumulation: the industrialization of southern Europe?', in A. Williams (ed.) *Southern Europe*, 179–207, Harper and Row, London.

Hughes, J.D. (1994) *Pan's Travail. Environmental Problems of the Ancient Greeks and Romans*. The Johns Hopkins University Press, Baltimore, MD.

Hughes, M.K. (1988) 'Ice-layer dating of eruption at Santorini', *Nature* 335, 211–212.

Hughes, M.K. and Diaz, H.F. (1994) 'Was there a "Mediaeval Warm Period", and if so, when?', *Climatic Change*, 26, 109–142.

Hunt, C.O. (1995) 'The natural landscape and its evolution', in G. Barker (ed.) *A Mediterranean Valley. Landscape Archaeology and Annales History in the Biferno Valley*, 62–83, Leicester University Press, Leicester.

Huntley, B. (1990) 'European vegetation history: palaeovegetation maps from pollen data – 13000 BP to present', *Journal of Quaternary Science* 5, 103–122.

Huntley, B. (1993) 'The use of climate response surfaces to reconstruct palaeoclimate from Quaternary pollen and plant macrofossil data', *Philosophical Transactions of the Royal Society B* 341, 215–223.

Huntley, B. and I. Prentice (1988) 'July temperatures in Europe, 6000 years before present', *Science* 241, 687–690.

Huxley, A. and W. Taylor (1989) *Flowers of Greece and the Aegean*. The Hogarth Press, London.

ICONA (1979) *Precipitaciones Maximas en Espana*. Monografias 21, Ministerio de Agricultura, Madrid.

Ilan, D. (1995) 'The dawn of internationalism – the middle Bronze Age', in T.E. Levy (ed.) *The Archaeology of Society in the Holy Land*, 297–317, Leicester University Press, Leicester.

Imeson, A.C. and M. Bakker (1998) 'Effects of grazing on land degradation', in C. Kosmas (ed.) *MEDALUS III Meeting, Lesvos Field Guide*, 28–60, Agricultural University of Athens, Athens.

Imeson, A.C. and I.M. Emmer (1992) 'Implications of climatic change on land degradation in the Mediterranean', in L. Jeftić, J.D. Milliman and G. Sestini (eds) *Climate Change and the Mediterranean*. Volume 1, 95–128, Arnold, London.

Imeson, A.C., F.J.P.M. Kwaad and J.M. Verstraten (1982) 'The relationship of soil physical and chemical properties to the development of badlands in Morocco', in R.B. Bryan and A. Yair (eds) *Badland Geomorphology and Piping*, 47–70, Geobooks, Norwich.

Imeson, A.C. and H. Lavee (1998) 'Soil erosion and climate change: the transect approach and the influence of scale', *Geomorphology* 23(2–4), 219–227.

Imeson, A.C., J.M. Verstraten, E.J. van Mulligen and J. Sevink (1992) 'The effects of fire and water repellency on infiltration and runoff under Mediterranean type forest', *Catena* 19, 345–361.

Inbar, M. (1992) 'Rates of fluvial erosion in basins with a Mediterranean type climate', *Catena* 19, 393–409.

Isager, S. and J.E. Skydsgaard (1992) *Ancient Greek Agriculture*. Routledge, London.

Ispikoudis, I., G. Lyrintzis and S. Kyriakakis (1993) 'Impact of human activities on Mediterranean landscapes in western Crete', *Landscape and Urban Planning* 24, 259–271.

Issar, A.S. and H.J. Bruins (1983) 'Special climatological conditions in the deserts of Sinai and the Negev during the latest Pleistocene', *Palaeogeography, Palaeoclimatology, Palaeoecology* 43, 63–72.

Issar, A., H. Tsoar and D. Levin (1989) 'Climatic changes in Israel during historical times and their impact on hydrological, pedological and socio-economic systems', in M. Leinen and M. Sarnthein (eds) *Palaeoclimatology and Palaeometeorology: Modern and Past Patterns of Global Atmospheric Transport*, 359–383, Kluwer Academic Publishers, Dordrecht.

Jackson, M.L., R.N. Clayton, A. Violante and P. Violante (1982) 'Eolian influence on terra rossa soils of Italy traced by quartz oxygen isotope ratios', in H. van Olpen and F. Veniale (eds) *International Clay Conference*, 293–301, Elsevier, Amsterdam.

Jalut, G., J. Monserrat Marti, M. Fontugne, G. Delibrias, J.M. Vilaplana and R. Julià (1992) 'Glacial to interglacial vegetation changes in the northern and southern Pyrénées: deglaciation, vegetation cover and chronology', *Quaternary Science Reviews* 11, 449–480.

Jameson, M.H., C.N. Runnels and T.H. van Andel (1994) *A Greek Countryside: The Southern Argolid from Prehistory to the Present Day*. Stanford University Press, Stanford, CA.

Jarman, M.R., G.N. Bailey and H.N. Jarman (eds) (1982) *Early European Agriculture: Its Foundation and Development*. Cambridge University Press, Cambridge.

Jeffrey, P. and M. Lemon (1996) 'Understanding the dynamics of sustainable communities: stochasts, cartesians and social networks', Paper presented at the inaugural conference of the European branch of the International Society for Ecological Economics. Université de Versailles-Saint Quentin en Yvelines. Paris, France. 23–25 May 1996.

Jeftić, L., S. Kečkeš and J.C. Pernetta (eds) (1996) *Climate Change and the Mediterranean*. Volume 2. Arnold, London.

Jeftić, L., J.D. Milliman and G. Sestini (eds) (1992) *Climate Change and the Mediterranean*. Volume 1. Arnold, London.

Jenkins, R. (1979) *The Road to Alto: An Account of Peasants, Capitalists and the Soil in the Mountains of Southern Portugal*. Pluto Press, London.

Jenkyns, H.C. (1980) 'Tethys: past and present', *Proceedings of the Geologists' Association* 91, 107–118.

Jennings, J.N. (1985) *Karst Geomorphology*. Basil Blackwell, Oxford.

Johnsen, S.J., H.B. Clausen, W. Dansgaard, K. Fuhrer, N. Gundestrup, C.U. Hammer, P. Iversen, J. Jouzel, B. Stauffer and J.P. Steffensen (1992) 'Irregular glacial interstadials recorded in a new Greenland ice core', *Nature* 359, 311–313.

Johnson, R.G. (1997) 'Climate control requires a dam at the strait of Gibraltar', *EOS* 27, 280–281.

Jones, A.R. (1984) 'Agriculture: organization, reform and the EEC', in A. Williams (ed.) *Southern Europe*, 236–267, Harper and Row, London.

Jones, J.A.A. (1997) *Global Hydrology: Processes, Resources and Environmental Management*. Longman, Harlow.

Jorda, M. (1992) 'Morphogénèse et fluctuations climatiques dans les Alpes françaises du sud de l'Âge du Bronze au Haut Moyen Âge', *Les Nouvelles d'Archéologie* 50, 95–103.

Jorda, M., M. Provensal and R. Royet (1990) 'L'histoire "naturelle" d'un site de l'Age du Fer sur le piémont méridional des Alpilles. Le domaine de Servanne (Bouches-du-Rhône)', *Gallia* 47, 57–66.

Judson, S. (1963) 'Erosion and deposition of Italian stream valleys during historic time', *Science* 140, 898–899.

Judson, S. and A. Kahane (1963) 'Underground drainageways in southern Etruria and northern Latium', *Papers of the British School at Rome* 31, 74–99.

Julià, R., J. Negendank and G. Seret (1994) *Origin and Evolution of Desertification in the Mediterranean Environment in Spain*. Final Report on Contract EV5V-CT91-0037, EU, Brussels.

Junta de Andalucía, Consejo de Medio Ambiente (1996) *Guía del Parque Natural Sierra Nevada*. Junta de Andalucía, Granada.

Kafri, U. and B. Lang (1987) 'New data on the Late Quaternary fill of the Hula Basin, Israel', *Israel Journal of Earth Sciences* 36, 73–81.

Kalantaridis, C. and L. Labrianidis (1999) 'Family production and the global market: rural industrial growth in Greece', *Sociologia Ruralis* 39(2), 146–164.

Karouzis, G. (1980) *Report on Aspects of Land Tenure in Cyprus*. Ministry of Agriculture and Natural Resources, Land Consolidation Authority, Nicosia.

Kasapidis, P. and G.M. Tsiourlis (2000) 'Preliminary results on the structure of ecosystems and the desertification process in relation to overgrazing in the Asteroussia Mountains (Crete, Greece)', in P. Balabanis, D. Peter, A. Ghazi and M. Tsoges (eds) *Mediterranean Desertification, Research Results and Policy Implications*, European Commission, EUR 19303, Brussels.

Katsoulis, B.D. (1987) 'Indications of change of climate from the analysis of air-temperature time-series in Athens, Greece', *Climatic Change* 10, 67–79.

Katsouyanni, K. (1995) 'Health effects of air-pollution in Southern Europe – are there interacting factors?', *Environmental Health Perspectives* 103, 23–27.

Katsouyanni, K., A. Pantazopoulou, G. Touloumi, I. Tselepidaki, K. Moustris, D. Asimakopoulos, G. Poulopoulou and D. Trichopoulos (1993) 'Evidence for interaction between air-pollution and high-temperature in the causation of excess mortality', *Archives of Environmental Health* 48(4), 235–242.

Kaufman, B., S. Richards, D.A. Dierig, R. Parish and D.C. Funnell (1999) 'Climate change in mountain regions: some possible consequences in the Moroccan High Atlas', *Global Environmental Change* 9(1), 45–58.

Kaufman, D. (1992) 'Hunter-gatherers in the Levantine Epipalaeolithic: the socioecological origins of sedentism', *Journal of Mediterranean Archaeology* 5(2), 165–201.

Keeble, D. (1989) 'Core-periphery disparities, recession and new regional dynamisms in the European Community', *Geography* 74, 1–11.

Keigwin, L.D. and R.C. Thunnell (1979) 'Middle Pliocene climatic change in the western Mediterranean from faunal and oxygen isotope trends', *Nature* 282, 292–296.

Keller, J., W.B.F. Ryan, D. Ninkovitch and R. Altherr (1978) 'Explosive activity in the Mediterranean over the past 200,000 yr as recorded in deep sea sediments', *Geological Society of America Bulletin* 89, 591–604.

Kemp, B.J. (1989) *Ancient Egypt: Anatomy of a Civilization*. Routledge, London.

Kenna, M.E. (1993) 'Return migrants and tourism development: an example from the Cylcades', *Journal of Modern Greek Studies* 11, 75–95.

Kennedy, D.L. and D. Riley (1990) *Rome's Desert Frontier: From the Air*. Batsford, London.

Kenyon, K. (1957) *Digging up Jericho*. Ernest Benn, London.

Kieffer, G. and J.-C. Tanguy (1994) 'Risques volcaniques et sismiques à l'Etna (Sicile)', *Bulletin de la Société Géologique de la France* 165(1), 37–47.

King, G.P.C. and D. Sturdy (1994) 'Tectonics', in S.E. van der Leeuw (ed.) *Understanding the Natural and Anthropogenic Causes of Soil Degradation and Desertification in the Mediterranean Basin*. Volume 1: *Land Degradation in Epirus*, 13–45, Final Report on Contract EV5V-CT91-0021, EU, Brussels.

King, R. (1984) 'Population mobility: emigration, return migration and internal migration', in A. Williams (ed.) *Southern Europe*, 147–178, Harper and Row, London.

King, R. (ed.) (1997) *The Mediterranean: Environment and Society*. Arnold, London.

King, R. and S. Burton (1989) 'Land ownership values and rural structural change in Cyprus', *Journal of Rural Studies* 5, 267–277.

Kipnis, B.A. (1997) 'Dynamics and potentials of Israel's megalopolitan processes', *Urban Studies* 34(3), 489–501.

Kirkby, M.J. (in press) 'Effects of climate change on water erosion at different temporal and spatial scales', in J. Boardman and D. Favis Mortlock (eds) *Climate Change and Soil Erosion*. Oxford University Press, Oxford.

Kirkby, M.J., A.J. Baird, S.M. Diamond, J.G. Lockwood, M.L. McMahon, P.L. Michell, J. Shao, J.E. Sheehy, J.B. Thornes and F.I. Woodward (1996) 'The MEDALUS slope catena model: a physically based process model for hydrology, ecology and land-degradation interactions', in C.J. Brandt and J.B. Thornes (eds) *Mediterranean Desertification and Land Use*, 303–354, John Wiley and Sons, Chichester.

Kirkby, S.J. (1996) 'Recreation and the quality of Spanish coastal waters', in M. Barke, J. Towner and M.T. Newton (eds) *Tourism in Spain: Critical Issues*, 189–211, CAB International, Wallingford.

Kosakevitch, A., F. Garcia Palomero, X. Leca, J.-M. Leistel, N. Lenotre and F. Sobol (1993) 'Contrôles climatique et géomorphologique de la concentration de l'or dans les chapeaux de fer de Rio Tinto (Province de Huelva, Espagne)', *Comptes Rendus de l'Académie des Sciences de Paris, Série II*, 316, 85–90.

Kosmas, C.S. (1993) 'Field site investigations: Spata, Athens, Greece', in J.B. Thornes (ed.) *MEDALUS I Final Report*, 581–607, Final Report on Contract EPOC-CT90-0014-(SMA), EU, Brussels.

Kosmas, C.S., M. Bakker, G. Bergkamp, V. Detsis, J. Diamantopoulos, St Gerontidis, A.C. Imeson, O. Levelt, M. Maranthianou, R. Oortwijn, D. Oustwoud Wijdnes, J. Poesen, L. Vandevkkerckhove and Th. Zaphirou (1998) *MEDALUS Lesvos Field Guide. MEDALUS III Meeting, Lesvos, 24–28 April 1998*. Laboratory of Soils and Agricultural Chemistry, Agricultural University of Athens.

Kosmas, C.S. and N.G. Danalatos (1994) 'Climate change, desertification and the Mediterranean region', in M.D.A. Rounsevell and P.J. Loveland (eds) *Soil Responses to Climate Change*, 25–38, NATO ASI Series, Vol. 123, Springer-Verlag, Berlin.

Kosmas, C.S., N. Danalatos, L.II. Cammeraat, M. Chabart, J. Diamantopoulis, R. Farand, L. Gutierrez, A. Jacob, H. Marques, J. Martinez-Fernandez, A. Mizara, N. Moustakas, J.M. Nicolau, C. Oliveros, G. Pinna, R. Puddu, J. Puigdefábregas, M. Roxo, A. Simao, G. Stamou, N. Tomasi, D. Usai and A. Vacca (1997) 'The effect of land use on runoff and soil erosion rates under Mediterranean conditions', *Catena* 29, 45–59.

Kosmas, C.S., N.G. Danalatos and S. Gerontidis (2000) 'The effect of land parameters on vegetation performance and degree of erosion under Mediterranean conditions', *Catena* 40(1), 3–17.

Kosmas, C.S., N. Moustakas, N.G. Danalatos and N. Yassoglou (1996) 'The Spata field site', in C.J. Brandt and J.B. Thornes (eds) *Mediterranean Desertification and Land Use*, 207–228, John Wiley and Sons, Chichester.

Kotarba, A. (1980) 'Splash transport in the steppe zone of Mongolia', *Zeitschrift für Geomorphologie Supplementband* 35, 92–102.

Kotb, T.H.S., T. Watanabe, Y. Ogino and K.K. Tanji (2000) 'Soil salinization in the Nile Delta and related policy issues in Egypt', *Agricultural Water Management* 43, 239–261.

Krämer, L. (2000) *EC Environmental Law*. 4th edn. Sweet and Maxwell, London.

Kromer, B. and B. Becker (1993) 'German oak and pine C^{14} calibration, 7200–9439 BC', *Radiocarbon* 35, 125–135.

Kronfield, J., J.C. Vogel and A. Rosenthal (1988) 'Natural isotopes and water stratification in the Jordan valley aquifers', *Israel Journal of Earth Sciences* 39, 71–76.

Kubilay, N.N., A.C. Saydam, S. Yemenicioglu, G. Kelling, S. Kapur, A. Karaman and E. Akça (1997) 'Seasonal chemical and mineralogical variability of atmospheric particles in the coastal region of the northeast Mediterranean', *Catena* 28, 313–328.

Kujit, I. (1994) 'Pre-Pottery Neolithic A settlement variability: evidence for sociopolitical developments in the southern Levant', *Journal of Mediterranean Archaeology* 7, 165–192.

Kujit, I. and O. Bar-Yosef (1994) 'Radiocarbon chronology for the Levantine Neolithic: observations and data', in O. Bar-Yosef and R.S. Kra (eds) *Late Quaternary Chronology and Paleoclimates of the Eastern Mediterranean*, 227–245, Radiocarbon, Tucson, AZ.

Kuniholm, P.I., B. Kromer, S.W. Manning, M. Newton, C.E. Latini and M.J. Bruce (1996) 'Anatolian tree rings and the absolute chronology of the eastern Mediterranean, 2220–718 BC', *Nature* 381, 780–783.

Kuniholm, P.I. and M.W. Newton (1996) 'Çatal Höyük: A Preliminary Report', in I. Hodder (ed.) *On the Surface: Çatalhöyük 1993–1995*, 345–347, McDonald Institute for Archaeological Research, Cambridge.

Kuniholm, P.I. and C.L. Striker (1987) 'Dendrochronological investigations in the Aegean and neighboring regions, 1983–1986', *Journal of Field Archaeology* 14, 385–398.

Kutiel, H. and P.A. Kay (1992) 'Recent variations in 700 hPa geopotential heights in summer over Europe and the Middle East, and their influence on other meteorological factors', *Theoretical and Applied Meteorology* 46, 99–108.

Kutiel, P. and M. Inbar (1993) 'Fire impacts on soil nutrients and soil erosion in a Mediterranean pine forest plantation', *Catena* 20, 129–139.

Kutiel, P., H. Kutiel and H. Lavee (2000) 'Vegetation response to possible scenarios of rainfall variations along a Mediterranean-extreme arid climatic transect', *Journal of Arid Environments* 44, 277–290.

Kutiel, P., H. Lavee and M. Shoshany (1995) 'Influence of climatic transect upon vegetation dynamics along a Mediterranean – arid transect', *Journal of Biogeography* 22, 1060–1071.

Kutzbach, J., P.J. Guetter, P.J. Behling and R. Selin (1993) 'Simulated climatic changes: results of the COHMAP climate-model experiments', in H.E. Wright, J.E. Kutzbach, T. Webb III, W.F. Ruddiman, F.A. Street-Perrott and P.J. Bartlein (eds) *Global Climates Since the Last Glacial Maximum*, 24–93, University of Minnesota Press, Minneapolis.

Lamarche Jr, V.C. and K.K. Hirschboek (1984) 'Frost rings in trees as records of major volcanic eruptions', *Nature* 307, 121–126.

Lamb, H.F., U. Eicher and V.R. Switsur (1989) 'An 18,000-year record of vegetation, lake-level and climatic change from Tigalmamine, Middle Atlas, Morocco', *Journal of Biogeography* 16, 65–74.

Lamb, H.F., F. Gasse, A. Benkaddour, N. El Hamouti, S. van der Kars, W.T. Perkins, N.J. Pearce and C.N. Roberts (1995) 'Relation between century-scale Holocene arid intervals in tropical and temperate zones', *Nature* 373, 134–137.

Lamb, H.H. (1970) 'Volcanic dust in the atmosphere; with a chronology and assessment of its meteorological significance', *Proceedings of the Royal Society of London* 266A, 425–533.

Lamb, H.H. (1972) *Climate: Present, Past and Future*. Vol. 1: *Fundamentals and Climate Now*. Methuen, London.

Lamb, H.H. (1977) *Climate: Present, Past and Future.* Vol. 2: *Climatic History and the Future.* Methuen, London.

Lambeck, K. (1996) 'Sea-level change and shore-line evolution in Aegean Greece since Upper Palaeolithic time', *Antiquity* 70, 588–611.

Lancaster, N. (1995) *Geomorphology of Desert Dunes.* Routledge, London.

Lancaster, N. and W.G. Nickling (1994) 'Aeolian sediment transport', in A.D. Abrahams and A.J. Parsons (eds) *Geomorphology of Desert Environments*, 447–473, Chapman and Hall, London.

Lancel, S. (1992) *Carthage.* Fayard, Paris.

Langbein, W.B. and S.A. Schumm (1958) 'Yield of sediment in relation to mean annual precipitation', *Transactions of the American Geophysical Union* 39, 1076–1084.

Laubscher, H. and D. Bernouilli (1982) 'History and deformation of the Alps', in K.J. Hsü (ed.) *Mountain Building Processes*, 169–180, Academic Press, London.

Lautensach, H. (1971) *La Precipitation en la Peninsula Iberica.* Madrid, Servicio Meteorologico Nacional, Centro de Analisís y Predicción. Notas de Meteorólogica Sinoptica, 25, March.

Laval, H.J. Medus and M. Roux (1991) 'Palynological and sedimentological records of Holocene human impact from the Étang de Berre, southeastern France', *The Holocene* 1, 269–272.

Lavee, H., P. Kutiel, M. Segev and Y. Benyamini (1995) 'Effect of surface roughness on runoff and erosion in a Mediterranean ecosystem: the role of fire', *Geomorphology* 11, 227–234.

Lavee, H., A.C. Imeson and P. Sarah (1998) 'The impact of climate change on geomorphology and desertification along a Mediterranean-arid transect', *Land Degradation and Development* 9(5), 407–422.

Leblanc, M. (1976) 'Oceanic crust at Bou Azzer', *Nature* 261, 34–35.

Le Brun, A. (1981) *Un Site Néolithique Précéramique en Chypre: Cap Andreos Castros.* Recherches sur les Grandes Civilisations, Mémoire no. 5, Éditions ADPF, Paris.

Le Brun, A. (1988) *Fouilles Récentes à Khirokitia (Chypre) 1983–1986.* Éditions Recherches sur les Civilisations Mémoire no. 81, Paris.

Lecolle, P. (1984) 'Influence de l'altitude en climat Méditerranéen sur les teneurs en oxygène 18 et carbone 13 des coquilles de Gastéropodes terrestres', *Comptes Rendues de l'Académie de Science de Paris*, Série II 294, 211–214.

Legge, A.J. (1994) 'Animal remains and their interpretation', in R.J. Harrison, G.C. Moreno López and A.J. Legge (eds) *Moncín: Un Poblado de la Edad del Bronce*, 453–482, Departamento de Educación y Cultura, Zaragoza.

Lehavy, Y.M. (1989) 'Dhali-*Agridi*: the Neolithic by the river', in L.E. Stager and A.M. Walker (eds) *American Expedition to Idalion, Cyprus, 1973–1980*, 203–218, Oriental Institute Communications no. 24, The Oriental Institute of the University of Chicago, Chicago.

Le Houérou, H.N. (1973) 'Fire and vegetation in the Mediterranean Basin', in *Proceedings Annual Tall Timbers Fire Ecology Conference 13*, 237–277, Tall Timbers Research Station, Tallahassee, Florida.

Le Houérou, H.N. (1981) 'Impact of man and his animals', in F. Di Castri, D.W. Goodall, and R.I. Specht (eds) *Mediterranean-type Shrublands*, 479–521, Elsevier Scientific, Amsterdam.

Le Houérou, H.N. (1987) 'Vegetation wildfires in the Mediterranean Basin: evolution and trends', *Ecologia Mediterranea* XIII(4), 12.

Le Houérou, H.N. (1992) 'Vegetation and land use in the Mediterranean Basin by the year 2050: a prospective study', in L. Jeftić, J.D. Milliman and G. Sestini (eds) *Climate Change and the Mediterranean.* Volume 1, 175–232, Arnold, London.

Lemon, M., C. Blatsou and R. Seaton (1995a) 'Environmental degradation and intensive citrus production in the Argolid Valley, Greece', in *L'Homme et la Dégradation de l'Environnement. XVe Rencontres Internationales d'Archéologie et d'Histoire d'Antibes*, 417–433, Éditions APDCA, Juan-les-Pins.

Lemon, M., R. Seaton, C. Blatsou and N. Calamaras (1995b) 'Agriculture, policy and environmental degradation: the case of the Argolid Valley', *Medit* 95(4), 26–33.

Lemon, M., R. Seaton and J. Park (1994) 'Social enquiry and the measurement of natural phenomena: the degradation of irrigation water in the Argolid Plain, Greece', *International Journal of Sustainable Development and World Ecology* 1, 206–220.

Leontidou, L. (1990) *The Mediterranean City in Transition: Social Change and Urban Transition.* Cambridge University Press, Cambridge.

Leopold, L.B. and S.A. Schumn (1958) 'Yield of sediment in relation to mean annual precipitation', *Transactions of the American Geophysical Union* 39,1076–1084.

Le Pichon, X. (1980) 'La lithosphère océanique, fondement de la tectonique globale', *Mémoire hors série de la Société Géologique de la France* 10, 339–350.

Le Pichon, X. (1982) 'Land-locked ocean basins and continental collision: the Eastern Mediterranean as a case example', in K.J. Hsü (ed.) *Mountain Building Processes*, 201–211, Academic Press, London.

Leroi-Gourhan, A. and F. Darmon (1987) 'Analyses palynologiques de sites archéologiques du pléistocène final dans la Vallée du Jourdain', *Israel Journal of Earth Sciences* 36, 65–72.

Le Roy Ladurie, E. (1969) *Les Paysans de Languedoc*. Champs Flammarion, Paris.

Le Roy Ladurie, E. (1983) *Histoire du Climat depuis l'An Mil*. 2nd edn. Champs Flammarion, Paris (Abridged translation of first edition, 1972, as *Times of Feast, Times of Famine*, Allen and Unwin, London).

Leroy, S.A.G., S. Giralt, P. Francus and G. Seret (1996) 'The high sensitivity of the palynological record in the Vico Maar lacustrine sequence (Latium, Italy) highlights the climatic gradient through Europe for the last 90 ka', *Quaternary Science Reviews* 15, 189–201.

Letolle, R., H. De Lumley and C. Vergnaud-Grazzini (1971) 'Composition isotopique de carbonates organogènes quaternaires de Méditerranée occidentale: essai d'interprétation climatique', *Comptes Rendues de l'Académie de Science de Paris, Série D* 273, 2225–2228.

Levin, N. and A. Horowitz (1987) 'Palynostratigraphy of the Early Pleistocene QI palynozone in the Jordan–Dead Sea Rift, Israel', *Israel Journal of Earth Sciences* 36, 45–58.

Lewin, J., M.G. Macklin and J.C. Woodward (eds) (1995) *Mediterranean Quaternary River Environments*. A.A.Balkema, Rotterdam.

Lewthwaite, J.G. (1982) 'Acorns for the ancestors: the prehistoric exploitation of woodland in the West Mediterranean', in S. Limbrey and M. Bell (eds) *Archaeological Aspects of Woodland Ecology*, 217–230, BAR International Series 146, Oxford.

Lewthwaite, J. (1985) 'The Neolithic of Corsica', in C. Scarre (ed.) *Ancient France*, 146–183, Edinburgh University Press, Edinburgh.

Lewthwaite, J.G. (1989) 'Isolating the residuals: the Mesolithic basis of man-animal relationships on the Mediterranean islands', in C. Bonsall (ed.) *The Mesolithic in Europe*, 541–555, John Donald Publishers Ltd., Edinburgh.

Lewuillon, L. (1991) 'Les murs de pierre sèche en milieu rural', in J. Guilaine (ed.) *Pour Une Archéologie Agraire*, 193–221, Armand Colin, Paris.

Lézine, A.-M. and J. Casanova (1991) 'Correlated oceanic and continental records demonstrate past climate and hydrology of North Africa', *Geology* 19, 307–310.

Lightfoot, D.R. (1996) 'Syrian qanat Romani: history, ecology, abandonment', *Journal of Arid Environments* 33, 321–336.

Lindh, G. (1992) 'Hydrological and water resources impact of climate change', in L. Jeftić, J.D. Milliman and G. Sestini (eds) *Climate Change and the Mediterranean*. Volume 1, 58–93, Arnold, London.

Lindzen, R.S. (1994) 'On the scientific basis for global warming scenarios', *Environmental Pollution* 83(1–2), 125–134.

Lindzen, R.S. (1997) 'Can increasing carbon dioxide cause climate change?' *Proceedings of the National Academy of Sciences of the United States of America* 94(16), 8335–8342.

Linés Escardó, A. (1970) 'The climate of the Iberian peninsula', in C.C. Wallén (ed.) *Climates of Northern and Western Europe*. 195–239, Elsevier, Amsterdam.

Linick, T.W., A. Long, P.E. Damon and C.W. Ferguson (1986) 'High-precision radiocarbon dating of bristlecone pine from 6554 to 5350 BC', *Radiocarbon* 28, 943–953.

Liphschitz, N. (1986) 'Overview of the dendrochronological and dendroarchaeological research in Israel', *Dendrochronologia* 4, 37–58.

Liritzis, Y. (1985) 'Archaeomagnetism, Santorini volcanic eruptions and fired destruction levels on Crete', *Nature* 313, 75–76.

Liritzis, Y. and P. Petropoulos (1992) 'A preliminary study of the relationship between large earthquakes and precipitation in the region of Athens, Greece', *Earth, Moon and Planets* 57, 13–21.

Liski, J., T. Karjalainen, A. Pussinen, G.-J. Nabuurs and P. Kauppi (2000) 'Trees as carbon sinks and sources in the European Union', *Environmental Science and Policy* 3, 91–97.

Llamas, R.M. (1997) 'Transboundary water resources in the Iberian Peninsula', in N.P. Gleditsch (ed.) *Conflict and the Environment*, 335–353, Kluwer Academic Publishers, Dordrecht.

Llasat, M.C. and M. Puigcerver (1994) 'Meteorological factors associated with floods in the north-eastern part of the Iberian peninsula', *Natural Hazards* 9, 81–93.

Llongueras i Campana, M. (1987) 'Los antecedentes y el proceso de neolitización en Catalunya', in J. Guilaine, J. Courtin, J.-L. Roudil and J.-L. Vernet (eds) *Premières Communautés Paysannes en Méditerranée Occidentale*, 593–597, Éditions du CNRS, Paris.

Llorens, P., J. Latron and F. Gallart (1992) 'Analysis of the role of agricultural abandoned terraces on the hydrology and sediment dynamics in a small mountainous basin (High Llobregat, Eastern Pyrenees)', *Pirineos* 139, 27–49.

Llorens, P., R. Poch, D. Rabada and F. Gallart (1995) 'Study of the changes of hydrological processes induced by afforestation in Mediterranean mountainous abandoned fields', *Physics and Chemistry of the Earth* 20, 375–383.

López Bermúdez, F. (1976) *La Vega Alta de Segura*. Geography Department, University of Murcia.

López Bermúdez, F. (1979) 'Inundaciones catastróficas, precipitaciones torrenciales y erosión en la provincia de Murcia', *Papeles del Departamento de Geografía, Universidad de Murcia*, 49–91.

López Bermúdez, F. (1993) 'Field site investigations: El Ardal, Murcia, Spain', in J.B. Thornes (ed.) *MEDALUS I Final Report*, 433–460, Final Report on Contract EPOC-CT90-0014-(SMA), EU, Brussels.

López Bermúdez, F. (1998) 'The Guadalentín Basin, Murcia, Spain', in P. Mairota, J.B. Thornes and N. Geeson (eds) *Atlas of Mediterranean Environments in Europe: The Desertification Context*, 136–143, John Wiley and Sons, Chichester.

López Bermúdez, F., M.A. Romero Diaz, J. Martinez Fernandez and F. Belmonte Serrato (1995) 'Field site: Murcia, Spain', in J.B. Thornes (ed.) *MEDALUS II: Managing Desertification*, 198–222, Final Report on Contract EV5V-CT92-0165, EU, Brussels.

López García, M.J. (1991) *La Temperatura del Mar Balear a Partir de Imagenes de Satelite*, Universidad de Valencia, Departamento de Geografía.

López Rodriguez, J.J. (1991) *Estudio de los Parametros Climatologicos y Edafologicas en Distinctos Metodos de Mantenimiento del Suelo*. Escuela Tecnica Superior de Ingenieros Agronimos, Universidad de Cordoba, Cordoba.

Louis, A. (1995) 'Description of the soil associations', in Tavernier, R. (ed.) *1995 Soil Map of European Communities. 1:1,000,000*, 30–54, Commission of the European Communities, Luxembourg.

Lowe, J.J. (1992) 'Lateglacial and early Holocene lake sediments from the northern Apennines, Italy – pollen stratigraphy and radiocarbon dating', *Boreas* 21, 193–208.

Lowe, J.J., C.A. Accorsi, M.B. Mazzanti, A. Bishop, S. van der Kaars, L. Forlani, A.M. Mercuri, C. Rivalenti, P. Torri and C. Watson (1996) 'Pollen stratigraphy of sediment sequences from lakes Albano and Nemi (near Rome) and from the central Adriatic, spanning the interval from oxygen isotope stage 2 to the present day', *Memorie del Istituto Italiano Idrobiologico* 55, 71–98.

Lowe, J.J. and M.J.C. Walker (1997) *Reconstructing Quaternary Environments*. 2nd edn. Longman, London.

Loy, W.G. and H.E. Wright Jr (1972) 'The physical setting', in W.A. McDonald and G.R. Rapp Jr, *The Minnesota Messinia Expedition: Reconstructing a Bronze Age Regional Environment*, 36–46, University of Minnesota Press, Minneapolis.

Loyé-Pilot, M.D., J.M. Martin and J. Morelli (1986) 'Influence of Saharan dust on the rain acidity and atmospheric input to the Mediterranean', *Nature* 321, 427–428.

Lubell, D., F.A. Hassan, A. Gautier and J.-L. Ballais (1976) 'The Capsian escargotières. An interdisciplinary study elucidates Holocene ecology and subsistence in north Africa', *Science* 191, 910–919.

Lukey, B.T., J. Sheffield, J.C. Bathurst, R.A. Hiley and N. Mathys (2000) 'Test of the SHETRAN technology for modelling the impact of reforestation on badlands runoff and sediment yield at Draix, France', *Journal of Hydrology* 235, 44–62.

Lumley, H. de and Y. Boone (1976) 'Les structures d'habitat au paléolithique inférieur', in H. de Lumley (ed.) *La Préhistoire Française 1*, 625–676, Société Préhistorique Française, Paris.

Lumley, M.A. de (1976) 'Les anténéanderthals du sud', in H. de Lumley (ed.) *La Préhistoire Française 1*, 547–560, Société Préhistorique Française, Paris.

Lytras, C. (1993) 'Devloping water resources', in J. Charalambous and G. Georghallides (eds) *A Survey on the History of Cyprus in its European Context: The Achievements of Cyprus since Independence and Future Prospects*, 123–137, University of North London Press, London.

McCorriston, J. (1994) 'Acorn eating and agricultural origins: California ethnographies as analogies for the ancient Near East', *Antiquity* 68, 97–107.

McCoy, F.W. (1980) 'The Upper Thera (Minoan) ash in deep-sea sediments: distribution and comparison with other ash layers', in C. Doumas (ed.) *Thera and the Aegean World II. Papers and Proceedings of the Second International Scientific Congress, Santorini, Greece, August 1978*, 57–78, Thera and the Aegean World, London.

McDonald, A.T. and D. Kay (1988) *Water Resources, Issues and Strategies*. Longman, Harlow.

MacDonald, A., J. Candela and H.L. Bryden (1995) 'An estimate of the net heat transport through the Strait of Gibraltar', in P.E. La Violette (ed.) *Seasonal and Interannual Variability of the Western Mediterranean*, 13–32, American Geophysical Union, Washington, DC.

MacDonald, D., J.R. Crabtree, G. Wiesinger, T. Dax, N. Stamou, P. Fleury, J.G. Lazpita and A. Gibon (2000) 'Agricultural abandonment in mountain areas of Europe: Environmental consequences and policy response', *Journal of Environmental Management* 59, 47–69.

McDonald, W.A. and G.R. Rapp Jr (1972) *The Minnesota Messinia Expedition. Reconstructing a Bronze Age Regional Environment*. University of Minnesota Press, Minneapolis.

McDonald, W.A. and R.H. Simpson (1972) 'Archaeological exploration', in W.A. McDonald and G.R. Rapp Jr (eds) *The Minnesota Messinia Expedition. Reconstructing a Bronze Age Regional Environment*, 117–147, University of Minnesota Press, Minneapolis.

McGuire, W.J., R.J. Howarth, C.R. Firth, A.R. Solow, A.D. Pullen, S.J. Saunders, I.S. Stewart and C. Vita Finzi (1997) 'Correlation between rate of sea-level change and frequency of explosive volcanism in the Mediterranean', *Nature* 389, 473–476.

MacKenzie, D. (1992) 'Pundits pontificate while Etna erupts', *New Scientist* 25 April, 5.

MacKenzie, D. (1998) 'Doñana damned: pollution could threaten Spain's natural resource treasure for decades', *New Scientist* 2 May, 12.

McKenzie, D.P. (1970) 'Plate tectonics of the Mediterranean region', *Nature* 226, 239–243.

MacLeod, D.A. (1980) 'The origin of the red Mediterranean soils in Epirus, Greece', *Journal of Soil Science* 31, 125–136.

McNeill, J.R. (1992) *The Mountains of the Mediterranean World: An Environmental History*. Cambridge University Press, Cambridge.

McPherson, C.B. and F. Sarda (1984) 'Fishes and fishermen: the exploitable trophic levels', in R. Margalef (ed.) *Key Environments: Western Mediterranean*, 296–316, Pergamon, Oxford.

Maddox, J. (1984) 'From Santorini to Armageddon', *Nature* 307, 107.

Magaritz, M. and G.A. Goodfriend (1987) 'Movement of the desert boundary in the Levant from latest Pleistocene to early Holocene', in W.H. Berger and L.D. Labeyrie (eds) *Abrupt Climatic Change*, 173–183, Reidel, Dordrecht.

Magaritz, M. and J. Heller (1982) 'Annual cycle of $^{18}O/^{16}O$ and $^{13}C/^{12}C$ isotope ratios in landsnail shells', *Isotope Geoscience* 1: 243–255.

Magaritz, M., J. Heller and M. Volokita (1981) 'Land–air boundary environment as recorded by the $^{18}O/^{16}O$ and $^{13}C/^{12}C$ isotope ratios in the shells of land snails', *Earth & Planetary Science Letters* 52, 101–106.

Maheras, P. (1989) 'Principal component analysis of western Mediterranean air-temperature variations 1866-1985', *Theoretical and Applied Climatology* 39, 137–145.

Maheras, P., C. Blafoutis and M. Vafiadis (1992) 'Precipitation in the central Mediterranean during the last century', *Theoretical and Applied Climatology* 45, 209–216.

Maheras, P. and H. Kutiel (1999) 'Spatial and temporal variations in the temperature regime in the Mediterranean and their relationship with circulation during the last century', *International Journal of Climatology* 19, 745–764.

Mainguet, M. (1991) *Desertification: Natural Background and Human Mismanagement*. Springer-Verlag, Berlin.

Mairota, P. (1988) 'Agricultural changes and land degradation in the Agri Basin', in P. Mairota, J.B.

Thornes and N. Geeson (eds) *Atlas of Mediterranean Environments in Europe: the Desertification Context*, John Wiley and Sons, Chichester.

Mairota, P., J.B. Thornes and N. Geeson (eds) (1998) *Atlas of Mediterranean Environments in Europe: The Desertification Context*. John Wiley and Sons, Chichester.

Malamud, B.D. and D.L. Turcotte (1999) 'Self-affine time series: 1. Generation and analyses', *Advances in Geophysics* 40, 1–90.

Maldonado, A. (1983) 'Evolution of the Mediterranean basins and a detailed reconstruction of Canozoic palaeoceanography', in R. Margelef (ed.) *The Mediterranean*, 17–59, Pergamon Press, Oxford.

Maldonado, A. and D.J. Stanley (1979) 'Depositional patterns and late Quaternary evolution of two Mediterranean submarine fans: a comparison', *Marine Geology* 31, 215–250.

Malone, C., S. Stoddart and R. Whitehouse (1994) 'The Bronze Age of southern Italy, Sicily and Malta *c*.2000–800 BC', in C. Mathers and S. Stoddart (eds) *Development and Decline in the Mediterranean Bronze Age*, 167–195, J.R. Collis Publications, Sheffield.

Makhzoumi, J.M. (1997) 'The changing role of rural landscapes: olive and carob multi-use tree plantations in the semiarid Mediterranean', *Landscape and Urban Planning* 37, 115–122.

Manabe, S. and R.J. Stouffer (1995) 'Simulation of abrupt climate change induced by freshwater input to the North Atlantic Ocean', *Nature* 378, 165.

Manning, S.W. (1988) 'Dating of the Santorini eruption', *Nature* 333, 401.

Manning, S.W. (1994) 'The emergence of divergence: development and decline on Bronze Age Crete and the Cyclades', in C. Mathers and S. Stoddart (eds) *Development and Decline in the Mediterranean Bronze Age*, 221–270, J.R. Collis Publications, Sheffield.

Mansanet Terol, C.M. (1987) *Incendios Forestales en Alicante. Estudio de la Evolución de la Vegetación Quemada*. Publicaciones de la Caja de Ahorros Provincial, Alicante.

Manzella, G.M.R. (1995) 'The seasonal variability of the water masses and transport through the Straits of Sicily', in P. La Violette (ed.) *Seasonal and Interannual Variability of the Western Mediterranean Sea*, 33–46, American Geophysical Union, Washington, DC.

Mardones, M. and G. Jalut (1983) 'La tourbière de Biscaye (alt. 409 m, Hautes Pyrénées): approche paléoécologique des 45.000 dernières années', *Pollen et Spores* XXV, 163–212.

Margalef, R. (1984) 'Introduction to the Mediterranean', in R. Margalef (ed.) *Key Environments: Western Mediterranean*, 1–16, Pergamon, Oxford.

Margaris, N.S. and A.T. Grove (1993) 'Landscape dynamics in relation to desertification', in J.B. Thornes (ed.) *MEDALUS I Final Report*, 264–80, Final Report on Contract EPOC-CT90-0014-(SMA), EU, Brussels.

Margaris, N.S., E. Koutsidou and C. Giourga (1996) 'Changes in traditional Mediterranean land-use systems', in C.J. Brandt and J.B. Thornes (eds) *Mediterranean Desertification and Land Use*, 29–42, John Wiley and Sons, Chichester.

Margaris, N.S., E.J. Koutsidou and C.E. Giourga (1998) 'Agricultural transformations', in P. Mairota, J.B. Thornes and N. Geeson (eds) *Atlas of Mediterranean Environments in Europe: The Desertification Context*, 82–84, John Wiley and Sons, Chichester.

Mariolakos, I. (1991) 'Prediction of natural mass movement in tectonically active areas', in M.E. Almeida-Teixeira, R. Fantechi, R. Oliveira and A. Gomes Coelho (eds) *Natural Hazards and Engineering Geology. Prevention and Control of Landslides and Other Mass Movements*, 69–81, Commission of the European Communities, EUR 12918, Brussels.

Marqués, M.A. (1991) 'Soil erosion research: experimental plots on agricultural and burnt environments near Barcelona', in M. Sala, J.L. Rubio and J.M. García-Ruiz (eds) *Soil Erosion Studies in Spain*, 153–164, Geoforma Ediciones, Logroño.

Marqués, M.A. and J. Roca (1987) 'Soil loss measurements in an agricultural area of north-east Spain', in V. Gardiner (ed.) *Internation Geomorphology 1986. Proceedings of the 1st Conference, Volume II*, 483–493, John Wiley and Sons, Chichester.

Martin Penela, A.J. (1994) 'Pipe and gully systems – development in the Almanzora basin (Southeast Spain)', *Zeitschrift für Geomorphologie* 38, 207–222.

Martinelli, N., O. Pignatelli and M. Romagnoli (1994) 'Primo contributo allo studio dendroclimatologico del Cerro (*Quercus cerris* L.) in Sicilia', *Dendrochronologia* 12, 61–76.

Martinson, D.G., N.G. Pisias, J.D. Hays, J. Imbrie, T.C. Moore and N.J. Shackleton (1987) 'Age dating and the orbital theory of the ice ages: development of a high-resolution 0 to 300,000-year chronostratigraphy', *Quaternary Research* 27, 1–29.

Martyn, D. (1992) *Climates of the World*. Elsevier, Amsterdam.

Mas-Pla, J., J. Bach, E. Viñals, J. Trilla and J. Estralrich (1999) 'Salinization processes in a coastal leaky aquifer system (Alt Empordà, NE Spain)', *Physics and Chemistry of the Earth (B)* 24, 337–341.

Masson, E. (1993) 'Vallée des Merveilles. Un berceau de la pensée religieuse européenne', *Les Dossiers de l'Archéologie* 181, 145pp.

Mathers, C. (1994) 'Goodbye to all that? Contrasting patterns of change in the south-east Iberian Bronze Age', in C. Mathers and S. Stoddart (eds) *Development and Decline in the Mediterranean Bronze Age*, 21–72, J.R. Collis Publications, Sheffield.

Mathers, C. and S. Stoddart (eds) (1994) *Development and Decline in the Mediterranean Bronze Age*. J.R. Collis Publications, Sheffield.

Matos, S. de (1993) 'Glacial and periglacial geomorphology and present-day climatic conditions in Serra da Estrela', in *3ª Reunião do Quaternario Ibérico, Coimbra, Portugal*, 165–170, GTPEQ AEQUA.

Mattingly, D.J. (1995) *Tripolitania*. Batsford, London.

Mazzarello, A. and F. Palumbo (1992) 'Rainfall fluctuations over Italy and their association with solar-activity', *Theoretical and Applied Climatology* 45, 201–207.

Mazzarello, A. and A. Palumbo (1994) 'The lunar modal-induced signal in climatic and oceanic data over the Western Mediterranean area and on its bistable phasing', *Theoretical and Applied Climatology* 50, 93–102.

Medforum (2000) Tourism in the Mediterranean. http://www.medforum.org/ulixes21/in/proj_elt_in.htm. (accessed May 2000).

Mediterranean Hydrological Cycle Observing System (Med-HYCOS) (2000 http://www.medhycos.mpl.ird.fr (accessed October 2000).

Mee, C. (1991) 'Rural settlement change in the Methana peninsula, Greece', in G. Barker and J. Lloyd (eds) *Roman Landscapes: Archaeological Survey in the Mediterranean Region*, 223–232, Archaeological Monographs of the British School at Rome, no. 2, British School at Rome, London.

Meiggs, R. (1982) *Trees and Timber in the Ancient Mediterranean World*. Clarendon Press, Oxford.

Melentis, J.K. (1978) 'The Dinaric and Aegean Arcs: Greece and the Aegean Sea', in A.E.M. Nairn, W.H. Kanes and F.G. Stehli (eds) *The Ocean Basins and Margins*. Volume 4A: *The Eastern Mediterranean*, 263–275, Plenum Press, New York.

Mellaart, J. (1967) *Çatal Hüyük: A Neolithic Town in Anatolia*. Thames and Hudson, London.

Mellaart, J. (1975) *Neolithic of the Near East*. Thames and Hudson, London.

Melmoth, W. (1915) *Pliny the Younger: Letters*. Loeb Classical Library, London.

Métaillie, J.-P. (1981) *Le Feu Pastoral dans les Pyrénées Centrales (Barousse, Oueil, Larboust)*. Éditions du CNRS, Paris.

Métaillie, J.-P. (1986) 'Photographie et histoire du paysage: un exemple dans les Pyrénées luchonnaises', *Revue Géographique des Pyrénées et du Sud-Ouest* 57(2), 179–208.

Métaillie, J.P. (1987) 'The degradation of the Pyrenees in the nineteenth century – an erosion crisis?', in V. Gardiner (ed.) *Internation Geomorphology 1986. Proceedings of the 1st Conference, Volume II*, 533–543, John Wiley and Sons, Chichester.

Metaxas, D.A., A. Bartzokas and A. Vitsas (1991) 'Temperature fluctuations in the Mediterranean area during the last 120 years', *International Journal of Climatology* 11, 897–908.

Michaelides, K. and J. Wainwright (2002) 'Modelling the Effects of Hillslope-Channel Coupling on Catchment Hydrological Response', *Earth Surface Processes and Landforms* 27, 1441–1458.

Michaels, P.J., R.C. Balling, R.S. Vose and P.C. Knappenberger (1998) 'Analysis of trends in the variability of daily and monthly historical temperature measurements', *Climate Research* 10(1), 27–33.

Milliman, J.D. (1992) 'Sea-level response to climate change and tectonics in the Mediterranean Sea', in L. Jeftić, J.D. Milliman and G. Sestini (eds) *Climate Change and the Mediterranean*. Volume 1, 45–57, Arnold, London.

Milliman, J.D., L. Jeftić and G. Sestini (1992) 'The Mediterranean Sea and climate change – an overview', in L. Jeftić, J.D. Milliman and G. Sestini (eds) *Climate Change and the Mediterranean*. Volume 1, 1–14, Arnold, London.

Misgav, A. (2000) 'Visual preference of the public for vegetation groups in Israel', *Landscape and Urban Planning* 48, 143–159.

Miskovsky, J-C. (1974) 'Le Quaternaire du midi Méditerranéen. Stratigraphie et paléoclimatologie d'après l'étude sédimentologique du remplissage des grottes et abris sous roche (Ligurie, Provence, Languedoc Méditerranéen, Rousillon, Catalogne)', *Études Quaternaires Mémoire no. 3*, Université de Provence, Marseille.

Mitchell, T. (1999) 'Dreamland, the neoliberalism of your desires', *Middle East Report* 210, 31.

Miyamura, S. (1988) 'Some remarks on historical seismograms and the microfilming project', in W.H.K. Lee, H. Meyers and K. Shimazaki (eds) *Historical Seismograms and Earthquakes of the World*, 401–419, Academic Press, London.

Mommersteeg, H.J.P.M., M.F. Loutre, R. Young, T.A. Wijmstra and H. Hooghiemstra (1995) 'Orbital forced frequencies in the 975000 year pollen record from Tenaghi Philippon (Greece)', *Climate Dynamics* 11, 4–24.

Monaghan, J.J., P.J. Bicknell and R.J. Humble (1994) 'Volcanoes, tsunamis and the demise of the Minoans', *Physica D* 77, 217–228.

Monnié, A. (1931) 'La régime du Tarn à Albi', *Revue Géographique des Pyrénées et du Sud-Ouest* 2, 162–189 and 337–358.

Montanari, A. (1995) 'Tourism and the environment: limitations and contradictions in the EC's Mediterranean area', *Tijdschrift voor Economische en Sociale Geographie* 86(1), 32–41.

Morell, I., E. Giménez and M.V. Esteller (1996) 'Application of principal components analysis to the study of salinization on the Castellón Plain (Spain)', *The Science of the Total Environment* 177, 161–171.

Moresi, M. and G. Mongelli (1988) 'The relation between the terra rossa and the carbonate-free residue of the underlying limestones and dolostones in Apulia, Italy', *Clay Minerals* 23, 439–446.

Morgan, R.P.C. (1995) *Soil Erosion and Conservation.* 2nd edn. Longman, Harlow.

Morris, A. (1992) 'Spain's new economic geography – the Mediterranean axis', *Scottish Geographical Magazine* 108(2), 92–98.

Morris, A. (1996) 'Tourism and local awareness: Costa Brava, Spain', in G.K. Priestley, J.A. Edwards and H. Coccossis (eds) *Sustainable Tourism? European Experiences*, 70–85, CAB International, Wallingford.

Mueller, S. (1982) 'Deep structure and recent dynamics in the Alps', in K.J. Hsü (ed.) *Mountain Building Processes*, 181–199, Academic Press, London.

Muhly, J.D. (1984) 'New evidence for sources of and trade in Bronze-Age tin', in A.D. Franklin, J.S. Olin and T.A. Wertime (eds) *The Search for Ancient Tin*, 43–48, Smithsonian Institute, Washington, DC.

Mulligan, M. (1996) 'Modelling the complexity of land-surface response to climatic variability in Mediterranean environments', in M.G. Anderson and S.M. Brooks (eds) *Advances in Hillslope Processes.* Volume 2, 1099–1149, John Wiley and Sons, Chichester.

Muñoz, D. and A. Udias, (1988) 'Evaluation of damage and source parameters of the Málaga earthquake of 9 October 1680', in W.H.K. Lee, H. Meyers and K. Shimazaki (eds) *Historical Seismograms and Earthquakes of the World*, 208–221, Academic Press, London.

Musgrave, P. (1992) *Land and Economy in Baroque Italy. Valpolicella, 1630–1797.* Leicester University Press, Leicester.

Musset, D., F.-X. Emery, P. Coste, N. Coulet, C. de Villeneuve-Bargemont, T. Schippers, F. Prévost and H. Germain (1986) *Histoire et Actualité de la Transhumance en Provence*, Les Alpes de Lumière 95/96, Edisud, Aix-en-Provence.

Naguib, M.K. (1970) 'Precipitation in the UAR in relation to different synoptic patterns', *Meteorological Research Bulletin* 2, 207–221.

Narcisi, B. (1996) 'Tephrochronology of a Late Quaternary lacustrine record from the Monticchio Maar (Vulture volcano, southern Italy)', *Quaternary Science Reviews* 15, 155–165.

Nativ, R., E. Adar, O. Dahan and I. Nassim (1997) 'Water salinization in arid regions – observations from the Negev desert, Israel', *Journal of Hydrology* 196, 271–296.

Naveh, Z. (1974) 'The ecological management of non-arable Mediterranean uplands', *Journal of Environmental Management* 2, 35.

Naveh, Z. (1975) 'The evolutionary significance of fire in the Mediterranean region', *Vegetatio* 29, 199–209.

Naveh, Z. (1994) 'The rôle of fire and its management in the conservation of Mediterranean ecosystems and landscapes', in J.M. Moreno and W.C. Oechel (eds) *The Rôle of Fire in Mediterranean-Type Ecosystems*, 163–185, Springer-Verlag, New York.

Naylon, J. (1992) 'Ascent and decline in the Spanish regional system', *Geography* 77, 46–62.

Neboit, R. (1977) 'Un exemple de morphologenèse accélérée dans l'antiquité: les valées de Basento et du Cavone en Lucanie (Italie)', *Méditerranée* 30, 39–50.

Neboit, R. (1984) 'Érosion des sols et colonisation grecque en Sicilie et en Grande Grèce', *Bulletin de l'Association de Géographes Français* 499, 5–13.

Negev, A. (1966) *Cities of the Desert*. E. Lewin-Epstein, Tel-Aviv.

Neumann, J. (1973) 'Sea and land breezes in the classical Greek literature', *Bulletin of the American Meteorological Society* 54, 5–8.

Neumann, J. (1986) 'Recent climatic fluctuations in Israel as indicated by Jerusalem meteorological data', *Israel Journal of Earth Sciences* 35, 51–53.

Niccolai, M., L. Petkov, F. Meneguzzo and P. Vignaroli (1995) 'Abruzzo region: climatology, remote sensing and socio-economics', in J.B. Thornes (ed.) *MEDALUS II. Research and Policy Interfacing in Selected Regions*, 404–448, Final Report on Contract EV5V-CT92-0166, EU, Brussels.

Nicholls, R.J. and F.M.J. Hoozemans (1996) 'The Mediterranean: vulnerability to coastal implications of climate change', *Ocean and Coastal Management* 31(2–3), 105–132.

Nicholls, R.J., F.M.J. Hoozemans and M. Marchand (1999) 'Increasing flood risk and wetland losses due to global sea-level rise: regional and global analyses', *Global Environmental Change* 9, S69–S87.

Nicholson, S.E. (1989) 'African drought: characteristics, causal theories and global teleconnections', in A. Berger, R.E. Dickinson and J.W. Kidson (eds) *Understanding Climate Change*, 79–100, American Geophysical Union, Washington, DC.

Nicol-Pichard, S. (1982) 'Analyse pollinique de sédiments littoraux post-glaciaires de l'embouchure de Paillon (Nice)', *Ecologia Mediterranea* 8(4), 87–95.

Nicol-Pichard, S. (1987) 'Analyse pollinique d'une séquence tardi- et postglaciaire à Tourves (Var, France)', *Ecologia Mediterranea* 13, 29–42.

Niklewski, J. and W. van Zeist (1970) 'A late Quaternary pollen diagram from northwestern Syria', *Acta Botanica Neerlandica* 19, 737–754.

Ninkovich, D. and J.D. Hays (1972) 'Mediterranean island arcs and origin of high potash volcanics', *Earth and Planetary Science Letters* 16, 331–345.

NOAA (1984) 'FAO/UNESCO UNEP gridded FAO/UNESCO soil units', http://www.ngdc.noaa.gov/seg/eco/cdroms/gedii_a/datasets/a16/fao.htm. (Accessed August 2001).

Nola, P. (1994) 'A dendroecological study of larch at timberline in the central Italian Alps', *Dendrochronologia* 12, 77–91.

Noy, T., A.J. Legge and E.S. Higgs (1973) 'Excavations at Nahel Oren, Israel', *Proceedings of the Prehistoric Society* 39, 75–99.

Noy-Meir, I. (1973) 'Desert ecosystems: environment and producers', *Annual Review of Ecology and Sytematics* 4, 25–51.

Nur, A. and E.H. Cline (2000) 'Poseidon's horses: plate tectonics and earthquake storms in the Late Bronze Age Aegean and eastern Mediterranean', *Journal of Archaeological Science* 27, 43–63.

Oates, D. and J. Oates (1976) *The Rise of Civilization*. Phaidon, Oxford.

Obled, C. (1988) 'Rainfall studies and flash flood problems in the French Mediterranean Region: recent achievements and current problems', in F. Siccardi and R.L. Bras (eds) *Selected Papers from Workshop on Natural Disasters in European Mediterranean Countries, Perugia, Italy, 1988*, 375–395, US National Science Foundation, Washington, DC.

Oke, T.R. (1987) *Boundary Layer Meteorology*. 2nd edn. Routledge, London.

Oldfield, F. (1996) 'The PALICLAS project: synthesis and overview', *Memorie del Istituto Italiano Idrobiologico* 55, 329–357.

Oostwoud Wijdenes, D., J. Poesen, L. Vandekerckhove and E. de Luna (1997) 'Chiseling effects on the vertical distribution of rock fragments in the tilled layer of a Mediterranean soil', *Soil and Tillage Research* 44, 55–66.

O'Riordan, T. and Voisey, H. (eds) (1998) *The Transition to Sustainability: The Politics of Agenda 21 in Europe*, Earthscan, London.

O'Rourke, E. (1999) 'Changing identities, changing landscapes: human–land relations in transition in the Aspre, Roussillon', *Ecumene* 6(1), 29–50.

Ortloff, C.R. and D.P. Crouch (1998) 'Hydraulic analysis of a self-cleaning drainage outlet at the Hellenistic city of Priene', *Journal of Archaeological Science* 25, 1211–1220.

Osborne, C.P., P.L. Mitchell, J.E. Sheehy and F.I. Woodward (2000) 'Modelling the recent historical impacts of atmospheric CO_2 and climate change on Mediterranean vegetation', *Global Change Biology* 6, 445–458.

Overpeck, J.T., L.C.Peterson, N. Kipp, J. Imbrie and D. Rind (1989) 'Climate change in the circum-North Atlantic region during the last deglaciation', *Nature* 338, 553–557.

Owens, E.J. (1991) *The City in the Greek and Roman World*. Routledge, London.

Paepe, R. (1986) 'Landscape changes in Greece as a result of changing climate during the Quaternary', in R. Fantechi and N.S. Margaris (eds) *Desertification in Europe*, 49–58, Reidel, Dordrecht.

Palumbo, A. (1986) 'Lunar daily variations in rainfall', *Journal of Atmospheric and Terrestrial Physics* 48, 145–148.

Palutikof, J.P., M. Conte, J. Casimiro Mendes, C.M. Goodess and F. Espirito Santo (1996) 'Climate and climatic change', in C.J. Brandt and J.B. Thornes (eds) *Mediterranean Desertification and Land Use*, 43–86, John Wiley and Sons, Chichester.

Palutikof, J.P., C.M. Goodess and X. Guo (1994) 'Climate change, potential evapotranspiration and moisture availability in the Mediterranean Basin', *International Journal of Climatology* 14, 853–869.

Palutikof, J.P., X. Guo, T.M.L. Wigley and J.M. Gregory (1992) Regional changes in climate in the Mediterranean Basin due to global greenhouse warming. *MAP Technical Report Series* 66, Mediterranean Action Plan/UNEP, Athens.

Palutikof, J.P., R.M. Trigo and S.T. Adcock (1999) 'Scenarios of future rainfall over the Mediterranean: Is the region drying?', in P. Balabanis, D. Peter, A. Ghazi and M. Tsoges (eds) *Mediterranean Desertification, Research Results and Policy Implications*, 33–39, European Commission, EUR 19303, Brussels.

Palutikof, J.P. and T.M.L. Wigley (1996) 'Developing climate change scenarios for the Mediterranean region', in L. Jeftić, S. Kečkeš and J.C. Pernetta (eds) *Climate Change and the Mediterranean*. Volume 2, 27–54, Arnold, London.

Pannicucci, M. (1972) 'Richerche orientative sui fenomeni erosivi nei terreni argillosi', *Annali del Istituto Sperimentale Studio e Difesa Suolo* 3, 131–146.

Papanastasis, V. (1977) 'Fire ecology and management of phrygana communities in Greece', in H.A. Mooney and C.E. Conrad (eds) *Proceedings of the Symposium on Environmental Consequences of Fire and Fuel Management in Mediterranean Ecosystems*, 476–482, USDA Forest Service, General Technical Report WO-3.

Papanastasis, V. (1993) 'Legal status of land tenure and use and its implication for open landscapes of western Crete', *Landscape and Urban Planning* 24, 273–277.

Papanastasis, V. and G. Lyrintzis (1995) 'Effect of human activities on desertification of Psilorites mountain with emphasis on livestock husbandry', in M. Dubost (ed.) *MEDIMONT: A Multinational, Multidisciplinary Research Programme on the Rôle and Place of the Mountains in the Desertification of the Mediterranean Mountain Regions. Second and Final Scientific Report*, Final Report on Contract EV5V-CT91-0045, EU, Brussels.

Pardé, M. (1933a) 'Le régime de l'Aude', *Revue Géographique des Pyrénées et du Sud-Ouest* 4, 5–29.

Pardé, M. (1933b) 'Les crues de Décembre 1932 dans le Languedoc et le Rousillon', *Revue Géographique des Pyrénées et du Sud-Ouest* 4, 499–512.

Pardé, M. (1953) 'Sur les inondations en Aquitaine, spécialement dans le bassin de la Garonne. A propos de la grande crue de Février 1952', *Revue Géographique des Pyrénées et du Sud-Ouest* 24, 163–57 and 25, 5–38.

Parsons, A.J., A.D. Abrahams and J.R. Simanton (1992) 'Microtopography and soil-surface materials on semi-arid pedimont hillslopes', *Journal of Arid Environments* 22, 107–115.

Parsons, A.J., A.D. Abrahams and J. Wainwright (1996) 'Responses of interrill runoff and erosion rates to vegetation change in Southern Arizona', *Geomorphology* 14, 311–317.

Parsons, A.J. and J. Wainwright (*in prep.*) 'Mechanisms of rill initiation', *Journal of Geophysical Research – Surface Processes.*

Parsons, A.J., J. Wainwright and A.D. Abrahams (1993) 'Tracing sediment movement on semi-arid grassland using magnetic susceptibility', *Earth Surface Processes and Landforms* 18, 721–732.

Parsons, A.J., J. Wainwright, D.M. Powell and R. Brazier (in press) 'A new conceptual model for understanding and predicting erosion by water', *Earth Surface Processes and Landforms.*

Pasquale, V., M. Verdoya, P. Chiozzi and J. Safanda (2000) 'Evidence of climate warming from underground temperatures in NW Italy', *Global and Planetary Change* 25(3), 215–222.

Paterne, M., F. Guichard and J. Labeyrie (1988) 'Explosive activity of the south Italian volcanoes during the past 80,000 years as determined by marine tephrachronology', *Journal of Volcanology and Geothermal Research* 34, 153–172.

Paterne, M., J. Labeyrie, F. Guichard, A. Mazaud and F. Maître (1990) 'Fluctuations of the Campanian explosive volcanic activity (south Italy) during the past 190,000 years, as determined by marine tephrochronology', *Earth and Planetary Science Letters* 98, 166–174.

Pavese, M.P., V. Banzon, M. Colacino, G.P. Gregori and M. Pasqua (1992) 'Three historical data series on floods and anomalous climatic events in Italy', in R.S. Bradley and P.D. Jones (eds) *Climate Since A.D. 1500*, 155–170, Routledge, London.

Payne, S. (1985) 'Animal bones from Aşıklı Hüyük', *Anatolian Studies* 35, 109–122.

Peacock, D.P.S. (1992) *Rome in the Desert: A Symbol of Power.* University of Southampton, Southampton.

Pearce, F. (1997) 'Greenhouse wars', *New Scientist* 19 July, 38–43.

Pearson, G.W., B. Becker and F. Qua (1993) 'High-precision C^{14} measurement of German and Irish oaks to show the natural C^{14} variations from 7890 to 5000 BC', *Radiocarbon* 35, 93–104.

Pearson, G.W. and M. Stuiver (1993) 'High-precision bidecadal calibration of the radiocarbon time scale, 500–2500 BC', *Radiocarbon* 35, 25–33.

Pellicer, M. and P. Acosta (1983) 'El neolitico antiguo en Andalucia Occidental', in *Actes du Colloque International de Préhistoire, Montpellier, 1981: Le Néolithique Ancien Méditerranéen*, 49–60, La Fédération Archéologique de l'Hérault, Montpellier.

Pérez Cueva, A.J. and R. Armengot Serrano (1983) 'El temporal de Octubre de 1982 en el marco de las lluvias torrenciales en la cuenca baja del Júcar', *Cuadernos de Geografía* 32–33, 61–86.

Pérez Cueva, A. and A. Calvo (1984) 'Lluvias torrenciales y cambios geomorfólogicos en una pequena cuenca de montana: el barranco de la Cuesta de la Vega (Valencia)', *Cuadernos de Investigacion Geografica* 10, 169–182.

Pérez-Obiol, R. (1988) 'Histoire tardiglaciaire et holocène de la végétation de la région volcanique d'Olot (N.E. Péninsule Ibérique)', *Pollen et Spores* XXX, 189–202.

Pérez-Obiol, R. and R. Julià (1994) 'Climatic change on the Iberian Peninsula recorded in a 30,000-yr pollen record from Lake Banyoles', *Quaternary Research* 41, 91–98.

Perissoratis, C., D. Georgas, M.C. Alexiadou, G. Dikaiakos, A. Lascaratos, S. Leontaris, N. Margaris and K. Tsakiri (1996) 'Implications of expected climatic changes for the Island of Rhodes', in L. Jeftić, S. Kečkeš and J.C. Pernetta (eds) *Climate Change and the Mediterranean.* Volume 2, 57–142, Arnold, London.

Peterson, J.W.M. (1992) 'Computer-aided projection of part of the Orange B cadastre to the Cèze Valley', *Dialogues d'Histoire Ancienne*, 18, 169–176.

Petley, D. (1996) 'The mechanics and landforms of deep-seated landslides', in M.G. Anderson and S.M. Brooks (eds) *Advances in Hillslope Processes.* Volume 2, 826–836, John Wiley and Sons, Chichester.

Pfeifer, K. (1999) 'How Tunisia, Morocco, Jordan and even Egypt became IMF "success stories" in the 1990s', *Middle East Report* 210, 23.

Phillips, C.P. (1998) 'The Crete Senesi, Tuscany: a vanishing landscape?', *Landscape and Urban Planning* 41(1), 19–26.

Pichler, H. and W.L. Friedrich (1980) 'Mechanism of the Minoan eruption of Santorini', in C. Doumas (ed.) *Thera and the Aegean World II: Papers and Proceedings of the Second International Scientific Congress, Santorini, Greece, August 1978*, 15–30, Thera and the Aegean World, London.

Pignatti, S. (1978) 'Evolutionary trends in Mediterranean flora and vegetation', *Vegetatio* 37(3), 175–185.

Pigou, A.C. (1920) *The Economics of Welfare*. 1st edn. Macmillan, London.

Pinol, J., J. Terradas and F. Lloret (1998) 'Climate warming, wildfire hazard, and wildfire occurrence in coastal eastern Spain', *Climatic Change* 38(3), 345–357.

Pirazzoli, P.A. (1989) 'Present and near-future sea-level changes', *Palaeogeography, Palaeoclimatology, Palaeoecology* 75, 241–258.

Pirazzoli, P.A. (1991) *World Atlas of Holocene Sea-Level Changes*. Elsevier, Amsterdam.

Pizzolotto, R. and P. Brandmayr (1995) 'The assessment of human influence and desertification processes in the landscape of the Ionian side of Calabria (Southern Italy)', in M. Dubost (ed.) *MEDIMONT: A Multinational, Multidisciplinary Research Programme on the Rôle and Place of the Mountains in the Desertification of the Mediterranean Mountain Regions. Second and Final Scientific Report*, Final Report on Contract EV5V-CT91-0045, EU, Brussels.

Plana Castelvi, A. (1989) 'Usos y abusos del agua: el caso del Rio Besos', in A. Gil Olcina and A. Morales Gil (eds) *Avenidas y Inundaciones en la Cuenca del Mediterraneo*, 153–166, Caja de Ahorros Mediterraneos, Alicante.

Planchais, N. (1982) 'Palynologie lagunaire de l'Étang de Mauguio. Paléoenvironnement végétal et évolution anthropique', *Pollen et Spores* XXIV, 93–118.

Planchais, N. (1985) 'Analyses polliniques du remplissage holocène de la lagune de Canet (plaine de Rousillon, département des Pyrénées Orientales)', *Ecologia Mediterranea* 11, 117–127.

Planchais, N. and I. Parra Vergara (1984) 'Analyses polliniques de sédiments lagunaires et côtiers en Languedoc, en Roussillon et dans la province de Castellon (Espagne): bioclimatologie', *Bulletin de la Société Botanique Française* 131, *Actualités Botaniques* 1984(2/3/4), 97–105.

Poesen, J.W.A. (1992) 'Mechanisms of overland flow generation and sediment production on loamy and sandy soils with and without rock fragments', in A.J. Parsons and A.D. Abrahams (eds) *Overland Flow Hydraulics and Erosion Mechanics*, 275–305, UCL Press, London.

Poesen, J.W.A. (1995) 'Soil erosion in Mediterranean environments', in R. Fantechi, D. Peter, P. Balabanis and J.L. Rubio (eds) *Desertification in a European Context: Physical and Socio-economic Aspects*, 123–152, European Commission Report EUR 15415, EU, Brussels.

Poesen, J. and K. Bunte (1996) 'Effects of rock fragments on desertification processes in Mediterranean environments', in J.B. Thornes and C.J. Brandt (eds) *Mediterranean Desertification and Land Use*, 247–269, John Wiley and Sons, Chichester.

Poesen, J.W.A. and J.M. Hooke (1997) 'Erosion, flooding and channel management in Mediterranean environments of southern Europe', *Progress in Physical Geography* 21(2), 157–199.

Poesen, J.W.A. and H. Lavee (1994) 'Rock fragments in top soils: significance and processes', *Catena* 23, 1–28.

Poesen, J.W.A. and H. Lavee (1997) 'How efficient were ancient rainwater harvesting systems in the Negev Desert, Israel?', *Bull. Séanc. Acad. R. Sci. Outre-Mer* 43, 405–419.

Poesen, J.W.A., K. Vandaele and B. van Wesemael (1998) 'Gully erosion: importance and model implications', in J. Boardman and D. Favis-Mortlock (eds) *Modelling Soil Erosion by Water*, 285–311, Springer-Verlag, Berlin.

Poesen, J.W.A., B. van Wesemael, G. Govers, J. Martinez-Fernandez, P. Desmet, K. Vandaele, T. Quine and G. Degraer (1997) 'Patterns of rock fragment cover generated by tillage erosion', *Geomorphology* 18, 183–197.

Pollard, J. and R. Dominguez Rodriguez (1995) 'Unconstrained growth. The development of a Spanish resort', *Geography* 80(1), 33–44.

Polunin, O. and A. Huxley (1965) *Flowers of the Mediterranean*. Chatto and Windus, London.

Polunin, O. and B.E. Smythies (1973) *Flowers of South-West Europe: A Field Guide*. Oxford University Press, Oxford.

Pons, A., J.-L. de Beaulieu, J. Guiot and M. Reille (1987) 'The Younger Dryas in Southwestern Europe: an abrupt climatic change as evidenced from pollen records', in W.H. Berger and L.D. Labeyrie (eds) *Abrupt Climatic Change*, 195–208, Reidel, Dordrecht.

Pons, A., J. Guiot, J.-L. de Beaulieu and M. Reille (1992) 'Recent contributions to the climatology of the last glacial–interglacial cycle based on French pollen sequences', *Quaternary Science Reviews* 11, 439–448.

Pons, A. and P. Quézel (1985) 'The history of the flora and vegetation and past and present human disturbance in the Mediterranean region', in C. Gomez-Campo (ed.) *Plant Conservation in the Mediterranean Area*, 25–43, Dr W Junk Publishers, Dordrecht.

Pons, A. and M. Reille (1988) 'The Holocene- and Upper Pleistocene pollen record from Padul (Granada, Spain): a new study', *Palaeogeography, Palaeoclimatology, Palaeoecology* 66, 243–263.

Pons, A., C. Toni and H. Triat (1979) 'Edification de la Camargue et histoire holocène de sa végétation', *Terre Vie, Revue Ecologique* supplement 2, 13–30.

Portères, P.R. (1979) 'Présence et utilisation des végétaux identifiés', in C. Roubet (ed.) *Économie Pastorale Préagricole en Algérie Orientale: Le Néolithique de Tradition Capsienne. Exemple: L'Aurès*, 439–448, Éditions du CNRS, Paris.

Postiglione, L.F. Basso, M. Amato and F. Carone (1990) 'Effects of soil tillage methods on soil losses, on soil characteristics and on crop production in a hilly area of southern Italy', *Agricoltura Mediterranea* 120, 148–158.

Potter, C. (1998) *Against the Grain: Agri-Environmental Reform in the United States and the European Union*. Wallingford, CAB International.

Potter, T.W. (1979) *The Changing Landscape of South Etruria*. Paul Elek, London.

Poulovassilis, A., T. Mimides, P. Giannoulopoulos, N. Psyhoyou and S. Alexandris (1994) 'Understanding the water table in the Argolid', in S.E. van der Leeuw (ed.) *Understanding the Natural and Anthropogenic Causes of Soil Degradation and Desertification in the Mediterranean Basin*. Volume 5, *Part II: Agricultural Production and Water Quality in the Argolid, Greece*, 5–88, Final Report on Contract EV5V-CT91-0021, EU, Brussels.

Prat, N. and C. Ibañez (1995) 'Effects of water transfers projected in the Spanish National Hydrological Plan on the ecology of the lower River Ebro (N.E. Spain) and its delta', *Water Science and Technology*, 31, 79–86.

Preiss, E., J.-L. Martin and M. Debussche (1997) 'Rural depopulation and recent landscape changes in a Mediterranean region: consequences to the breeding avifauna', *Landscape Ecology* 12, 51–61.

Price, C., S. Michaelides, S. Pashiardis and P. Alpert (1999) 'Long term changes in diurnal temperature range in Cyprus', *Atmospheric Research* 51(2), 85–98.

Pridham, G. (1999) 'Towards sustainable tourism in the Mediterranean? Policy and practice in Italy, Spain and Greece', *Environmental Politics* 8(2), 97–116.

Priestley, G.K. (1996) 'Structural dynamics of tourism and recreation-related development: the Catalan coast', in G.K. Priestley, J.A. Edwards and H. Coccossis (eds) *Sustainable Tourism? European Experiences*, 99–119, CAB International, Wallingford.

Priestley, G.K., J.A. Edwards and H. Coccossis (eds) (1996) *Sustainable Tourism? European Experiences*, CAB International, Wallingford.

Provansal, M. (1995) 'The role of climate in landscape morphogenesis since the Bronze Age in Provence, southeastern France', *The Holocene* 5, 348–353.

Provansal, M., L. Bertucchi and M. Pelissier (1994) 'The swampy grounds in western Provence, indicators of the Holocene morphogenesis', *Zeitschrift für Geomorphologie* 38(2), 185–205.

Puigcerver, M., S. Alonso, J. Lorente, M.C. Llasat, A. Redaño, A. Burgueño and E. Vilar (1986) 'Preliminary aspects of rainfall rates in the northeast of Spain', *Theoretical and Applied Climatology* 37, 97–109.

Puigdefábregas, J. (1993) 'Field site investigations: Tabernas, Almería, Spain', in J.B. Thornes (ed.) *MEDALUS I Final Report*, 461–503, Final Report on Contract EPOC-CT90-0014-(SMA), EU, Brussels.

Puigdefábregas, J., C. Aguliera, A.J. Brenner, S.C. Clark, M. Cueto, L. Delgado, F. Domingo, L. Gutierrez, L.D. Incoll, R. Lazaro, J.M. Nicolau, G. Sanchez, A. Sole and S. Vidal (1996) 'The Rambla Honda field site: interactions of soils and vegetation along a catena in semi-arid south-east Spain', in C.J. Brandt and J.B. Thornes (eds) *Mediterranean Desertification and Land Use*, 137–168, John Wiley and Sons, Chichester.

Puigdefábregas, J., M. Cueto, F. Domingo, L. Gutierrez, G. Sanchez and A. Sole (1998) 'Rambla Honda, Tabernas, Almería, Spain', in P. Mairota, J.B. Thornes and N. Geeson (eds) *Atlas of Mediterranean Environments in Europe: The Desertification Context*, 110–112, John Wiley and Sons, Chichester.

Puigdefábregas, J. and T. Mendizabal (1998) 'Perspectives on desertification: western Mediterranean', *Journal of Arid Environments* 39, 209–224.

Pungetti, G. (1995) 'Anthropological approach to agricultural landscape history in Sardinia', *Landscape and Urban Planning* 31(1–3), 47–56.

Purcell, N. (1990) 'Mobility and the polis', in O. Murray and S. Price (eds) *The Greek City from Homer to Alexander*, 29–58, Clarendon Press, Oxford.

Pyatt, F.B., G. Gilmore, J.P. Grattan, C.O. Hunt and S. McLaren (2000) 'An imperial legacy? An exploration of the environmental impact of ancient metal mining and smelting in southern Jordan', *Journal of Archaeological Science* 27, 771–778.

Pye, K. (1992) 'Aeolian dust transport and deposition over Crete and adjacent parts of the Mediterranean Sea', *Earth Surface Processes and Landforms* 17, 271–288.

Pye, K. and H. Tsaor (1990) *Aeolian Sand and Sand Dunes*. Unwin Hyman, London.

Pyle, D.M. (1990) 'New estimates for the volume of the Minoan eruption', in D. Hardy, V.P. Galanopoulos, N.C. Flemming and T.H. Druitt (eds) *Thera and the Aegean World III. Proceedings of the Third International Congress, Santorini, Greece, 3–9 September 1989*, 113–121, The Thera Foundation, London.

Quirantes, J., E. Barahona and A. Iriarte (1991) 'Soil degradation and erosion in southeastern Spain. Contributions of the Zaidin Experimental Station CSIC (Granada, Spain)', in M. Sala, J.L. Rubio and J.M. García-Ruiz (eds) *Soil Erosion Studies in Spain*, 211–217, Geoforma Ediciones, Logroño.

Rambal, S. (1987) 'Evolution de l'occupation des terres et ressources en eau en region méditerranéenne karstique', *Journal of Hydrology* 93, 339–357.

Ramis, D. and P. Bover (2001) 'A review of the evidence for domestication of *Myotragus balearicus* Bate (1909 (Artiodactyla, Caprinae)) in the Balearic Islands', *Journal of Archaeological Science* 28, 265–282.

Rampino, M.R., S. Self and R.W. Fairbridge (1979) 'Can rapid climatic change cause volcanic eruptions?', *Science* 206, 826–829.

Randić, A., A. Abramić, D. Balenović, B. Biondić, R. Cimerman, G. Dorčić, E. Draganović, F. Gašparović, N. Karajić, N. Kozelićki, M. Mastrović, K. Pandžić, M. Rukavina, N. Smodlaka and S. Vidić (1996) 'Implications of expected climatic changes for the Cres–Lošinj Islands', in L. Jeftić, S. Kečkeš and J.C. Pernetta (eds) *Climate Change and the Mediterranean*. Volume 2, 431–548, Arnold, London.

Rauws, G. and G. Govers (1988) 'Hydraulic and soil mechanical aspects of rill generation on agricultural soils', *Journal of Soil Science* 39, 111–124.

Reed, C.A. (1977) *The Origins of Agriculture*. Mouton, The Hague.

Reij, C., I. Scoones and C. Toulmin (1996) *Sustaining the Soil: Soil and Water in Sub-Saharan Africa: Issues and Options*. IFAD & CDCS – Free University Amsterdam, Amsterdam.

Reille, M. (1984) 'Origine de la végétation actuelle de la Corse sud-orientale; analyse pollinique de cinq marais côtiers', *Pollen et Spores* XXVI, 43–60.

Reille, M. and J.L. de Beaulieu (1988) 'History of the Würm and Holocene vegetation in western Velay (Massif Central, France): a comparison of pollen analysis from three corings at Lac du Bouchet', *Review of Palaeobotany and Palynology* 54, 233–248.

Reille, M. and J.L. de Beaulieu (1990) 'Pollen analysis of a long upper Pleistocene continental sequence in a Velay maar (Massif Central, France)', *Palaeogeography, Palaeoclimatology, Palaeoecology* 80, 35–48.

Reille, M. and J.J. Lowe (1993) 'A re-evaluation of the vegetation history of the eastern Pyrenees (France) from the end of the last glacial to the present', *Quaternary Science Reviews* 12, 47–77.

Renard, K.G., G.R. Foster, G.A. Weesies and J.P Porter (1991) 'RUSLE: Revised Universal Soil Loss Equation', *Journal of Soil and Water Conservation* 46(1), 30–33.

Rendell, H. (1982) 'Clay hillslope erosion rates in the Basento Valley, S. Italy', *Geografiska Annaler* 64A, 141–147.

Rendell, H.M. (1986) 'Soil erosion and land degradation in southern Italy', in R. Fantechi and N.S. Margaris (eds) *Desertification in Europe*, 184–193, Reidel, Dordrecht.

Renfrew, A.C. (1972) *Emergence of Civilization: The Cyclades and the Aegean in the Third Millennium*. Methuen, London.

Renfrew, A.C., J.R. Cann and J.E. Dixon (1965) 'Obsidian in the Aegean', *Annual of the British School at Athens* 60, 225–247.

Ribeiro, T. and V. Rodrigues (1998) 'The evolution of sustainable development strategies in Portugal', in T. O'Riordan and H. Voisey (eds) *The Transition to Sustainability: The Politics of Agenda 21 in Europe*, 202–213, Earthscan, London.

Richez, G. (1996) 'Sustaining local cultural identity: social unrest and tourism in Corsica', in G.K. Priestley, J.A. Edwards and H. Coccossis (eds) *Sustainable Tourism? European Experiences*, 176–188, CAB International, Wallingford.

Richter, I. and K. Strobach (1978) 'Benioff zones of the Aegean Arc', in H. Closs, D. Roeder and K. Schmidt (eds) *Alps, Apennines and Hellenides*, 410–414, Inter-Union Commision on Geodynamics Scientific Report No. 38, E. Schweizerbart'sche Verlagsbuchhandlung, Stuttgart.

Richter, K. and D. Eckstein (1990) 'A proxy summer rainfall record for southeast Spain derived from living and historic pine trees', *Dendrochronologia* 8, 67–82.

Ricou, L.-E. (1994) 'Tethys reconstructed: plates, continental fragments and their boundaries since 260 Ma from Central America to South-eastern Asia', *Geodinamica Acta* 7(4), 169–218.

Ritchie, J.C. (1984) 'Analyse pollinique de sédiments holocènes supérieurs des hauts plateaux du Maghreb oriental', *Pollen et Spores* XXVI, 489–496.

Ritchie, J.C., C.H. Eyles and C.V. Haynes (1985) 'Sediment and pollen evidence for an early to mid-Holocene humid period in the eastern Sahara', *Nature* 314, 352–355.

Roberts, B.R. (1992) *Land Care Manual*. University of New South Wales Press, Sidney, Australia.

Roberts, N. (1990) 'Human-induced landscape change in south and southwest Turkey during the later Holocene', in S. Bottema, G. Entjes-Nieborg and W. van Zeist (eds) *Man's Role in the Shaping of the Eastern Mediterranean Landscape*, 53–67, Balkema, Rotterdam.

Roberts, N., P. Boyer and R. Parish (1996) 'Preliminary results of geoarchaeological investigations at Çatalhöyük', in I. Hodder (ed.) *On the Surface: Çatalhöyük 1993–95*, 19–40, McDonald Institute for Archaeological Research, Cambridge.

Robertson, A.H.F., P.D. Clift, P.J. Degnan and G. Jones (1991) 'Palaeogeographic and palaeotectonic evolution of the Eastern Mediterranean Neotethys', *Palaeogeography, Palaeoclimatology, Palaeoecology* 87, 289–343.

Robertson, A.H.F., S. Eaton, E.J. Follows and A.S. Payne (1995) Depositional processes and basin analysis of Messinian evaporites in Cyprus', *Terra Nova* 7, 233–253.

Robertson, A.H.F. and M. Grasso (1995) 'Overview of the Late Tertiary–Recent tectonic and palaeo-environmental development of the Mediterranean region', *Terra Nova* 7, 114–127.

Robinson, A.R. and Golnaraghi, M. (1994) 'The physical and dynamical oceanography of the Mediterranean Sea', in P. Malanotte-Rizzoli and A.R. Robinson (eds) *Ocean Processes in Climate Dynamics: Global and Mediterranean Examples*, 255–306, Kluwer Academic, Dordrecht.

Robinson, O.F. (1992) *Ancient Rome. City Planning and Administration*. Routledge, London.

Rodrigo, F.S., M.J. Esteban-Parra and Y. Castro-Diez (1994) 'An attempt to reconstruct the rainfall régime of Andalusia (southern Spain) from 1601 A.D. to 1650 A.D. using historical documents', *Climatic Change* 27, 397–418.

Rodriguez, R., M.-C. Llasat and D. Wheeler (1999) 'Analysis of the Barcelona precipitation series 1850–1991', *International Journal of Climatology* 19, 787–801.

Rognon, P. (1987) 'Late Quaternary reconstruction for the Maghreb (North Africa)', *Palaeogeography, Palaeoclimatology, Palaeoecology* 58, 11–34.

Rojo, T. (1994) 'Scenarios for the industrialization of the western Mediterranean', *Futures* 26(5), 467–489.

Romano, R. (1982) 'Succession of the volcanic activity in the Etnean area', *Memorie della Società Geologica Italiana* XXIII, 27–48.

Romero Diaz, M.A., F. López Bermúdez, J.B. Thornes, C. Francis and G.C. Fisher (1988) 'Variability of overland erosion rates in a semi-arid Mediterranean environment under matorral cover', *Catena Supplement* 13, 1–11.

Rosa Attolini, M., M. Galli, T. Nanni, L. Ruggiero and F. Zuanni (1988) 'Preliminary observations of the fossil forest of Dunarobba (Italy) as a potential archive of paleoclimatic information', *Dendrochronologia* 6, 33–48.

Rossi, F and F. Siccardi (1989) 'Coping with floods: the research policy of the Italian Group for Prevention of Hydrogeological Disasters', in F. Siccardi and R.L. Bras (eds) *Selected Papers from Workshop on Natural Disasters in European Mediterranean Countries, Perugia, Italy, 1988*, 395–415, US National Science Foundation, Washington, DC.

Rossignol, M. (1962) 'Analyse pollinique de sédiments marins Quaternaires en Israël. II Sédiments Pleistocènes', *Pollen et Spores* IV, 121–148.

Rossignol-Strick, M., N. Planchais, M. Paterne and D. Duzet (1992) 'Vegetation dynamics and climate during the deglaciation in the south Adriatic basin from a marine record', *Quaternary Science Reviews* 11, 415–423.

Rother, W., B.B. Manca, B. Klein, D. Bregant, D. Georgopoulos, V. Beitzel, V. Kovacevic and A. Luchetta (1996) 'Recent changes in Eastern Mediterranean deep waters', *Science* 271, 333–335.

Roubet, C. (1980) *Économie Pastorale Préagricole en Algérie Orientale: Le Néolithique de Tradition Capsienne*. Éditions du CNRS, Paris.

Roure Nolla, J.M., R. Pérez-Obiol, J. Belmonte Soler, E. Yll and J.-L. Cano (1995) *Palynological Study on Desertification in South-Western Europe: Timing, Natural Trends and Human Impact. Spain*. Final Report on Contract EV5V-CT91-0027, EU, Brussels.

Roxo, M.J. (1993) 'Field site investigations: lower Alentejo, Beja and Mértola, Portugal', in J.B. Thornes (ed.) *MEDALUS I Final Report*, 406–432, Final Report on Contract EPOC-CT90-0014-(SMA), EU, Brussels.

Roxo, M.J. (1994) 'Accão Antropica no Processo de Degração de Solos – A Serra de Serpa e Mertola', doctoral thesis, Nova Universidade de Lisboa.

Rubio, J.L., V. Andreu and R. Cerni (1990) 'Degradacion del suelo por erosion hidrica: diseño experimental y resultados preliminares', in J. Albaladejo, M.A. Stocking and E. Díaz (eds) *Soil Degradation and Rehabilitation in Mediterranean Environmental Conditions*, 216–235, CSIC, Madrid.

Rumney, G.R. (1968) *Climatology and the World's Climates*. Macmillan, New York.

Rymer, H., J. Cassidy, C.A. Locke and J.B. Murray (1995) 'Magma movements in Etna volcano associated with the major 1991–1993 lava eruption: evidence from gravity and deformation', *Bulletin of Volcanology* 57, 451–461.

Saad, S.I. and S. Sami (1967) 'Pollen and spores of the Nile Delta', *Pollen et Spores* IX, 467–503.

Saez, M., J. Sunyer, J. Castellsague, C. Murillo and J.M. Anto (1995) 'Relationship between weather temperature and mortality – a time-series analysis approach in Barcelona', *International Journal of Epidemiology* 24(3), 576–582.

Safar, W., F. Serre-Bachet and L. Tessier (1992) 'Les plus vieux pins d'alep vivants connus', *Dendrochronologia* 10, 41–52.

Sahsamanoglou, H.S. and T.J. Makrogiannis (1992) 'Temperature trends over the Mediterranean region, 1950–88', *Theoretical and Applied Climatology* 45, 183–192.

Sala, M. (1982) 'Datos cuantitativos de los procesos geomorfológicos fluviales actuales en la cuenca de la riera de Fuirosos (Mont-Negre, mazico litoral catalán)', *Cuadernos de Investigaciones Geográficos* 7, 53–70.

Sala, M. and A. Calvo (1990) 'Response of four different Mediterranean vegetation types to runoff and erosion', in J.B. Thornes (ed.) *Vegetation and Erosion*, 347–362, John Wiley and Sons, Chichester.

Sala, O.E., F.S. Chapin III, J.J. Armesto and 16 others (2000) 'Global biodiversity scenarios for the year 2100', *Science* 287, 1770–1774.

Sallares, R. (1991) *The Ecology of the Ancient Greek World*. Duckworth, London.

Sánchez, G. and J. Puigdefábregas (1994) 'Interactions of plant growth and sediment movement on slopes in a semi-arid environment', *Geomorphology* 9, 243–260.

Sánchez, J., R. Boluda, C. Morell, J.C. Colomer and A. Artigao (1998) 'Degradation index of desertification threatened soils in the Mediterranean region (first approximation). Application in Castilla-La Mancha (Spain)', in A. Rodriguez, C.C. Jímenez and M.L. Tejedor (eds) *The Soil as a Strategic Resource: Degradation Processes and Conservation Measures*, 441–448, Geoforma Ediciones, Logroño.

Sánchez, J.R., V.J. Mangas, C. Ortiz and J. Bellot (1994) 'Forest fire effect on soil chemical properties and runoff', in M. Sala and J.L. Rubio (eds) *Soil Erosion and Degradation as a Consequence of Forest Fires*, 53–66, Geoforma Ediciones, Logroño.

Sánchez-Arcilla, A., J.A. Jiménez, M.J.F. Stive, C. Ibañez, N. Pratt, J.W. Day Jr and M. Capobianco

(1996) 'Impacts of sea-level rise on the Ebro Delta: a first approach', *Ocean and Coastal Management* 30(2–3), 197–216.

Sánchez Soler, M.J. and A.M. Fernandez del Rincon (1991) 'El parque nacional de las Tablas de Daimiel', in J.A. Gonzalez Martin and A. Vazquez Gonzalez (eds) *Guia de Castilla la Mancha, Espacios Naturales*, 481–498, Servicio de Communicaciones: Castilla la Mancha, Ciudad Real.

Sapelli, G. (1995) *Southern Europe since 1945: Tradition and Modernity in Portugal, Spain, Italy, Greece and Turkey*. Longman, Harlow.

Savat, J. (1981) 'Work done by splash – laboratory experiments', *Earth Surface Processes and Landforms* 6, 275–283.

Savat, J. and J.W.A. Poesen (1981) 'Detachment and transportation of loose sediment by raindrop splash. Part I The calculation of absolute data on detachability and transportability', *Catena* 8, 1–17.

Savostin, L.A., J.-C. Sibuet, L.P. Zonenshain, X. Le Pichon and M.-J. Roulet (1986) 'Kinematic evolution of the Tethys belt from the Atlantic Ocean to the Pamirs since the Triassic', *Tectonophysics* 123, 1–35.

Sawkins, F.J. (1990) *Metal Deposits in Relation to Plate Tectonics*. Springer-Verlag, New York.

Scheele, M. (1996) 'The agri-environmental measures in the context of CAP reform', in M. Whitby (ed.) *The European Environment and CAP Reform*, 39–45, CAB International, Wallingford.

Schell, C., S. Black and K.A. Hudson-Edwards (2001) 'Sediment source and transport characteristics of the Rio Tinto, Huelva, SW Spain', in I.D.F. Foster (ed.) *Tracers in Geomorphology*, John Wiley and Sons, Chichester.

Scherm, H. (2000) 'Simulating uncertainty in climate-pest models with fuzzy numbers', *Environmental Pollution* 108, 373–379.

Schick, A. (1977) 'A tentative sediment budget for an extremely arid watershed in the southern Negev', in D.O. Doehring (ed.) *Geomorphology in Arid Regions*, 139–163, Proceedings of the 8th Binghampton Symposium, Binghampton, NY.

Schmalz, R.F. (1991) 'The Mediterranean salinity crisis: alternative hypothesis', *Carbonates and Evaporites* 6, 121–126.

Schneider, R. and K. Tobolski (1983) 'Palynologische und stratigraphische Untersuchungen im Lago di Ganna (Varese, Italien)', *Botanica Helvetica* 93, 115–122.

Schuldenrein, J. (1986) 'Palaeoenvironment, prehistory and accelerated slope erosion along the central Israeli coastal plain: a geoarcheological case study', *Geoarcheology* 1(1), 61–81.

Schulte, L. (1996a) 'Quaternary morphodynamic and climatic change in the middle and lower Aguas', in P.V. Castro, R.W. Chapman, S. Gili, V. Lull, R. Micó, C. Rihuete, R. Risch and E. Sanahula Yll (eds) *Aguas Project: Palaeoclimatic Reconstruction and the Dynamics of Human Settlement and Land Use in the Area of the Middle Aguas (Almería) of the South-east of the Iberian Peninsula*, 38–42, Final Report on Contract EV5V-CT94-0487, EU, Brussels.

Schulte, L. (1996b) 'Morfogenesis cuaternaria en el curso inferior del rio de Aguas (Cuenca de Vera, Provincia de Almeria)', in A. Grandal d'Anglade and J. Pagés Valcarlos (eds) *IV Reunión de Geomorfología*, 223–234, Sociedad Española de Geomorfología, O Castro, Coruña.

Schultz, H.D. (1983) 'Zur Lage holozäner Küsten in den Mündungsgebieten des Río de Vélez und des Río Algarrobo (Málaga)', *Madrider Mitteilungen* 24, 59–64.

Scoging, H.M. (1982) 'Spatial variations in infiltration, runoff and erosion on hillslopes in semi-arid Spain', in R.B. Bryan and A. Yair (eds) *Badland Geomorphology and Piping*, 89–112, GeoBooks, Norwich.

Scoging, H.M. and J.B. Thornes (1979) 'Infiltration characteristics in a semiarid environment', in *The Hydrology of Areas of Low Precipitation, Proceedings of the Canberra Symposium, December 1979*, 159–167, IAHS Publication no. 128, IAHS Press, Wallingford.

Scora, G.A., R.W. Scora, M.A. Mimikou, S.P. Kanellopoulou and E.A. Baltas (1999) 'Human implication of changes in the hydrological regime due to climate change in Northern Greece', *Global Environmental Change* 9(2), 139–156.

Scotese, C.R. (1991) 'Jurassic and Cretaceous plate tectonic reconstructions', *Palaeogeography, Palaeoclimatology, Palaeoecology* 87, 493–501.

Selby, M.J. (1993) *Hillslope Materials and Processes*. Oxford University Press, Oxford.

Selli, R. (1973) 'An outline of the Italian Messinian', in C.W. Drooger (ed.) *Messinian Events in the Mediterranean*, 150–175, North Holland, Amsterdam.

Sempere Torres, D., C. Salles, J.D. Creutin and G. Delrieu (1992) 'Quantification of soil detachment by raindrop impact: performance of classical formulae of kinetic energy in Mediterranean storms', in *Erosion and Sediment Transport Monitoring Programmes in River Basins (Proceedings of the Oslo Symposium, August 1992)*, 115–124, IAHS Publication No. 210, IAHS Press, Wallingford.

Semple, E.C. (1932) *The Geography of the Mediterranean Region: Its Relation to Ancient History*. Constable and Co. Ltd., London.

Sengör, A.M.C. (1985) 'The story of Tethys: how many wives did Okeanos have?', *Episodes* 8, 3–12.

Šercelj, A. (1966) 'Pelodne analize Pleistocenskih in Holocenskih sedimentov Ljubljanskega Barje', *Slovenska Akademija Znanosti in Umentnosta Classis IV Dissertationes* 9, 431–472.

Serre, F. (1978) 'The dendroclimatological value of the European larch (*Larix decidua* Mill.) in the French Maritime Alps', *Tree-Ring Bulletin* 38, 25–34.

Serre-Bachet, F. (1985) 'Une chronologie pluriséculaire du sud de l'Italie', *Dendrochronologia* 3, 45–66.

Serre-Bachet, F. (1986) 'Une chronologie maîtresse du sapin (*Abies alba* Mill.) du Mont Ventoux (France)', *Dendrochronologia* 4, 87–96.

Serre-Bachet, F. (1991) 'Tree-rings in the Mediterranean area', in B. Frenzel, A. Pons, and B. Gläser (eds) *Evaluation of Climate Proxy Data in Relation to the European Holocene*, 133–147, Gustav Fischer Verlag, Stuttgart.

Serre-Bachet, F. and J. Guiot (1987) 'Summer temperature changes from tree rings in the Mediterranean area during the last 800 years', in W.H. Berger and L.D. Labeyrie (eds) *Abrupt Climatic Change*, 89–97, Reidel, Dordrecht.

Serre-Bachet, F., J. Guiot and L. Tessier (1992) 'Dendroclimatic evidence from southwestern Europe and northwestern Africa', in R.S. Bradley and P. Jones (eds) *Climate since 1500 AD*, 349–365, Routledge, London.

Sestini, G. (1992a) 'Implications of climatic changes for the Po Delta and Venice Lagoon', in L. Jeftić, J.D. Milliman and G. Sestini (eds) *Climate Change and the Mediterranean*. Volume 1, 428–494, Arnold, London.

Sestini, G. (1992b) 'Implications of climatic change for the Nile Delta', in L. Jeftić, J.D. Milliman and G. Sestini (eds) *Climate Change and the Mediterranean. Volume 1*, 535–601, Arnold, London.

Shackleton, N.J. and N.D. Opdyke (1977) 'Oxygen isotope and palaeomagnetic evidence for early Northern Hemisphere glaciation', *Nature* 270, 216–219.

Shackley, M. (1980) *Neanderthal Man*. Duckworth, London.

Shakesby, R.A., C. de O.A. Coelho, A.D. Ferreira, J.P. Terry and R.P.D. Walsh (1994) 'Fire, post-burn land management practice and soil erosion response curves in eucalyptus and pine forests, north-central Portugal', in M. Sala and J.L. Rubio (eds) *Soil Erosion and Degradation as a Consequence of Forest Fires*, 111–132, Geoforma Ediciones, Logroño.

Shams el Din, M.I. (1970) 'On the occurrence of thunderstorm in early autumn along the western Mediterranean coast of UAR', *Meteorological Research Bulletin*, 2, 223–244.

Shannon, J., W.R.R. Richardson and J.B. Thornes (2002) 'Modelling event-based fluxes in ephemeral streams', in L.J. Bull and M.J. Kirkby (eds) *Dryland Rivers*. John Wiley and Sons, Chichester.

Shapira, A. (1983) 'A probabilistic approach for evaluating earthquake risks, with application to the Afro-Eurasian junction', *Tectonophysics* 91, 321–324.

Shaw, E.M. (1983) *Hydrology in Practice*. Van Nostrand Reinhold, Wokingham.

Shelton, B.B. (1994) 'Massalia and colonization in the north-western Mediterranean', in G.R. Tsetskhadze and F. de Angelis (eds) *The Archaeology of Greek Colonization: Essays Dedicated to Sir John Boardman*, 61–87, Oxbow, Oxford.

Shepherd, R. (1980) *Prehistoric Mining and Allied Industries*. Academic Press, London.

Sherratt, A. (1981) 'Plough and pastoralism: aspects of the secondary products revolution', in I. Hodder, G. Isaac and N. Hammond (eds) *Pattern of the Past*, 261–305, Cambridge University Press, Cambridge.

Siccardi, F. and R.L. Bras (1989) *Selected Papers from Workshop on Natural Disasters in European Mediterranean Countries, Perugia, Italy, 1988*. US National Science Foundation, Washington, DC.

Siegel, F.R. (1996) *Natural and Anthropogenic Hazards in Development Planning*. Academic Press, London.

Sifeddine, A., P. Bertrand, E. Lallier-Vergès and A.J. Patience (1996) 'Lacustrine organic fluxes and palaeoclimatic variations during the last 15ka: Lac du Bouchet (Massif Central, France)', *Quaternary Science Reviews* 15, 203–211.

Simkin, T., L. Siebert, L. McClelland, D. Bridge, C. Newhall and J.H. Latter (1981) *Volcanoes of the World: A Regional Directory, Gazetteer and Chronology of Volcanism during the Last 10,000 Years*. Hutchinson Ross Publishing Company, Stroudsberg, PA.

Simmons, A.H. (1991) 'Humans, island colonization and Pleistocene extinctions in the Mediterranean: the view from Akrotiri *Aetokremnos*, Cyprus', *Antiquity* 65, 857–869.

Simpson, J. (1995) *Spanish Agriculture: The Long Siesta, 1765–1965*. Cambridge University Press, Cambridge.

Sivall, T. (1957) 'Sirocco in the Levant', *Geografiska Annaler* 39, 114–142.

Slavkoff, E. (1999) 'European regional policy and practice for combating desertification – results and prospects', in P. Balabanis, D. Peter, A. Ghazi and M. Tsoges (eds) *Mediterranean Desertification. Research Results and Policy Implications*, 255–264, European Commission, EUR 19303, Brussels.

Slimani, M and T. Lebel (1986) 'Comparison of three methods to estimate rainfall frequency parameters according to duration of accumulation', in V.P. Singh (ed.) *Hydrologic Frequency Modelling*, 277–291, D. Reidel, Dordrecht.

Smart, P.L. and P.D. Frances (1991) *Quaternary Dating Methods: A User's Guide*. Quaternary Research Association Technical Guide No. 4, Cambridge.

Smit, A. and T.A. Wijmstra (1970) 'Application of transmission electron microscope analysis in the reconstruction of former vegetation', *Acta Botanica Neerlandica* 19, 867–876.

Smith, A.G. and N.H. Woodcock (1982) 'Tectonic syntheses of the Alpine-Mediterranean region: a review', in H. Berckhemer and K.J. Hsü (eds) *Alpine-Mediterranean Geodynamics*, 15–38, American Geophysical Union, Washington, DC.

Smithers, J. and B. Smit (1997) 'Human adaptation to climatic variability and change', *Global Environmental Change* 7, 129–146.

Solé, A., R. Josa, G. Pardini, R. Aringhieri, F. Plana and F. Gallart (1992) 'How mudrock and soil physical properties influence badland formation at Vallcebre (Pre-Pyrenees, NE Spain)', *Catena* 19, 287–300.

Soler, M. and M. Sala (1992) 'Effects of fire and of clearing in a Mediterranean *Quercus ilex* woodland: an experimental approach', *Catena* 19, 321–332.

Soler, M., M. Sala and F. Gallart (1994) 'Post fire evolution of runoff and erosion during an eighteen month period', in M. Sala and J.L. Rubio (eds) *Soil Erosion and Degradation as a Consequence of Forest Fires*, 149–161, Geoforma Ediciones, Logroño.

Sonntag, C., U. Thorweihe, J. Rudolph, E.P. Löhnert, C. Junghans, K.O. Münnich, E. Klitzsch, E.M. El Shazly and F.M. Swailem (1980) 'Isotopic identification of Saharan groundwaters, groundwater formation in the past', in E.M. van Zinderen Bakker and J.M. Coetzee (eds) *Palaeoecology of Africa*, 159–171, A.A. Balkema, Rotterdam.

Sordinas, A. (1969) 'Radiocarbon dates from Corfu, Greece', *Antiquity* XLI, 64.

Sorriso Valvo, M. (1991) 'Mass movement and tectonics', in M.E. Almeida-Teixeira, R. Fantechi, R. Oliveira and A. Gomes Coelho (eds) *Natural Hazards and Engineering Geology. Prevention and Control of Landslides and Other Mass Movements*, 127–138, Commission of the European Communities, EUR 12918, Brussels.

Spanish Government (1998) *El Libro Blanco de Agua en España*. Spanish Government, Madrid.

Sparks, R.S.J. (1985) 'Archaeomagnetism, Santorini volcanic eruptions and fired destruction levels on Crete', *Nature* 313, 74–75.

Sparks, R.S.J., S. Brazier, T.C. Huang and D. Muerdter (1983) 'Sedimentology of the Minoan deep-sea tephra layer in the Aegean and Eastern Mediterranean', *Marine Geology* 54, 131–167.

Specht, R.L. (1969) 'A comparison of sclerophyllous vegetation characteristics of Mediterranean type climates in France, California and southern Australia', *Australian Journal of Botany* 17, 293–308.

Spivey, N. and S. Stoddart (1990) *Etruscan Italy: An Archaeological History*. Batsford, London.

Sprengel, U. (1975) 'La pastorizia transumante nell'Italia centro-meridionale', *Annali del Mezzogiorno* 15, 271–327.

Stampfli, G., J. Marcoux and A. Baud (1991) 'Tethyan margins in space and time', *Palaeogeography, Palaeoclimatology, Palaeoecology* 87, 373–409.

Stančic, Z., T. Podobnikar, K. Oštir and V. Gaffney (1996) 'From scientific research projects to local decision makers' education', in *The European Co-operation Network for Education and Research in Land Information Systems (EUROLIS). Fifth Seminar: European Land Information Systems ELIS'96, Warsaw, 19–22 June, 1996*, 93–102.

Stanley, D.J. (1977) 'Post-Miocene depositional patterns and structural displacement in the Mediterranean', in A.E.M. Nairn, W.H. Kanes and P.G. Stehli (eds) *The Ocean Basins and Their Margins: The Eastern Mediterranean*, Vol. 4A, 77–150, Plenum Press, New York.

Stanley, D.J. and Sheng, H. (1986) 'Volcanic shards from Santorini (Upper Minoan ash) in the Nile Delta, Egypt', *Nature* 320, 733–735.

Stebbing, E.P. (1938) 'The advance of the Sahara', *Geographical Journal* 85, 506–519.

Steen, E. (1998) 'Tunisia, A Mediterranean country with dry-area problems', *Ambio* 27(3), 238–243.

Stevenson, A.C. and R.J. Harrison (1992) 'Ancient forests in Spain: a model for land-use and dry forest management in south-west Spain from 4000 BC to 1900 AD', *Proceedings of the Prehistoric Society* 58, 227–247.

Stoiber, R.E., S.N. Williams and B. Huebert (1987) 'Annual contribution of sulphur dioxide to the atmosphere by volcanoes', *Journal of Volcanology and Geothermal Research* 33, 1–8.

Striem, H.L. (1967) 'A comparative study of rainfall frequency-distribution spectra for Jerusalem and Rome', *Israel Journal of Earth Sciences* 16, 22–29.

Stringer, C. and C. Gamble (1993) *In Search of the Neanderthals: Solving the Puzzle of Human Origins*. Thames and Hudson, London.

Stuiver, M. and G.W. Pearson (1993) 'High-precision bidecadal calibration of the radiocarbon time scale, AD 1950–500 BC and 2500–6000 BC', *Radiocarbon* 35, 1–23.

Stuiver, M. and P.J. Reimer (1993) 'Extended C^{14} data-base and revised Calib 3.0 C^{14} age calibration program', *Radiocarbon* 35, 215–230.

Suc, J.-P. (1978) 'Analyse pollinique de dépôts plio-pléistocènes du sud du massif basaltique de l'Escandorgue (site de Bernasso, Lunas, Hérault, France)', *Pollen et Spores* XX, 497–512.

Suc, J.-P. (1982) 'Palynostratigraphie et paléoclimatologie du Pliocène et du Pléistocène inférieur en Méditerranée nord-occidentale', *Comptes Rendus de l'Académie des Sciences de Paris*, Série II 294, 1003–1008.

Suc, J.-P. (1984) 'Origin and evolution of the Mediterranean vegetation and climate Europe', *Nature* 307, 429–432.

Suc, J.-P. (1989) 'Latitudinal and altitudinal organization of Upper Cainozoic vegetation in western Mediterranean region', *Bulletin de la Société Géologique de France* 5, 541–550.

Suc, J.-P. and Bessais, E. (1990) 'Pérennité d'un climat thermo-xériquie en Sicile avant, pendant, après la crise de salinité messinienne', *Comptes Rendus de l'Académie des Sciences de Paris*, Série II 310, 1701–1707.

Suc, J.-P. and W.H. Zagwijn (1983) 'Plio-Pleistocene correlations between the northwestern Mediterranean region and northwestern Europe according to recent biostratigraphic and palaeoclimatic data', *Boreas* 12, 153–166.

Sullivan, D.G. (1988) 'The discovery of Santorini Minoan tephra in western Turkey', *Nature* 333, 552–554.

Sweeting, M.M. (1972) *Karst Landforms*. Macmillan, London.

Szaboles, I. (1990) 'Effects of predicted climate change on European soils, with particular regard to salinization', in M. Boer and R.S. de Groot (eds) *Landscape Ecological Impact of Climatic Change*, 177–193, IOS, Amsterdam.

Tachau, F. (1984) *Turkey, The Politics of Authority, Democracy and Development*. Praeger Publishers, New York.

Tamari, A. (1976) 'Climatic fluctuations in the eastern basin of the Mediterranean based on dendrochronological analysis', unpublished PhD thesis, University of Tel Aviv.

Tavernier, R. (1985) *Soil Map of the European Communities 1:1,000,000*. CEC Publication EUR 8982, Luxembourg.

Tavernier, R. (ed.) (1995) *Soil Map of the European Communities 1:1,000,000*. Commission of the European Communities, Luxembourg.

Taylor, R.E. (1987) *Radiocarbon Dating: An Archaeological Perspective*. Academic Press, London.

Tchernov, E. (1987) 'The age of the 'Ubeidiya formation, an Early Pleistocene hominid site in the Jordan valley, Israel', *Israel Journal of Earth Sciences* 36, 3–30.

Tchernov, E. (1991) 'Biological evidence for human sedentism in southwest Asia during the Natufian', in O. Bar-Yosef and F. Valla (eds) *The Natufian Culture in the Levant*, 315–340, International Monographs in Prehistory, Ann Arbor, MI.

Telelis, I. and E. Chrysos (1992) 'The Byzantine sources as documentary evidence for the reconstruction of historical climate', in B. Frenzel, C. Pfister and B. Gläser (eds) *European Climate Reconstructed from Documentary Data: Methods and Results*, 17–31, Gustav Fischer Verlag, Stuttgart.

Terlouw, K. (1996) 'A general perspective on the regional development of Europe from 1300 to 1850', *Journal of Historical Geography* 22(2), 129–146.

Ternan, J.L., A.G. Williams and M. Gonzalez del Tanago (1994) 'Soil properties and gully erosion in the Guadalajara Province, Central Spain', in R.J. Rickson (ed.) *Conserving Soil Resources: European Perspectives*, 56–69, CAB International, Wallingford.

Terral, J.-F. and G. Arnold-Simard (1996) 'Beginnings of olive cultivation in eastern Spain in relation to Holocene bioclimatic changes', *Quaternary Research* 46, 176–185.

Terry, J.P. (1994) 'Soil loss from erosion plots of differing post-fire forest cover, Portugal', in M. Sala and J.L. Rubio (eds) *Soil Erosion and Degradation as a Consequence of Forest Fires*, 133–148, Geoforma Ediciones, Logroño.

Tessier, L. (1986) 'Chronologie de mélèzes des Alpes et petit age glaciaire', *Dendrochronologia* 6, 97–113.

Thiébault, S. (1988) *L'Homme et le Milieu Végétal: Analyses Anthracologiques de Six Gisements des Préalpes au Tardi et au Postglaciaire*. Documents d'Archéologie Française 15, Paris.

Thiel, C.C. (1976) 'Earthquake prediction: opportunity to avert disaster', *USGS Circular* 729.

Thirgood, J.V. (1981) *Man and the Mediterranean Forest: A History of Resource Depletion*. Academic Press, London.

Thomas, R.G. (1993) 'Rome rainfall and sunspot numbers', *Journal of Atmospheric and Terrestrial Physics* 55, 155–164.

Thomas, D.S.G. (1997) 'Arid environments: their nature and extent', in D.S.G. Thomas (ed.) *Arid Zone Geomorphology. Process, Form and Change in Drylands*, 1–12, John Wiley and Sons, Chichester.

Thomas, D.S.G. and N.J. Middleton (1994) *Desertification: Exploding the Myth*. John Wiley and Sons, Chichester.

Thompson, M. and S. Rayner (1998) 'Cultural discourses', in S. Rayner and E.L. Malone (eds) *Human Choice and Climate Change*, 265–343, Battelle Press, Columbus, OH.

Thornes, J.B. (1974) 'The rain in Spain', *Geographical Magazine* XLVI, 337–343.

Thornes, J.B. (1976) *Semi-Arid Erosional Systems*, Geographical Paper No. 7, London School of Economics.

Thornes, J.B. (1977) 'Channel changes in ephemeral streams: observations, problems and models', in K.J. Gregory (ed.) *River Channel Changes*, 317–355, John Wiley and Sons, Chichester.

Thornes, J.B. (1985) 'The ecology of erosion', *Geography* 70, 222–236.

Thornes, J.B. (1988) 'Erosional equilibria under grazing', in J. Bintliff, D. Davidson and E. Grant (eds) *Conceptual Issues in Environmental Archaeology*, 193–210, Edinburgh University Press, Edinburgh.

Thornes, J.B. (1990) 'The interaction of erosional and vegetational dynamics in land degradation: spatial outcomes', in J.B. Thornes (ed.) *Vegetation and Erosion*, 41–53, John Wiley and Sons, Chichester.

Thornes, J.B. (1993a) 'Catchment and channel hydrology', in A.D. Abrahams and A.J. Parsons (eds) *Geomorphology of Desert Environments*, 257–288, Chapman and Hall, London.

Thornes, J.B. (1993b) 'Channel processes and forms', in A.D. Abrahams and A.J. Parsons (eds) *Geomorphology of Desert Environments*, 288–318, Chapman and Hall, London.

Thornes, J.B. (1998) 'Mediterranean desertification and Di Castri's Fifth Dimension', *Mediterraneo* 12/13, 149–167.

Thornes, J.B. (1999) 'Modelling ephemeral channel behaviour in the context of desertification', in G. Enne, Ch. Zanolla and D. Peter (eds) *Desertification in Europe: Mitigation, Strategies, Land-Use Planning*, 96–107, Directorate General for Research Environment and Climate Programme, EUR19390, Brussels.

Thornes, J.B. (2000) 'Mediterranean desertification: the issues', in P. Balabanis, D. Peter, A. Ghazi and M. Tsogas (eds) *Mediterranean Desertification: Research Results and Policy Implications*, 9–17, European Commission, EUR 19303, Brussels.

Thornes, J.B. (in press) 'Stability and instability in the management of Mediterranean desertification', in J. Wainwright and M. Mulligan (eds) *Environmental Modelling: Finding Simplicity in Complexity*, John Wiley and Sons, Chichester.

Thornes, J.B. and I. Alcántara-Ayala (1998) 'Modelling mass failure in a Mediterranean mountain environment: climatic, geological, topographical and erosional controls', *Geomorphology* 24, 87–100.

Thornes, J.B. and J.C. Brandt (1993) 'Erosion–vegetation competition in an environment undergoing climatic change with stochastic rainfall variations', in A.C. Millington and K.T. Pye (eds) *Environmental Change in the Drylands. Biogeographical and Geomorphological Responses*, 306–320, John Wiley and Sons, Chichester.

Thornes, J.B. and A. Gilman (1983) 'Potential and actual erosion around archaeological sites in southeast Spain', in J. de Ploey (ed.) *Rainfall Simulation, Runoff and Soil Erosion*, 91–113, Catena Supplement No. 4, Catena, Cremlingen.

Thornes, J.B., J. Shao, S. Diamond, M. McMahon and J.C. Hawkes (1996) 'Testing the MEDALUS model', *Catena* 26, 106–156.

Thunell, R.C. (1979) 'Climatic evolution of the Mediterranean Sea during the last 5.0 million years', *Sedimentary Geology* 23, 67–79.

Thunell, R.C. and D.F. Williams (1983) 'Paleotemperature and paleosalinity history of the eastern Mediterranean during the late Quaternary', *Palaeogeography, Palaeoclimatology, Palaeoecology* 44, 23–39.

Thunell, R.C. and D.F. Williams (1989) 'Glacial-Holocene salinity changes in the Mediterranean Sea: hydrographic and depositional effects', *Nature* 338, 493–496.

Thunell, R.C., D.F. Williams, E. Tappa, D. Rio and I. Raffi (1990) 'Pliocene-Pleistocene stable isotope record for Ocean Drilling Program Site 653, Tyrrhenian Basin: implications for the palaeoenvironmental history of the Mediterranean Sea', in K.A. Kastens, J. Mascle *et al.* (eds) *Proceedings of the Ocean Drilling Program*. Volume 107: *Scientific Results. Tyrrhenian Sea*, 387–399, Ocean Drilling Program, College Station, TX.

Till, C. and J. Guiot (1990) 'Reconstruction of precipitation in Morocco since 1100 A.D. based on *Cedrus atlantica* tree-ring widths', *Quaternary Research* 33, 337–351.

Tilling, R.I. (1989) 'Volcanic hazards and their mitigation: progress and problems', *Reviews in Geophysics* 27, 237–269.

Tiné, S. (1983) *Passo di Corvo e la Civiltà Neolitica del Tavoliere*. Sagep, Genoa.

Tivy, J. (1993) *Biogeography: A Study of Plants on the Ecosphere*. 3rd edn. Longman Scientific and Technical, Harlow.

Tlemcani, R. (1999) *Etat, Bazar et Globalisation: L'Aventure de l'Intifah en Algerie*. Les Editions el Hikma, Algiers.

Todd, I. (1987) *Vasilikos Valley Project 6: Excavations at Kalavassos-Tenta I*, Studies in Mediterranean Archaeology LXXI, 6, Paul Åström, Goteborg.

Tomadin, L., R. Lenaz, V. Landuzzi, A. Mazzucotelli and R. Vannucci (1984) 'Wind-blown dusts over the central Mediterranean', *Oceanologica Acta* 7, 13–23.

Tomadin, L., R. Lenaz and R. Sartori (1990) 'Clay minerals as natural tracers in sediments, water column, and lower atmosphere of the Tyrrhenian basin (Mediterranean Sea)', *Mineralogica Petrologica Acta* XXXIII, 81–91.

Torri, D. and J.W.A. Poesen (1992) 'The effect of soil surface slope on raindrop detachment', *Catena* 19, 561–578.

Torri, D., M. Sfalanga and M. del Sette (1981) 'Splash detachment: runoff depth and soil cohesion', *Catena* 14, 149–155.

Torri, D., M. Sfalanga and G. Chisci (1987) 'Threshold conditions for incipient rilling', *Catena* 14, 97–105.

Toulmin, C. (1999) 'International experience in desertification policy: the global picture', in P. Balabanis, D. Peter, A. Ghazi and M. Tsogas (eds) *Mediterranean Desertification, Research Results and Policy Implications*, 331–340, European Commission, EUR 19303, EU, Brussels.

Tourneux, F.-P. (1994) 'Analyse de l'environnement des sites', in S.E. van der Leeuw (ed.) *Understanding the Natural and Anthropogenic Causes of Soil Degradation and Desertification in the Mediterranean Basin. Volume 3: Dégradation et Impact Humain dans la Moyenne et Basse Vallée du Rhône dans l'Antiquité* (Part II), 143–168, Final Report on Contract EV5V-CT91-0021, EU, Brussels.

Tout, D. (1990) 'The horticulture industry of Almería Province, Spain', *The Geographical Journal* 156, 304–312.

Tout, D.G. and M.V. Kemp (1985) 'The named winds of Spain', *Weather* 40, 322–329.

Trabaud, L. (1994) 'Postfire plant community dynamics in the Mediterranean Basin', in J.M. Moreno and W.C. Oechel (eds) *The Role of Fire in Mediterranean-Type Ecosystems*, 1–15, Springer-Verlag, Berlin.

Tracy, M. (1989) *Government and Agriculture in Western Europe 1880–1988*. 3rd edn. Harvester Wheatsheaf, London.

Trenberth, K.E. (ed.) (1992) *Climate System Modeling*. Cambridge University Press, Cambridge.

Triat, H. (1975) 'Analyse pollinique de la tourbière de Fos-sur-Mer (Bouches-du-Rhône)', *Ecologia Mediterranea* 1, 109–121.

Triat-Laval, H. (1982) 'Pollenanalyse de sediments quaternaires récents du pourtour de l'Étang de Berre', *Ecologia Mediterranea* 8, 97–115.

Trigger, B. (1972) 'Determinants of urban growth in pre-industrial societies', in P.J. Ucko, R. Tringham and G.W. Dimbleby (eds) *Man, Settlement and Urbanism*, 601–637, Duckworth, London.

Trimble, S.W. and P. Crosson (2000) 'U.S. soil erosion rates – myth and reality', *Science* 289, 248–250.

Trinkaus, E. (ed.) (1989) *The Emergence of Modern Humans: Biocultural Adaptations in the Later Pleistocene*. Cambridge University Press, Cambridge.

Tropeano, D. (1983) 'Soil erosion on vineyards in the Tertiary Piedmontese basin (Northwestern Italy). Studies on experimental areas', in J. De Ploey (ed.) *Rainfall Simulation, Runoff and Soil Erosion*, 115–127, Catena Supplement 4, Braunschweig.

Tropeano, D. (1984) 'Rate of soil-erosion processes on vineyards in central Piedmont (NW Italy)', *Earth Surface Processes and Landforms* 9, 253–266.

Tropeano, D. and P. Olive (1989) 'Vitesse de la sédimentation holocène dans la plaine occidentale du Pô (Italie)', *Bulletin de l'Association Française pour l'Étude du Quaternaire* 26, 65–71.

Truman, C.C. and J.M. Bradford (1990) 'Effect of antecedent soil moisture on splash detachment under simulated rainfall', *Soil Science* 150, 787–798.

Trump, D. (1976) *Skorba*. Reports of the Research Committee of the Society of Antiquaries of London XXII, Oxford.

Trump, D. (1984) 'The Bonu Iginu project: results and prospects', in W.H. Waldren, R. Chapman, J. Lewthwaite and R.-C. Kennard (eds) *The Deya Conference of Prehistory. Early Settlement in the Western Mediterranean Islands and their Peripheral Areas*, 511–532, BAR, Oxford.

Trümpy, R. (1982) 'Alpine palaeogeography: a reappraisal', in K.J. Hsü (ed.) *Mountain Building Processes*, 149–156, Academic Press, London.

Tsoar, H. and A. Karnieli (1996) 'What determines the spectral reflectance of the Negev-Sinai sand dunes', *International Journal of Remote Sensing* 17, 513–525 and 3319–3321.

Turkelbloom, F., J. Poesen, I. Ohler, K. van Keer, S. Ongprasert and K. Vlassak (1997) 'Assessment of tillage erosion rates on steep slopes in northern Thailand', *Catena* 29, 29–44.

Turner, J. (1978) 'The vegetation of Greece during prehistoric times – the palynological evidence', in C. Doumas (ed.) *Thera and the Aegean World II. Papers and Proceedings of the Second International Scientific Congress, Santorini, Greece, August 1978*, 765–773, Thera and the Aegean World, London.

Turner, J. and J.R.A. Greig (1975) 'Some Holocene pollen diagrams from Greece', *Review of Palaeobotany and Palynology* 20, 171–204.

Turon, J.-L. (1984) 'Direct land-sea correlations in the last interglacial complex', *Nature* 309, 673–676.

Tusa, S. (1985) 'The beginnings of early farming communities in Sicily: the evidence of the Uzzo cave', in C. Malone and S. Stoddart (eds) *Papers in Italian Archaeology IV. The Cambridge Conference. Part ii: Prehistory*, 61–82, BAR, Oxford.

Tykot, R.H. (1998) 'Mediterranean islands and multiple flows. The sources and exploitation of Sardinian obsidian', in M.S. Shackley (ed.) *Archaeological Obsidian Studies. Method and Theory*, 67–82, Plenum Press, New York.

Tzedakis, P.C. (1993) 'Long-term tree populations in northwest Greece through multiple Quaternary climatic cycles', *Nature* 364, 437–440.

Tzedakis, P.C. (1994) 'Vegetation change through glacial–interglacial cycles: a long-term pollen-sequence perspective', *Philosophical Transactions of the Royal Society of London* 345B, 403–432.

Udias, A. (1982) 'Seismicity and seismotectonic stress field in the Alpine-Mediterranean region', in H. Berckhemer and K.J. Hsü (eds) *Alpine-Mediterranean Geodynamics*, 75–82, American Geophysical Union, Washington, DC.

UNEP (1990) 'Global assessment of land degradation/desertification – GAP II', *Desertification Control Bulletin* 18, 24–25.

UNEP (1992) *Earth Summit 1992: The UN Conference on Environment and Development*, Rio de Janeiro.

UNEP (1994) *United Nations Convention to Combat Desertification in Countries Experiencing Serious Drought and/or Desertification*, signed June 1994.

Vafeidis, A. (2001) 'Modelling Impacts of Fires on Soil Erosion', unpublished PhD thesis, King's College London.

Vafeidis, A., N.A. Drake and J. Wainwright (1997) 'The contribution of fire to erosion in Greece: a remote sensing and GIS analysis', *Proceedings of RSS '97 Observations and Interactions, 2–4 September 1997, Reading*, 178, Remote Sensing Society, Nottingham.

Vai, G.B. (1991) 'Palaeozoic strike-slip rift pulses and palaeogeography in the circum-Mediterranean Tethyan realm', *Palaeogeography, Palaeoclimatology, Palaeoecology* 87, 223–252.

Valderrabano, J. and L. Torrano (2000) 'The potential for using goats to control *Genista scorpius* shrubs in European black pine stands', *Forest Ecology and Management* 126, 377–383.

Van Andel, T.H. (1998) 'Middle and Upper Palaeolithic environments and the calibration of ^{14}C dates beyond 10,000 BP', *Antiquity* 72, 26–33.

Van Andel, T.H. and C. Runnels (1987) *Beyond the Acropolis: A Rural Greek Past*. Stanford University Press, Stanford, CA.

Van Andel, T.H. and C.N. Runnels (1988) 'An essay on the 'Emergence of Civilization' in the Aegean world', *Antiquity* 62, 234–247.

Van Andel, T. and C.N. Runnels (1995) 'The earliest farmers in Europe', *Antiquity* 69, 481–500.

Van Andel, T., C.N. Runnels and K.O. Pope (1986) 'Five thousand years of land use and abuse in the southern Argolid, Greece', *Hesperia* 55, 103–128.

Van Andel, T.H. and E. Zangger (1990) 'Landscape stability and stabilisation in the prehistory of Greece', in S. Bottema, G. Entjes-Nieborg and W. van Zeist (eds) *Man's Role in the Shaping of the Eastern Mediterranean Landscape*, 139–157, Balkema, Rotterdam.

Van Andel, T.H., E. Zangger and A. Demitrack (1990) 'Land use and soil erosion in prehistoric and historical Greece', *Journal of Field Archaeology* 17, 379–396.

Van Asch, Th.W.J. (1980) 'Water erosion on slopes and landsliding in a Mediterranean landscape', *Utrechtse Geografische Studies* 20, Utrecht.

Van Asch, Th.W.J. (1986) 'Hazard mapping as a tool for landslide prevention in Mediterranean areas', in R. Fantechi and N.S. Margaris (eds) *Desertification in Europe*, 126–135, Reidel, Dordrecht.

Van den Brink, L.M. and C.R. Janssen (1985) 'The effect of human activities during cultural phases on the development of montane vegetation in the Sierra de Estrela, Portugal', *Review of Palaeobotany and Palynology* 44, 193–215.

Van der Leeuw, S.E. (ed.) (1998) *The Archaeomedes Project: Understanding the Natural and Anthropogenic Causes of Land Degradation and Desertification in the Mediterranean Basin*. Office for Official Publications of the European Communities, EUR 18181, Luxembourg.

Van der Leeuw, S.E., M. Lemon and R.A.F. Seaton (1995) 'Agricultural policy and desertification', in R. Fantechi, D. Peter, P. Balabanis and J.L. Rubio (eds) *Desertification in a European Context: Physical and Socio-Economic Aspects*, 197–212, European Commission, EUR15415EN, Luxembourg.

Van der Wiel, A.M. and T.A. Wijmstra (1987a) 'Palynology of the lower part (78–120m) of the core Tenaghi Philippon II, Middle Pleistocene of Macedonia, Greece', *Review of Palaeobotany and Palynology* 52, 73–88.

Van der Wiel, A.M. and T.A. Wijmstra (1987b) 'Palynology of the 112.8–197.8 m interval of the core Tenaghi Philippon III, Middle Pleistocene of Macedonia', *Review of Palaeobotany and Palynology* 52, 89–117.

Van der Zwaan, G.J. and L. Gudjonsson (1986) 'Middle Miocene–Pliocene stable isotope stratigraphy and palaeoceanography of the Mediterranean', *Marine Micropalaeontology* 10, 71–90.

Van Wersch, H.J. (1972) 'The agricultural economy', in W.A. McDonald and G.R. Rapp Jr, *The Minnesota Messinia Expedition. Reconstructing a Bronze Age Regional Environment*, 117–147, University of Minnesota Press, Minneapolis.

Van Zeist, W., H. Woldring and D. Stapert (1975) 'Late Quaternary vegetation and climate of southwestern Turkey', *Palaeohistoria* 17, 54–143.

Van Zeist, W. and S. Bottema (1971) 'Plant husbandry in early Neolithic Nea Nikomedia, Greece', *Acta Bottanica Neerlandica* 20, 524–528.

Van Zuidam, R.A. (1975) 'Geomorphology and archaeology: evidences of interrelation at historical sites in the Zaragoza region, Spain', *Zeitschrift für Geomorphologie* 19, 319–328.

Van Zuidam, R.A. (1976) *Geomorphological Development of the Zaragoza Region, Spain. Processes and Landforms Related to Climatic Changes in a Large Mediterranean River Basin.* ITC, Enschede.

Vaquer, J. and M. Barbaza (1987) 'Cueillette ou horticulture mésolithique: la Balma de l'Abeurador', in J. Guilaine, J. Courtin, J.-L. Roudil and J.-L. Vernet (eds) *Premières Communautés Paysannes en Méditerranée Occidentale*, 231–242, Éditions du CNRS, Paris.

Várallyay, G. (1994) 'Climate change, soil salinity and alkalinity', in M.D.A. Rounsevell and P.J. Loveland (eds) *Soil Responses to Climate Change*, 39–111, NATO ASI Series, Vol. 123, Springer-Verlag, Berlin.

Vassale, R. (1994) 'The use of explosive for the diversion of the 1992 Mt. Etna lava flow', *Acta Vulcanologica* 4, 173–177.

Vega, J.A. and F. Díaz-Fierros (1987) 'Wildfire effects on soil erosion', *Ecologia Mediterranea* XIII, 119–125.

Vergnaud-Grazzini, C., J.F. Saliège, M.J. Urrutiager and A. Iannace (1990) 'Oxygen and carbon isotope stratigraphy of ODP Hole 653A and site 654: the Pliocene-Pleistocene glacial history recorded in the Tyrrhenian Basin (West Mediterranean)', in K.A. Kastens, J. Mascle *et al.* (eds) *Proceedings of the Ocean Drilling Program. Volume 107: Scientific Results. Tyrrhenian Sea*, 361–386, Ocean Drilling Program, College Station, TX.

Verhagen, P. (1994) 'The implementation of GIS in the ARCHAEOMEDES project (Lower Rhône valley)', in *Table-ronde du Service de l'Archéologie: La Représentation Cartographique – Lyon, 12/4/1994.*

Verhagen, Ph., F. Favory and S.E. van der Leeuw (1994) 'Traitement des données dans le système d'information géographique', in S.E. van der Leeuw (ed.) *Understanding the Natural and Anthropogenic Causes of Soil Degradation and Desertification in the Mediterranean Basin. Volume 3: Dégradation et Impact Humain dans la Moyenne et Basse Vallée du Rhône dans l'Antiquité (Part II)*, 115–141, Final Report on Contract EV5V-CT91-0021, EU, Brussels.

Vermeersch, P.M., E. Paulissen, S. Stokes, C. Charlier, P. Van Peer, C. Stringer and W. Lindsay (1998) 'A Middle Palaeolithic burial of a modern human at Taramsa Hill, Egypt', *Antiquity* 72, 475–484.

Vernet, J-L. (1972) 'Contribution à l'Histoire de la Végétation du Sud-Est de la France au Quaternaire. Étude de Macroflores, de Charbon de Bois Principalement', Thèse de Docteur ès Sciences Naturelles, Université de Montpellier.

Vernet, J.-L. (1980) 'Le végétation du bassin de l'Aude, entre Pyrénées et Massif Central, au Tardiglaciaire et au postglaciaire d'après l'analyse anthracologique', *Review of Palaeobotany and Palynology* 30, 33–55.

Vernet, J.-L. and S. Thiébault (1987) 'An approach to northwestern Mediterranean recent prehistoric vegetation and ecologic implications', *Journal of Biogeography* 14, 117–127.

Veyne, P. (1990) *Bread and Circuses: Historical Sociology and Political Pluralism.* Allen Lane, London.

Vezzoli, L. (1991) 'Tephra layers in Bannock Basin (Eastern Mediterranean)', *Marine Geology* 100, 21–34.

Vigotti, M.A., G. Rossi, L. Bisanti, A. Zanobetti and J. Schwartz (1996) 'Short term effects of urban air pollution on respiratory health in Milan, Italy, 1980–89', *Journal of Epidemiology and Community Health* 50, S71–S75.

Viguier, J.M. (1993) 'Mesure et modélisation de l'érosion pluviale. Application au vignoble de Vidauban (Var, France)', unpublished PhD thesis, Université d'Aix-Marseille II.

Vita Finzi, C. (1969) *The Mediterranean Valleys. Geological Changes in Historical Times*. Cambridge University Press, Cambridge.

Vita Finzi, C. (1976) 'Diachronism in Old World alluvial sequences', *Nature* 262, 218–219.

Vita Finzi, C. (1978) *Archaeological Sites in their Setting*. Thames and Hudson, London.

Vita Finzi, C. (1986) *Recent Earth Movements. An Introduction to Neotectonics*. Academic Press, London.

Vita Finzi, C. and E.S. Higgs (1970) 'Prehistoric economy in the Mount Carmel area of Palestine: site catchment analysis', *Proceedings of the Prehistoric Society* 36, 1–37.

Wagstaff, J.M. (1981) 'Buried assumptions: some problems in the interpretation of the "Younger Fill" raised by recent data from Greece', *Journal of Archaeological Science* 8, 247–264.

Wainwright, J. (1991) 'Erosion of semi-arid archaeological sites: a case study in natural formation processes', unpublished PhD thesis, University of Bristol.

Wainwright, J. (1992) 'Assessing the impact of erosion on semi-arid archaeological sites', in M. Bell and J. Boardman (eds) *Past and Present Soil Erosion*, 228–241, Oxbow Books, Oxford.

Wainwright, J. (1994) 'Anthropogenic factors in the degradation of semi-arid regions: a prehistoric case study in southern France', in A.C. Millington and K. Pye (eds) *Effects of Environmental Change on Drylands*, 285–304, John Wiley and Sons, Chichester.

Wainwright, J. (1996a) 'Infiltration, runoff and erosion characteristics of agricultural land in extreme storm events, SE France', *Catena* 26, 27–47.

Wainwright, J. (1996b) 'A comparison of the infiltration, runoff and erosion characteristics of two contrasting 'badland' areas in S. France', *Zeitschrift für Geomorphologie Supplementband* 106, 183–198.

Wainwright, J. (1996c) 'Hillslope response to extreme storm events: the example of the Vaison-la-Romaine event', in M.G. Anderson and S.M. Brooks (eds) *Advances in Hillslope Processes*, Volume 2: 997–1026, John Wiley and Sons, Chichester.

Wainwright, J. (2000) 'Contextes géomorphologiques et géoarchéologiques des habitats de l'Age du Bronze en Méditerranée Occidentale', in *Proceedings, XXIVe Congrès Préhistorique de France*, 11–26.

Wainwright, J. (2003) 'History and evolution of Mediterranean desertification', *Advances in Environmental Monitoring and Modelling*.

Wainwright, J., M. Mulligan and J.B. Thornes (1999a) 'Plants and water in drylands', in A.J. Baird and R.L. Wilby (eds) *Ecohydrology*, 78–126, Routledge, London.

Wainwright, J., A.J. Parsons and A.D. Abrahams (1995) 'Simulation of raindrop erosion and the development of desert pavements', *Earth Surface Processes and Landforms* 20, 277–291.

Wainwright, J., A.J. Parsons and A.D. Abrahams (1999b) 'Rainfall energy under creosotebush', *Journal of Arid Environments* 43, 111–120.

Wainwright, J., A.J. Parsons and A.D. Abrahams (2000) 'Plot-scale studies of vegetation, overland flow and erosion interactions: case studies from Arizona and New Mexico', *Hydrological Processes* 14, 2921–2943.

Wainwright, J., A.J. Parsons, D.M. Powell and R. Brazier (2001) 'A new conceptual framework for understanding and predicting erosion by water from hillslopes and catchments', in J.C. Ascough II and D.C. Flanagan (eds) *Soil Erosion Research for the 21st Century. Proceedings of the International Symposium*, 607–610, American Society of Agricultural Engineers, St Joseph, MI.

Waldren, W.H. (1986) *The Balearic Pentapartite Division of Prehistory. Radiocarbon and Other Age Determinations*. BAR, Oxford.

Walker, D.S. (1960) *The Mediterranean Lands*. Methuen, London.

Wallmann, P.C.B., G.A.Mahood and D.D. Pollard (1988) 'Mechanical models for correlation of ring-fracture eruptions at Pantelleria, Strait of Sicily, with glacial sea-level drawdown', *Bulletin of Volcanology* 50, 327–339.

Wansard, G. (1996) 'Quantification of paleotemperature changes during isotopic stage 2 in the La Draga continental sequence (NE Spain) based on the Mg/Ca ratio of freshwater ostracods', *Quaternary Science Reviews* 15, 237–245.

Ward, R.C. and M. Robinson (1999) *Principles of Hydrology*. 4th edn. McGraw-Hill, New York.

Warren, P. (1972) *Myrtos: An Early Bronze Age Settlement in Crete*. British School at Athens, London.

Warren, P. (1984) 'Absolute dating of the Bronze Age eruption of Thera (Santorini)', *Nature* 308, 492–493.

Wasson, R.J. and P.M. Naninga (1986) 'Estimating wind transport of sand on vegetated surfaces', *Earth Surface Processes and Landforms* 11, 505–514.

Water Framework Directive (2000) 'The EU Water Framework Directive', http://europa.eu.int/water/water-framework/index_en.html.

Watson, A.M. (1983) *Agricultural Innovation in the Early Islamic World: The Diffusion of Crops and Farming Techniques*. Cambridge University Press, Cambridge.

Watts, W.A. (1973) 'Rates of change and stability in vegetation in the perspective of long periods of time', in H.J.B. Birks and R.G. West (eds) *Quaternary Plant Ecology: The 14th Symposium of the British Ecological Society, University of Cambridge, 28–30 March 1972*. Blackwell Scientific Publications, Oxford.

Watts, W.A., J.R.M. Allen, B. Huntley and S.C. Fritz (1996a) 'Vegetation history and climate of the last 15,000 years at Laghi di Monticchio, southern Italy', *Quaternary Science Reviews* 15, 113–132.

Watts, W.A., J.R.M. Allen and B. Huntley (1996b) 'Vegetation history and palaeoclimate of the last glacial period at Lago Grande di Monticchio, southern Italy', *Quaternary Science Reviews* 15, 133–153.

Weijermars, R. (1988) 'Neogene tectonics in the western Mediterranean may have caused the Messinian Salinity Crisis and an associated glacial event', *Tectonophysics* 148, 211–219.

Weinstein-Evron, M. (1987) 'Paleoclimate reconstruction of the Late Pleistocene in the Hula Basin', *Israel Journal of Earth Sciences* 36, 59–64.

Weisgerber, G. and E. Pernicka (1995) 'Ore mining in prehistoric Europe: an overview', in G. Morteani and J.P. Northover (eds) *Prehistoric Gold in Europe: Mines, Metallurgy and Manufacture*, 159–182, Kluwer Academic Publishers, Amsterdam.

Weiss, B. (1982) 'The decline of Late Bronze Age civilization as a possible response to climatic change', *Climatic Change* 4, 173–198.

Weiss, H., M.-A. Courty, W. Wetterstrom, F. Guichard, L. Senior, R. Meadow and A. Curnow (1993) 'The genesis and collapse of Third Millennium North Mesopotamian civilization', *Science* 261, 995–1004.

Wendorf, F. and R. Schild (1984) 'The emergence of food production in the Egyptian Sahara', in J.D. Clark and S.A. Brandt (eds) *From Hunters to Farmers: The Causes and Caonsequences of Food Production in Africa*, 93–101, UCLA Press, Berkeley, CA.

Wetterstrom, N. (1993) 'Foraging and farming in Egypt. The transition from hunting and gathering to horticulture in the Nile Valley', in T. Shaw, P. Sinclair, B. Andah and A. Opoko (eds) *The Archaeology of Africa: Foods, Metals and Tools* 165–226, Routledge, London.

Wheatley, P. (1972) 'The concept of urbanism', in P.J. Ucko, R. Tringham and G.W. Dimbleby (eds) *Man, Settlement and Urbanism*, 601–637, Duckworth, London.

White, K., N. Drake, A. Millington and S. Stokes (1996) 'Constraining the timing of alluvial fan response to Late Quaternary changes, southern Tunisia', *Geomorphology* 17, 295–304.

White, K.D. (1970) *Roman Farming*. Thames and Hudson, London.

White, K.D. (1984) *Greek and Roman Technology*. Thames and Hudson, London.

White, P.E. (1995) 'Modelling rural population change in the Cilento region of southern Italy', *Environment and Planning A* 17, 1401–1413.

Whitehouse, R. (1987) 'The first farmers in the Adriatic and their position in the Neolithic of the Mediterranean', in J. Guilaine, J. Courtin, J.-L. Roudil and J.-L. Vernet (eds) *Premières Communautés Paysannes en Méditerranée Occidentale*, 357–366, Éditions du CNRS, Paris.

Whittle, A. (1996) *Europe in the Neolithic: The Creation of New Worlds*. Cambridge University Press, Cambridge.

Wigley, T.M.L. (1992) 'Future climate of the Mediterranean Basin with particular emphasis on changes in precipitation', in L. Jeftić, J.D. Milliman and G. Sestini (eds) *Climatic Change and the Mediterranean: Environmental and Societal Impacts of Climatic Change and Sea-Level Rise in the Mediterranean Region*, 15–44, Arnold, London.

Wijmstra, T.A. (1969) 'Palynology of the first 30 metres of a 120 m deep section in northern Greece', *Acta Botanica Neerlandica* 18, 511–527.

Wijmstra, T.A. and A. Smit (1976) 'Palynology of the middle part (30–78 metres) of the 120 m deep section in northern Greece (Macedonia)', *Acta Botanica Neerlandica* 25, 297–312.

Wilkinson, A. (1998) 'Steps towards integrating the environment into other EU policy sectors', in T. O'Riordan and H. Voisey (eds) *The Transition to Sustainability: the politics of Agenda 21 in Europe*, 113–129, Earthscan, London.

Williams, D.F., R.C. Thunell, E. Tappa, D. Rio and I. Raffi (1988) 'Chronology of the oxygen isotope record, 0–1.88 million years before present', *Palaeogeography, Palaeoclimatology and Palaeoecology* 64, 221–240.

Williams, T., N. Thouveny and K.M. Creer (1996) 'Palaeoclimatic significance of the 300 ka mineral magnetic record from the sediments of Lac du Bouchet, France', *Quaternary Science Reviews* 15, 223–235.

Williams, W. and E.M. Papamichael (1995) 'Tourism and tradition: local control versus outside interests in Greece', in M.-F. Lanfant, J.B. Allcock and E.M. Bruner (eds) *International Tourism*, 127–142, Sage, London.

Williams-Thorpe, O. (1995) 'Obsidian in the Mediterranean and the Near East: a provenancing success story', *Archaeometry* 37(2), 217–248.

Willis, K.J. (1992a) 'The Late Quaternary vegetation history of Greece: I Lake Gramousti', *New Phytologist* 121, 101–117.

Willis, K.J. (1992b) 'The Late Quaternary vegetation history of Greece: II Rezina marsh', *New Phytologist* 121, 19–138.

Willis, K.J. (1992c) 'The Late Quaternary vegetation history of Greece: III A comparative study of two contrasting sites', *New Phytologist* 121, 139–156.

Willis, K.J. (1994) 'The vegetational history of the Balkans', *Quaternary Science Reviews* 13, 769–788.

Wilson, L. (1980) 'Energetics of the Minoan eruption: some revisions', in C. Doumas (ed.) *Thera and the Aegean World II. Papers and Proceedings of the Second International Scientific Congress, Santorini, Greece, August 1978*, 31–35, Thera and the Aegean World, London.

Wilvert, C. (1994) 'Spain – Europe California', *Journal of Geography* 93, 74–79.

Wischmeier, W. (1976) 'Use and misuse of the Universal Soil Loss Equation', *Journal of Soil and Water Conservation* 31, 5–9.

Wittlinger, G. and H. Haessler (1978) 'Aftershocks of the Friuli 1976 earthquakes', in H. Closs, D. Roeder and K. Schmidt (eds) *Alps, Apennines and Hellenides*, 178–180, Inter-Union Commission on Geodynamics Scientific Report No. 38, E. Schweizerbart'sche Verlagsbuchhandlung, Stuttgart.

Wright, H.E., J.E. Kutzbach, T. Webb III, W.F. Ruddiman, F.A. Street-Perrott and P.J. Bartlein (eds) (1993) *Global Climates Since the Last Glacial Maximum*. University of Minnesota Press, Minneapolis.

WWF-Spain (2001) 'Spain's National Hydrological Plan: what is it and what is happening to it?', http://www.panda.org/europe/freshwater/regional/spain-nhp.html. (Accessed July 2001).

Wymer, J. (1982) *The Palaeolithic Age*. Croom Helm, London.

Yaalon, D. (1997) 'Soils in the Mediterranean: what makes them different?', *Catena* 28, 157–169.

Yaalon, D.H. and E. Ganor (1973) 'The influence of dust on soils during the Quaternary', *Soil Science* 116, 146–155.

Yair, A. (1994) 'The ambiguous impact of climate change at a desert fringe: northern Negev, Israel', in A. Millington and K. Pye (eds) *Environmental Change in Drylands: Biogeographical and Geomorphological Perspectives*, 199–227, John Wiley and Sons, Chichester.

Yair, A., P. Goldberg and B. Brimer (1982) 'Long term denudation rates in the Zin-Havarim badlands, northern Negev, Israel', in R. Bryan and A. Yair (eds) *Badland Geomorphology and Piping*, 279–291, GeoBooks, Norwich.

Yair, A. and M. Klein (1973) 'The influence of surface properties on flow and erosion processes on debris covered slopes in an arid area', *Catena* 1, 1–18.

Yakar, J. (1991) *Prehistoric Anatolia. The Neolithic Transformation and the Early Chalcolithic Period*. Monograph Series of the Institute of Archaeology, University of Tel Aviv, No. 9, Tel Aviv.

Yassoglou, N., C. Kosmas and M. Moustakas (1997) 'The red soils, their origin, properties, use and management in Greece', *Catena* 28, 261–278.

Yiakoulaki, M.D. and A.S. Nastis (1995) 'Intake by goats grazing kermes oak shrublands with varying cover in northern Greece', *Small Ruminant Research* 17, 223–228.

Yiakoulaki, M.D. and A.S. Nastis (1996) 'Effect of stocking rate on intake and weight gain of goats grazing in kermes oak shrubland', in *The Optimal Exploitation of Marginal Mediterranean Areas by Extensive Ruminant Production Systems, Proceedings of an International Symposium Organized by HSAP and EAAP*, 243–247, EAAP Publication No. 83, Thessaloniki.

Zanchi, C. and D. Torri (1980) 'Evaluation of rainfall energy in central Italy', in M. de Boodt and D. Gabriels (eds) *Assessment of Erosion*, 133–142, John Wiley and Sons, Chichester.

Zerbini, S., H.-P. Plag, T. Baker, M. Becker, H. Billiris, B. Bürki, H.-G. Kahle, I. Marson, L. Pezzoli, B. Richter, C. Romagnoli, M. Sztobryn, P. Tomasi, M. Tsimplis, G. Veis and G. Verrone (1996) 'Sea level in the Mediterranean: a first step towards separating crustal movements and absolute sea-level variations', *Global and Planetary Change* 14, 1–48.

Zêzere, J.L., A. de Brum Ferreira and M.L. Rodrigues (1999) 'The role of conditioning and triggering factors in the occurrence of landslides: a case study in the area north of Lisbon (Portugal)', *Geomorphology* 30(1–2), 133–146.

Zilhão, J. (1988) 'Nouvelles datations pour la préhistoire ancienne du Portugal', *Bulletin de la Société Préhistorique Française* 85(8), 247–250.

Zilhão, J. (1993) 'The spread of agro-pastoral economies across Mediterranean Europe: a view from the far west', *Journal of Mediterranean Archaeology* 6, 5–63.

Zinyowera, M.C., B.P. Jallow, R.S. Maya and H.W.O. Okoth-Ogendo (eds) (1998) 'Africa', in R.T. Watson, M.C. Zinyowera, R.H. Moss and D.J. Dokken (eds) *The Regional Impacts of Climate Change: An Assessment of Vulnerability*, 29–84, IPCC/Cambridge University Press, Cambridge.

Zohary, D. and M. Hopf (1975) 'Domestication of pulses in the old world', *Science* 187, 319–326.

Zohary, D. and M. Hopf (1988) *Domestication of Plants in the Old World*. Clarendon Press, Oxford.

Zohary, D. and P. Spiegel-Roy (1975) 'Beginnings of fruit growing in the old world', *Science* 187, 319–326.

Zolitschka, B. and J.F.W. Negendank (1996) 'Sedimentology, dating and palaeoclimatic interpretation of a 76.3 ka record from Lago Grande di Monticchio, southern Italy', *Quaternary Science Reviews* 15, 101–112.

Index